EMBEDDED MICROCONTROLLERS

Todd D. Morton
Western Washington University

Upper Saddle River, New Jersey
Columbus, Ohio

Library of Congress Cataloging-in-Publication Data
Morton, Todd D.
   Embedded microcontrollers / Todd D. Morton.
     p. cm.
   ISBN 0-13-907577-1
    1. Embedded computer systems. 2. Microcomputer programming.
   3. Real-time programming.
   I. Title.
   TK7895.E42 M67 2001
   621.39'16—dc21

00-042760
CIP

Vice President and Publisher: Dave Garza
Editor in Chief: Stephen Helba
Assistant Vice President and Publisher: Charles E. Stewart, Jr.
Production Editor: Alexandrina Benedicto Wolf
Design Coordinator: Robin G. Chukes
Production Coordination: Custom Editorial Productions, Inc.
Cover Designer: Thomas Mack
Cover Photo: Marjorie Dressler
Production Manager: Matt Ottenweller
Marketing Manager: Ben Leonard

This book was set in Times Roman by Custom Editorial Productions, Inc., and was printed and bound by R. R. Donnelley & Sons Company. The cover was printed by Phoenix Color Corp.

10 9 8 7 6 5 4 3 2
ISBN: 0-13-907577-1

Preface

This book is intended for anyone who wants to design small- to medium-sized embedded systems. It was written primarily for electronics engineering technology students, but should also be appropriate for most engineering students and practicing engineers. It is currently being used for two one-quarter courses in electronics engineering technology. The first is a required junior-level course on microprocessor-based applications, and the second is a senior elective course on embedded systems. The first course covers assembly language only, and the second course uses C.

Prerequisites for this book include a background in electronic circuits, basic digital logic, and general C programming. The C programming course is required only for Parts 4 and 5. A background in C and/or C++ is required because this book does not cover C language programming basics. It only addresses the concepts and techniques required for using C in a microcontroller-based embedded system.

► SOFTWARE AND HARDWARE USED IN THE BOOK

When writing a book on embedded systems, it is difficult to maintain the balance between being too specific to one hardware/software set and too general to be practical. I have tried to keep this balance by focusing on the concepts, processes, conventions, and techniques used in design and debugging.

The book focuses on the M68HC12 microcontroller from Motorola, but has also been successfully used in courses that use the M68HC11 family. It is intended to supplement, not replace, vendor documentation. I expect every student to have the complete vendor documentation for both the MCU and the development board he or she is using. For example, for the M68HC12 MCU the student should have the *CPU12 Reference Manual* and the *Technical Specifications* for the specific part. If the M68HC11 is used, the famous "pink books" are a requirement.

The development hardware used throughout the text is the Motorola 68HC912B32 EVB. In the first part of the text, only a single board is required, and all code is loaded into RAM. In the second half of the book, the background debug system is used, and the code is loaded into the target Flash ROM. This requires either two EVBs or one EVB as a target and a 68HC12 BDM pod such as the Noral 68HC12 BDM debugger. The concepts regarding the debug process and testing should apply to most modern development systems.

The development software used throughout is the Introl-CODE development system. Except for the sections that specifically address development using the Introl-CODE system, the C code is all ANSI-C, and therefore it would be reasonable to use another compiler.

The real-time kernel covered in this text is MicroC/OS-II. It is available in source form and is widely used. Many of the concepts apply to other kernels, especially the applications for typical kernel services.

► ## CHAPTER DESCRIPTIONS

The book is divided into five parts. The first half of the book emphasizes assembly code and the second half focuses on C. Hardware is covered throughout the text, especially in Part 3. Because of the use of pseudo-C from Chapter 6 on, it is reasonable to cover that material with an emphasis on C instead of assembly.

Part 1   Introduction.   This part introduces the reader to the background and perspective required in learning about embedded systems.

Part 2   Assembly Language Programming.   This part introduces programming in assembly language and covers the CPU12 programming model and program design. Students should be able to build a prewritten program after completing Chapter 3, and write complete programs that are executed by the D-Bug12 monitor in RAM after completing Chapter 6. Chapter 7 covers some basic applications that are appropriate for assembly code.

Part 3   Microcontroller Hardware and I/O.   Real-time concepts and I/O hardware are introduced, including interrupts and basic multitasking. All of the 68HC912B32 I/O resources are covered with the exception of the BDLC. MCU configuration for stand-alone systems along with bus expansion are covered in Chapters 10 and 11.

Part 4   Programming Microcontrollers in C.   Concepts in C for programming real-time embedded systems are covered. Emphasis is placed on memory usage and program efficiency appropriate for small MCUs.

Part 5   Real-Time Multitasking Kernels.   In this part basic multitasking design is covered along with using MicroC/OS-II, an off-the-shelf kernel.

► ACKNOWLEDGMENTS

Without the help and patience of the following people, this text would not have been possible: Rich Pennington at Introl Corp, Jim Sibigtroth at Motorola University and Austin Community College, Jean Labrosse at Micrium, Tony Plutino and Dave Hyder at Motorola, Marsh Faber and Mel Downs at Hewlett-Packard, Phil Meek and Harry Erickson at Noral Micrologics, Dave Garza and staff at Prentice-Hall, Kathleen Kitto and Andrew Pace at WWU, George Sweiss at ITT Technical Institute, Malvern Phillips at British Columbia Institute of Technology, the contributors to the Motorola 68HC11 and 68HC12 listserv, and, of course, all of my students. You will finally get those labs and homework assignments graded.

To Edye, Jeva, and Perry

Contents

▶ **PART 1**   **Introduction**

1   **Introduction to Microcontrollers**   **1**

    1.1   The Microcomputer   3
    1.2   The 68HC11 and 68HC12 Microcontrollers   14
    1.3   Historical Context   14
    1.4   Software and Hardware Development   15
           Summary   22
           Exercises   22

▶ **PART 2**   **Assembly Language Programming**

2   **Programming Basics**   **24**

    2.1   Programming Languages   24
    2.2   Types of Program Segments   30
    2.3   Software Construction   31
           Summary   34
           Exercises   34

3   **Simple Assembly Code Construction**   **35**

    3.1   Assembly Source Code   37
    3.2   A Basic Build Process   47

3.3   Run-Time Debugging—A Tutorial 52
        Summary 59
        Exercises 60

4   CPU12 Programming Model   61

4.1   The CPU Register Set 61
4.2   CPU12 Addressing Modes 64
4.3   The CPU12 Instruction Set 77
        Summary 78
        Exercises 78

5   Basic Assembly Programming Techniques   80

5.1   Data Transfer 80
5.2   Using the Stack 86
5.3   Basic Arithmetic Programming 92
5.4   Shifting and Rotating 105
5.5   Boolean Logic, Bit Testing, and Bit Manipulation 106
5.6   Branches and Jumps 113
5.7   Subroutines 120
5.8   Position Independence 128
        Summary 132
        Exercises 132

6   Assembly Program Design and Structure   137

6.1   Design and Documentation Tools 138
6.2   Structured Control Constructs 143
6.3   Data Storage 155
6.4   Program Structure 162
6.5   Passing Parameters 163
        Summary 172
        Exercises 173

7   Assembly Applications   174

7.1   Software Delay Routines 174
7.2   I/O Data Conversions 180
7.3   Basic I/O Routines 194
7.4   Fixed-Point Arithmetic 208
        Summary 229
        Exercises 229

▶ PART 3   Microcontroller Hardware and I/O

8   Introduction to Real-Time I/O and Multitasking   231

8.1   Real-Time Systems 231
8.2   CPU Loads 233

8.3   I/O Detection and Response   234

8.4   Basic Cooperative Multitasking   261

8.5   Using CPU12 Interrupts   274

8.6   Basic Real-Time Debugging   289

      Summary   293

      Exercises   293

9   Microcontroller I/O Resources   295

9.1   General Purpose I/O   296

9.2   Timers   306

9.3   Serial I/O   347

9.4   A-to-D Conversion   366

      Summary   376

      Exercises   376

10   The Final Product   378

10.1   MCU Hardware Design   379

10.2   Reset Exceptions   386

10.3   M68HC912B32 Operating Modes   394

10.4   Configuration and Start-Up Code   400

10.5   Final Product Development   406

      Summary   416

      Exercises   417

11   System Expansion   418

11.1   The Bus Cycle   419

11.2   Chip-Select Logic   421

11.3   Bus Timing Analysis   434

      Summary   437

      Exercises   438

▶ PART 4   Programming Microcontrollers in C

12   Modular and C Code Construction   439

12.1   C Source Code   440

12.2   The Modular Build Process   454

12.3   Source-Level Debugging   475

      Summary   481

      Exercises   481

13   Creating and Accessing Data in C   483

13.1   Introduction to Data Types   484

13.2   ANSI-C Data Types   485

13.3   Variables and Stored Constants   494

13.4 Pointers 498
13.5 Arrays and Strings 503
13.6 Structures 507
13.7 Enumerated Types 509
13.8 Bit Operations 509
Summary 516
Exercises 517

14 C Program Structures 518

14.1 Control Structures 518
14.2 Functions 536
14.3 Modules 548
14.4 Start-Up and Initialization 557
Summary 564
Exercises 564

▶ **PART 5 Real-Time Multitasking Kernels**

15 Real-Time Multitasking in C 566

15.1 Real-Time Programming Review 566
15.2 Real-Time Kernel Overview 570
15.3 Cooperative Kernel Design 574
Summary 591
Exercises 592

16 Using the MicroC/OS-II Preemptive Kernel 594

16.1 Overview 595
16.2 Tasks and Task Switching 603
16.3 Interrupt Service Routines 609
16.4 Timers 610
16.5 Intertask Communication 614
16.6 µC/OS-Based Stopwatch Program 638
Summary 645
Exercises 646

▶ **APPENDICES**

A Programming Conventions 648

B Basic I/O 651

C uC/OS Reference 672

▶ **REFERENCES 685**

▶ **INDEX 687**

Introduction to Microcontrollers

We have all heard the stories about how the microprocessor has revolutionized many aspects of our everyday lives. The most visible examples are desktop computer systems and the Internet. Another part of this revolution that we do not often hear about is embedded systems. *Embedded systems* are electronic systems that contain a microprocessor or microcontroller, but we do not think of them as computers—the computer is hidden, or embedded, in the system. Examples of embedded systems include automobiles, industrial controllers, instrumentation, network routers, and household appliances, now even including rice cookers and toasters. Homes in the United States have an average of 30 to 40 microprocessors each, yet only 45% of these homes have a desktop computer. The rest of these processors are used in embedded applications.

In this book we will concentrate on the largest segment of the embedded systems market—the small systems. These are systems that require 8- or 16-bit microprocessors or microcontrollers. Figure 1.1 shows a typical system, a digital thermometer. It is made up of a temperature sensor connected to an ADC, a microprocessor (CPU), RAM, ROM, chip select logic, and an LCD module.

If there are two essential design characteristics for embedded systems they are cost sensitivity and diversity. It does not make sense to use the same system for an infrared remote control as an unmanned spacecraft. Of course this is obvious, but the wide range of complexity is what has guided the evolution of the technology used for embedded systems. In

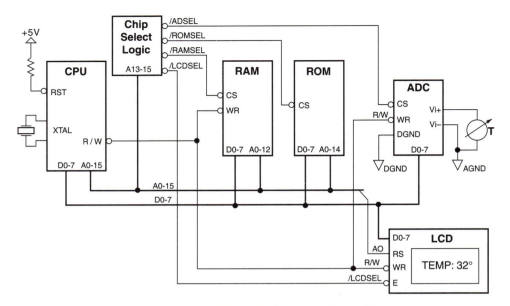

Figure 1.1   A Typical Small Embedded System—Digital Thermometer

addition to this diversity is the requirement that costs must be reduced as much as possible. A video game system will not sell if it costs $1,000. Yet the processor technology used in a video game is on a par with that used in desktop PCs. Therefore there cannot be a *one-system-fits-all* approach to embedded systems design. Embedded system designs require *application-specific* hardware and software, hardware and software designed for each different application.

Another characteristic of embedded systems development is the wide range of skills required to design a system. In the simple example shown in Figure 1.1 there are several technologies represented, as explained in the following list:

- The CPU requires skills in software design and microprocessor interfacing.
- The chip select logic requires familiarity with digital logic.
- The ADC and temperature sensor require skills in analog design and sampling theory. There may also be some knowledge of basic physics required to understand the characteristics of the sensor.
- The LCD requires an understanding of user interfaces and the optical characteristics of the LCD.

In addition, the system may be networked, may run on alternate power sources, or may be placed in a harsh environment. Some embedded system designs are large enough so that it is practical to break the design tasks down into specialties. However, it is most beneficial to understand or at least be interested in learning all the technologies involved. Because of the diverse skills required, this can be a fascinating and rewarding field for an engineer or technician.

▶ 1.1 THE MICROCOMPUTER

We will start by covering the heart of all embedded systems—the *microcomputer*. As shown in Figure 1.2, the microcomputer is made up of the CPU, memory devices, I/O devices, and the bus system.

The CPU or central processing unit is another name for a *microprocessor*. It controls the system and processes data, the memory stores the CPU's programs and data, the I/O devices provide an interface with the outside world, and the bus system provides for a flexible interconnection system. The thermometer in Figure 1.1 is a microcomputer. It has a CPU, its memory includes a RAM and a ROM device, and the I/O consists of an ADC connected to a temperature sensor and an LCD module. The microcomputer is a very flexible system. It allows the designer to include only the devices required for the specific application, which is especially important for embedded systems.

If a microcomputer system is combined into a single integrated circuit (IC), it is called a *single-chip microcomputer* or a *microcontroller (MCU)*. These terms are often used interchangeably but they really represent two different devices designed for different applications. A single-chip microcomputer contains resources typically used for computer systems, such as a memory management unit and a disk controller. The microcontroller, on the other hand, contains resources typically used for embedded systems, such as timers and ADCs. We will be focusing on microcontrollers in this text.

To see the effect of using a single-chip microcontroller, let's go back to our simple embedded system example in Figure 1.1. There are several ways to implement this design. We can use a microprocessor-based system, we can use a microcontroller with an external bus, or we can use a microcontroller in single-chip mode.

If the design is implemented using a microprocessor, all the blocks shown in Figure 1.1 are separate ICs. This means that at least five ICs are required in addition to the temperature sensor and LCD module. Since most of the ICs are connected to the bus, they are large ICs with a large number of pins. This would result in a relatively expensive large printed circuit board.

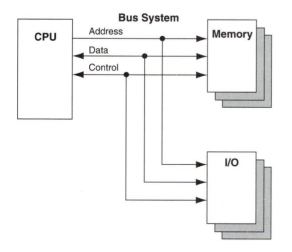

Figure 1.2   The Microcomputer

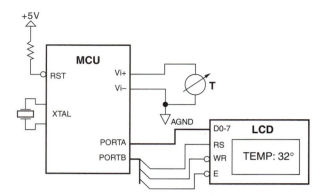

Figure 1.3 A Microcontroller-Based System

The next option is to use a microcontroller that is configured in expanded mode. *Expanded mode* means the bus system is available outside of the microcontroller IC. If the microcontroller contains the chip select logic, the RAM, and the ADC, then the system is down to two ICs, the MCU and the ROM. The PCB size is reduced and typically the cost will go down. This is a compromise design. The number of ICs has been reduced and, the flexibility provided by the external bus still exists. It still requires a large ROM device and the power consumption is still relatively high.

The last option is to implement the design with a microcontroller in single-chip mode. An MCU in single-chip mode does not provide the bus for external connections. The extra pins vacated by removing the bus can be used to reduce the size of the IC package or to add extra I/O. As shown in Figure 1.3 this option results in a single-chip solution. Only the LCD module and temperature sensor circuits are required, so this can be an extremely cost effective and compact design.

1.1.1 The Microprocessor

The microprocessor or CPU is the controller for the microcomputer system. It controls all the bus activity, performs calculations, and makes decisions. The microprocessor is programmable—its operations are controlled by a sequence of instructions. These instructions include three general types: data transfer instructions, arithmetic and logic instructions, and program control instructions. A sequence of microprocessor instructions is called a *program* or *software*.

The combination of a programmable CPU and the bus system results in an extremely flexible system that can easily be customized for a given application. In embedded systems this flexibility is used to create application-specific hardware that runs a single application-specific program.

The microprocessors used for embedded systems are relatively simple when compared with microprocessors designed for desktop computers. Currently the highest volume microcontrollers are those with 8-bit CPUs. Small packages and cost effectiveness are the guiding factors when selecting a microcontroller for an embedded system. It would not make sense to use an expensive 32-bit CPU in a television remote control or a toaster. There are embedded systems on the market that do have powerful 32-bit CPUs, however.

These devices are used in systems that require moving a large amount of data at high speed, or systems that require a large number of complex calculations. Common applications for these microcontrollers include video games, laser printers, network routers, or automotive engine control systems. In this book we focus on the intermediate range of 8- and 16-bit microcontrollers and the design constraints typical of these devices. Some of the material is certainly applicable to the smallest 4- and 8-bit designs, and some of the material also is applicable to the larger 32-bit designs.

1.1.2 The Bus System

The *bus system* for a microcomputer provides a flexible means to transfer data among the CPU, the memory, and the I/O devices. It is flexible because it is a shared bus. To add a memory or peripheral device to the system, you simply have to connect them to the bus system and add the required decoding logic. The CPU controls the bus system by providing a device address on the address bus and bus control signals on the control bus for direction and timing. It then either provides data (writes) or samples data (reads) on the data bus.

The Address Bus. The address bus is made up of CPU outputs that contain the source or destination location for a data transfer. Access to specific locations is controlled on two levels. Chip select logic decodes the address bus to determine which memory or peripheral device to access. Then the device address decoding logic decodes the address to determine the specific location within the device. This is analogous to the postal system in which the mail is directed first to an area post office based on the address. The post office then directs the mail to the specific mailbox based on the address.

Most small microcontrollers have a linear address space. A *linear address space* is one in which each address referred to by an instruction directly corresponds to that location in memory. It is the easiest type of addressing to use but it can result in inefficiencies in the CPU. Paged memory systems can increase the CPU efficiency but can be difficult to work with. In a paged system the address bus contains the location information within a current *page* of addresses. Another CPU register must be used to select the current page.

The size of the address bus determines the total number of locations that are directly accessible by the CPU. For each location to have a unique address you can only have as many locations as you have unique combinations in the address word. Therefore, there are 2^N possible addresses for an *N*-bit address bus.

EXAMPLE 1.1

Addressable Space for a 16-bit Address Bus

The 68HC11 microcontroller has a 16-bit address bus, which is the typical size for a small microcontroller. What is the maximum number of directly accessible locations?

Solution

For a 16-bit address bus there are $2^{16} = 65,536$ locations. For this case we would normally say there are 64K-bytes of memory space (1K-byte = 2^{10} = 1,024 bytes). Note this is not the same as 10^3!

This type of notation is commonly used for addressable space or memory size designations. Some similar designators are also used:

$$1\text{M-byte} = 2^{20} = 1{,}048{,}576 \text{ bytes}$$

$$1\text{G-byte} = 2^{30} = 1{,}073{,}741{,}824 \text{ bytes}$$

You have to be careful with these definitions. 1M-byte is 2^{20} when the context is an addressable device but may be 10^6 in other contexts. For example when referring to the size of a disk drive the 10^6 definition is normally used—but this is not always the case.

The Data Bus.   The data bus is a set of signals that contain the data to be transferred between the CPU and a memory or I/O device. The data bus must be bidirectional because the CPU must be able to read and write data. When the CPU reads data from a peripheral, its data lines are set as inputs. When it writes data to a peripheral, its data lines are set as outputs.

When we use the term "n-bit microcontroller," n is the size or width of the data bus. For example the 68HC11 has an 8-bit data bus size, so it is referred to as an "8-bit microcontroller." The 68HC12 is a 16-bit microcontroller because its internal data bus is 16 bits wide.

The size of the data bus determines how much data can be transferred in a single bus cycle. Desktop computer systems, which must transfer very large amounts of data, require 32- or 64-bit microprocessors. Most embedded systems, however, do not require a large amount of data to be transferred so a smaller data bus is acceptable and more cost effective.

The small systems covered in this book will require only an 8-bit CPU. There are times, however, when a 16-bit CPU may be required even if there is very little data to be moved. This is because the 16-bit bus enables the CPU to fetch larger instructions or an instruction and an argument in a single cycle. Consequently it takes fewer cycles to execute the instructions on a 16-bit data bus than on an 8-bit bus.

Bus Control Signals.   The rest of the bus signals are used for bus control. At a minimum, these signals provide the following functions:

- Data direction. For Motorola microcontrollers, this is the $R/\overline{W}$ signal.
- Bus synchronization. For the 68HC11 and 68HC12, this is the E-clock.

There may also be other control signals such as the following examples:

- An additional bus latch signal for multiplexed buses. In the 68HC11 family, this is the *AS* signal.
- Memory and I/O request signals for separate memory and I/O spaces. These are common on older Intel-type processors.
- Bus sharing signals such as *bus request* and *bus acknowledge.* These are used for shared bus systems in which more than one processor or peripheral share control over the bus.

- Data acknowledge signals for asynchronous bus systems. This is */DTACK* on the 68000 family of microprocessors.
- Low-byte and high-byte signals for 16-bit data transfers from 8-bit external devices.

Sharing the Bus.   Since the data bus is bidirectional and can be connected to multiple devices, there is a possibility of a bus collision. If two devices try to put data on the bus at the same time, a *bus collision* occurs. At the logic level a bus collision means there are two outputs connected together. If one output is high and the other is low, a low resistance path is created between power and ground, and one of the device outputs will eventually fail. Because of this only devices with tristate outputs can be connected to the data bus. A tristate output is one that can be placed in a high impedance state, effectively disconnecting the output from the bus. Figure 1.4 shows the detail of the output stages of three devices, the CPU, a RAM, and an EPROM. To simplify the figure only the connection to bit-0 of the data bus, *D0,* is shown. Notice that all three devices have tristate outputs connected to *D0.*

The CPU output is enabled for all write cycles ($R/\overline{W} = 0$). The RAM output is enabled if the RAM is selected (*/RAMSEL* = 0) and if it is a read cycle ($R/\overline{W} = 1$). The EPROM's output is enabled if the EPROM is selected (*/EPROMSEL* = 0).

The *chip select logic* is a decoding circuit that selects the appropriate device based on the current address on the address bus. It is up to the chip select logic to make sure that there are not two devices selected at one time. The logic may be implemented externally using digital logic devices or be internal to a microcontroller in the form of *programmable chip selects.* Microcontrollers that have internal chip selects are designed for expanded mode operation. Two examples are the MC68HC11F1 and the MC68HC812A4.

The Memory Map.   A memory map documents the addressable space of the CPU. It shows how the space is divided into blocks. Each block corresponds to a different device.

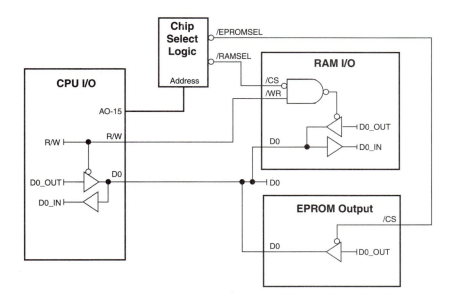

Figure 1.4   Data Bus Access Output Logic Detail

It should include the block location, block size, device type, and usage. Following are a few rules for creating concise memory maps:

- Every byte must be accounted for in the map, even if it is unused or makes up a mirror image.
- The starting and ending address for every device block must be shown.
- The memory map should start with address $0000 on top and end with $FFFF at the end.
- You should not attempt to show blocks to scale.

The chip select logic is what determines the contents of the memory map. Therefore, in order to change the memory map, the chip select logic must be changed or, in the case of internal chip selects, reprogrammed. Table 1.1 shows the memory map of the M68EVB912B32 development board. In this system the chip select logic is internal to the microcontroller and all the memory and I/O are internal resources.

Multiplexed Bus Systems. A bus system contains a large number of signals that require a large number of IC pins. Increasing the number of pins on an IC package in order to make a bus available to external devices can significantly increase the cost and size of the IC. For a typical 8-bit system with a 16-bit address bus, at least 26 pins are required for the bus alone.

For low-cost, simple applications that require small packages, a microcontroller will be used in single-chip mode. The bus system is still used internally by the microcontroller to access its on-chip resources, but the bus is not available outside the microcontroller. This means either a smaller package can be used or more I/O signals can be made available for a given package.

A compromise between having the bus system with a pin for every bus signal and a microcontroller with no external bus system is the multiplexed bus system. *Multiplexed bus systems* use time-division multiplexing to share IC pins between the data bus and part or all of the address bus. This compromise reduces the cost of the microcontroller or CPU by reducing the number of pins required for the package. It requires additional external logic to demultiplex the bus signals, however. This technique is used in many systems in which the number of interconnections can become prohibitive. The PCI computer bus uses a multiplexed bus system to reduce the size of the bus connectors. The M68HC11E series

TABLE 1.1   M68EVB912B32 MEMORY MAP

Address Range	Device Type	Usage	Address Range
$0000–$01FF	Internal Registers	Internal Registers	$0000–$01FF
$0200–$07FF	Unused	Unused	$0200–$07FF
$0800–$0BFF	On-Chip RAM	User Code/Data	$0800–$09FF
		D-Bug12 RAM	$0A00–$0BFF
$0C00–$0CFF	Unused	Unused	$0C00–$0CFF
$0D00–$0FFF	Byte-Erasable EEPROM	User Code/Data	$0D00–$0FFF
$1000–$7FFF	Unused	Unused	$1000–$7FFF
$8000–$FFFF	Flash EEPROM	D-Bug12 Program Code	$8000–$FFFF

microcontrollers have the 8-bit data bus multiplexed with the least-significant byte of the address bus. During the first half of a bus cycle, the pins contain the lower byte of the address. During the second half of the bus cycle, the pins contain the data. This saves eight pins on the IC.

There are two problems with multiplexed bus systems: (1) external circuitry is required to demultiplex the bus signals, and (2) given the same technology, a multiplexed bus system has a lower maximum bus speed than a nonmultiplexed bus system.

1.1.3 Memory Types and Applications

Memory devices are used to store the microprocessor's program and data. There are several types of memory devices with applications in embedded systems. Memory that is connected to the bus system is called *primary memory*. Let's look at the different types of primary memory to see how they are used within the microcomputer system.

The Four "Abilities."   To understand how and why different memory devices are used, we must look at four important memory characteristics:

- Accessibility
- Writability
- Volatility
- Programmability

Accessibility.   Obviously all memory devices must be readable; otherwise the contents would not be accessible. There are different methods used to access the contents of a memory device, however.

Virtually all memory devices connected to the microcomputer bus system use random access. *Random access* refers to a memory device that allows direct access to any location. The access time in random access memories is approximately the same for every location.

At this point it would seem that all memory would be random access. What other methods could be used? An example of a memory device that is not a random access device is a FIFO (First-In-First-Out) memory. In a FIFO memory the data must be read in the same order that they were stored. You cannot randomly read any location within the device. In fact a FIFO memory does not even have addressing pins on the device.

Writability.   Although all memory devices are readable, not all memory devices are writable. A memory device is said to be writable if the contents can be changed by a CPU write cycle with or without wait states. This means a single CPU *store* instruction can be used to change the contents of a location.

The most common writable memory is the *RAM,* Random Access Memory. This is an unfortunate name. As mentioned above almost all memory devices connected to a microcomputer bus system are random access devices. A much better name for RAM would have been *Read-Write Memory.*

Devices that cannot be written to are called *Read-Only Memories* or *ROMs*. The contents of a ROM cannot be changed in a single CPU write cycle. This includes in-circuit programmable devices like EEPROM and Flash EEPROM.

Volatility. A memory device is *nonvolatile* if its contents are retained when its power source is removed. A *volatile* memory device loses its contents with the loss of power. To the programmer volatility affects the persistence of the data in the memory. If the contents must survive when there is no power, a nonvolatile memory device must be used. In general RAM devices are volatile and ROM devices are nonvolatile.

If a particluar application requires both writability and nonvolatility, a battery-backed RAM can be used. A *battery-backed RAM* or *BBRAM* is a hybrid device that contains a RAM, a battery, and power management circuitry. When the normal power source is turned off, the power management circuit switches the RAM power to the battery, which holds the contents of the RAM until the normal power source is turned back on. BBRAM is extremely useful during the development cycle of a project.

Nonvolatile is not the same as nonchangeable. EEPROM contents can be changed in circuit, but they are nonvolatile devices because the contents remain with a loss of power.

Programmability. The process of changing the contents of a ROM device is called *programming*. Some ROMs can be programmed only one time. This means that if the program or data contained in the ROM must be changed, then a new device is required. There are three different methods for programming ROMs: (1) mask programming by the manufacturer, (2) user programming using a device programmer, and (3) user programming in circuit. Table 1.2 shows the basic ROM types in use today.

In order to examine how these memory devices are used in a microcomputer, two types of systems will be described—stand-alone systems and development systems.

Stand-alone Systems. Stand-alone systems are the embedded system final products— the actual products that are manufactured and sold. The stand-alone embedded system runs a single program when power is applied. This program must be available to the CPU when the system is turned on, so it must be stored in a nonvolatile memory device. These programs are referred to as *firmware* because they cannot be changed without replacing or reprogramming the ROM device.

TABLE 1.2 ROM TYPES AND CHARACTERISTICS

ROM Type	Description	Programming Method	Erase Method
ROM	Mask ROM	Manufacturer, one time	None
PROM	Programmable ROM	Device Programmer, one time	None
EPROM	Erasable PROM	Device Programmer	UV light source
OTP	One-Time Programmable	Device Programmer, one time	None
EEPROM	Electrically Erasable EPROM	Device Programmer, in circuit	Byte-wide erase
Flash EEPROM	Electrically Erasable EPROM	Device Programmer, in circuit	Device or block erase

TABLE 1.3 MEMORY USES

Contents	Stand-alone Systems	Computer/Development Systems
Variables, Stack	RAM	RAM
Program, Constants	Masked ROM PROM EPROM/OTP EEPROM Flash EEPROM	RAM BBRAM
Nonvolatile Parameters	BBRAM EEPROM Flash EEPROM	RAM BBRAM
Computer BIOS, Debug Monitor		Masked ROM PROM EPROM/OTP EEPROM Flash EEPROM

Development Systems and Computer Systems.   Both of these systems are characterized by the fact that they run many different programs at different times. Consequently the program code must be changed easily. In computer systems the operating system loads a program from a disk drive before it is executed. In development systems a program is loaded into memory, usually from a workstation, before execution. Once loaded, the program is executed. This means the programs must be loaded into a memory device that is writable or in-circuit programmable. Normally there is no need for the program to be restricted to nonvolatile memory types. If the development system is powered down, the user can simply reload the program from the workstation.

Of course computer systems and development systems must also execute a program when power is first applied. Computer systems have firmware called a *bootloader* or *BIOS* program. Development systems have firmware consisting of a monitor program.

Table 1.3 shows the memory types normally used for these two systems. Notice that the stand-alone device must have its program and constants in nonvolatile memory, but during development the program and constants are stored in RAM. This enables us to rapidly develop the program because a ROM device does not have to be programmed for each code-debug cycle.

Remember some EEPROM devices can be reprogrammed in circuit. This means we may be able to write a program that changes the contents of these devices. This property is ideal for nonvolatile parameters—variables that are persistent through power cycles. When firmware is contained in an EEPROM, however, it must be done with care. After all, if we can write a program that reprograms an EEPROM device, we can also create a bug that does the same thing. This would result in a hard and possibly catastrophic failure. It would not even recover after a *reset* or power cycle. As far as the user is concerned, the system would be dead. Most modern EEPROM devices and microcontrollers have built-in protections to guard against accidental reprogramming.

1.1.4 I/O Devices

I/O devices include all resources that provide an interface between the microcomputer and the outside world. These interfaces can vary widely from obvious human-based interfaces like a keypad or display to a network or machine interface that has no human interaction at all.

The set of I/O resources available on a microcontroller often determines whether that microcontroller will work with an application. They provide the necessary hardware interface to an external system, and they can reduce the load on the microprocessor.

General Purpose I/O. General purpose I/O, GPIO, is the most basic form of I/O. It is a set of input pins or output pins that can be accessed by the CPU. Some GPIO pins can be programmed to work in either direction. Another traditional term for GPIO is *parallel I/O*. Figure 1.5 shows a simplified functional logic diagram for a bidirectional GPIO port. Only GPIO bit 0 is shown to simplify the diagram. There are two registers in the diagram, the data register *PORT* and the data direction register *DDR*.

The data direction register sets the direction of the port. If its output is a one, then the port is an output, and if its output is a zero, then the port is an input. The *DDR* state can be changed by writing to the *DDR*, which is a mapped location in the microcontroller address space. In this case, if this *DDR* is to be changed, the correct value is placed on bit-0 of the data bus *D0* and the *WR_DDR* signal is activated. The *DDR* state can also be read by reading the location of the *DDR*, which activates the *RD_DDR* signal.

If the *PORT* pin is set to be an output, the *PORT* register controls its state. If it is an input, the logic level on the input pin controls its state. A write to the *PORT* register, which is also mapped in the microcontroller address space, will activate the *WR_PORT* signal. Note that writing to the *PORT* register has no effect on the pin if the port is set to be an input. A read of the *PORT* register always results in the state of the pin, regardless of port direction. Source 1.1 shows the typical software required to configure the M68HC912B32's *PORTP*, bit 0, *PP0*.

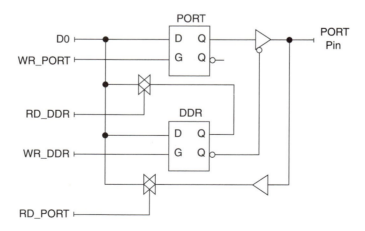

Figure 1.5 Bidirectional GPIO Functional Logic Diagram

Source 1.1 GPIO Software

```
bset PORTP,BIT0    ;Preset PORTP data
bset DDRP,BIT0     ;Configure PORTP, BIT0 as an output
```

In this example *PORTP,* bit 0 is set first. This may seem backward because we have not configured *PP0* to be an output yet. But by presetting the data register, we can avoid momentary glitches on the output. *PP0* then is configured as an output by writing a one to bit 0 of the data direction register *DDRP.* Once the port is set to an output, the preset port data are connected to the output pin.

Timers.   Timer circuits are the second most widely used I/O resource. Timers include both timing and counting functions and may be either inputs or outputs. Timer circuits are used to reduce the load on the CPU and to create real-time events. For example a common application of a timer circuit is to produce periodic interrupts. Periodic interrupts are very common in embedded system programs as they provide a timed "heartbeat" for the real-time program execution.

ADCs and DACs.   ADCs and DACs are widely used I/O resources. Many embedded systems must digitize analog signals for digital displays or transmission over a serial port. As the microcontrollers become more powerful, however, more signal processing occurs after the signal is digitized. By using digital signal processing techniques, overall system performance and cost can be improved over traditional analog processing techniques.

DACs are commonly used for automated control. A system that requires an analog signal for control would traditionally require a user to turn a dial. With a DAC a microcontroller circuit can replace the dial and make hands-off automated control possible.

Serial Ports.   Serial ports are another very popular type of I/O resource. In general there are two types of serial ports available in microcontrollers, synchronous serial ports and asynchronous serial ports.

Asynchronous serial ports, such as the SCI on Motorola MCUs, are usually used to communicate with computers over an RS232 interface. Providing connectivity to a computer system not only enhances the application but also provides an invaluable service tool. Many embedded systems that do not require computer communications during normal operation have serial connectors added for servicing and testing. Asynchronous serial ports can also be used to create small LANs for distributed control or processing.

Synchronous serial ports are characterized by providing their own synchronization clock signal. These ports are generally designed for in-system communications. This includes microcontroller-to-microcontroller or microcontroller-to-peripheral communications. Motorola's SPI port is an example of a synchronous serial port. With the advent of single-chip microcontrollers, the synchronous serial ports have become much more important because they allow the system to be expanded without requiring a bus system.

There are several different synchronous serial protocols. Motorola's is called the SPI, Serial Peripheral Interface. Others include Dallas's one-wire bus, National's Microwire, and the I²C bus. Depending on the required level of functionality, the SPI can usually communicate with devices designed for the other synchronous standards.

▶ 1.2 THE 68HC11 AND 68HC12 MICROCONTROLLERS

In this text, we focus on three Motorola microcontrollers. The primary focus is on the 68HC12 family devices—the M68HC912B32 and the M68HC812A4. When 68HC11 examples are introduced, they will be designed for the M68HC11F1.

These three microcontrollers represent two different CPUs. We will refer to the 68HC12 family CPU as the *CPU12* and the 68HC11 CPU as the *CPU11*. The software model is the same for a given CPU so code designed for one *CPU11* device normally works on any *CPU11* device.

Although most of the examples and code in this text are designed for the 68HC12 family microcontrollers, with small changes they can be applied to a 68HC11 device.

▶ 1.3 HISTORICAL CONTEXT

Computer systems are those systems we think of as computers, such as portable PCs, desktop computers, workstations, and mainframes. These systems are general computing devices because they can be used for many different applications based on the software installed. The hardware for computer systems is designed to store and move large amounts of data and execute a wide variety of software as fast as possible.

We have all heard the stories about the computers that used to fill rooms in academic and defense labs. Maybe you have seen an early calculator about the same size as today's desktop computer. The advent of the microprocessor dramatically changed the design of computer systems. With the microprocessor, computer systems evolved quickly into what we have today. Computer systems are making a dramatic impact on our society in what has been called the "Information Age."

The advent of the microprocessor also has resulted in a dramatic change in the design of embedded systems. Some of the most important requirements for embedded systems are small size, low cost, and low power consumption. These requirements typically are more important than the processing power. Microprocessors and now microcontrollers have made these systems more realizable.

The evolution of embedded systems is also quietly making a major impact on the world in which we live today. There are far more embedded system designs than computer designs and a far greater volume of units sold. Currently the microprocessor or microcontroller with the highest volume sales is the Motorola 6805 family of microcontrollers, which are inexpensive and simple 8-bit processors that are about as different as you can get from the microprocessors designed for today's computer systems.

To illustrate this let's look into the American household. Currently about 45% of the households in America have a personal computer. That is a large number of microprocessors. However, the typical American household now averages between 30 and 40 microprocessors, and the number is rising. Only about five of these are in your PC. All the other processors are used in embedded systems. Look around you to see whether you can determine the number of microprocessors in your home, car, or dorm room.

Before the microprocessor, a designer was restricted to standard digital logic or custom integrated circuits when designing an embedded system. The standard logic designs tended to be large and expensive, and they consumed relatively large amounts of power. They generally were used in systems that required very few logic states, or registers.

The custom IC was technologically a perfect solution for embedded applications because it could be designed to meet the specific requirements for an application. There are an endless number of embedded applications, however, and that would mean an endless number of custom ICs. Unfortunately custom ICs are expensive to develop, and once completed they cannot be altered for a different application. Only large companies with a high volume of products or military and space organizations could take this route, designing custom ICs for each application.

The microprocessor was the solution to the problem of developing less-expensive, programmable controllers. A microprocessor is a single IC that can be programmed to perform the functions required for the given application. Microprocessors also can be made in large volumes because they can meet the needs of many applications.

For many years the microprocessors that were used for computer systems were the same as those used for embedded systems. Processors like the Motorola 6809, Mostek 6502, Zilog Z80, and Intel 8085 were popular for embedded designs and were used in the earliest PCs. Some embedded designs, however, demanded lower size and power consumption. For these applications, the microcontroller was developed.

The early microcontrollers like the Intel 8048, Motorola 6801, and National COP parts again were limited in their applications. Because of the IC technology at the time, the microcontrollers had a very limited set of resources. It was difficult to find a microcontroller that fit the resource requirements of a particular application. If an application did not work with the available microcontrollers, a microprocessor-based system had to be used. In the early days of the microcontroller, the microprocessor-based system was still more widely used because of its flexibility.

Today's application-specific IC design technology has changed the situation. Manufacturers can now easily design a new version of a microcontroller with a different set of resources and make the part available to everyone. They can also include more resources on a single chip, making the microcontroller more versatile. Consequently some microcontroller families have hundreds of different versions. Examples include the 8051 family, the 6805 family, and the 68HC11 family. The different microcontrollers in these families each contain the same CPU, so software is easily ported from one family member to another. The only limitation is the manufacturer's ability to produce the different parts. IC fabrication facilities are in high demand, so some of the lower-demand parts do not receive priority. Now most embedded systems are designed using microcontrollers because the designer can usually find a microcontroller with resources that fit the application.

▶ 1.4 SOFTWARE AND HARDWARE DEVELOPMENT

In this section we examine the development cycle involved in the design of an embedded system. The development of an embedded system not only involves creating schematics and writing code, it also includes the tasks involved during the complete lifetime of the product. In fact many of the considerations that must be made in today's competitive environment are actually based on the reuse of existing designs. So the project lifetime of an embedded system continues after production and includes product maintenance, bug reporting, and archiving.

One unique characteristic of an embedded system is the close interaction between hardware and software. Because of this interaction the software and hardware should never be designed in isolation from each other. When the project is small and a single designer is responsible for both hardware and software, this may not be a problem. For larger projects in which there may be separate programmers and hardware designers, however, it is essential that they work closely together. The embedded programmers must understand the hardware, and the hardware designers must understand the software. One way to distribute tasks to designers is by using functional blocks in which each designer is responsible for both the software and hardware for a function or set of functions. This may be a better way to break up the tasks, but it requires designers who are capable of designing both hardware and software, and it requires careful coordination of the system-wide software integration and hardware constraints.

Figure 1.6 shows the development cycle for a small embedded system from concept to the final prototype construction and testing. It shows the major stages of the development cycle along with some of the tasks, processes, and outcomes associated with each stage.

Following are short descriptions of each of the stages in the development cycle. Depending on a company's environment and the type of product, the process used may be more or less formal than the process described here.

1.4.1 Concept/Problem Definition

The first step in any project development is to define the problem to be solved by the project. The outcome should be a simple problem definition from the user's point of view. It should describe the problem, not address the solution. A simple example of a problem might be, "The speaker's movement is encumbered by the microphone wire," whereas the statement "A wireless microphone needs to be created" is not a problem definition because it indicates a solution to a problem. It is important for the end users to be closely involved in creating the problem definition and for market analysis to be completed to see what other products address the problem.

1.4.2 Requirements/Specifications

This is the process of developing a solution to the problem on paper in the form of a requirements document or specifications. Requirements define what the design will do to solve the problem, but they do not address the methods used to implement the solution.

The complexity and formality of the requirements depend on the type of company and products. If the product is large and complex and must be solved by many people, the requirements document becomes more important. It may be used to formally define the design tasks for each person or group of people.

The requirements document is also used to create the test plan. When the project is complete, it must be tested to verify that it meets the requirements. The requirements document defines what needs to be tested. For large projects a separate test engineer may independently create a test plan based on the requirements. When the designers are finished with the product, it is then passed on to the tester.

Some informal *proof-of-concept* design work may be done at this point to verify that the conceptual ideas and specifications are actually possible. There is rarely much time to perform proof-of-concept designs at this stage, so it is usually preferable to reuse designs

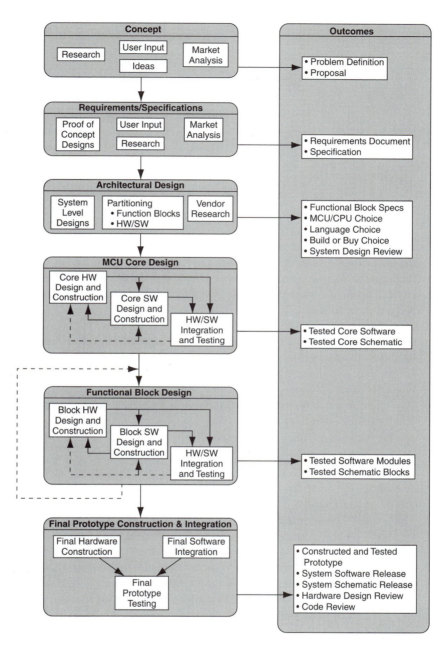

Figure 1.6 Small Embedded Systems Development Cycle

from past projects. Reuse of a proven design as a platform that can be built on confidently for a new solution is a good way to ensure the success of a new project design. If a radical change is required in the design, however, it is important to start testing the new concepts as soon as possible.

1.4.3 Architectural Design

Architectural design, also called *system design,* is the top layer of the design process in which the design is partitioned into functional blocks and appropriate technologies. For embedded systems the following are some of the architectural design decisions that must be made:

Hardware versus Software
Processor Choice
Language Choice
Major Hardware and Software Blocks
Buy versus Build

Hardware versus Software.   Because it can have a major impact on the overall design, the technologies used for each function block must be determined at the system level. This includes the decision to use software or hardware. Based on the requirements, the designer can determine the most appropriate technology for a function. Should the function be implemented in software, in hardware, or in both?

Some functions can be implemented in either hardware or software. If we want to use software, an analysis of the overall CPU load must be made to determine whether it can handle the extra task. Some functions cannot be handled by the CPU even if it is the only task assigned to it. If the CPU cannot handle the task, then extra hardware must be used. What type of hardware is needed: an MCU resource, an ASIC, or a purchased subsystem?

It is important to remember that hardware represents a recurring cost. A part must be purchased for every unit sold. Software represents a nonrecurring cost. It must be developed up front, but it does not represent a per-unit cost unless there are licensing fees.

The Processor Choice.   The choice of processor or at least the family of processors must be made at this early stage. This includes the choice of using a microcontroller or a microprocessor-based design. If it is a microcontroller, will it use single-chip or expanded mode? The processor choice not only impacts the CPU load but also the available on-chip resources, the type of source code, and the software development tools required for the design.

If the company has already made an investment in development tools and reusable software libraries, it makes sense to use a compatible processor. Other processors should only be considered when there is no alternative or there is a potential for significant product differentiation. When it really comes down to it, the processor choice for most projects has already been made. That is why it is extremely important to do research when you actually do select a new processor. You and everyone else in your company may be living with it for a long time.

Modern microcontroller families offer some flexibility. Since the CPU is the same for a given microcontroller family, the investment in software and software tools is not lost by choosing another member of that family. Therefore it is a safer decision to choose a microcontroller from a large family so there will be more flexibility in future project decisions. Do not assume, however, because an MCU shows up in the manufacturer's selection guide that it is actually available within your production schedule or price constraints. The selection of a specific MCU is normally a two-way process—narrow down the choice to the MCUs that are actually available, find the MCUs with the best fit for your project, and then look more closely at cost and delivery. Often you will have to make a tough decision—use the common, older part with lower cost and ensured delivery or use a rarer, possibly newer part that is a better fit for your design. It all depends on the project. You should always balance the lure of glitzy new parts with the confidence and stability involved in older proven parts.

When choosing a processor it is also important to avoid "religious" wars. Yes, some processors are faster than others for a given application, and some have very irritating instruction sets. Nevertheless the bottom line is, "Can it do the job and can you get the parts for an appropriate cost?" Beyond that, there are good and bad aspects to every processor family.

The Language Choice.   The language or languages used for the project needs to be selected early because the choice can affect the selection of processor, memory, and development tools. Another important issue is reusability. Is there already software available for a similar system that can be reused? What language would be preferable for future designs so the software designed for this project can be reused?

Again language selection is an area filled with religious wars. Try to avoid these wars and select the language objectively. Many times the choice comes down to the language that is most familiar to the programmer. It is only after learning multiple languages, however, that the choice of language can be based on technical and economical benefits and not on the programmer's bias.

Normally the language decision is between assembly language or C. Most of the complaints about either choice are based on poor programming structure or style, not on the languages themselves. It is easy to write a poor program in C and it is even easier to write a poor program in assembly. In addition many of the attributes of high-level languages are actually attributable to modular programming techniques and the use of software libraries. These methods can also be used with assembly code.

What drives the language choice today is portability, readability, and reliability. These factors are all based on the growing code size and today's market pressures. Technical merits of a language are just not as important. The bottom line is to know your language, know your design, and know your compiler.

Major Hardware and Software Blocks.   Major functional block decisions must be made at this point, such as the selection of a real-time kernel, major algorithms and data structures in software, or the power source or networking protocol used in hardware.

For microcontroller-based designs, this is the point at which you need to keep track of the required MCU resources such as GPIO ports, timers, and ADCs. It is a good practice to make a checklist of the required MCU resources. It can be very embarrassing to be late in a design stage only to find you are short one GPIO pin. You can also do this to some extent for the software tasks by estimating the CPU load for a given task and keeping a running total.

To Build or Buy.   The decision to build or buy software and hardware components is important. It can make or break the project. Buying software or hardware subsystems can be more cost effective if it is a considerable task to design your own. It is a good way to avoid "reinventing the wheel."

Typical software functions that can be purchased include real-time kernels, networking drivers, or math libraries. Hardware functions that typically are purchased include power supplies, data acquisition systems, or display modules.

Considerable problems can occur when purchasing software or hardware. No matter how much a component costs, there is no guarantee it will work, in time, for the project. The purchased software or hardware may not be a good fit for the project. If software source code is not available, you will need to rely on the manufacturer to revise the code or fix bugs. The same goes for hardware that has defects. With this in mind, remember that the manufacturer's priorities may not be the same as your own.

For software, the best choice may be open source code. With open source software, you can fix bugs and revise the code yourself. When you acquire the code, the bulk of the work is already done for you, but you also have the flexibility to customize the code to meet your needs and provide product differentiation.

System Design Outcomes.   The outcomes of the system design stage are primarily the system-wide design decisions already mentioned. Another outcome may be a document defining the requirements for each functional block. This document can be important to define the requirements for individual designers or design groups that will be responsible for that function.

1.4.4 Detailed Design and Construction

The detailed design and construction phase of the project is the primary focus of this text. It is the point at which you actually design, construct, and test the system. For software this means detailed design, writing, and debugging the code. For hardware this means detailed design, prototyping, and testing circuits.

MCU Core Design.   In Figure 1.6, the detailed design stage is shown as consisting of multiple steps. It starts out with the construction of the MCU core hardware and software. The reason the MCU core block is shown first is because it is the heart of the system. Once it is complete, the rest of the functions can be added and software can be written and tested.

Functional Block Design.   Once the MCU core is completed, the other functional blocks are constructed and tested. This is shown in Figure 1.6 as a repeating loop to emphasize the technique of breaking the system down into manageable functional blocks and working on each block separately. If only one person is working on the design, then the function blocks are completed sequentially as shown in the figure. If there are multiple designers, they can be completed in parallel.

The tasks involved for each functional block design include detailed hardware design and construction, detailed software design and construction, and hardware-software integration and testing. Note that Figure 1.6 shows there may be interdependencies between the hardware and software design. The hardware may not be testable until some of the software is complete and the software may not be testable until part of the hardware is complete.

Typically the hardware is constructed and minimally tested first. Then the majority of the work goes into writing the software and testing the software with the hardware.

1.4.5 Final Prototype Construction and Integration

The final prototype and integration is the last phase shown in Figure 1.6. It is the point at which all the functional blocks come together and a final hardware prototype is constructed. The software modules are also combined, if they have not been already, and the code is revised to run as a stand-alone application on the prototype. Before this point some of the software and hardware may have been constructed and tested on a development system. Depending on the development system used and the complexity of the project, this may be a very simple step or it may be a major undertaking. It is important not to underestimate the hours required to complete this stage. This is a critical step, which can contain some rather obscure bugs. It is not the type of work to be done at midnight before a major product demonstration.

1.4.6 Reviews

Reviews are demonstrations, schematic inspections, or code inspections. The main idea is to have other people familiar with the project and technologies assist in error detection. Reviews have been found to be an extremely productive way to detect errors. There are several reasons for this. First, when the designer prepares for a review, he tends to look at his design differently. By simply trying to articulate how the design works to others, the designer may catch errors. You have probably had the experience of trying to get help from someone only to find the solution yourself during the process of explaining the problem.

Having others inspect your design also provides a different perspective. Everybody's mind works differently, and when someone is not familiar with the design, he or she may examine it more carefully. It is common for the original designer to miss a serious error in a simple piece of code. The designer may look at the code and think, "I know that part works," and skip over it without a thorough review.

As shown in Figure 1.6, there may be reviews after the architectural design stage and after the hardware and software detailed design stages. The hardware design review usually involves the original designer presenting the schematic to the reviewers. The software review is often referred to as a *code review*. It involves the original designer presenting the source code to the reviewers. It is essential that the review materials be provided to the reviewers before the actual presentation. The importance and effectiveness of reviews cannot be overemphasized.

1.4.7 Unit Testing

The difference between the testing performed during the construction phases and unit testing is twofold. First, the unit testing is typically done by an independent tester who tests the product to verify that it meets the specifications in the requirements document. Second, errors detected at this stage tend to be very expensive to fix. It is important to realize that unit testing should not be the place to catch mistakes. You should provide the tester with a design that you fully expect to be bug free.

Another testing technique that is, for better or for worse, more popular now is public testing. Public testing involves the release of a preliminary design to select customers and having them provide feedback and perform error detection. This is also referred to as *beta testing*. Public testing can provide invaluable feedback with respect to error detection and the functional design. Customer feedback may not only result in code changes but also specification changes. Public testing should be a part of most test plans. It should never take the place of internal reviews and testing, however. Any code released to customers, whether beta or production, creates a lasting impression on the quality of your products. Releasing bug-prone betas not only creates the general impression that your product quality is poor, it can sometimes result in customers feeling that you are trying to save money by replacing your testing group with them.

1.4.8 Release

The process of releasing a design to production is not a simple matter of handing a floppy disk and a stack of paper to the production folks. Careful coordination and revision control is needed before a design is actually released. This is especially true with complex projects that have multiple programmers working together on different functional blocks. A released version of code or schematic package must be archived so it is always accessible. These functions can be provided by *revision control software*. Revision control software not only archives different revisions and generates revision numbers, it also locks files so they cannot be changed accidentally.

▶ ## SUMMARY

In this chapter we covered the structure, development, and applications for embedded systems. It is intended to give a broad view and context for the rest of the material in this text.

We introduced the microcomputer and its structure including different design implementations. We saw that a microcontroller includes all four components of a microcomputer in a single IC. We primarily use the 68HC12 family of microcontrollers, in the examples in this book.

We also looked at the development process for an embedded system. We focus on the detailed design tasks in this text, but we always have to keep the complete process in mind because it is the context in which we will always be working.

At this point we should also recognize the differences between embedded systems and computer systems. Although they both are microprocessor-based systems, they have very different requirements and markets.

EXERCISES

1. The 68000 microprocessor has a 24-bit linear address bus. What is the maximum number of directly accessible locations?
2. The circuit in Figure 1.4 shows an EPROM device output connected to bit 0 of the data bus. As the circuit is shown, what happens if the user accidentally writes to the EPROM? What signal could be used to improve the EPROM chip select logic?

3. Show the memory map for a 16-bit address space that has a 32K-byte EPROM starting at address $8000, a 2K-byte RAM starting at $2000, and a 32-byte register block starting at address $1000.

4. The following program is used to initialize *PORTP,* bit 0 to a high output:

```
bset DDRP,BIT0
bset PORTP,BIT0
```

Assuming the pin for this bit has a pull-up resistor, sketch and describe a timing diagram that shows the potential low glitch that can occur on the pin. Refer to Figure 1.5 and assume it starts in the reset state.

PART 2
Assembly Language Programming

Programming Basics

This chapter provides an overview of the basic concepts involved in programming embedded microcontrollers. The details are covered throughout the rest of the text. We examine different programming languages and their relationships to each other as well as the software construction process.

The important issues in today's development environment include reusability and maintainability. Programming languages, programming structure or style, and the software construction process all have a significant impact on reusability and maintainability of a program design.

▶ 2.1 PROGRAMMING LANGUAGES

Fundamentally programming is the design, coding, and debugging of a sequence of instructions or algorithms. In an ideal environment the programmer would only have to concentrate on the design of these algorithms and not be concerned with the implementation details. There is a fundamental problem with this ideal, however. The programmer communicates in a human language, such as English, but the CPU only understands binary.

In this section we look at three of the language types used to bridge the CPU–human language gap. These three languages are

Machine language   The binary language used directly by the CPU. This is the lowest-level language.

24

Assembly language	A low-level language made up of instruction mnemonics that map directly to machine instructions.
C	A high-level language. It is used as an abstraction layer to hide the implementation details of the CPU.

In general, the higher the *level* of the language, the more abstract it is from the actual CPU operation. This abstraction results in a more efficient programming process, but as we will see, this programming efficiency comes at the cost of lower machine code efficiency and some loss of control over the CPU operation.

What Is Source Code?   The term *source* is used to represent the original programming code generated by the programmer. If the programmer writes in assembly language, the assembly language is the source. If the programmer writes the program in C, the C code is the source code. The machine code could even be the source code if the programmer were to write the program in binary.

What Is Object Code?   The *object* code is the resulting code from the software build process. When programming microcontrollers, the object is always machine language.

2.1.1 Machine Language

Remember that the CPU is a digital circuit. The only language it understands is a language of ones and zeros—binary. Figure 2.1 shows a small sequence of instructions for the CPU12 in binary machine language. This is what the CPU reads, and it bases its operations on these instructions. The program in this form is called the *binary machine code.* It may also be referred to as *machine code, binary code,* or *object code.* It represents the actual ones and zeros in the CPU's program memory.

The machine language consists of CPU opcodes and operands. An *opcode* is one or more bytes that make up a single machine instruction. The *operands* are the binary instruction arguments. For example in Figure 2.1 the contents of addresses $8000, $8003,

Contents	Address
11110110	8000
00001000	8001
00000001	8002
11111011	8003
00001000	8004
00000000	8005
01111011	8006
00001000	8007
00000010	8008

Figure 2.1   Program Memory Contents/Binary Machine Code

and \$8006 are instruction opcodes. The rest of the memory locations are operands. There is no way to tell what is an opcode and what is an operand just by looking at the machine code. In fact the CPU cannot tell the difference. It simply assumes that the first byte it gets to is an opcode and goes from there.

Binary machine code is required by the CPU, but for a human it is just a bunch of ones and zeros. For even the simplest program, we cannot look at this code and understand what the instruction sequence is. There needs to be another language that we can read and understand. Then there needs to be a way to translate that language to binary machine code.

Hex Machine Code. The first and simplest translation that can be done is not a language translation at all. In order to make it easier for the programmer to read the machine code, it can be represented in hexadecimal instead of binary code. Not only is the binary machine code hard to read, it is also very susceptible to errors if editing is required. In the example in Figure 2.1, 72 bits make up the complete program sequence. If a single bit is incorrect, the program will not work. Note that the addresses in Figure 2.1 are already represented in hexadecimal. You can imagine how impractical it would be to try to work with addresses in binary. By converting the binary numbers to hexadecimal, it is much easier to edit or analyze the program.

Notational Note

Throughout this text you will find hex values displayed with the Motorola hex prefix "\$." There will be times, however, when the data are assumed to be in hex and the "\$" is not shown. This will include figures, tables, and listings that show the contents of memory. For example Figure 2.1 shows all addresses in hex but does not use the prefix "\$." Figure 2.2 also shows all addresses and memory contents in hex without the prefix. It should always be clear, by context, whether the value is in hexadecimal, decimal, or binary. The "\$" prefix will always be used in source code and in the body of the text. Later in the text, when we are covering C source code, the "0x" prefix will be used to represent a hex number.

Figure 2.2 shows the memory display generated by Motorola's D-Bug12 monitor command, *MD*. The address and the data are shown in hex. The first four characters contain the address of the first byte in the line. In this case the starting address is \$8000, which contains an \$F6. The rest of the line shows the contents of the 16 addresses \$8000 through \$800F. You should be able to see that contents of addresses \$8000–\$8008 shown in Figure 2.2 agree with the memory contents shown in Figure 2.1.

By using a hexadecimal representation of the machine code, it is easier to read and edit. In the past it was common to perform all memory editing and loading by entering hex codes with a hex keypad. As you can imagine, it would only be practical for very small programs. Today some minor memory editing may be done using hex during the debugging stage, but program loading is automated by using a desktop computer connected to the target CPU board.

```
8000   F6 08 01 FB - 08 00 7B 08 - 02 00 00 00 - 00 00 00 00
```

Figure 2.2 Hex Machine Code as Displayed by D-Bug12

2.1.2 Assembly Language

In order to make the machine instructions readable to the programmer, the CPU manufacturers define an assembly language for each CPU. The assembly language is made up of a set of mnemonics along with syntax rules for the arguments. Each mnemonic corresponds directly to a CPU machine instruction or opcode. Source 2.1 is the CPU12 assembly program for the machine code shown in Figure 2.1. The assembly code is much easier to read and understand. In this case we can see that the program loads *ACCB* with *var2,* adds that to *var1,* and then stores the result in *var3.* It is not important right now to know these instructions. It should be obvious though that this program would be easier for the programmer to work with than machine code.

Source 2.1 A CPU12 Assembly Language Program

```
ldab    var2
addb    var1
stab    var3
```

The translation from assembly language to machine code involves a one-to-one mapping process from the instruction mnemonic and argument to the binary opcode and operand. The translation can be done manually by using the instruction descriptions provided by the manufacturer. Since it is a simple lookup process, however, it is an ideal application for a computer. Assembling code by hand is a waste of the programmer's time. However, it is important that the programmer is able to make the translation as required.

The Assembler.   The computer program that performs the assembly language to machine language translation is called an *assembler.* If the computer that runs the assembler is not the same machine that will run the machine code, it is called a *cross-assembler.* For embedded systems programming we normally use cross-assemblers; however, throughout this text, cross-assemblers will simply be referred to as assemblers.

Figure 2.3 shows the data flow diagram for the assembly process. In addition to the binary machine code, the assembler may generate a program listing. Listing 2.1 shows the program listing generated by the Introl-CODE assembler. It shows the line number, the address, the machine code, and the original assembly code. The program listing is used for program debugging.

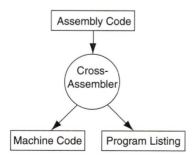

Figure 2.3   Assembler Flow Diagram

Listing 2.1 Program Listing

```
14   00008000 f60801          ldab    var2
15   00008003 fb0800          addb    var1
16   00008006 7b0802          stab    var3
```

Each line in the listing contains the information for a single instruction and its arguments. The address shown is the address of the first byte for that line. Following the address are the opcode and operands that the assembler generated from the instruction. For example the *addb* opcode, $FB, is at address $8003. If you compare Listing 2.1 with Figure 2.2, Source 2.1, and Figure 2.1, you should be able to see how the machine code and assembly code correlate.

The most important characteristic of assembly language is that it is a precise representation of the actual CPU operations. Each assembly instruction corresponds directly to a CPU operation. When an application requires this precision, assembly language is the language of choice. There are also some other important considerations to be made, however.

First of all, assembly language is CPU dependent and therefore it is not portable because the code would have to be rewritten to work on a different CPU. This affects the reusability of the program. Although assembly languages for different CPUs are very similar, this rewrite still can represent a substantial task, which in turn can affect the time to market and reliability of a design.

In addition, in some respects the precise nature of assembly language can be a negative characteristic. Assembly language precisely represents the CPU operation, which not only includes the desired high-level algorithm of the program but also all the implementation details. For example the *ldab var2* instruction in Source 2.1 is an implementation detail. It has nothing to do with the original intent of the program, which is to add *var1* and *var2*.

Since assembly language includes all the CPU operations, it is not only CPU dependent but is also cluttered with the implementation details, which may hide the original intent of the program. For these reasons it may be desirable to add an abstraction layer to hide these details. The program would be more to the point and independent of the processor if you were able to say, "Set *var3* equal to *var1* plus *var2*." This is where high-level languages come in.

2.1.3 C, A High-Level Language

The contents of a high-level language more closely represent the function or algorithm required rather than the CPU implementation details. Source 2.2 shows how the example used previously would be programmed in C.

Source 2.2 A High-Level C Program

```
var3 = var1 + var2;
```

The program is now reduced to a single line that shows only the addition function. There are no implementation details for the CPU operation. Therefore the program is shorter, easier to read, and independent of the CPU.

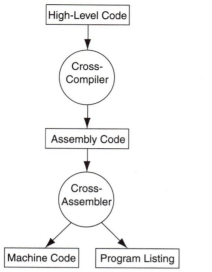

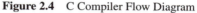

Figure 2.4 C Compiler Flow Diagram

The Compiler.   Another translator is required to translate the high-level language to machine code. As shown in Figure 2.4, the process actually performs the translation in two steps. First the high-level language is translated into assembly code. Then an assembler is used to translate the assembly code into machine code.

The process used to translate a high-level language to assembly language is called a *compiler.* If the compiler generates assembly code for a different CPU than it is executed on, it is a *cross-compiler.* Again, most compilers used for embedded microcontrollers are cross-compilers, but throughout this text they will be referred to as compilers.

In general, using a high-level language can result in a program that is independent of the CPU. Because it is CPU independent, the program is more portable. To change to a different CPU, we would only have to change the compiler and assembler to generate code for the new CPU. In actual practice, especially for embedded systems, many other considerations must be made before the code is completely portable. Using a high-level language makes it possible to write portable code, but a high-level program is not necessarily portable. We will discuss these portability issues later in the chapters on C programming.

We have seen that using a high-level language improves the readability of the program, makes the programming process more efficient, and makes it possible to write portable code. The downside to a high-level language, however, is that there is no longer a one-to-one correspondence between an instruction and the actual CPU operation. The compiler generates the assembly code and therefore plays a large role in determining the actual CPU operation. Because compilers are not usually as smart as programmers are, the machine code generated by a compiler is typically larger and less efficient than the machine code generated from assembly source code. This may be disconcerting to some programmers, but it is not a good enough reason to avoid high-level languages. Instead it means that you must know your compiler and know how the different parts of your C code will be implemented. These details are usually left out of a beginning course in C programming.

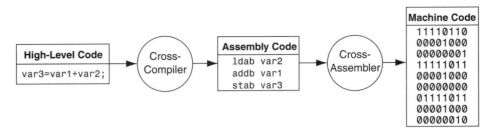

Figure 2.5 Summary of the Program Translation Process

To summarize, Figure 2.5 shows the three versions of our program together with the compiler and assembler processes. It shows the processes used and the code generated from a C source to the machine object. This overall process is called the *software build process*. We will look at different software build processes in more detail in Chapters 3 and 12.

▶ 2.2 TYPES OF PROGRAM SEGMENTS

Before we go on to the software construction process, let's define some terms for different program segments we may be asked to design or analyze. These segments range from a single line of source code to a full stand-alone program designed for a final product.

2.2.1 Program Segment Terms

Code Snippets. A code snippet is simply one or more lines of code that are part of a larger program. Code snippets are used to illustrate programming concepts. Since a code snippet is part of a larger program, it may not be possible to execute the snippet by itself. It would require a debugging monitor with software breakpoints and instruction tracing to actually see the effects of the code when executed.

Macros. Macros are code snippets that are named. You can include a macro in a program by referring to it by name. Macros do not use a subroutine call or return. Like snippets, macros cannot be executed alone without a debugging monitor.

Routines. A routine is a block of code that performs a single function. Routines are always executed with a subroutine call and return with a return-from-subroutine. In assembly a routine is called a *subroutine*. In C a routine is called a *function*. Routines again cannot run by themselves. There must be a calling program or a debug monitor to call the routine. Routines are the basic building blocks for larger programs.

Modules. A module is a single file that contains a collection of related routines. Modules are used for information hiding, portability, and improving the overall program organization. Modules may be dependent on a resident monitor, a multitasking kernel, or a main program. We will cover program modules in Chapter 12.

Monitor-Dependent Programs.   This is a complete program with the exception of the MCU initialization and *RESET* code. It requires a debugging monitor to perform all the required MCU initialization. Traditionally most of a project's development work is completed with the program in this form.

Stand-Alone Programs.   This is the complete program used in the final product. It includes everything needed for the system to run once power is applied. Stand-alone programs for embedded systems must be contained in nonvolatile memory devices if they are to run after cycling power. Very few of the examples shown in this text are stand-alone programs, however; we will look at the design and construction of stand-alone programs in Chapter 10.

▶ 2.3 SOFTWARE CONSTRUCTION

Software construction is the part of the software development cycle in which you actually write and debug the code. It consists of detailed design, coding, a build process, and debugging. Figure 2.6 shows the software construction cycle. It is an iterative process, in which errors are found, the source is changed, and the process is repeated until the code performs as required.

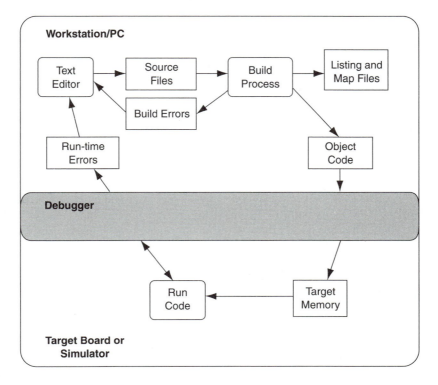

Figure 2.6   The General Software Construction Process

The first process shown is a text editor. This is the program used to create all source files. The software build process then creates the object files from the source files. The build process may include a compiler, an assembler, a linker, and other utility programs to build the object file, listings, and map files. Throughout this text we will look at different build processes with varying degrees of complexity. Once the object file is generated, it is loaded into the MCU memory or simulator memory and executed. The process used to communicate and control the CPU operation is called a *debugger*. Debuggers may reside on the target or may consist of a PC application. We also look at different debugging tools throughout this text.

During software construction there are two classes of errors, build errors and run-time errors. Build errors include compiler, assembler, or linker errors. They are usually the result of syntax or configuration errors in the source files. Run-time errors are errors in the actual operation of the program. The source of run-time errors may be a software bug or a misinterpretation of the software requirements. When errors are found, the programmer edits the source files and repeats the process. It is perfectly normal and expected that this is an iterative process. Unless the software is extremely simple or your brain works like a computer, you will make errors. It is unreasonable to expect otherwise.

The number of iterations through this process can vary tremendously, partly due to programming style and partly due to the complexity of the problem. Following are examples of two extremes in programming styles.

2.3.1 The Code Hacker

A code hacker is a programmer who cannot wait to type and run some new code. A code hacker can usually be found by following a trail of broken keyboards and Mountain Dew empties. The hacker makes no effort to find the actual source of programming errors or to design a good solution. The first line of code that has any remote relationship to the problem is changed, the file is saved (sometimes), and the process is repeated. Usually this haphazard approach adds more errors than it fixes, but if the code hacker works on it long enough, a working solution will eventually appear. Once the program works the code hacker will either send the code (let's hope it is the last version) to someone else or call the instructor over and announce aloud, "I'm done."

The code hacker's final code tends to be buggy and has a limited life span. As soon as the EPROM is burnt or the instructor arrives, another bug emerges and the program crashes. The code hacker then wakes up long enough to blame the hardware.

2.3.2 The Computer-Shy Perfectionist

Some programmers actually take error messages personally. They will try to write a complete program in a single iteration of the construction process. This type of programmer is terrified of the keyboard and will avoid it until he or she can prove that the program design will work. The program design has pages of flow diagrams, state diagrams, and even source code snippets written on paper. The keyboard still has not been touched, however. These programmers usually cannot be found—they are in seclusion until they are positive their program works or until the assignment is due. Even then these programmers show up with a fist full of papers and no code.

Be careful. This type of programmer can be mistaken for an excellent designer. The only problem is the missing code. Do not believe it when one of these programmers says, "It works. All I have to do is type it in. It will only be a few minutes."

2.3.3 Becoming a Good Programmer

Both of these extremes in programming style are missing some important points. The most obvious point the hacker is missing is the value of careful design. Careful design is needed not only to avoid introducing new bugs but also to make sure the code actually meets the requirements. The computer-shy designer, on the other hand, is missing the value of using computer tools. The process of using a text editor and building and running code with a debugger can cut hours of misguided analysis and research. The tools are there to help the programmer gain insight into the operation of the system. If they are not used, many hours can be wasted getting nowhere.

The debugging process should be looked at as part of a learning process in which the debugging tools are used to observe and gather information about how the program is actually working. You do not need to find a bug to get value out of the debugging process. The hacker's guessing and the perfectionist's research can never make up for the information gathered from good debugging tools. A good programmer not only understands that her program works, she also understands why it works and how it works.

2.3.4 Construction Times

It is helpful to define specific times within the software construction cycle. Understanding the time within the cycle when an event or process occurs helps us understand a lot more about the characteristics and limitations placed on those events and processes.

Build Time.   *Build time* is the time in which the program is built, that is, assembled, compiled, or linked. It can be defined in more detail as *compile time, assemble time,* and *link time.*

Some of the things that happen at build time include program definition, determination of constant values, and definition of static memory locations.

In general things that happen at build time cannot be changed during load time or run time. The exceptions would be program memory locations for relocatable code or parameter values stored in EEPROM.

Load Time.   During program development, this is the point at which the program is loaded into CPU memory. On the final product, this corresponds to the state of memory when system power is applied. There are only a few things that can happen at load time but not at program time.

One of the most common is a memory checksum or CRC value that must be stored in memory for a power-up memory test. This is because we do not know the memory contents until we build the program. If we do not know the memory contents, then we cannot calculate the checksum or CRC. This is also related to another thing that happens at load time—filling the unused memory. It would be a waste of time to define constants to fill all remaining memory during build time. Normally the memory is filled with the desired opcode by a device programmer or monitor and the program is loaded on top of that. Some development systems can also automatically produce S-Record files that contain the fill character.

Run Time.   Run time is the time when the CPU is actually running your program. Anything that must change during operation must be changeable at run time. Variables are the most obvious item that must be set during run time. Since variables must be changeable, they must be contained in RAM. Since they are in RAM, their values are unknown at system power-up (load time). So they must be initialized at run time. Initializing variables is actually an interesting problem. We just said that variables must be initialized at run time. We need to know the initial value, however, so the initial value must be defined at build time. How do we get something that must be defined at build time into something that can only be set at run time? The answer is that we use both a constant and a variable. The constant is set equal to the initial value at build time. At run time the value of the constant is copied into the variable during the initialization process.

▶ SUMMARY

In this chapter we looked at some of the basic concepts for programming and software construction. Machine code, assembly code, and high-level languages were covered along with a summary of each language's advantages and disadvantages. We also looked at program construction and programming styles.

The material in this chapter presented the big picture in software construction and programming techniques. In the next few chapters we begin to dig into this material in more detail.

EXERCISES

1. What are some of the advantages and disadvantages of using assembly source versus C source code?
2. Convert the following binary numbers to hexadecimal:
   (a) 00000000
   (b) 01011010
   (c) 11110000
   (d) 10000001
3. By looking in the *CPU12 Reference Manual*, verify the opcodes given in Listing 2.1. Rewrite the listing if the *addb* instruction is replaced with the *subb* instruction.
4. What is the difference between a routine and a module?
5. Are the contents of the following objects defined at build time, load time, or run time?
   (a) program code
   (b) the program stack space
   (c) the stack pointer
   (d) variables
   (e) constants
6. If a program needs to use the results of two constants multiplied together, should the multiplication be performed at build time or run time? Why?

3

Simple Assembly Code Construction

Because there are so many topics that need to be covered to write embedded programs, it can be difficult to get started. Yet the only way to really learn this material is by practice. Reading this book alone will not make anyone a good programmer. You must write code and debug it to gain the required insight and experience. So how do we practice if there are so many things to learn? Let's start by looking at a list of topics that need to be covered to write basic assembly programs:

The Software Construction Process	An Editor and/or IDE
	The Assembler
	Debugging Commands
	Debugging Hardware Configuration
The Microcontroller Software Model	CPU Registers
	CPU Instruction Set
General Software Design	Structured Programming
	Program Design Techniques
Microcontroller Hardware	MCU Configuration
	MCU On-Chip Resources
	MCU Operating Modes and Exceptions
Real-Time I/O Techniques	Event Detection
	Real-Time Multitasking

In this chapter we will simplify the task of programming by developing only the basic tools needed to construct a nonmodular assembly language program. This will allow us to run the examples and test the concepts covered in the next few chapters. Let's first look at the list of topics again and reduce it to the items that are essential to this task. The rest can be covered later.

Software Construction Process.   We must have a software construction process in order to write software. At this point, however, we do not need a high-end system that is designed for large, complex real-time designs. Let's start by introducing a very simple construction process. The process described is practical for monitor-dependent or stand-alone programs up to about 500–1K-bytes. Once the code grows larger, it is best to move up to the more complex modular build process described in Chapter 12. Although the process described here can be used for stand-alone programs, we will only be creating monitor-dependent programs in this chapter.

All the code will be contained in a single file and will have absolute locations. This way we do not need to use a linker or learn about the syntax requirements for modular programs. The program location will always be defined using the *org* assembler directive.

The debug process will use a single development board with a serial port connected to a desktop computer. It uses a monitor program resident on the board, which has commands to load code into memory, display registers, and memory locations. We will also use the monitor to set software breakpoints and trace instruction execution. The programs will normally be loaded into RAM.

With this system we will need to know how to use the editor, the assembler, the development board, and the monitor commands. We will also need to learn the basic assembler syntax and directives.

The CPU Programming Model.   The microprocessor's programming model is introduced in the next two chapters, in which we cover the CPU12 registers and most of the instruction set. We will need to know that material before writing our own code. In this chapter we demonstrate the basic software construction process by using a prewritten demonstration program.

MCU Hardware.   The hardware design concepts are kept to a minimum until Chapters 9 through 11. We can put off the hardware design because most of the hardware is already contained in the microcontroller. Once we can write simple programs, we will look at hardware issues such as accessing on-chip resources and designing external circuitry. The initialization code required for hardware configuration will also be avoided at this point. We will start with monitor-dependent programs only. The monitor program will always handle all the initialization tasks. The only hardware-related material we need to know now is the memory map of the development board. The memory map is required so we know where to load our code.

There are some examples that use general purpose I/O, GPIO. It may help to review the section on GPIO in Chapter 1 and for more detail, read the section on GPIO in Chapter 9.

Software Design.   Software design techniques also will be covered in more detail later. The time that should be spent on software design goes up as the complexity of the program goes up. Since we are starting with very simple programs, there is little formal design required. Once we have covered the microcontroller software model and basic programming techniques, we will cover software design in Chapter 6.

Real-Time Techniques.   We cannot get very far with our embedded software without using real-time techniques. We can, however, cover many programming techniques before introducing the complexity of real-time programming. This too will be delayed until Chapter 8, after the basic assembly programming, design, and applications are covered.

Documentation.   The material covered in this chapter is not intended to replace the user manuals from the assembler and development board manufacturers. It is expected that the manufacturer's user manuals will be available and will be used as a reference for details. This text covers the overall concepts and processes.

▶ 3.1 ASSEMBLY SOURCE CODE

In this section we look at a basic source program called *demo1*. The complete program is contained in a single module (file), so a linker and modular programming syntax are not required. The source for *demo1* is shown in Source 3.1. This program generates 60 one-second pulses, but more importantly, the source code contains most of the pieces of a typical assembly source program.

3.1.1 Program Contents and Organization

Program organization is the order and contents of the source file. The program in Source 3.1 is used to illustrate the organization used throughout this text. Appendix A also covers the coding conventions used in this text. There are different conventions for organizing the pieces of an assembly program. If you do not follow the organization described here, you should at least define and follow your own conventions and follow them consistently.

Comments.   There are two different types of comments: comments that make up complete lines and comments added to the end of a source line. In general your programs should have a combination of these types. The complete line comments are used as headers to describe what is to follow, and the comments at the end of a line describe a more specific operation.

Source 3.1 An Assembly Code Example, demo1.s12

```
;*************************************************************************
; A simple demonstration program to see the HC12 run.
; It generates 60 1-second, active-low, pulses on PP0.
;   MCU: 68HC912B32EVB, E=8MHz
;   Monitor: D-Bug12
;
; 11/4/99 Todd Morton
;*************************************************************************
; Equates
```

```
;***********************************************************************
DDRP              equ $57                 ;Port definitions
PORTP             equ $56
BIT0              equ %00000001
TC1MS             equ 1996                ;Delay count for 1ms
;***********************************************************************
; Program
;***********************************************************************
                  org $0800

main              bset PORTP,BIT0         ;Initialize PORTP, BIT0
                  bset DDRP,BIT0
                  movb InitCnt,CurCnt     ;Initialize pulse counter


;***********************************************************************
; Main loop for output pulse generation
;***********************************************************************
pulse             bclr PORTP,BIT0         ;Turn pulse on.
                  ldd #250                ;Wait 250ms
                  jsr WaitDms
                  bset PORTP,BIT0         ;Turn pulse off.
                  ldd #750                ;Wait 750ms
                  jsr WaitDms
                  dec CurCnt              ;Count pulses?
                  bne pulse               ;  No: Another pulse
                  swi                     ;  Yes: Return to monitor


;***********************************************************************
; Subroutine WaitDms - A programmable delay in ms.
;     Arguments: The number of ms is passed in ACCD.
;     Registers: preserves all registers except CCR.
;     Stack Reqs: 6 bytes stack space
;     Req. Subs: Dly1ms
;***********************************************************************
WaitDms           pshd                    ;preserve ACCD
msdlp             jsr Dly1ms              ;execute 1ms ACCD times
                  subd #1
                  bne msdlp
                  puld                    ;recover ACCD
                  rts


;***********************************************************************
; Subroutine Dly1ms - 1ms delay loop.
;     MCU: 68HC12, E=8MHz, no clock stretching, 16-bit bus
;     Registers: preserves all registers except CCR.
;     Stack Reqs: 2 bytes stack space
;***********************************************************************
Dly1ms            pshx                    ;preserve IX
                  ldx #TC1MS              ;execute loop TC1MS times
```

```
d1mslp          dex
                bne d1mslp
                pulx                    ;recover IX
                rts

;*********************************************************************
; Constants
;*********************************************************************
InitCnt         fcb 60                  ;Initial pulse count
;*********************************************************************
; Variables
;*********************************************************************
CurCnt          rmb 1                   ;LED flash counter
;*********************************************************************
```

Comment headers are required for the program, each subroutine, and each interrrupt service routine.

The program header should include a functional description of the program, list the program dependencies, and of course, include the programmer's name. There also may be a revision number and revision date. In the case of Source 3.1, there is no revision number, but the date of the last revision is shown.

Comments are very important. Remember that most of the code that goes into a project is reused. If the code is poorly commented, it will take more time to understand how the code works, increase the chance of a bug, and decrease the mood level of the programmer trying to understand the code. One thing to note here is that we can also have too many comments. If the program is filled with obvious or inaccurate comments, the reader will start ignoring all comments. One common mistake is to repeat the source instruction in the comment. A comment should never describe a single source instruction. Look at the following example:

```
    ldaa var                    ;Load ACCA with var
```

The comment on this line is worthless. It simply repeats the instruction. Comments should always be written at a higher level to describe the function of the code or should contain a special note or programmer warning. They should be written in the appropriate native language, such as English, but complete sentences are not necessary. Sometimes it is also appropriate to write comment headers in a pseudocode like pseudo-C to describe the function at a higher level.

Equates.   A section of program equates should follow the program header. In *demo1* this is the section labeled *Equates*. Equates are constant definitions for the assembler. By using equates you can use symbol names to make your code more readable. Most assemblers require that equates are defined first.

Main Program.   After equates there should be an origin directive to define the starting location of the program code. The main program follows the origin directive. It is called the main program because it is the program executed from the monitor program. Unlike C programs the main program does not have to be called main. Any label can be used at the start of the main program.

By placing the main program before constant and variable definitions, the address in the origin statement can always be used to run the program regardless of future code changes. For example to run *demo1* the following monitor command would always be used:

```
>g 0800
```

This is because the main program for *demo1* will always be located at $0800.

Subroutines.   All subroutines and interrupt service routines should follow the main program. In Source 3.1 *WaitDms* and *Dly1ms* are subroutines. Subroutines themselves should never be contained within the main program code or another subroutine. Of course you can call a subroutine from another subroutine, but the code itself should never be nested.

Stored Constants.   The stored constants are defined after all the program code. These constants are actually placed in the MCU memory so they must follow an origin statement. In Source 3.1 there is not a new origin statement for the constants so they will immediately follow the program code in memory. There is one stored constant in Source 3.1, *InitCnt*.

Variables.   Variables normally are placed at the end of the program. One thing unique about variables is that they must be placed in RAM. During development the whole program is placed in RAM, so there is not another origin statement for the variables. If, however, you place the program in ROM, there has to be a separate origin statement for the variables. There is one variable in Source 3.1, *CurCnt*.

Again, you should follow the organization shown in Source 3.1. If you do not, at least be consistent with your own organization. Now let's look at the assembler syntax in more detail. Remember that you also can reference the user manual for your assembler for details.

3.1.2 Assembler Syntax

First the assembler expects its input to be a text-only (ASCII) file, which contains assembler text only. This means the file cannot contain any embedded word processor formatting, html tags, or e-mail headers. A text editor is normally used to generate the file, but a word processor can also be used if the file is saved as "text only." Realize that an editor designed for program development is quite different from a word processor. When writing programs, it is preferable to use a text editor.

Source Line Syntax.   There are three possible types of lines in a source file—blank lines, comment lines, and source lines. The blank lines are obvious. They simply serve to provide spacing for readability. If the assembler does not accept blank lines, we can add a comment character at the beginning of the line.

If a line begins with an asterisk or semicolon, everything that follows in the line is a comment. Most assemblers accept either asterisks or semicolons as comment indicators.

Comments may also be placed at the end of a source line, separated from the operation or operand field by either white space or a semicolon. Since some assemblers require the semicolon, it is good practice to always use a semicolon to start a comment at the end of a line.

If the line does not start with a comment character, it is a source line. Source lines may contain a source code instruction, a directive for the assembler, or a label. Each source line must have the following format:

```
label:       operation operand,operand         ;comment
```

To separate the fields either space(s) or tab(s) are used as white space. Spaces are preferable because they will result in the correct spacing for any monospaced font in any editor or display. Tabs are more convenient, however. A good programming text editor allows you to use tabs but instead of inserting tab characters, it inserts the correct number of spaces for the same effect. It also helps to have an editor that allows auto-indenting. With auto-indenting, you do not have to enter the white space that starts most lines.

Labels. The first item on a source line is an optional label. A label is a symbol that is defined in the first column of the source code. The value of the label is set equal to the address of the first byte defined on that line. Labels are used as references for jump, call, and branch instructions.

The colon placed after the label is optional. When a colon does follow a label, references to that label do not include the colon. The colon is used to make it easier for a search tool to find the label definition. For example in Source 3.1 the subroutine *Dly1ms* is defined with a colon following the label. When this subroutine is called using *jsr Dly1ms,* however, the colon is not used. If we then needed to find the subroutine *Dly1ms,* we would search for *Dly1ms:,* and only the label definition would be found. Since most assembly programs are not very large this function is not very useful. In addition the Introl-CODE assembler uses the colon to export the label for use in other files. If you only have one source file, this is not a problem. If you have multiple source files, however, you may get undesirable results.

After the label is the operation. The operation can be either an assembly mnemonic or an assembler directive. An assembler directive is an instruction for the assembler. The microcontroller does not see the instructions or understand them. Following the operation is the operand or operands separated by commas. Note that some assemblers require arguments separated by white space instead of commas.

Number Syntax. Assemblers designed for Motorola processors use the following format to determine the base of a number:

```
[radix]n
```

where the radixes are

Radix	Type
$	Hexadecimal
@	Octal
%	Binary
&	Decimal

and *n* is 0...9, A...F, or a...f. If no radix is given, most assemblers default to decimal. Beware, however, some assemblers out there default to hexadecimal.

Symbols.   Symbols are the names given to equated constants and labels. Different assemblers have different rules governing legal symbols. Things to watch out for include the maximum length of a symbol and whether the assembler is case sensitive. Symbols in the Introl-CODE assembler are made up of letters (a...z, A...Z), digits (0...9), ?, $, _, and periods (.). They must begin with a letter, a period, or a question mark, and can be any length. Although symbols can be any length, only the first 16 characters are printed in the listing file. Therefore the practical limit is 16 characters for a symbol name. This is long enough to create good self-documenting symbols and labels. If your assembler does not accept at least eight characters for symbol names, it is time for a new assembler. The Introl-CODE assembler is also case sensitive so *ABc* is not the same as *ABC*. This is another important attribute of a good assembler because again, it allows more readable, self-documenting symbol names.

Special $ and * Characters.   When the characters $ or * are used alone in an expression, they are converted to the address of the beginning of that line. Most assemblers accept one or the other or both of these characters for this function. Following is a typical example:

```
        bra *
```

This is equivalent to

```
trap    bra trap
```

Readability Odds and Ends.   It is very important to produce readable code. Look at the formatting of Source 3.1 and take note of some techniques that make the code more readable. All instructions should line up in a single column. This may be difficult when long label names are used, but for most cases it is easy to do and well worth it. Some programmers also prefer that arguments line up in the same column. This is a good idea, but you should avoid shifting the argument too far away from the instruction.

3.1.3 Assembler Directives

Assembler directives are instructions directed to the assembler, not the microcontroller. The following descriptions cover the basic directives needed for the simple assembly process described in this chapter. Other directives are described later in the text as needed. You should be able to write any assembly program contained in a single module with these directives alone. Again, these directives are specifically for the Introl-CODE assembler. Most other 68HC11 and CPU12 assemblers follow roughly the same rules, however.

Origin, *org*.   The origin directive defines a starting address for the assembler. It tells the assembler where to place the following code. Following is the syntax:

```
org abs_expr
```

where *abs_expr* is an absolute expression consisting only of constants and symbols previously defined within the current input file.

Source 3.1 contains a single *org* statement immediately before the main program. In that case it sets the start of the program to be at address $0800. Since it is the only *org* statement in the program, all other code follows in order. Unless there are separate memory devices used or special locations required for different sections of the program, there should be only one *org* statement.

For the simple assembler process described here, the *org* directive should always be used for portability to other common assemblers. Later, once we start writing modular programs, a linker is required and we will replace the *org* directive with the *section* directive.

Equate, *equ*.   Equates are used to set a symbol equal to a value. The syntax is

```
symbol   equ abs_expr
```

where *abs_expr* is an absolute expression consisting only of constant values and symbols previously defined within the current input file.

Equates are important for both code readability and revisability. By using symbols instead of numbers, the programmer knows the purpose of the value. For example the program in Source 3.1 has the following two equates:

```
PORTP    equ $56            ;address of PORTP
BIT0     equ %00000001      ;bit location of bit0
```

The assembler will replace every occurrence of *PORTP* with $56 and *BIT0* with %00000001. The second line in the program uses these two symbols:

```
   bset PORTP,BIT0
```

By using these symbols, the intention of the code is very clear. This line sets bit 0 of *PORTP.* It is much more understandable than

```
   bset $56,%00000001
```

Or even worse

```
   bset 86,1
```

Equates are used to define constant values, but these constants are not stored in MCU memory. Equated constants are used only by the assembler.

Stored Constants.   Stored constants are actually stored in the MCU memory. They are normally used to create lookup tables and initial values for initialized variables. There are several directives that can be used to define stored constants including *fcb, fdb, fcc,* and *dc*. First we will cover the traditional Motorola-type directives—*fcb, fdb,* and *fcc*—and then the general define constant directive, *dc*.

Form Constant Byte, *fcb*. *fcb* is used to define one or more bytes of constant storage. The syntax is

```
symbol    fcb expr{,expr}
```

where *expr* is an expression limited to a one byte result. There may be more than one *expr* separated by commas. For example

```
NUM       equ 10
          org $0100
TABLE     fcb $10,17,NUM+6
```

The *fcb* directive above has three entries. Therefore it creates three constants. Each constant will be stored in one memory byte. Because of the *org* statement on the second line, the first byte will be stored at address $0100. Following are the resulting memory contents in hex:

Contents	Address
10	0100
11	0101
10	0102

Form Double Byte, *fdb*. *fdb* is used to define one or more two-byte constants. The syntax is

```
symbol    fdb expr{,expr}
```

where *expr* is an expression limited to a 16-bit result. There may be more than one *expr* separated by commas. For example

```
          org $2000
LIST      fdb $C13,$1000,LIST+$FF,50
```

The *fdb* directive above creates eight bytes to be stored in memory. The first two bytes will be stored at address $2000 because of the *org* directive on the preceding line. Following are the resulting memory contents in hex:

Contents	Address
0C	2000
13	2001
10	2002
00	2003
20	2004
FF	2005
00	2006
32	2007

The first two expressions in the list are obvious and result in the values $0C13 and $1000 being stored in memory locations $2000–$2003. The third item on the list is a little tricky. It uses the symbol *LIST*. Since *LIST* is in column 1 of the *fdb* line, it is a label, which is set equal to the address of that line. In this case the address is $2000, so *LIST + $FF* is equal to $20FF. The last argument in the list is only a single byte. Notice the *fdb* directive extends the byte and uses two bytes of memory.

Multiple Byte Storage.   Notice the order in which the bytes are stored by the *fdb* directive. The most-significant byte is always stored in the lower address location. This is called *big-endian* storage. In general, Motorola-type processors always use big-endian storage. Intel-type processors use *little-endian* storage in which the byte order is reversed. If you were using a little-endian processor, the constants would be stored as follows:

Contents	Address
13	2000
0C	2001
00	2002
10	2003
FF	2004
20	2005
32	2006
00	2007

If you are learning your first assembly language, big-endian storage is easy to understand and seems more natural. Little-endian-type processors take a little time to get used to.

Form Constant Character, *fcc*.   *fcc* is used to define one byte of constant storage for each ASCII character in a string. The syntax is

```
symbol  fcc delimited string
```

where the *delimited string* is a character string enclosed with two delimiting characters. The first character following the *fcc* command is the delimiting character. Typical delimiting characters are the backslash (/), single quotation mark (') and double-quotation mark ("). For example

```
        org $1000
STRING  fcc /Hello/
```

In this case the delimiting character is a backslash (/). Try to be consistent by always using the same delimiter. Sometimes it is not possible, however, such as when the delimiting character is contained in the string.

The *fcc* directive in the example creates five bytes to be stored in memory. The first byte is stored at address $1000 because of the *org* statement on the preceding line. Following are the resulting memory contents in hex and the corresponding ASCII characters:

ASCII Character	Memory Contents	Memory Address
H	48	1000
e	65	1001
l	6C	1002
l	6C	1003
o	6F	1004

Define Constant, *dc*.   The *define constant* directive is more general than *fcb, fdb,* and *fcc* and is becoming more widely used than the Motorola-type directives. The Introl-CODE assembler accepts both forms. Check with your assembler documentation to see which form you can use. The syntax is

```
[symbol]    dc[.fsize] expr{,expr}
```

where the size of the constant is defined by the *fsize* modifier. The two most common size modifiers are *b* (byte) and *w* (16-bit word). *dc.b* can replace *fcb* and *dc.w* can replace *fdb*. For strings the *expr* must contain the string delimited with single or double quotation mark characters. The following example will store an ASCII string, which is compatible with both the BUFFALO monitor utility routines and C:

```
hello   dc.b 'Hello',$04,0
```

Following are the resulting memory contents in hex:

ASCII Character	Memory Contents	Memory Address
H	48	1000
e	65	1001
l	6C	1002
l	6C	1003
o	6F	1004
EOT	04	1005
NUL	00	1006

The $04 is the ASCII end-of-text character that is used by the BUFFALO routines and the $00 at the end is a NULL for the termination of a C string.

Variable Storage.   Static variables require memory space for storage. The space must reside in RAM because variables must be writable. When we tell the assembler to reserve memory space for a variable, the assembler will simply skip that space without putting anything into it. This means nothing gets loaded into that memory location, and there is no reference to the location in the binary file that is loaded into CPU memory. At power up the contents of RAM are unknown, so some variables may require initialization. The other function of the variable storage directives is to set a label equal to the address of the memory reserved. This label is then used throughout the program as the variable name. Again there is a Motorola-type directive, *rmb,* and a more general directive, *ds.*

Reserve Memory Byte, *rmb.* Reserve memory byte reserves the number of bytes indicated for a variable or variables. The syntax is

```
[symbol]   rmb abs_expr
```

where *abs_expr* is the number of bytes to be reserved.

Define Storage, *ds.* Define storage works the same way as *rmb* except a size modifier is added to determine the number of bytes to reserve for each entry.

```
[symbol]   ds{.size} abs_expr
```

allocates *abs_expr* of size *size* in the current location. The following two snippets of code are equivalent:

```
* Motorola-type:
Var     rmb 1
Pointer rmb 2

* General type:
Var     ds.b 1
Pointer ds.w 1
```

Var is a label set equal to the address of an 8-bit variable, and *Pointer* is set equal to the address of a 16-bit variable.

There are several other directives, but the ones introduced here are all that are required for simple, nonmodular assembly programs. Other assemblers also may have different directives for the same purposes. Read your assembler documentation and make sure you understand the directive syntax for origin, equates, stored constants, and variable definitions.

▶ 3.2 A BASIC BUILD PROCESS

The software build process is the part of software construction that translates the source code to object code. It also generates other files required for debugging. Now we will look more closely at a simple build process for assembling a single source file. Figure 3.1 shows the data flow diagram for the build process. This process uses the Introl-CODE assembler and S-Record utility. It assumes these processes are called from the command line.

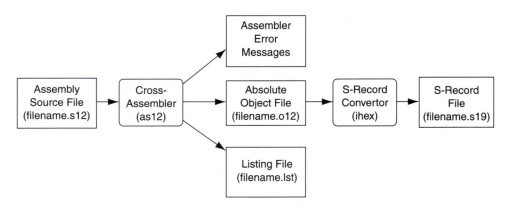

Figure 3.1   A Simple Assembly Build Process

The Introl-CODE assembler, *as12,* translates the assembly source file, *filename.s12,* into binary machine code, *filename.o12*. The assembler will also generate a listing file, *filename.lst*. If there are assembler errors, the assembler reports these errors with messages to the console. In order to load the code to a development board, the object must be converted into an S-Record file. This is done by the Introl-CODE utility *ihex*.

To build the *demo1* program shown in Source 3.1, we would use the following command line sequence:

```
% as12 –ghrtq demo1.s12
% ihex demo1.o12
```

Optionally a shell script could be written to execute both of these commands with a single build command. It does not get much simpler than this.

Note

The *as12* command line options are there to save paper and disable some automatic features in the assembler. Remember that they are optional. It is important to refer to the Introl-CODE manual to determine the command line options required in your environment.

After the processes are run, there are four files: the source file, the binary object file, the listing file, and the S-Record file. The listing file generated by the Introl-CODE assembler for *demo1* is shown in Listing 3.1. We will use this for debugging in Section 3.3. The first part of the file shows the original source code along with line numbers, addresses, and the object code. Following the code is a *section synopsis*. It lists the number of bytes in the complete program. In this case there are $3A, or 58, bytes used. The last part of the listing file is the *symbol table*, which shows the definitions of the symbols used in the program. For example the symbol *WaitDms* is set to $0824. This agrees with the address on line 44 where *WaitDms* is defined.

Listing 3.1 Listing File from the Source Code in Source 3.1

```
 1                              ;********************************************************
 2                              ; A simple demonstration program to see the HC12 run.
 3                              ; It generates 60 1-second, active-low, pulses on PP0.
 4                              ;   MCU: 68HC912B32EVB, E=8MHz
 5                              ;   Monitor: D-Bug12
 6                              ;
 7                              ;   11/4/99 Todd Morton
 8                              ;********************************************************
 9                              ; Equates
10                              ;********************************************************
11 00000057         DDRP            equ $57
12 00000056         PORTP           equ $56
13 00000001         BIT0            equ %00000001
14 000007cc         TC1MS           equ 1996            ;Delay count for
                                                         1ms
15                              ;********************************************************
16                              ; Program
17                              ;********************************************************
18                                              org $0800
19
20 00000800  4c5601  main            bset PORTP,BIT0     ;Initialize PORTP,
                                                          BIT0
21 00000803  4c5701                  bset DDRP,BIT0
22 00000806  180c08380839            movb InitCnt,CurCnt ;Initialize pulse
                                                          counter
23
24                              ;********************************************************
25                              ; Main loop for output pulse generation
26                              ;********************************************************
27 0000080c  4d5601  pulse           bclr PORTP,BIT0     ;Turn pulse on.
28 0000080f  cc00fa                  ldd #250            ;Wait 250ms
29 00000812  160824                  jsr WaitDms
30 00000815  4c5601                  bset PORTP,BIT0     ;Turn pulse off.
31 00000818  cc02ee                  ldd #750            ;Wait 750ms
32 0000081b  160824                  jsr WaitDms
33 0000081e  730839                  dec CurCnt          ;Count pulses?
34 00000821  26e9                    bne pulse           ;  No: Another
                                                             pulse
35 00000823  3f                      swi                 ;  Yes: Return to
                                                             monitor
36
37                              ;********************************************************
38                              ; Subroutine WaitDms - A programmable delay in ms.
```

```
39                               ;      Arguments: The number of ms is passed in ACCD.
40                               ;      Registers: preserves all registers except CCR.
41                               ;      Stack Reqs: 6 bytes stack space
42                               ;      Req. Subs: Dly1ms
43                               ;****************************************************
44  00000824  3b       WaitDms      pshd                      ;preserve ACCD
45  00000825  16082f   msdlp        jsr Dly1ms                ;execute 1ms ACCD
                                                                times
46  00000828  830001                subd #1
47  0000082b  26f8                  bne msdlp
48  0000082d  3a                    puld                      ;recover ACCD
49  0000082e  3d                    rts
50
51                               ;****************************************************
52                               ; Subroutine Dly1ms - 1ms delay loop.
53                               ;      MCU: 68HC12, E=8MHz, no clock streching, 16-bit
                                            bus
54                               ;      Registers: preserves all registers except CCR.
55                               ;      Stack Reqs: 2 bytes stack space
56                               ;****************************************************
57  0000082f  34       Dly1ms       pshx                      ;preserve IX
58  00000830  ce07cc                ldx #TC1ms                ;execute loop TC1ms
                                                                times
59  00000833  09       d1mslp       dex
60  00000834  26fd                  bne d1mslp
61  00000836  30                    pulx                      ;recover IX
62  00000837  3d                    rts
63
64                               ;****************************************************
65                               ; Constants
66                               ;****************************************************
67  00000838  3c       InitCnt      fcb 60        ;Initial pulse count
68                               ;****************************************************
69                               ; Variables
70                               ;****************************************************
71  00000839           CurCnt       rmb 1         ;LED flash counter
72                               ;****************************************************
```

Section synopsis

    1 0000003a (58)

Symbol table

            1 00000800 | Dly1ms   1 0000082f | WaitDms   1 00000824 | main
            1 00000800 |

```
msdlp       1 00000825 | CurCnt     1 00000839 | InitCnt    1 00000838 | d1mslp
            1 00000833 |
```

Symbol cross-reference

```
CurCnt             22      33    *71
Dly1ms             45     *57
InitCnt            22     *67
WaitDms            29      32    *44
d1mslp            *59      60
main              *20
msdlp             *45      47
pulse             *27      34
```

The S-Record file is shown in Figure 3.2. This file will be sent to the development board. For this process the binary file is an intermediate file used to generate the S-Record. Once the S-Record file has been generated, the binary file can be deleted. S-Record is a standard text format for sending binary code to programmers or development boards. The reasons we use a format like this are as follows:

- The file format follows a known standard, so any device that meets the standard can receive it regardless of the contents.
- The file is a text file so you can look at it on the terminal and see what is being sent to the development board.
- The destination address for the data is given.
- A checksum is included at the end of each line for error detection. Since we are sending these files over a serial link, error detection is important.

The first two characters in each line indicate the type of record. Following are some of the possibilities:

S Character	Description
S1	16-bit Address
S2	24-bit Address
S3	32-bit Address
S9	Termination Record

In this case we have 16-bit addresses and a termination record at the end. The next two characters on each line list the record length in bytes. Next is the address field. In this case

```
S12308004C56014C5701180C083808394D5601CC00FA1608244C5601CC02EE160824730820
S11D08203926E93F3B16082F83000126F83A3D34CE07CC0926FD303D3C00E8
S9030000FC
```

Figure 3.2   S-Record from Program in Source 3.1

the addresses are 16 bits or four hex digits. You can see that the starting address of the first line is $0800. The data field is after the address field. The last two digits for each line contain an inverted checksum for that line. It is easy to see the checksum in the last line. All the data are $00 except for the byte count, which is $03. Therefore the checksum is $03, and if it is inverted we get $FC, as shown at the end of the line.

▶ 3.3 RUN-TIME DEBUGGING—A TUTORIAL

So far in this chapter, we have looked at the content and syntax of an assembly source file and a simple build process to generate machine code in S-Record format. In this section we cover the last piece of the software construction process, debugging. As with the rest of this chapter, the debugging process covered here is a basic process for use with small programs that do not have significant real-time requirements. It will be expanded on in later chapters, as the programs become larger and more complex.

3.3.1 Debugging Hardware Configuration

The debugging system we concentrate on here is based on a development board with a resident monitor program. Other names for development boards include evaluation boards (EVBs) or trainer boards. The EVB term comes from industry and refers to boards produced by IC vendors so customers can evaluate a particular IC. There are EVBs for everything from microcontrollers to sensors. The EVBs that Motorola produced for the 68HC11 microcontrollers turned out to be a very cost-effective platform for developing small products and consequently became very popular. Evaluation boards for most microcontrollers are now available from the microcontroller manufacturer. Because of their cost-effectiveness, they are not only used by industry but also by schools and hobbyists. There are also many development boards produced by other manufacturers and educational programs. If a monitor program is included, they all support the basic debugging operations described in this section. In this text we will concentrate on the Motorola M68EVB912B32 evaluation board, which is based on the M68HC912B32 microcontroller and the D-Bug12 monitor. For those programmers familiar with Motorola's BUFFALO monitor, D-Bug12 will look very familiar.

In this section, we do not use the background debug system included in the M68HC912B32 microcontroller. We only use one development board in *EVB Mode* connected to the PC via a serial port, as shown in Figure 3.3.

This is the least expensive and simplest option. If the background debug system is used, two development boards or one development board and a BDM pod are required. We look at debugging using the background debug system in Chapter 10.

The PC can be any computer with a serial port, a terminal emulator, and access to the S-Record file. In this case a Windows-based PC is described. To communicate to the EVB from the PC, a terminal emulator program is required. There are many terminal programs for the PC. We use HyperTerminal in this text because it comes with most Win95, Win98, and WinNT systems. It also has the capability to insert character and line delays, which will be required later, when EEPROM or Flash EEPROM is programmed.

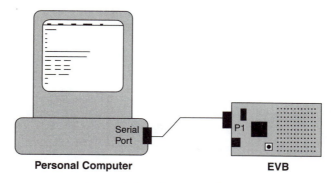

Figure 3.3 A Debugging System

3.3.2 The Debug Monitor

The debugging monitor program is a program that contains utilities for loading the S-Record file from the PC, examining memory and registers, and controlling execution of the program. The monitor program that comes with the M68EVB912B32 evaluation board is D-Bug12. It is a typical debugging monitor and is available from Motorola's Web site in S-Record form. You should refer to the *M68EVB912B32 Users' Manual* for the full documentation on the EVB and D-Bug12. There is also additional documentation on the Motorola Web site.

The D-Bug12 monitor is located in the M68HC912B32 Flash EEPROM when purchased from Motorola. In EVB mode the EVB board will *reset* into D-Bug12. The following display shows the D-Bug12 monitor prompt when the EVB *RESET* button is pressed or power is turned on:

```
D-Bug12 v2.0.2
Copyright 1996 - 1997 Motorola Semiconductor
For Commands type "Help"

>
```

The complete memory map for the EVB is shown in Table 3.1. Since D-Bug12 is located in the Flash EEPROM, the programs we are developing can only be loaded into RAM or the byte-erasable EEPROM. D-Bug12 also uses some of the on-chip RAM, so we are left with 512 bytes of RAM at $0800–$09FF and 768 bytes of byte-erasable EEPROM at $0D00–$0FFF. Most of our programs will be loaded into RAM, which places a serious limit on the size of our code. It will be more than adequate for the simple programs required in the next few chapters, however. For example the *demo1* program uses only 58 bytes.

Notice from Source 3.1 that the origin address in *demo1* is set to $0800, which is the correct starting address for this EVB. We use the *demo1* program as an example as we go through the following debugging steps.

TABLE 3.1 THE M68EVB912B32 MEMORY MAP

Address Range	Device Type	Usage	Address Range
$0000–$01FF	Internal Registers	Internal Registers	$0000–$01FF
$0200–$07FF	Unused	Unused	$0200–$07FF
$0800–$0BFF	On-Chip RAM	User Code/Data	$0800–$09FF
		D-Bug12 RAM	$0A00–$0BFF
$0C00–$0CFF	Unused	Unused	$0C00–$0CFF
$0D00–$0FFF	Byte-Erasable EEPROM	User Code/Data	$0D00–$0FFF
$1000–$7FFF	Unused	Unused	$1000–$7FFF
$8000–$FFFF	Flash EEPROM	D-Bug12 Program Code	$8000–$FFFF

3.3.3 Loading S-Record Files

Once the program has been assembled and the S-Record file has been generated, we need to load it into the EVB memory. The EVB is connected to a PC COM port, so we use HyperTerminal for communication between the PC and the EVB. The D-Bug12 monitor program then decodes the S-Record and puts the binary object code into microcontroller memory. To perform this step we use the D-Bug12 *LOAD* command. At the EVB prompt type

```
>load
```

The board will now wait for S-Record formatted data to be sent from the PC. There are a couple of ways to do this. If you already have the S-Record file on the PC, you can use *Send Text File* from the *Transfer* menu on HyperTerminal. Alternatively if you already have the S-Record displayed in another window, you can select all the S-Record, copy it, and paste it in the HyperTerminal window. Once the S-Record has been received, the D-Bug12 monitor will respond with a new prompt. If the S-Record is long enough, D-Bug12 will also display asterisks to indicate that it is in the process of receiving the S-Record.

Once the program has been loaded into memory, it can be executed. To run programs under D-Bug12, use the *G* command as follows:

```
>g 0800
```

If you downloaded *demo1*, you should get 60 active-low pulses on PORTP, bit 0. After the 60 pulses, control is passed back to D-Bug12, and it will display the register contents and a new prompt.

If you did use *demo1*, everything should have worked as expected. The program is a working program and contains no errors. When you start writing your own programs, however, this is usually not the case. Typically there are errors that you need to find, or you want to verify that the code is working as expected. In order to make more detailed observations of the MCU operation, we need to be able to display memory contents and registers, stop program execution at various points in the program, and run the program one instruction at a time. Following are some examples of using D-Bug12 for these operations. It is important to learn how to use these commands. They provide the feedback required to find bugs or to provide insight into how your program operates.

3.3.4 Register and Memory Display and Modify

The first important debugging commands are those used for displaying and editing memory and CPU registers. To display memory the *MD* command is used. For example

```
>md 0800 083f

0800  4C 57 01 4C - 56 01 18 0C - 08 38 08 39 - 4D 56 01 CC   LW.LV....8.9MV..
0810  00 FA 16 08 - 24 4C 56 01 - CC 02 EE 16 - 08 24 73 08   ....$LV......$s.
0820  39 26 E9 3F - 3B 16 08 2F - 83 00 01 26 - F8 3A 3D 34   9&.?;../...&.:=4
0830  CE 07 CC 09 - 26 FD 30 3D - 3C 00 81 0D - 00 00 00 80   ....&.0=<.......
```

This command displayed memory locations $0800 through $083f. Since this is where *demo1* is located, the contents should agree with the listing and S-Record shown in Listing 3.1 and Figure 3.1. To modify a memory location, the *MM* command can be used.

To display the register contents, use *RD*. This results in the following display:

```
>rd

PC    SP    X     Y     D = A:B   CCR = SXHI NZVC
0827  0A00  0000  0000    00:00         1001 0100
```

It displays the contents of each CPU register. To modify register contents, use *RM*.

The most common use of these four commands is to use *MD* to display program memory in order to check on the value of a variable or to check the integrity of the program.

3.3.5 Software Breakpoints

When we run a program, the CPU executes the program until control is passed back to the monitor with either an *swi* instruction or an EVB *RESET*. If the program is not an endless loop or does not end with an *swi*, the CPU will continue to execute beyond the program code until it crashes.

If we start a program, and it does not return to the monitor without a *RESET*, or it is not performing the expected function, there is a bug. We have no way of knowing where the bug occurred, however; we just know that the program failed somewhere. A breakpoint is a useful utility that allows us to stop execution at a certain point in the program. When we get to the breakpoint, we can look at the registers and memory contents to see whether there are any clues that will lead us to the bug. The other possibility is that we never reach the breakpoint, in which case we know the bug occurred earlier on in the program.

A software *breakpoint* is the command used to stop execution at a certain point in the program. It is called a software breakpoint because the monitor actually replaces the opcode at the desired address with an *swi*, which transfers control to D-Bug12 and displays the next instruction to be executed and the current register contents. Because the software breakpoint must replace an opcode, the address of the breakpoint must be the address of the first byte of an opcode. In effect this limits you to the addresses shown on the listing file. In addition since the monitor program has to replace the opcode temporarily, the program must be running in RAM. This is a significant limitation to software breakpoints, especially now that a lot of development is being done in Flash EEPROM.

For example, let's say we want to stop *demo1* at the start of each pulse. If we look at the listing in Listing 3.1, we can see that each pulse starts at line 27. The address for that line is $080C. Let's set this breakpoint and run the program.

```
>br 080c
Breakpoints: 080C
>g 0800
User Breakpoint Encountered

  PC    SP    X     Y     D = A:B    CCR = SXHI NZVC
 080C  0A00  0000  0000    3C:00           1001 0000
 080C  4D5601        BCLR  $0056,#$01
```

The program ran until it reached address $080C, and then it stopped and displayed the registers and the next instruction to be executed. At this point it may be interesting to look at the contents of the variable, *CurCnt*. By looking at the listing file, we see that *CurCnt* is at address $0839. Let's display the memory location with *MD:*

```
>md 0839

0830  CE 07 CC 09 - 26 FD 30 3D - 3C 3C 81 0D - 00 00 00 80    ....&.0=<<......
```

The *MD* command always shows a complete line, so we have to find location $0839. In this case we can see that the contents of address $0839 is $3C, which is 60 decimal. This count should decrement after each pulse. To verify that it works correctly, let's run the program until it reaches the start of the next pulse. The breakpoint is already set to the correct address, so we have to tell the monitor to continue to run the program from the current value of the program counter. To do this we use the *G* command without an argument.

```
>g
User Breakpoint Encountered

  PC    SP    X     Y     D = A:B    CCR = SXHI NZVC
 080C  0A00  0000  0000    02:EE           1101 0000
 080C  4D5601        BCLR  $0056, $01
```

The program ran until it returned to the start of the pulse and reached the breakpoint again. Note that the program should go through this loop 60 times. We can now check to see whether the program did decrement *CurCnt* by using the *MD* command.

```
>md 0839

0830  CE 07 CC 09 - 26 FD 30 3D - 3C 3B 81 0D - 00 00 00 80    ....&.0=<;......
```

You can see that *CurCnt*, address $0839, is now $3B or 59 decimal, so it did decrement correctly.

3.3.6 Instruction Tracing

There are also times when you want to see the results of each instruction. The D-Bug12 *TRACE* command allows us to run one instruction, then check the register values and memory contents, and then run the next instruction. Another name for this process is *single stepping*.

We could do this by using breakpoints, but it would require us to set a new breakpoint after each instruction. The D-Bug12 *TRACE* command, *T,* automates this process. We can use the trace command to see the microcontroller decrement *CurCnt* and make the decision whether to jump back to create another pulse or return to the monitor. From Listing 3.1 we can see that *CurCnt* is decremented at line 33. We first set a breakpoint at $081E and then use the *TRACE* command to see the step-by-step operation of the CPU.

```
>br 81e
Breakpoints: 081E
>g 0800
User Breakpoint Encountered


 PC    SP    X     Y     D = A:B   CCR = SXHI NZVC
081E  0A00  0000  0000    02:EE          1101 0100
081E  730839        DEC   $0839
```

We are now at the decrement instruction. After we run it, *CurCnt,* address $0839, should be one less. Before we run the next instruction, let's look at the value of *CurCnt.*

```
>md 0839

0830  CE 07 CC 09 - 26 FD 30 3D - 3C 3C 81 0D - 00 00 00 80    ....&.0=<<......
```

CurCnt is equal to $3C. Now let's run the instruction using the *TRACE* command and look at the value of *CurCnt* again.

```
>t

 PC   SP    X     Y     D = A:B   CCR = SXHI NZVC
0821  0C00  0000  0000    02:EE          1101 0000
0821  26E9        BNE   $080C
>md 0839

0830  CE 07 CC 09 - 26 FD 30 3D - 3C 3B 81 0D - 00 00 00 80    ....&.0=<;......
```

You can see that *CurCnt* is now $3B, and the processor stopped at a conditional branch instruction. This conditional branch will not branch if *CurCnt* was decremented to zero. Since *CurCnt* is not zero, it should branch back to $080C. Let's run another *TRACE* and see it happen.

```
S>t

 PC   SP    X     Y     D = A:B   CCR = SXHI NZVC
080C  0C00  0000  0000    02:EE          1101 0000
080C  4D5601        BCLR  $0056,#$01
```

Simple Assembly Code Construction

Sure enough, the program counter went to $080C, which means we are back to the start of the pulse loop to create another pulse.

This process of predicting what will happen when an instruction is executed and then verifying it with the *TRACE* command can be called "playing computer." It is an excellent way to learn how the CPU and its instructions work. If you are learning a new instruction set, or if this is your first experience writing programs for a microprocessor, you should make use of this method. Make sure to try to predict the outcome before running the instruction. Otherwise you may find yourself tracing through instructions blindly, skipping over a bug or not catching an important detail.

3.3.7 Other Debugging Tools

When debugging embedded microcontrollers, many other tools should be used. The most obvious are the normal electronic test instruments like the oscilloscope. Other software development tools include simulators and in-circuit emulators. Let's look at some of these tools in more detail.

Oscilloscope. If there is one instrument that is essential to embedded systems development, it is the oscilloscope. Because of the nature of embedded systems, there is typically a combination of digital and analog signals. Oscilloscopes are normally thought of as tools for debugging hardware. Scopes can also be useful for software debugging and analysis, however. For example the easiest way to test the timing of the pulses generated by *demo1* is to connect a scope to *PORTP* bit 0 and measure the pulse widths. Later in this text we look at the use of an oscilloscope for debugging real-time software.

Logic Analyzers. Here is another indispensable instrument for embedded systems development. The logic analyzer is similar to an oscilloscope because it shows signal waveforms. The logic analyzer, however, only shows logic levels, not the continuous analog waveform. It also has many channels. In order to debug a typical 8-bit microprocessor bus, at least 27 channels are required. The signals can then be displayed as timing waveforms or as logic states. We can use the logic analyzer to capture real-time bus activity if the bus system is available externally. We cannot, however, stop the processor operation or look at CPU registers. The logic analyzer combined with a debugging monitor can be effective for debugging most small systems.

Mixed-Signal Oscilloscope. Because embedded systems typically involve both analog and digital signals, we need an oscilloscope and a logic analyzer. Many newer instruments have the capability to do both, however. Many of the expensive logic analyzers allow an oscilloscope expansion board to be added. A less expensive solution is a mixed-signal oscilloscope like the Hewlett-Packard 54645D. It combines two analog oscilloscope channels with 16 digital channels, an ideal combination for single-chip microcontroller designs.

Simulators. Another way to debug programs is by using a simulator. Simulators are computer programs that simulate the operation of the CPU. A program is passed to the simulator and the simulator runs the program. Simulators are typically easy to use, and a development board is not required. They have built-in monitor functions, so you can control execution as well as examine and edit memory and registers. The problem with simulators, especially for an embedded system, is that they cannot simulate the I/O signals and surrounding hardware well. Embedded systems by their nature are tied to I/O devices. If the simulator cannot simulate the device, we cannot test that part of the code. The simulator would never catch many hardware problems. For example the M68HC912B32 does not operate correctly if the supply voltage comes up too slowly. Simulators will not catch these types of problems. Simulators are getting better, along with other digital and analog EDA tools. When the processor simulator is integrated with general EDA tools, a more complete simulation of the system can be done. At that point simulators are an excellent tool for debugging microcontrollers.

Because EVBs are easy to use, you do not gain much by using a simulator. Simulators should only be used when the hardware is not accessible or while debugging complicated real-time software. In a simulator you can not only stop program execution, you can also stop the simulated "real time." This allows you to examine program parameters without missing real-time events.

In-Circuit Emulators.   In-circuit emulators, ICEs, are a combination hardware and software system that is used to emulate the CPU. Typically these systems are plugged into the CPU socket of the target board. In-circuit emulators can provide all of the functionality of a monitor program in real time, along with some of the functionality of a logic analyzer. They are also the most expensive solution. If it means getting a more reliable product to market quickly, however, it may be well worth the investment.

The problem with in-circuit emulators is the ability of the ICE manufacturers to keep up with the CPU and MCU manufacturers. New MCUs are introduced more quickly and have higher clock speeds, making them more difficult to emulate. Consequently some MCU and CPU manufacturers are putting debug circuitry on the IC and providing a way to access it. Later we look at a debugging system that uses the background debug system on the M68HC12 to provide in-circuit emulator-like functionality. The background debug system reduces the hardware cost of the emulator while providing most the functionality.

▶   SUMMARY

In this chapter we covered enough material to build and debug a simple assembly language program with the Introl-CODE assembler and the M68EVB912B32 evaluation board. We also covered assembler syntax and directives. We cannot write our own program yet; that is covered in the next few chapters. Given a program, however, we now can assemble it and test it.

With this in mind, it is time to practice. A great way to do this is to create the *demo1* program and go through the steps covered in this chapter. Another great way to learn how to use the debugger and at the same time verify your solutions is to actually build and if necessary run the exercises at the end of the chapter.

EXERCISES

1. A 68HC11 EVB requires that the user program is located between $0000 and $02FF. Show the directive to locate a program for this board.

2. Add the necessary origin directives to *demo1.s12* to locate the program code and constants in the byte-erasable EEPROM and the variables in user RAM on the M68EVB912B32.

3. How many bytes of memory are used for each of the following directives?

    a.   `fcc /Good luck/`
    b. `PTR` `equ $0002`
    c.   `rmb 64`
    d.   `dc.b 10`
    e.   `ds.w 2`
    f. `TBL` `fdb $0100,0,%0101`

4. Show the values that will be loaded into memory for each of the directives in Exercise 3. If the contents are unknown, enter a question mark.

5. Given the following assembler directives:

```
          org        $1000
Const     fcb        $3b,$b,21,'b'
          fdb        500,Const+10,$af24
var       rmb        1
```

show the locations and the contents of each memory byte affected. If the contents are unknown, enter a question mark.

CPU12 Programming Model

The next step in writing embedded programs in assembly is to learn the CPU programming model. The programming model defines the register set, addressing modes, and instruction set for a given CPU. The programming model is entirely a function of the CPU. Therefore in this text we refer to CPU11 and CPU12 instead of a specific member of the MCU family.

In this chapter we start looking at the CPU12's programming model by examining the register set, addressing modes, and instruction execution time. In Chapter 5 we complete the programming model by covering the instruction set.

Because the CPU12's programming model is based on the CPU11, we also look at the CPU11 in order to compare the two CPUs. Again, we only cover the general concepts and techniques for using the CPU12's programming model in this chapter. This is not intended to be complete documentation for the CPU12. For detailed descriptions of the CPU12, you should refer to the *CPU12 Reference Manual* from Motorola. For detailed documentation for the CPU11, refer to the *M68HC11 Reference Manual* from Motorola.

▶ 4.1 THE CPU REGISTER SET

The type of register set in a CPU has a major impact on programming techniques and style. There are two general types of register sets—accumulator-based and register-based.

In an accumulator-based register set, there are typically just a few registers, so arguments are more often contained in memory than in registers. Because arguments tend to be contained in memory, accumulator-based CPUs must have a complete set of addressing modes to efficiently access memory. If there are addressing modes missing, programs can become inefficient, and you will find yourself moving data into and out of registers more often than should be necessary.

In the register-based CPUs, there are several general-purpose registers. The idea is that operations on registers tend to be faster than operations on memory. If you have many registers, you can work with data while in the CPU registers without having to move them to memory.

The important thing to watch out for in a register-based CPU is that the register set is symmetrical. A symmetrical register set is one that follows the same rules for register operations for all registers. In other words all registers should work identically with all instructions. This presents a trade-off. The more general-purpose registers in a CPU, the more opcodes required. Therefore, for a given set of instructions, the instruction size must be larger in CPUs with more registers. If the instruction size is increased, the execution speed may go down, which can offset the efficiency provided by having many registers.

4.1.1 The CPU12 and CPU11 CPU Register Sets

The CPU12's register set is accumulator-based. It is identical to the CPU11's register set, which in turn was based on the 6800's register set. Although this register set is somewhat limiting, Motorola stuck with it in order to make the CPU12 source code–compatible with the CPU11. Luckily they added additional addressing modes to make the CPU12 code more efficient and more enjoyable to write. Figure 4.1 shows the CPU12 and CPU11 register set. There is one 16-bit accumulator, two 16-bit index registers, a 16-bit stack pointer, a 16-bit program counter, and an 8-bit condition code register. Following is a quick description of the CPU12 registers.

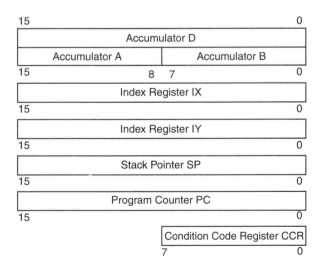

Figure 4.1 68HC11 and 68HC12 Register Set

Accumulator. The CPU12 has a single 16-bit accumulator that can be accessed as a 16-bit register, ACCD, or two 8-bit registers, ACCA and ACCB. ACCA corresponds to the most-significant byte of ACCD, and ACCB corresponds to the least-significant byte of ACCD. You can use either access method at any time. The accumulator is the destination for most arithmetic instructions and is normally used as a general-purpose register.

Index Registers. There are two 16-bit index registers, IX and IY. Index registers are typically used as pointers. They are also used as accumulators for instructions that have more than a single 16-bit result. On the CPU12 the two index registers can be used interchangeably without execution speed or code size penalties.

Stack Pointer. A single 16-bit stack pointer, SP, is used for both the system and user stack operations. There is a difference between the CPU12's stack pointer and the CPU11's stack pointer. The CPU12 stack pointer points to the last item placed on the stack, whereas the CPU11's stack pointer points to the next available location on the stack. The CPU12's stack pointer has been made more useful because it can also be used as an indexing register in all the indexed addressing modes.

Program Counter. The 16-bit program counter, PC, always points to the next instruction in the instruction sequence. It can be changed by branch, call, or jump instructions. The CPU12's program counter can also be used in most indexed addressing modes. This in effect adds PC relative addressing for the 68HC12 family and makes it possible to write relocatable machine code.

Condition Code Register. The 8-bit condition code register is shown in more detail in Figure 4.2. It contains five arithmetic flags, *H, N, Z, V,* and *C,* and three MCU control bits, *S, X,* and *I.* Following is a short description of these flags. Applications and examples that use these flags are covered in Chapter 5.

Half-Carry Flag, **H.** The *H* flag's primary purpose is to indicate a carry out of bit 3 for BCD addition. It is used by the *daa* instruction after an 8-bit addition. It is also changed by several other instructions.

Negative Flag, **N.** The primary purpose for the *N* flag is to indicate a negative number. It is set equal to bit-7 following an arithmetic or logical operation or a compare, load, or store operation. The *N* flag is often used as a general-purpose flag, especially to indicate an error condition.

Zero Flag, **Z.** The *Z* flag indicates a zero result. It is set to one if every bit in the result is zero after an arithmetic or logical operation or a compare, load, or store operation. It is also commonly used as a general-purpose flag.

S	X	H	I	N	Z	V	C
7	6	5	4	3	2	1	0

Figure 4.2 The 68HC11 and CPU12 Condition Code Register

Overflow Flag, V. The overflow flag indicates a two's-complement overflow following an arithmetic operation. It is also cleared by many other operations including loads and stores.

Carry Flag, C. The *C* flag indicates an unsigned carry or borrow. After an addition the *C* flag indicates a carry out of the most-significant bit. After a subtraction it indicates a borrow into the most-significant bit. It is also used by the multiply, divide, rotate, and shift commands.

Stop Control Bit, S. The *S* bit is used to disable the *stop* instruction. If it is set, the *stop* instruction is treated as a *nop*. If it is cleared, the *stop* instruction will shut down all MCU clocks, placing the microcontroller into a low-power mode.

Pseudo-Nonmaskable Interrupt Mask, X. The *X* bit is used to initially disable a pseudo-nonmaskable interrupt request from the */XIRQ* pin. If *X* is set, an *XIRQ* request is ignored. Once cleared it cannot be set again by using software. The *X* bit is cleared automatically when an *XIRQ* is serviced.

Global Interrupt Mask, I. The global interrupt mask, *I,* is used to disable all maskable interrupt sources. These sources include the */IRQ* pin and interrupt requests from the on-chip resources. It can be set or cleared at any time. It is set automatically when an interrupt is serviced.

▶ 4.2 CPU12 ADDRESSING MODES

An *addressing mode* refers to the process used by the CPU to determine the location of an argument. The location of the argument is called the *effective address, EA.* The CPU12 addressing modes are a superset of the CPU11 addressing modes. We will cover the modes in common with both CPUs first and then discuss the additional modes of the CPU12.

We will use some of the CPU12 instructions to illustrate the addressing modes. This is a good time to start looking at the instruction descriptions in Motorola's *CPU12 Reference Manual.*

There are six addressing modes in common with both the CPU11 and CPU12: inherent, immediate, extended/direct, indexed, and relative addressing. All of these modes are implemented the same in both processors except for the indexed addressing modes. For the indexed modes, we will look at both the CPU11 and CPU12 implementation.

4.2.1 Inherent Addressing

When using inherent addressing, the effective address is contained in the opcode itself. This addressing mode is also called *implied addressing* or *register addressing* because the argument is always located in a CPU register.

Source 4.1 shows some examples using inherent addressing. The first example clears ACCA by loading it with $00. The CPU knows the destination is ACCA because of the opcode, $87. The second example clears ACCB. Notice it has a different opcode, $C7. This is how the CPU determines which register to clear. You can see the effective address is actually a register implied by the opcode. The third example is a multiply instruction. The instruction mnemonic does not indicate the source or destination. If we look at the instruction description, however, the *mul* instruction always multiplies ACCA with ACCB and places the result into ACCD. In this case inherent addressing is used for both the source and destination arguments.

Source 4.1 Inherent Addressing Examples

```
CPU12
Opcodes          Instruction          Operation Description
   87               clra              ; $00 -> ACCA
   C7               clrb              ; $00 -> ACCB
   12               mul               ; ACCA x ACCB -> ACCD
```

4.2.2 Immediate Addressing

When using immediate addressing, the argument itself follows the opcode. Therefore the effective address is equal to the address immediately following the opcode ($EA =$ &opcode + 1). The syntax for immediate addressing is

```
operation #argument
```

Source 4.2 shows some examples using immediate addressing.

Source 4.2 Immediate Addressing Examples

```
        * Equates
        INIT         equ $02ff
        MAXCNT       equ 13


CPU12
Opcodes          Instruction          Operation Description
  860d             ldaa #MAXCNT        ; $0d -> ACCA
  cf02ff           lds #INITSP         ; $02ff -> SP
```

In the first line, *MAXCNT,* which is equal to 13, is loaded into ACCA. Notice the opcode for an immediate load to ACCA is $86. Immediately following the $86 is $0d, which is the argument itself. The second line is an example of an immediate load into the stack pointer, which is a 16-bit register. Again the value of the argument, *INITSP,* immediately follows the opcode, $CF.

When the source is assembled, the number of bytes following the opcode will always match the destination register size. Notice that since the destination register SP is 16 bits, the immediate operand is 16 bits, $02ff. This also would be the case if the actual value of the argument could be represented by only 8 bits. If the argument size is larger than the destination register size, we get an assembler error.

Immediate addressing is normally one of the fastest addressing modes, but notice that the argument must always be a constant because the assembler must know it at build time. If you need to load a register with a variable value, immediate addressing cannot be used. Of course an immediate value can never be a destination. Note there are no store instructions that use immediate addressing.

Remember the "#"!

Forgetting to include the "#" is probably the most common mistake made by CPU11 or CPU12 programmers, experienced and inexperienced alike. The usual symptoms include inconsistent argument values and failures that cause strange program counter or stack pointer values.

4.2.3 Extended and Direct Addressing

Both extended and direct addressing are forms of *absolute addressing*. When using absolute addressing the effective address follows the opcode. The difference between extended and direct addressing is the size of the effective address operand. Extended addressing uses a 16-bit operand and direct addressing uses an 8-bit operand.

Since extended addressing has a 16-bit operand, it is the most general form because it can access any location in a 64K-byte memory map. If you (or the assembler) do not know the location of an argument, extended addressing should be used.

To reduce code size and execution time, the direct addressing mode was introduced on the 6800 microprocesor. The 8-bit operand contains the least-significant byte of the effective address. The value of the most-significant byte is always $00. Therefore you can only access memory locations $0000–$00FF using direct addressing. This area of the memory map has traditionally been called the *direct page*.

With the introduction of the 68HC12, direct addressing has become much less important. The 16-bit data bus and instruction queue allow the CPU12 to execute most extended addressing instructions in the same number of clock cycles as direct addressing. The only instructions that execute faster using direct addressing are *brset, brclr,* and the store instructions. With all other instructions the only benefit of using direct addressing is to save one byte of code space.

For both extended and direct addressing modes, the syntax is

```
operation operand
```

To force the assembler to use one mode or the other, we can use forcing characters as follows:

```
operation >operand      ; force extended
operation <operand      ; force direct
```

Because the syntax for both direct addressing and extended addressing is the same, we need to be cognizant of the assembler's rules for selecting the addressing mode.

To optimize code size and execution speed, an assembler will typically choose direct addressing if it knows the effective address will fall on the direct page, $0000–$00FF. A two-pass assembler must make this decision on its first pass, however. On the first assembler pass, it will not know what the effective address will be if a forward referenced label is used. A *forward-referenced label* is a label that is defined after the instruction. Source 4.3 shows an example of a forward-referenced label.

On the first assembler pass the assembler reaches the *ldaa var1* instruction before the *rmb* line that defines *var1*. In this case the assembler does not know the effective address so it must use extended addressing. The assembler uses the extended addressing opcode, $B6, and reserves two bytes for the effective address.

Source 4.3 A Forward-Referenced Label

```
CPU12
Opcodes                Source
                         ...
b60020                 ldaa var1             ; (var1) -> ACCA
                                               ...
                       var1                  rmb 1
```

The assembler then continues through the program, finally reaching the label *var1* and resolving its address, which in this case turns out to be $0020. On the second pass the assembler reaches the *ldaa var1* line again and inserts $0020 into the two reserved bytes from the first pass. In this example the assembler had to use extended addressing despite the fact that the address turned out to be on the direct page. There are two ways to change this code so the assembler will use direct addressing.

In Source 4.4 the label *var1* is defined before it is used, so it is not a forward-referenced label. This time, during the first pass, the assembler reaches the *rmb* line to resolve *var1*. It then reaches the *ldaa var1* instruction. Since it already knows the address of *var1,* which we assume is again $0020, it uses direct addressing. It uses the direct addressing opcode $96 this time and places the least-significant byte of the effective address, $20, in the 1-byte operand.

Source 4.4 A Previously Defined Label

```
CPU12
Opcodes                Source
          var1         rmb 1
                         ...
9620                   ldaa var1             ;(var1) -> ACCA
                         ...
```

Notice we had to change the order of our code and place the variables before the program. We can do this, but it does not follow the program organization described in Chapter 3. The potential problem with putting the variables first is that if we use a single origin directive, the start of the program will change every time we add or remove a variable. It can be rather frustrating to spend several minutes debugging a working program when the only reason it crashed is because you keyed *g 0804* instead of *g 0805*—the new location of the program.

Source 4.5 shows how the addressing mode can be forced to either extended or direct. Forcing direct can be used if you are using forward-referenced labels and you know they will fall on the direct page.

Source 4.5 Forcing Direct Addressing

```
CPU12
Opcodes                Instruction           Operation Description

9620                   ldaa <var1            ; Force direct addressing
b60020                 ldaa >var1            ; Force extended addressing
```

Forcing extended may be needed if we must write code that is precisely timed regardless of the location of the variables. If we try to force direct and *var1* turns out to be greater than $00FF, we get an error message from the assembler.

Note

AS11 Phasing Errors

There is a rather nasty bug in the Motorola AS11 assembler that everyone eventually discovers. If a forward-referenced label is encountered during the first pass, the assembler will correctly assume extended addressing and save two bytes for the operand. On the second pass, however, if the effective address turned out to be on the direct page, the assembler will try to switch back to direct and only use a 1-byte operand. This bug results in a lot of phasing errors. The worst thing is that the phasing errors do not generally land on the line that actually caused the problem. You move the program (so the effective address is no longer on the direct page) and the errors go away! The only good way to fix this bug is to use forcing characters—or better yet, use a different assembler.

4.2.4 68HC11 Indexed Addressing

Before we look at the CPU12 indexed addressing modes, let's look first at the way the CPU11 implements its indexed addressing mode. When using the CPU11's indexed addressing, the effective address is calculated by adding the contents of the index register to an unsigned, 8-bit constant offset, *ff*. The syntax is

```
operation ff,x                          ;IX indexed addressing
operation ff,y                          ;IY indexed addressing
```

If there is no offset, you can use either of the following on most assemblers:

```
operation ,x
operation 0,x
```

In the CPU11, using IY as the index register results in an additional byte added to the opcode ($18). This added byte results in both a speed and code space penalty. This is an example of asymmetry in the CPU11's register set. Using one index register has different results than using the other index register.

Source 4.6 includes two lines that use the CPU11's indexed addressing. First the index registers are initialized. In this case they are initialized to point to *REGBASE,* $1000. The next line uses IX as the index register. It assembles to an opcode, $a6, followed by the 8-bit offset, $0a. When IY is used, there are two opcode bytes, $18a7, and the offset, $02.

Source 4.6 68HC11 Indexed Addressing

```
* Equates
REGBASE             equ $1000
PORTG               equ $02
PORTE               equ $0a
```

```
68HC11
Opcodes                  Instruction              Description
18ce1000                 ldy #REGBASE             ;$1000 -> IY
ce1000                   ldx #REGBASE             ;$1000 -> IX
a60a                     ldaa PORTE,x             ;($0a + IX) -> ACCA
18a702                   staa PORTG,y             ;ACCA -> ($02 + IY)
```

It is important to note that the CPU11 will only allow constant offsets. This can be a frustrating limitation when trying to address a table of values in which the table offset is variable. Source 4.7 shows how the *abx* instruction must be used to address variable locations in a table.

Source 4.7 Implementing a Variable Offset with CPU11

```
ldx #TABLE_BASE          ;TABLE_BASE -> IX
ldab OffsetVar           ;(OffsetVar) -> ACCB
abx                      ;$00:ACCB + IX -> IX
```

The problem with this method is that it changes the index register itself. If this code were in a loop, IX would have to be reinitialized to *TABLE_BASE* every time the table is accessed. This problem is solved in the CPU12 by the addition of the register offset indexed addressing modes.

4.2.5 CPU12 Indexed Addressing

In the CPU12 there are four different classes of indexed addressing. These four classes can be subdivided into seven different indexed addressing modes.

The addition of these addressing modes in the CPU12 is one of the most important improvements made over the CPU11. Because of these additional modes, the CPU12's code can be made relocatable and more efficient, especially when using local variables. Table 4.1 lists all the indexing modes and separates them into classes.

TABLE 4.1 CPU12 INDEXED ADDRESSING MODES

Class	Modes	Syntax	Indexing Registers
Constant Offset Indexing	5-bit constant offset	n,r	X,Y,SP,PC
	9-bit constant offset	n,r	X,Y,SP,PC
	16-bit constant offset	n,r	X,Y,SP,PC
Auto-Increment, Auto-Decrement Indexing	Preincrement	$n,+r$	X,Y,SP
	Postincrement	$n,r+$	X,Y,SP
	Predecrement	$n,-r$	X,Y,SP
	Postdecrement	$n,r-$	X,Y,SP
Register Offset Indexing	*A* contains offset	A,r	X,Y,SP,PC
	B contains offset	B,r	X,Y,SP,PC
	D contains offset	D,r	X,Y,SP,PC
Indirect Indexing	Constant offset indirect	$[n,r]$	X,Y,SP,PC
	Register *D* offset indirect	$[D,r]$	X,Y,SP,PC

All CPU12 indexed addressing modes assemble to one opcode byte, a postbyte, and zero to two additional bytes. The opcode indicates the instruction and the indexed addressing mode. The postbyte provides detailed information for the indexed mode. The additional bytes are used when the offset is larger than five bits.

4.2.6 Constant Offset Indexed Addressing

The CPU12's constant offset indexed addressing mode adds efficiency and flexibility to the 68HC11's indexed mode. Again, when using constant offset indexed addressing, the effective address is calculated by adding the contents of an index register to a constant offset. For source code compatibility, the syntax for the CPU12 constant offset indexed addressing is the same as the 68HC11's. The enhancements made to CPU12 are a more flexible offset, enhanced program coding, and the use of SP and PC as indexing registers. Following is the syntax:

```
operation ,r          ;No offset
operation 0,r         ;No offset
operation n,r         ;Positive Constant offset, n
operation -n,r        ;Negative Constant offset, -n
```

where

$$r = \text{the indexing register (IX, IY, SP, PC)}$$

$$n\ (-n) = \text{signed constant offset (5-bit, 9-bit, 16-bit)}$$

CPU12 has three different constant offset addressing modes: 5-bit, 9-bit, and 16-bit. The size of the offset determines the value of the postbyte and the number of additional bytes added to the instruction opcode.

5-Bit Constant Offset. The 5-bit constant offset mode assembles to only two bytes, the postbyte and the opcode. The contents of the postbyte are

rr0nnnnn

where

$$rr = \text{the indexing register (IX = 00, IY = 01, SP = 10, PC = 11)}$$

$$nnnnn = \text{the 5-bit offset, } n$$

Since the constant offset is a signed 5-bit number, this mode is limited to an offset range of

$$-16 \le n \le +15$$

The CPU12 code size for the 5-bit mode is always the same or better than the CPU11's. Moreover, because most offsets tend to be within the 5-bit range, the overall code is more efficient than the equivalent CPU11 code.

9-Bit Constant Offset. The 9-bit constant offset mode adds an additional byte that indicates the magnitude of the offset. The contents of the postbyte are

111rr00s

where

rr = the indexing register (IX = 00, IY = 01, SP = 10, PC = 11)

s = the sign of the offset (1 = negative, 0 = positive)

The 9-bit constant offset indexed mode is limited to an offset range of

$$-256 \leq n \leq +255$$

16-Bit Constant Offset. The 16-bit constant offset mode adds two additional bytes containing the offset. The contents of the postbyte are

111rr010

where

rr = the indexing register (IX = 00, IY = 01, SP = 10, PC = 11)

The 16-bit offset addressing mode has an unlimited range within the 64K-byte memory map.

Since the offsets are constant and must be known at assembly time, the assembler will choose the most efficient coding for the given offset. Source 4.8 shows an example of each constant offset mode.

Source 4.8 Constant Offset Indexed Addressing

```
CPU12
Opcodes         Instruction         Description
a604            ldaa $04,x          ;($04+IX) -> ACCA
ece810          ldd $10,y           ;($10+IY):($10+IY+1) -> ACCD
a6fa1000        ldaa $1000,pc       ;($1000+PC) -> ACCA
```

Check the CPU12 opcodes given in Source 4.8. Make sure you understand how the postbytes correlate with the descriptions given above. The first line in Source 4.8 uses a 5-bit constant offset because $04 is within the range of a 5-bit signed number. Its postbyte is $04. The second line uses a 9-bit constant offset with a postbyte of $E8. The last line uses 16-bit constant offset addressing because of the large offset and a postbyte of $FA.

Programming Note

The offsets shown in Source 4.8 are magic numbers. In a real program, they should all be equated symbols to make the program more readable. Numbers are used here to simplify the examples.

4.2.7 Auto-Increment/Decrement Indexed Addressing

It is common to have a pointer progress through an array of data objects during a program loop. Each time through the loop, the pointer must be incremented or decremented by the size of each object. For example if the object is simply a single byte, the pointer must be incremented or decremented by one each time through the loop. If the data object is a single 16-bit word, the pointer must be incremented or decremented by two each time through the loop. The auto-increment and auto-decrement addressing modes perform the incrementing and decrementing automatically.

When using auto-increment and auto-decrement addressing, there can be no offset, so the indexing register contains the effective address and points directly to the argument. As shown in Table 4.1 there are four different modes in this class: preincrement, postincrement, predecrement, and postdecrement. The syntax for these addressing modes is

```
operation n,+r          ;preincrement by n
operation n,r+          ;postincrement by n
operation n,-r          ;predecrement by n
operation n,r-          ;postdecrement by n
```

where

$$n = \text{the adjustment value, } 1 \leq n \leq 8$$

$$r = \text{the indexing register (IX, IY, SP)}$$

The program counter cannot be used as an indexing register for these modes. There would be no practical use for auto-incrementing or auto-decrementing the program counter anyway. The auto-increment and auto-decrement modes require one postbyte after the opcode. The postbyte is coded as follows:

```
rr1pnnnn
```

where

$$rr = \text{the indexing register (00 = IX, 01 = IY, 10 = SP)}$$

$$p = 0 \text{ for pre, 1 for post}$$

$$nnnn = \text{adjustment value coded as follows:}$$

$$0000 = +1...0111 = +8 \qquad \text{for increment adjustments}$$

$$1000 = -8...1111 = -1 \qquad \text{for decrement adjustments}$$

Note there is no zero and the code for the increment adjustments is

$$nnnn = n - 1$$

Source 4.9 shows a program that copies a 16-bit word from an array of 4-byte data objects to an array of 2-byte data objects. Assume IX is initialized to point to the source array, and IY is initialized to point to the destination array.

Source 4.9 Postincrement Addressing

```
CPU12
Opcodes    Instruction        Description
18023371   movw 4,x+,2,y+     ;(IX:IX+1)->(IY:IY+1),IX+4->IX,IY+2->IY
```

The *movw* opcode is two bytes, $1802. These two bytes are followed by two post-bytes—one for the source's addressing mode and one for the destination's addressing mode. In this case both source and destination use postincrement indexed addressing. The first postbyte, $33, is for postincrement by four of index register IX. The second postbyte, $71, is for postincrement by two of index register IY.

4.2.8 Register Offset Indexed Addressing

The effective address in register offset indexed addressing is calculated by adding the contents of an offset register to the contents of an indexing register. Register offset indexed addressing allows a variable offset value to be added to the indexing register without changing the value of the indexing register. To use this addressing mode the offset value must be placed in a register, either ACCA, ACCB, or ACCD. The syntax is

```
operation A,r              ;A contains offset
operation B,r              ;B contains offset
operation D,r              ;D contains offset
```

where

$$A, B, \text{ and } D = \text{the offset registers}$$

$$r = \text{the indexing register (IX, IY, SP, PC)}$$

The registered offset modes require one postbyte after the opcode. The postbyte is coded as follows:

111rr1aa

where

$$rr = \text{the indexing register } (00 = \text{IX}, 01 = \text{IY}, 10 = \text{SP}, 11 = \text{PC})$$

$$aa = \text{the offset register } (00 = A, 01 = B, 10 = D)$$

Source 4.10 shows a program that looks up a display code in *SEG_TBL* for the variable *DispVar*. It then outputs the result to *DISP_REG*.

Source 4.10 Registered Offset Indexed Addressing

```
CPU12
Opcodes          Instruction        Description
cd1000           ldy #SEG_TBL       ;SEG_TBL -> IY
b60835           ldaa DispVar       ;(DispVar) -> ACCA
a6ec             ldaa a,y           ;(IY+ACCA) -> ACCA
7a2000           staa DISP_REG      ;ACCA -> DISP_REG
```

In this example ACCA is used as the offset register and the destination register for the *ldaa a,y* instruction. This works correctly because the CPU uses the offset in ACCA before it is overwritten by the contents of the memory location pointed to by IY + ACCA.

This addressing mode offers a significant improvement over the CPU11's indexed mode when the program must access a variable entry in an array. As we saw in Source 4.7, when using the CPU11's constant offset indexing, the *abx* instruction had to be used to add a variable offset to the index register. Consequently the index register had to be reinitialized every time the array was accessed.

4.2.9 Indexed Indirect Addressing

When using indirect addressing the effective address is contained in the location pointed to by the sum of the indexing register and an offset. In other words the sum of the index register and the offset point to a pointer that points to the argument. The CPU12 indirect indexed addressing modes accept either a 16-bit constant offset or a 16-bit registered offset using ACCD. Following is the syntax:

```
operation [D,r]              ;ACCD contains offset
operation [n,r]              ;n is a 16-bit constant offset
```

where

$$D = \text{the register containing the offset}$$

$$n = \text{a 16-bit constant offset}$$

$$r = \text{the indexing register (IX, IY, SP, PC)}$$

This addressing mode requires a postbyte for the registered offset and, for the constant offset mode, two additional bytes for the 16-bit offset. The postbyte is coded as follows:

111rrx11

where

$$rr = \text{the indexing register } (00 = \text{IX}, 01 = \text{IY}, 10 = \text{SP}, 11 = \text{PC})$$

$$x = \text{the offset type } (0 = \text{constant offset}, 1 = \text{ACCD registered offset})$$

Source 4.11 shows how this addressing mode is used for accessing routines with addresses contained in a jump table.

Source 4.11 Indirect Indexed Addressing

```
            * Equates
            JMPTBL          equ $1000
            CMD1OFF         equ 2

CPU12
Opcodes                 Instruction             Description
ce1000                  ldx #JMPTBL             ;JMPTBL -> IX
15e30002                jsr [CMD1OFF,x]         ;((CMD1OFF + IX))  -> PC
15e7                    jsr [D,x]               ;((D + IX))-> PC
```

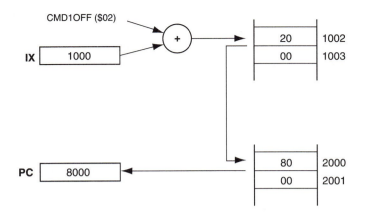

Figure 4.3 Indirect Addressing Flow Diagram

IX points to a table that contains the starting addresses for a set of command routines (an array of pointers to functions). In the second line a 16-bit constant offset, *CMD1OFF*, is used to point to the correct pointer. It requires four-bytes of machine code. The third line uses ACCD for a variable offset into the jump table. Notice it only requires two bytes of machine code. Of course this is not including the code required to load ACCD with the offset. Note that the two postbytes, $E3 and $E7, follow the description given previously.

The extra level of indirection in this mode can be confusing at first. Figure 4.3 shows the process used to perform the instruction on the second line of Source 4.11. In the example, assume *CMD1OFF* is $02, the content of IX is $1000, the content of addresses $1002 and $1003 is $2000 and, the content of addresses $2000 and $2001 is $8000.

First the contents of IX are added to the constant offset *CMD1OFF*. In this case the sum is $1002. The contents of addresses $1002 and $1003 are concatenated to form the effective address, which in this case is $2000. In order to complete the jump instruction, the contents of $2000 and $2001 are loaded into the program counter.

4.2.10 8-Bit Relative Addressing

Both the CPU11 and CPU12 use 8-bit relative addressing for branch instructions. When using relative addressing, the effective address is calculated by adding an offset, *rr*, to the current value of the program counter, PC. In this case the offset is a signed 8-bit offset so the range is limited to

$$-128 \leq rr \leq +127$$

The offset *rr* can be calculated by using

$$rr = A_D - PC$$

where

$$A_D = \text{the destination address}$$

$$PC = \text{contents of the program counter}$$

Normally we do not have to perform this calculation. We simply provide a label for the branch destination, and the assembler calculates the offset for us.

Source 4.12 shows an example of an endless trap using a relative branch. This is a common way to end a program. In this program we want to branch to the beginning of the same line so the program is trapped at the line. We use the label *trap* as the argument for the branch instruction. Note the label *trap* is set equal to the address at the beginning of the line, which is $0834. Yet, because relative addressing is used, the operand is not $0834. It is the relative offset, which has been calculated by the assembler. In this case it is

$$rr = A_D - PC = \$0834 - \$0836 = -2 = \$FE$$

You can see from the program listing that the destination address is $0834. The program counter always points to the next instruction, so it is equal to $0836. The offset is –2, which means the program counter jumps back two bytes. The signed 8-bit value for –2 is $FE.

Source 4.12 8-Bit Relative Addressing

Addresses	CPU12 Opcodes	Label	Instruction	Description
00000834	20fe	trap	bra trap	;PC-2 -> PC
00000836				

4.2.11 CPU12 Long Relative Addressing Mode

CPU12 also has long-branch instructions that use relative addressing with a 16-bit signed offset. Long branches are costly though. Compared to a regular branch a long branch adds two bytes of program code and one cycle of execution time.

Because the long branches are costly, you should use a regular branch if possible. The problem is that as the program grows, new range errors can occur from the regular branches. The normal culprit is the branch to subroutine. As you add more code, the subroutines end up farther away. Some assemblers will automatically perform branch size optimization. That is, the assembler will replace long branches with regular branches if they are within the 8-bit range. You should be able to turn this optimization off, because it is usually preferred to have the source file reflect the actual code. Branch optimization is most beneficial when the source code is a high-level language such as C.

4.2.12 Addressing More Than 64K-bytes

The CPU12 is set up to allow for more than 64K-bytes of memory space. The M68HC812A4 microcontroller is one version that lets you address up to 4M-bytes of memory. The CPU12 achieves this by using a paged addressing scheme so the complete address is made up of the contents of a page register, *PPAGE,* and the contents of the address bus. When a program is more than one page in size, subroutine calls must be made with the *call* instruction and returns made using the *rtc* instruction. The *call* and *rtc* instructions allow the page switch to be performed transparently. They will also work for subroutines within the same page.

▶ 4.3 THE CPU12 INSTRUCTION SET

The CPU instruction set defines all the CPU's assembly instructions. To fully define an instruction, we need the mnemonic description, argument description, the corresponding machine code, a detailed operational description including the effects on the condition codes, the available addressing modes, and a description of the cycle-by-cycle bus activity. It is not the intent of this text to provide a full description of the CPU12 instruction set. Instead we concentrate on programming techniques and applications using the instruction set. For a complete description of the CPU12 instructions set, you should refer to the Instruction Glossary in Motorola's *CPU12 Reference Manual.*

The CPU12 instruction set is a superset of the CPU11 instruction set, so CPU11 programs are source-code compatible with the CPU12. This means CPU11 assembly source can be reassembled for a CPU12. Of course for full compatibility there are many other considerations, such as peripheral configurations and the memory map.

A CPU instruction set can have a major impact on the efficiency of the program code. The instructions added to the CPU12, along with the addressing modes, play a large role in increasing the efficiency of CPU12 programs relative to the CPU11. Some of the most important instruction additions include a memory-to-memory move instruction, improved multiplication and divide instructions, long branches, load effective address instructions, and fuzzy logic support. When coupled with the additional addressing modes, CPU12 code can be significantly more efficient than CPU11 code. The techniques and applications of these instructions are covered in the next several chapters. Before we go on, let's take a look at how long it takes an instruction to execute.

4.3.1 Instruction Execution Time

Embedded systems are typically real-time systems, so it is very important to be able to determine the time it takes the CPU to execute instructions. The instruction execution time is determined by the number of bus cycles required for the instruction and the bus frequency of the MCU.

In general the number of bus cycles depends on many things, including the size of the opcodes, the number of arguments, the size of the data bus, and the complexity of the execution. The CPU12 executes programs much faster than the CPU11 for several reasons. First it has a 16-bit data bus, so it can fetch opcodes and arguments with fewer bus cycles. Second it can run at a higher bus frequency than the CPU11. And third it has added circuitry for complex instructions such as multiplies and divides, so the calculations are completed in fewer bus cycles. The CPU12 also has an instruction queue that allows it to fetch the next instruction while it is executing the current instruction.

The CPU12 instruction execution times are a function of several variables. These include clock stretching, data bus size to external devices, and code alignment. Considering all the variables, calculating the execution time of a CPU12 instruction can be complex. Because the CPU12 uses a queue and not a true pipeline, however, the number of cycles is always deterministic.

To simplify the task of calculating the execution time of an instruction, we will assume that there is no clock stretching, all accesses are either aligned or accessed from internal

RAM, and all accesses are over a 16-bit bus. Under these conditions we can use the cycle count given in the *CPU12 Reference Manual*. These are actually best-case conditions because they represent the lowest number of clock cycles required for each instruction.

To convert the number of cycles to time, we must know the bus frequency of the CPU. In the case of the CPU12, the bus frequency is based on the E-clock. For the M68HC912B32, the frequency of the E-clock, f_E, is one-half of the oscillator or crystal frequency found on the *XTAL* and *EXTAL* inputs. For the M68EVB912 EVB, the crystal frequency is 16MHz, so f_E = 8MHz. The total execution time is then

$$T_{ex} = N \times T_E$$

where

$$T_{ex} = \text{total execution time}$$
$$N = \text{the total number of clock cycles}$$
$$TE = \text{the E-clock period, } 1/f_E$$

Beware. This equation is for the MC68HC912B32 microcontroller. Different CPUs and MCUs have different crystal frequency to bus frequency relationships. For example the 68HC11 MCU has a bus frequency, f_E, equal to one-quarter of the crystal frequency, and the Z80 has a bus frequency that is equal to the crystal frequency. You must know this relationship before calculating time delays or evaluating the processing speed of a microprocessor or microcontroller.

▶ ## SUMMARY

In this chapter we covered the CPU's register set, the addressing modes, and the time it takes to execute an instruction. This is the first step required to write our own programs. In the next chapter we will finish the CPU12 programming model by looking at the techniques for using all the basic CPU12 instructions.

EXERCISES

1. Determine the object code for the following instructions, assuming a two-pass assembly process. Assume that *var1* is a forward-referenced label equal to $00E0.
   a. ldaa var1
   b. ldd 5,x
   c. adda $15
   d. ldd 5,x+
   e. ldd A,x
   f. ldd [D,pc]

2. Give the addressing mode and the effective addresses for the instructions in Exercise 1. Assume the following register and memory contents before each instruction:

```
8820   0D 26 E9 3F - 3B 16 08 2F - 83 00 01 26 - F8 3A 3D 34

  PC     SP     X      Y      D = A:B    CCR = SXHI NZVC
 0820   0900   0300   0000    80:00            1001 0100
```

3. For Exercise 1a, show the instruction syntax to force direct addressing.

4. Find the indexed addressing mode and the required number of additional bytes by decoding the following postbytes:
   a. 10011111
   b. 10100100
   c. 11111110
   d. 11111011
   e. 11110001
   f. 11101010

5. Implement the endless loop program shown in Source 4.12 using an *lbra* instruction. Show both the source and the machine code.

6. Calculate the execution time for the *ldaa* instruction for the following addressing modes. Assume f_{XTAL} = 16MHz and use the number of clock cycles listed in the *CPU12 Reference Manual*.
   a. Immediate
   b. Direct
   c. Extended
   d. Indexed, 5-bit constant offset
   e. Indexed, 16-bit constant offset
   f. Indexed indirect, constant offset
   g. Indexed, ACCD offset
   h. Indexed, auto-increment

7. Calculate the execution time of the *idiv* with f_{XTAL} = 4MHz.

Basic Assembly Programming Techniques

In this chapter we start writing assembly language programs for the 68HC11 and the 68HC12 microcontrollers. Basic techniques and examples for using most of the CPU12 instructions are covered. By going through this chapter and working the exercises, we can develop a basic set of programming skills that will be used later to construct complete programs. The theme of this chapter should be details, details, details. We need to know the details of the instruction operation. When and what flags are set, how many object code bytes are required, and how are relative offsets calculated?

The skills presented in this chapter cannot be absorbed by reading it one time. They will only become second nature with experience, so after going through this chapter once, use it as a reference. When using higher-level languages, these details will normally be hidden. This is one of the reasons we use high-level languages—you do not have to be bogged down with the details. However, these details are still there. A good programmer not only knows what the program does but how it does it. The *how* is contained in these details.

▶ 5.1 DATA TRANSFER

The first set of instructions we cover is the set of basic data transfer instructions. Data transfer instructions are the instructions we use to move data from one location to another. The data transfer instructions included here are immediate loads, memory-to-register loads, register-to-memory stores, register-to-register transfers and exchanges, and on the CPU12, memory-to-memory moves.

5.1.1 Data Transfers Are Copies

The first thing to realize is that whenever data are transferred, the CPU actually copies the source data to the destination. Therefore, with the exception of exchanges, the source data always remains intact. So the notation

(M) → ACCA

can be interpreted as "Copy the contents of *M* into ACCA," where *M* is a memory location and (*M*) is the contents of that location. We will also use (*M*) for immediate data, despite the fact that functionally, immediate data have no location.

If the source or destination register is a 16-bit register, two bytes of data are copied. These two bytes are assumed by the CPU12 to be stored in big-endian order. So

(M:M+1) → ACCD

can be interpreted as "Copy the contents of *M* into ACCA and the contents of *M* + 1 into ACCB."

5.1.2 Register Loads

To transfer an immediate value or a value from memory to a register, a load instruction is used. Table 5.1 shows the CPU12 register loads and the available addressing modes.

The CPU11 column in Table 5.1 indicates whether the instruction is available on a CPU11-based microcontroller. Use this as a reference if you are trying to write code that is compatible with a 68HC11 microcontroller. The addressing modes shown, however, are only for the CPU12. Refer to the *68HC11 Reference Manual* to find the addressing modes available for the 68HC11. The IDX column includes all direct indexed addressing modes. The [IDX] includes all the indirect indexed addressing modes.

The first six load instructions in Table 5.1 are straightforward loads. We have already looked at code examples in the previous chapter when we covered addressing modes.

TABLE 5.1 REGISTER LOADS

Mnemonic	Operation	CPU11	CPU12 Addressing Modes					
			INH	IMM	DIR	EXT	IDX	[IDX]
LDAA	(M) → ACCA	X		X	X	X	X	X
LDAB	(M) → ACCB	X		X	X	X	X	X
LDD	(M:M+1) → ACCD	X		X	X	X	X	X
LDS	(M:M+1) → SP	X		X	X	X	X	X
LDX	(M:M+1) → IX	X		X	X	X	X	X
LDY	(M:M+1) → IY	X		X	X	X	X	X
LEAS	*EA* → SP						X	
LEAX	*EA* → IX						X	
LEAY	*EA* → IY						X	

5.1.3 Load Effective Address

The last three instructions in Table 5.1 are *load effective address* instructions. Remember that the effective address is the address of an argument. Normally the CPU12 calculates the effective address in order to access the argument at that address. When using the load effective address instruction, the effective address itself is accessed. In effect it gives the programmer access to the CPU circuitry that performs the indexed addressing calculations. This function is useful for adding signed offsets to pointers. Source 5.1 shows two ways to add offsets to index registers used as pointers.

Source 5.1 Load Effective Address Examples

```
leas 4,sp          ;SP+4 -> SP
leay d,y           ;IY+ACCD -> IY
```

The first line adds four to the stack pointer. This is typically used to allocate or deallocate stack space for local variables or variables passed on the stack. We cover these topics later. If you tried to do the same thing with the CPU11, you would have to use four *ins* instructions. The second line in Source 5.1 shows how a register, in this case ACCD, can be used to add a variable offset to IY.

5.1.4 Register Store Instructions

To transfer data from a register to memory, a store instruction is used. Table 5.2 shows all the CPU12 register store instructions. Note that they are all available on the CPU11.

Of course there is not an immediate addressing mode for store instructions. Functionally, immediate data have no location. Immediate data are stored after the instruction opcode. It would not make sense to store data using immediate addressing because you would be storing the data in program space.

5.1.5 Transfers and Exchanges

To move data between two registers, transfer and exchange instructions are used. Table 5.3 shows all the CPU12 transfer and exchange instructions. The CPU12 added a

TABLE 5.2 REGISTER STORE INSTRUCTIONS

Mnemonic	Function	CPU11	CPU12 Addressing Modes					
			INH	IMM	DIR	EXT	IDX	[IDX]
STAA	ACCA → (M)	X			X	X	X	X
STAB	ACCB → (M)	X			X	X	X	X
STD	ACCD → (M:M+1)	X			X	X	X	X
STS	SP → (M:M+1)	X			X	X	X	X
STX	IX → (M:M+1)	X			X	X	X	X
STY	IY → (M:M+1)	X			X	X	X	X

TABLE 5.3 TRANSFER AND EXCHANGE INSTRUCTIONS

Mnemonic	Function	CPU11	CPU12 Addressing Modes					
			INH	IMM	DIR	EXT	IDX	[IDX]
TFR R₁,R₂	$R_1 \rightarrow R_2$		X					
SEX R₁,R₂	$R_1 \rightarrow R_2$		X					
TAB	ACCA → ACCB	X	X					
TAP	ACCA → CCR	X	X					
TBA	ACCB → ACCA	X	X					
TPA	CCR → ACCA	X	X					
TSX	SP → IX	X	X					
TSY	SP → IY	X	X					
TXS	IX → SP	X	X					
TYS	IY → SP	X	X					
EXG R₁,R₂	$R_1 \rightarrow R_2$		X					
XGDX	ACCD → IX	X	X					
XGDY	ACCD → IY	X	X					

general-purpose transfer and a general-purpose exchange to replace the CPU11 register-specific transfers and exchanges. The added instructions help make the CPU12 instruction set more symmetric.

First let's look at the transfer instructions. The old CPU11 instructions have a unique mnemonic for each register pair available for transfer. The new CPU12 transfer instruction uses a general transfer mnemonic with the registers indicated as arguments. The first argument is the source register, and the second argument is the destination register. For example

```
tfr a,b         ; ACCA → ACCB
tab             ; ACCA → ACCB
```

Both of these lines transfer data from ACCA to ACCB. The second line is compatible with the CPU11. There is a difference in operation between these two instructions, however. The *tab* instruction affects the CCR, whereas the *tfr* instruction has no effect on the CCR unless the CCR is a destination register.

The *tfr* instruction can transfer data from any register to any other register except the program counter. It can even transfer data to a destination register that is a different size than the source register is. We have to be careful, however, to see what the CPU12 will actually do when the register sizes are not the same.

If the transfer occurs from a 16-bit register to an 8-bit register, the least-significant byte of the source register is copied to the destination register. If a transfer occurs from an 8-bit register to a 16-bit register, CPU12 automatically performs a sign extension into the most-significant byte of the destination register. This is the appropriate action if the data being transfer are signed numbers. The ever-popular sign extension instruction, *sex,* is simply an alternate mnemonic for a *tfr* instruction with the source as an 8-bit register and the destination as a 16-bit register. So

```
tfr a,x
```

is identical in every way to

```
sex a,x
```

You would think that Motorola would make the *tfr* instruction an unsigned transfer and the *sex* instruction a signed transfer, but this is not the case.

It is not appropriate for sign extension to occur if the data are unsigned numbers such as an address or a counter. For these cases we have to revert to alternate methods to make the transfers. Source 5.2 shows some examples for making unsigned transfers from an 8-bit register to a 16-bit register.

Source 5.2 Unsigned Transfer Examples

```
CPU12
Opcodes          Instruction              Operation Description

*Transfer ACCA to ACCD, unsigned
b704             tfr a,d                  ;ACCA -> ACCD, sign -> ACCA
87               clra                     ;$00 -> ACCA

*Transfer ACCB to IX, unsigned, using leax.
ce0000           ldx #0                   ;$00 -> IX
1ae5             leax b,x                 ;IX+ACCB -> IX

*Transfer ACCB to IX, unsigned, using exchange and transfer.
b795             exg b,x                  ;$00:ACCB -> IX
b751             tfr x,b                  ;IXL -> ACCB

*Transfer CCR to IX, unsigned, using 16-bit transfer and ACCD.
b721             tfr cc,b                 ;CCR -> ACCB
87               clra                     ;$00 -> ACCA
b745             tfr d,x                  ;ACCD -> IX

*Transfer CCR to IX, unsigned, using the stack.
39               pshc                     ;SP-1 -> SP, CCR -> (SP)
1b9f             des                      ;SP-1 -> SP
6980             clr 0,sp                 ;$00 -> (SP)
30               pulx                     ;(SP:SP+1) -> IX, SP+2 -> SP
```

Because of a change between the CPU11 and CPU12 stack, the CPU12 stack pointer transfer instructions *tsx, tsy, txs,* and *tys* perform different operations than the CPU11 does. Table 5.4 shows the difference between the CPU11 and CPU12 stack transfer instructions. The change in the CPU12 stack will be discussed in detail in the next section.

5.1.6 Exchange Instructions

The CPU12 has a single exchange instruction that can be used with any two registers except the program counter. Exchange instructions involve an extra operation in order to temporarily store one of the registers. When exchanging between an 8-bit register and a

TABLE 5.4 STACK TRANSFER INSTRUCTIONS FOR THE CPU11 AND CPU12

Instruction	CPU11 Operation	CPU12 Operation
TSX	SP + 1 → IX	SP → IX
TSY	SP + 1 → IY	SP → IY
TXS	IX − 1 → SP	IX → SP
TYS	IY − 1 → SP	IY → SP

16-bit register, make sure to read the instruction description in the *CPU12 Reference Manual*. The least-significant byte of the 16-bit register is always transferred to the 8-bit register; however, the most-significant byte of the 16-bit register may be loaded with either a $00 or an $FF.

5.1.7 Clear Instructions

A clear instruction loads a register or a memory location with $00. The register clears are functionally the same as a load immediate with zero. The most useful clear instruction is the clear memory, because it will clear a memory location without first loading the memory location into a register. Table 5.5 shows the three clear instructions for the CPU11 and CPU12.

5.1.8 The CPU12 Move Instruction

The memory-to-memory move instruction is another new instruction that significantly improves code efficiency in the CPU12. It enables you to either move data from one memory location to another or move immediate data into a memory location without using a CPU register. Table 5.6 shows the two move instructions—one for single bytes and the other for 16-bit words.

TABLE 5.5 CLEAR INSTRUCTIONS

Mnemonic	Function	CPU11	CPU12 Addressing Modes					
			INH	IMM	DIR	EXT	IDX	[IDX]
CLRA	$00 → ACCA	X	X					
CLRB	$00 → ACCB	X	X					
CLR	$00 → (M)	X				X	X	X

TABLE 5.6 CPU12 MOVE INSTRUCTIONS

Mnemonic	Function	CPU11	CPU12 Addressing Modes					
			INH	IMM	DIR	EXT	IDX	[IDX]
MOVB	$(M_1) → (M_2)$			X		X	X	
MOVW	$(M_1:M_1+1) → (M_2:M_2+1)$			X		X	X	

Basic Assembly Programming Techniques

Source 5.3 shows two examples that use the move instructions. The first example uses the move to immediately store $0F into the address PORTT. It is a single instruction and does not use a CPU register. Following the move instruction is the equivalent CPU11 code to perform the same task. Note that we had to use ACCA, so if the data in ACCA needed to be preserved, we would have had to save them first.

In the second example indexed addressing was used to copy data from one memory location to another. Again the CPU11 equivalent would have required a CPU register.

Source 5.3 Move Instructions and the CPU11 Equivalents

```
************************************************************
        movb #$0f,PORTT         ;$0F -> PORTT

        ldaa #$0f               ;CPU11 equivalent
        staa PORTT


************************************************************
        movw 0,x,0,y            ;(IX:IX+1) -> (IY:IY+1)

        ldd 0,x                 ;CPU11 equivalent
        std 0,y
```

The first CPU12 example shown in Source 5.3 is one cycle faster than the CPU11 code, but it is one byte longer. In the second example, however, the *movw* instruction saves six clock cycles. This is a significant speed improvement, especially if the code is contained in a loop.

▶ 5.2 USING THE STACK

A stack is a last-in-first-out (LIFO) data structure. It is an ideal data structure for temporarily storing data from nested routines. Last-in-first-out means that the last item placed on the stack is the first item taken off the stack, just like a stack of plates. Many of you are already familiar with stacks if you use a Hewlett-Packard RPN calculator. A CPU stack is implemented in memory with the stack pointer SP and a set of stack instructions.

Note _____

Stack Direction Terminology

When describing a stack, it is very convenient to use terms such as *top, bottom, up, down, above,* or *below.* Because memory maps can be shown with $0000 on top or with $FFFF on top, however, these terms can be ambiguous.

In this text memory maps are always shown with $0000 on top. So the terms *up* or *above* refer to smaller addresses, and the terms *below* or *down* refer to larger addresses. This is also consistent with the stack of plates analogy. A plate is added to the top of the stack and the stack grows upward. *Top of the stack* always refers to the address of the last data stored on the stack and therefore should be the location pointed to by the stack pointer.

TABLE 5.7   CPU12 STACK INSTRUCTIONS

Mnemonic	Operation	CPU11
PSHA	SP–1 → SP, ACCA → (SP)	X
PSHB	SP–1 → SP, ACCB → (SP)	X
PSHC	SP–1 → SP, CCR → (SP)	
PSHD	SP–2 → SP, ACCD → (SP:SP+1)	
PSHX	SP-2 → SP, IX → (SP:SP+1)	X
PSHY	SP-2 → SP, IY → (SP:SP+1)	X
PULA	(SP) → ACCA, SP+1 → SP	X
PULB	(SP) → ACCB, SP+1 → SP	X
PULC	(SP) → CCR, SP+1 → SP	
PULD	(SP:SP+1) → ACCD, SP+2 → SP	
PULX	(SP:SP+1) → IX, SP+2 → SP	X
PULY	(SP:SP+1) → IY, SP+2 → SP	X

5.2.1 Stack Instructions

The stack instructions include push instructions and pull instructions. *Push* instructions place the contents of a register onto the stack. *Pull* instructions take data off the stack and copy them into a register. Some microprocessor manufacturers use the term *pop* instead of *pull* to refer to the pull operation. Table 5.7 shows all the CPU12 push and pull instructions. There is a push and pull instruction for all registers except the program counter and the stack pointer.

5.2.2 The CPU12 Stack Operation

Like most modern stacks, the CPU12 has a predecrement push and a postincrement pull. Source 5.4 shows a sequence of stack instructions, the operations that occur, and equivalent instructions using an index register as a stack pointer.

Source 5.4 Push and Pull Instruction Examples

```
        psha            ;SP-1->SP, ACCA->(SP)
        pshx            ;SP-2->SP, IX->(SP:SP+1)
        pulx            ;(SP:SP+1)->IX, SP+2->SP
        pula            ;(SP)->ACCA, SP+1->SP

* Equivalent instructions using IY as a stack pointer.

        staa 1,-y
        stx 2,-y
        ldx 2,y+
        ldaa 1,y+
```

The *psha* instruction places the contents of ACCA onto the stack. To do this it first decrements the stack pointer to point to the next available location. It then stores the contents

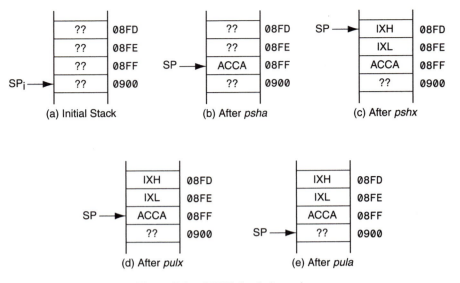

Figure 5.1 CPU12 Stack Operations

of ACCA in the location pointed to by the stack pointer. Figure 5.1 shows the contents of the stack and the stack pointer after each instruction. Initially the stack pointer contains $0900. Then after the *psha* instruction, the stack pointer contains $08FF, and the contents of ACCA have been stored at that location. As you can see, the stack grows toward lower addresses.

The *pshx* instruction places the contents of IX on the stack. Since this is a 16-bit register, it must decrement twice and then store IX. As shown in Figure 5.1c, the stack pointer ends up pointing to $08FD and IX has been stored at locations $08FD and $08FE. Notice that the IX bytes are in the big-endian order.

The *pulx* instruction then pulls the top two bytes off the stack into IX. In the instruction sequence given, this should copy the original data back into IX. To pull IX, IX is first loaded with the 16-bit word pointed to by the stack pointer, and then the stack pointer is incremented twice. As shown in Figure 5.1d, the stack pointer ends up at $08FF, which is the same location it was pointing to before the *pshx* instruction.

The last *pula* instruction pulls the byte on top of the stack into ACCA. ACCA is first loaded with the contents of the location pointed to by the stack pointer, and then the stack pointer is incremented. Again, because of the instruction sequence given, ACCA ends up with the same value it had initially. As you can see in Figure 5.1e, the stack pointer ends up with its initial value.

You can see in Figure 5.1 that the stack pointer always points to the last item placed onto the stack during every step in the instruction sequence. This characteristic of a predecrement push, postincrement pull stack is desirable because it allows the stack pointer to be used as a normal index register.

Source 5.4 also shows the equivalent instructions using an index register as a stack pointer. Although we can use an index register to implement a stack, the stack instructions are more convenient and efficient. In this case the last four lines are slower and result in twice as much code space as the normal stack instructions.

The CPU12 stack is used by the user program and automatically by the CPU for subroutine calls and interrupts. Therefore the CPU12 stack is both a *user stack* and the *system stack*. Some CPUs separate these two stacks. For example the 6809 CPU has a separate user stack and system stack. The intent was that the system stack would be used by the CPU only for subroutine and interrupt service routines, and the user stack would be used only by the user's code. Two separate stacks can make programming easier when writing assembly language programs. When writing code in a high-level language, the compiler keeps track of the stack, and not having two different stacks does not make a significant difference.

5.2.3 The CPU11 Stack Operation

One of the major changes made in the CPU12 is that it has a different stack structure than the CPU11 does. The 68HC11's CPU was based on the 6800's CPU, including the 6800's awkward postdecrement push, preincrement pull stack. Because of this, the CPU11 stack pointer always points to the next location on the stack, not to the last byte pushed onto the stack. Source 5.5 shows the same instruction sequence with the CPU11 operation descriptions.

Source 5.5 CPU11 Stack Instructions and Operations

```
psha              ; ACCA->(SP), SP-1->SP
pshx              ; IXL->(SP), SP-1->SP, IXH->(SP), SP-1->SP
pulx              ; SP+1->SP, (SP)->IXH, SP+1->SP, (SP)->IXL
pula              ; SP+1->SP, (SP)->ACCA
```

Notice the difference in operation between the CPU11 and CPU12. For a push instruction, the CPU decrements the stack pointer after the data are stored. For a pull instruction, it increments the stack pointer before the data are loaded. Figure 5.2 shows the stack

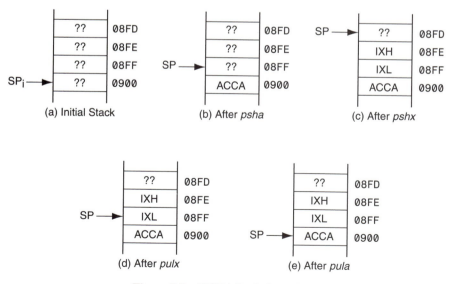

Figure 5.2 CPU11 Stack Operations

contents and stack pointer for the instruction sequence in Source 5.5. The CPU11 stack pointer always points to the next location on the stack and all the data are shifted to the next higher address relative to the CPU12 stack data shown in Figure 5.1.

Although the CPU11 stack is slightly different in its detailed operation, the overall function is still the same. ACCA was pushed onto the stack, IX was pushed onto the stack, IX was then pulled off the stack, and then ACCA was pulled off the stack. The results will be the same on either a CPU11-based or a CPU12-based microcontroller.

Although the overall function of the stack is the same for both the CPU11 and CPU12, there are rare cases in which the differences in the detailed operation can result in incompatible source code. So far we have seen that if only push and pull instructions are used, the CPU12 source is compatible with the CPU11. Now let's look at some other ways that the stack is accessed.

Besides the push and pull instructions, the second most common set of stack-related instructions are the transfer instructions involving the stack pointer. The instructions are *tsx, tsy, txs,* and *tys.* We saw in Table 5.4 that these instructions actually have different operations for the CPU11 and CPU12. The difference in operation is due to the difference between the two stacks. For example, if a *tsx* instruction is inserted after the two push instructions in the previous examples, IX would point to the last byte placed on the stack. For the CPU12 stack, IX would point to $08FD. For the CPU11 stack, IX would point to $08FE. Because the operations are adjusted to match the stack implementation, programs that use these instructions are also source-code compatible.

The potential for source-code incompatibility between the CPU11 and CPU12 exists, however, because of the differences in the stack. Source 5.6 shows some rather strange code that initializes a stack frame in a popular real-time kernel. It uses an index register to fill a stack frame with initial values. The problem with this code is that it does not use a transfer instruction to initialize the index register. Instead it initializes IX with the initial stack location and uses predecrement index addressing to fill the frame. This code works for the CPU12 stack, but it does not work for the CPU11 stack.

Source 5.6 Incompatible Stack Code

```
ldx #INIT_FSP
movw #INIT_PC,2,-x
movw #INIT_IY,2,-x
movw #INIT_IX,1,-x
movb #INIT_A,1,-x
movb #INIT_B,1,-x
movb #INIT_CCR,1,-x
```

At this point you may ask why anyone would write code like this anyway. This code is actually a variation of some code taken out of a popular real-time kernel for the 68HC12. The code in the kernel is used to create a stack frame filled with initial values. Creating initialized stack frames is one of the rare examples in which the stack differences between the CPU11 and CPU12 make it difficult to create compatible source code.

5.2.4 Stack Usage Rules

We will finish our discussion of stacks by covering five basic rules to follow when using the stack. Following these rules can prevent some of the most common and most disastrous software bugs.

Initialize the Stack Pointer.   The stack pointer must be initialized before it can be used. This may seem obvious, but it is common to forget to initialize the stack. This can happen specifically when converting code that was tested with a monitor program to stand-alone code, because most monitor programs initialize the stack before running the user program. So a typical user program may not have stack initialization, yet it would still work under the monitor.

Typically the stack pointer is initialized to the last byte of user RAM. This is because the stack grows into lower addresses. By placing the stack pointer at the last available RAM location, the stack has the most room to grow before it conflicts with variable locations. The CPU12 stack pointer can actually be initialized to the next address after the last byte of RAM, because the stack pointer is decremented first before storing the data pushed. It is common to initialize the stack pointer to the last byte so the code will be source-code compatible with the CPU11. The wasted byte may not be significant.

Remember that the stack is not only used by the user program. It is also used by the CPU for subroutines and interrupt service routines. This means the stack pointer must be initialized before any subroutine calls or serviced interrupts.

Balance the Stack.   Always make sure there are as many bytes pulled off the stack as there are bytes pushed onto the stack. Otherwise the stack can become unbalanced, and over time it may grow beyond the bounds of the stack space. You can also balance the stack by adding an offset to the stack pointer. There is an example of this in Source 5.6 where the *leas* instruction moves the stack pointer back to its initial value. An unbalanced stack can be an especially nasty bug. It may take a long time, with a specific combination of inputs, before the bug shows up.

Do Not Use Data above the Stack Pointer.   Notice that in Figure 5.1 when data are pulled off the stack, the data are actually copied like any other data transfer. Therefore the data remain in the memory locations until something else is pushed onto the stack at the same location. Beware! Never rely on the data above the stack pointer. If an interrupt occurs or a subroutine is called, the data will be destroyed. Even if there are no interrupts or subroutines, the data above the stack are still in danger from a context switch to a monitor program. BUFFALO and D-Bug 12 (resident on target) save the CPU registers on the stack whenever a software breakpoint occurs. So if you have a breakpoint, the data above the stack pointer are written over. This is also the case if a positive offset is added to the stack pointer to access data previously placed onto the stack. Again, the data above the stack pointer are not protected and can be changed.

Only Use Push, Pull, and Transfer Instructions for Stack Access.   This warning is important only when source-code compatibility between the CPU12 and CPU11 is required.

Use Standard Local Variable and Parameter Passing Techniques. Do not use the stack haphazardly for temporary storage. Try to stick to the standard techniques that are covered in Chapter 6 regarding parameter passing and local variables. Otherwise your pushes and pulls should always be close together in the same program structure. A common bug is caused by pushing data onto the stack, and then missing the pull because of a conditional branch, which in turn creates an unbalanced stack.

▶ 5.3 BASIC ARITHMETIC PROGRAMMING

In this section we look at basic arithmetic programming, including addition, subtraction, and compares. We are going to assume that you are already familiar with binary addition, BCD, and two's-complement numbers.

5.3.1 Addition Instructions

The CPU12 has the same addition instructions as the CPU11. These instructions are shown in Table 5.8. We can break these up into 8-bit additions (*aba, adda,* and *addb*), 8-bit additions with carry (*adca* and *adcb*), a 16-bit addition (*addd*), and two 8-bit to 16-bit additions (*abx* and *aby*). The *daa* instruction is used to adjust the result of a binary addition to a BCD result.

5.3.2 8-Bit Binary Addition

The first application is 8-bit binary addition. Let's look at an example and the detailed operation of the CPU, including the effect on the CCR flags. Let's start with the following example:

```
ldaa #$3a          ; $3a -> ACCA
adda #$7c          ; ACCA + $7c -> ACCA
```

TABLE 5.8   ADDITION INSTRUCTIONS

Mnemonic	Operation	CPU11	CPU12 Addressing Modes					
			INH	IMM	DIR	EXT	IDX	[IDX]
ABA	ACCA+ACCB → ACCA	X	X					
ABX	ACCB+IX → IX	X	X					
ABY	ACCB+IY → IY	X	X					
ADCA	(M)+ACCA+C → ACCA	X		X	X	X	X	X
ADCB	(M)+ACCB+C → ACCB	X		X	X	X	X	X
ADDA	(M)+ACCA → ACCA	X		X	X	X	X	X
ADDB	(M)+ACCB → ACCB	X		X	X	X	X	X
ADDD	(M:M+1)+ACCD → ACCD	X		X	X	X	X	X
DAA	Decimal Adjust ACCA	X	X					

The CPU Operation. Following is the CPU operation for this example:

```
              1111                    H  N  Z  V  C
    $3A       0011 1010              | 1| 1| 0| 1| 0|
 +  $7C    +  0111 1100
    $B6       1011 0110
```

To follow this operational description, refer to the *adda* instruction description in the *CPU12 Reference Manual*. The CPU operation is always a simple binary addition. The CPU does not know if the arguments are signed, unsigned, or BCD.

The flags are set according to the Boolean formulas in the *adda* instruction description. The half-carry flag, *H,* indicates a carry out of bit 3. If we look at the binary addition shown, we can see that there was a carry out of bit 3, so the *H* flag is set. The negative flag, *N,* indicates a negative result. A result is negative if bit 7 of the result is set. In this case bit 7 of the result is one, so the *N* flag is set. The zero flag, *Z,* indicates a result of zero. It is set if every bit in the result is zero. In this case the result is not zero, so the *Z* flag is cleared. The overflow flag, *V,* is set if the addition resulted in a two's-complement overflow. The easiest way to tell if there is an overflow is to look at the sign bits of the two arguments and the result. For addition operations, if the two arguments have the same sign and the result has the opposite sign, then the *V* flag is set. In this example the two arguments have a zero-sign bit, so they are both positive numbers. The result has a one-sign bit, so it is a negative number. There is no way to get a negative number when adding two positive numbers. Therefore a two's-complement overflow must have occurred, and the *V* flag was set. One thing to note is that if we are adding two arguments that have unlike signs, there is no possibility of getting a two's-complement overflow. Finally the carry flag, *C,* indicates a carry out of bit 7. In this case there was not a carry out of bit 7, so the *C* flag is cleared.

As mentioned earlier, the CPU does not know anything about the arguments. It simply performs an 8-bit binary addition on the arguments. Now it is up to us to interpret the results based on the actual data types.

Unsigned Interpretation. If the arguments are both unsigned numbers, the addition can be interpreted as follows:

```
    $3A      →        58
 +  $7C      →     +  124
    $B6      →       182
```

Eight-bit unsigned numbers can range from 0 to 255. If the result exceeds 255, then the *C* flag will be set, and the 8-bit result will be incorrect. In this case the result is 182, which does not exceed 255 and is correct. Now the question is how do we know if it is correct?

For unsigned numbers, the *C* flag indicates an incorrect 8-bit result. We saw previously that the *C* flag was cleared, so we know the result is correct. If the *C* flag were set, we would have to branch to an error routine to indicate that the result was greater than 255, or use the *C* flag to calculate a 16-bit result.

Signed Interpretation. If the arguments are both signed numbers, the addition is interpreted as follows:

$$
\begin{array}{rcl}
\$3A & \rightarrow & (+58) \\
+ \ \underline{\$7C} & \rightarrow & + \ \underline{(+124)} \\
\$B6 & \rightarrow & (\ -74)
\end{array}
$$

Eight-bit signed numbers can range from −128 to +127. If the result exceeds this range, the overflow flag, *V*, will be set. In this case the desired result, +182, exceeds the range for an 8-bit signed number. This error can be detected by testing the *V* flag. In this case the *V* flag is set, which indicates that the range was exceeded, and the result is incorrect.

To summarize, the CPU does the same operation, regardless of data type. It is up to the programmer to perform the correct tests to determine whether an error occurred. Based on the data type, the following tests are made:

Type	Test
Unsigned Numbers	*C* flag
Signed Numbers	*V* flag

5.3.3 Multiple-Byte Binary Addition

To add words larger than eight bits, we either have to use the *addd* instruction or perform multiple 8-bit additions using the add-with-carry instructions. Let's look at an *addd* example first.

```
ldd  #95          ;95 -> ACCD
addd var16        ;ACCD+(var16:var16+1)-> ACCD
std  var16        ;ACCD -> (var16:var16+1)

var16    rmb 2
```

This example adds 95 to the 16-bit global variable *var16*. The variable *var16* was defined using an *rmb* directive. Notice that it requires two bytes that will be at the locations *var16* and *var16+1*. Again, let's look at the CPU operation for the *addd* instruction. For this detail we will assume (*var16:var16+1*) initially contain $8001.

CPU Operation. Following is the CPU operation for this example:

```
                     11 111              H  N  Z  V  C
  $005f    0000 0000 0101 1111          ┌──┬──┬──┬──┬──┐
+ $8001  + 1000 0000 0000 0001          │ -│ 1│ 0│ 0│ 0│
  $8060    1000 0000 0110 0000          └──┴──┴──┴──┴──┘
```

The operation is the 16-bit equivalent of the 8-bit addition shown previously. The sign bit for 16-bit values is bit 15, so the negative flag and the overflow flag are controlled by bit 15 instead of bit 7. The same is true for the carry flag. Instead of indicating a carry out of bit 7, it indicates a carry out of bit 15. The half-carry flag, *H,* is not affected at all. Therefore the *addd* instruction cannot be used for BCD numbers. The signed and unsigned interpretations are

Unsigned Numbers			**Signed Numbers**		
$005f	→	95	$005f	→	(+ 95)
+ $8001	→	+ 32769	+ $8001	→	+ (−32767)
$8060	→	32864	$8060	→	(−32672)

In both cases, signed and unsigned, the results were correct, and both the overflow flag and the carry flag are cleared.

For word sizes 16 bits and greater, the add-with-carry instructions should be used as follows: The least-significant bytes are added with an 8-bit add instruction, and then the next most-significant bytes are added with an add-with-carry instruction. This step is repeated until the complete words are added. As an example, we can implement the *addd* example shown earlier using add-with-carry instructions.

```
ldd  #95        ;95 -> ACCD
addb var16+1    ;ACCB+(var+1)-> ACCB
adca var16      ;ACCA+(var)+C -> ACCA
```

The least-significant byte of ACCD, ACCB, is first added to the least-significant byte of *var16, (var16+1)*. This addition will set the carry flag if the result is greater than 255. Following is the CPU operation of the first addition:

```
          11 111         H  N  Z  V  C
$5f       0101 1111      ┌──┬──┬──┬──┬──┐
                         │1 │0 │0 │0 │0 │
+ $01   + 0000 0001      └──┴──┴──┴──┴──┘
$60       0110 0000
```

The most-significant byte of ACCD, ACCA, is then added to the most-significant byte of *var* using an add-with-carry instruction. The add-with-carry instruction will add the two bytes, and then add the current value of the carry flag. From the operation shown previously, we can see that for this example, the carry flag is cleared. Following is the CPU operation for the second add.

```
$00       0000 0000      H  N  Z  V  C
                         ┌──┬──┬──┬──┬──┐
+ $80   + 1000 0000      │0 │1 │0 │0 │0 │
$80       1000 0000      └──┴──┴──┴──┴──┘
+ (C)   +         0
$80       1000 0000
```

Notice the add-with-carry instruction also sets the flags. These flags can then be used by another add-with-carry instruction if the word was larger than 16 bits.

Unsigned Word Extension. When performing arithmetic operations, there are times when the two arguments are different sizes. The procedure for operating on mismatched arguments is to extend the smaller argument to the size of the larger argument first, and then perform the operation. The method to extend the argument size is different for unsigned numbers than for signed numbers. Let's look at a simple example that extends unsigned 8-bit data to 16-bits:

```
clra        ;0->ACCA
ldab var8   ;(var8)->ACCB
```

The 8-bit data is loaded into ACCB, and ACCA is cleared. The result is a 16-bit value in ACCD that is equal to the unsigned data in *var8*. This seems rather obvious. Let's look at what happens if we use this code for signed numbers, however. For this illustration, we will assume *(var8)* is $81, so after the code above is executed, ACCD contains $0081.

	8-bit, var8		16-bit, ACCD
Unsigned:	$81= 129	→	$0081=129
Signed:	$81=−127	→	$0081=129

The extension of the signed number changed the value of the signed number from −127 to 129.

Sign Extension.　When extending signed numbers, we have to perform *sign extension*. The process of sign extension copies the sign bit of the original data into all higher-order bits of the extended data. In the example given previously, since the sign bit of *var8* is a one, then ACCA must contain all ones. For the CPU11, we can extend *var8* into ACCD as follows:

```
        clra            ;$00 -> ACCA
        ldab var8       ;(var8) -> ACCB
        bpl pos_num     ;done if positive number
        ldaa #$ff       ;Extend sign if negative
pos_num
```

In this example a conditional branch is used to load ACCA with $FF if the contents of ACCB is not positive. Otherwise ACCA remains $00. On the CPU12 we can use the sign extension instruction *sex,* shown in Table 5.3. Since the *sex* instruction is a transfer operation, we need to first load *(var8)* into an 8-bit accumulator. Following is a CPU12 example:

```
ldab var8       ; (var8) -> ACCB
sex b,d         ; sign extend to ACCD
```

var8 is first loaded into ACCB, and then ACCB is sign extended into ACCD. To summarize, always use the simple extension of loading zeros in all higher-order bits for unsigned numbers, and always copy the sign bit into higher-order bits for signed numbers.

5.3.4 BCD Addition

There are some applications in which the basic data type is binary coded decimal (BCD) or ASCII, which can be easily converted to BCD. If arithmetic operations on the data are required, a decision must be made. We either can perform BCD arithmetic or convert the BCD data to binary and perform the normal binary arithmetic. For most applications, it is better to convert the BCD to binary and work with binary numbers. There are several reasons for this. The storage of binary numbers is much more efficient, binary arithmetic is much simpler, and the BCD-binary conversion can be done easily as data are passed to or from an external source.

For those occasional applications in which BCD arithmetic makes sense, most CPUs provide a *daa* instruction to adjust the result of a binary addition to BCD. It is important to realize that the operation of the *daa* instruction only adjusts a result of a binary addition or, for some CPUs, adjusts the result of a binary subtraction. It cannot be used on the CPU11 or CPU12 for BCD subtraction, and it cannot be used on any CPU to perform a binary to BCD conversion.

The basic operation of the *daa* instruction is

```
If(ACCA3..0 > 9 or H == 1) add $06
If(ACCA7..4 > 9 or C == 1) add $60
```

where $ACCA_{3..0}$ is the least-significant digit in ACCA, and $ACCA_{7..4}$ is the most-significant digit in ACCA. The two lines above are a sequence. The first line is performed first, followed by the second line.

Let's look at an example in which we add two packed-BCD numbers, 39 and 47:

```
ldaa #$39        ;$39 -> ACCD
adda #$47        ;ACCA+$47-> ACCA
daa              ;Decimal adjust ACCA
```

First take a look at the detailed *adda* operation:

```
              1111 111        H  N  Z  V  C
  $39         0011 1001      | 1| 1| 0| 1| 0|
+ $47       + 0100 0111
  $80         1000 0000
```

This is a binary addition, which produced a result of $80. This is obviously not the sum of 39 and 47 BCD. The *daa* instruction is used to adjust this result for the correct BCD result. First the least-significant digit of the result is checked to see if it is greater than 9, or if the half-carry flag has been set. The half-carry flag was set so the *daa* instruction adds $06.

```
  $80         1000 0000        H  N  Z  V  C
+ $06       + 0000 0110      | 0| 1| 0| ?| 0|
+ $86         1000 0110
```

Now the second part of the *daa* operation is performed. In this case the most-significant digit is not greater than nine and the *C* flag is not set, so $60 is not added, and the operation is complete. The result is 86, which is the correct sum of 39 and 47. The *C* flag is set correctly by the *daa* instruction to indicate a BCD carry. In this case there was no carry. If there was a carry, the *C* flag could be used for a carry into the next BCD digit.

5.3.5 Index Register Additions

There are two more addition instructions shown in Table 5.8 that are not really intended for arithmetic—*abx* and *aby*. These are special-purpose instructions for adding offsets to index registers. They add the 8-bit unsigned value in ACCB to an index register, IX or IY. The reason they are not good general-purpose addition instructions is that they have no effect on the flags. Therefore conditional branches based on the result cannot be used.

5.3.6 Subtraction Instructions

The CPU12 has the same subtraction instructions as the CPU11. These instructions are shown in Table 5.9. We can break these up into 8-bit subtractions (*sba, suba,* and *subb*), 8-bit subtractions with borrow (*sbca* and *sbcb*), and a 16-bit subtraction (*subd*). There are also three negate instructions (*neg, nega,* and *negb*) for performing a two's-complement conversion. These instructions are used to change the sign of the argument.

5.3.7 8-Bit Binary Subtraction

First let's look at the detail of an 8-bit subtraction. In this example we subtract $48 from $23.

```
ldaa #$23          ;$23 -> ACCA
suba #$48          ;ACCA-$48 -> ACCA
```

As we did with addition, let's look at the detailed CPU operation for a subtraction. First there is not a separate subtraction circuit in the CPU. Subtraction is accomplished by first negating the subtrahend then adding.

The CPU Operation. Following is the CPU operation for this example:

```
                                    1
  $23      0010 0011      →     0010 0011
- $48    - 0100 1000      →   + 1011 1000
  $DB                          1101 1011
```

H	N	Z	V	C
–	1	0	0	1

TABLE 5.9 SUBTRACTION INSTRUCTIONS

Mnemonic	Function	CPU11	INH	IMM	DIR	EXT	IDX	[IDX]
			CPU12 Addressing Modes					
SBA	ACCA–ACCB → ACCA	X	X					
SBCA	ACCA–(M)–C → ACCA	X		X	X	X	X	X
SBCB	ACCB–(M)–C → ACCB	X		X	X	X	X	X
SUBA	ACCA–(M) → ACCA	X		X	X	X	X	X
SUBB	ACCB–(M) → ACCB	X		X	X	X	X	X
SUBD	ACCD–(M:M+1) → ACCD	X		X	X	X	X	X
NEG	0–(M) → (M) X				X	X	X	
NEGA	0–ACCA → ACCA	X	X					
NEGB	0–ACCB → ACCB	X	X					

The flags are set according to the Boolean expressions given in the *suba* instruction description in the *CPU12 Reference Manual*. The half-carry flag is not affected by the subtraction instructions. It is only used for BCD addition. The negative flag is set to indicate a negative result. The zero flag is cleared because the result was not zero. The overflow flag indicates a two's-complement overflow. It has been cleared in this case. The *C* flag indicates a borrow. In this case it is set because the magnitude of the subtrahend is greater than the minuend. Just as with addition, the use of these flags depends on the interpretation of the subtraction problem with respect to the data type.

Unsigned Interpretation.   If the arguments are both unsigned numbers, then the subtraction can be interpreted as follows:

$$
\begin{array}{rcr}
\$23 & \rightarrow & 35 \\
- \ \underline{\$48} & \rightarrow & - \ \underline{72} \\
\$DB & \rightarrow & 219
\end{array}
$$

Unsigned numbers cannot be less than zero. Therefore the subtrahend cannot have a larger magnitude than the minuend. In this case the subtrahend, $48, is greater than the minuend, $23, so the expected result would be a negative number. The result is 219, which is obviously incorrect. Unsigned errors like this can be detected by checking the *C* flag to see if a borrow was required. In this case it was. If the *C* flag is set, we must either jump to an error routine or use the borrow to subtract one from the next most-significant byte.

Signed Interpretation.   If the arguments are both signed numbers, the subtraction is interpreted as follows:

$$
\begin{array}{rcr}
\$23 & \rightarrow & (+35) \\
- \ \underline{\$48} & \rightarrow & - \ \underline{(+72)} \\
\$DB & \rightarrow & (-37)
\end{array}
$$

The signed interpretation gives the correct results. This can be detected by checking the overflow flag. If the overflow flag was set, then there was a two's-complement overflow. When subtracting signed numbers, it is only possible to get an overflow when the arguments have opposite signs. The overflow flag is set when a negative number is subtracted from a positive number with a negative result, or if a positive number is subtracted from a negative number with a positive result.

Remember that the *C* flag is used for unsigned numbers, and the *V* flag is used for signed numbers.

5.3.8 Multiple-Byte Binary Subtraction

When word sizes are larger than eight bits, we must use either the *subd* instruction or one of the subtract-with-borrow instructions. Let's look at an example that uses the subtract-with-borrow instructions. When subtracting multiple-byte data using a subtract-with-borrow, we start at the least-significant bytes of the arguments and use an 8-bit subtraction, then go to the next most-significant bytes and use a subtract-with-borrow. In this example we will subtract 95 from the 16-bit data stored at *var16*.

```
        ldd   var16        ;(var16:var16+1) -> ACCD
        subb  #95          ;ACCB-95 -> ACCB
        sbca  #00          ;ACCA-00-C -> ACCA

var16   rmb 2
```

Notice that we performed the operation in a different order than we did in the addition example. Remember that the subtrahend is always the argument of the subtraction instruction. To look at the detailed operation of the CPU, we will again assume $8001 is stored in *(var16:var16+1)*. Following is the CPU operation of the first subtraction:

```
                                      1              H  N  Z  V  C
    $01        0000 0001    →     0000 0001         – │1 │0 │0 │1
 -  $5f     -  0101 1111    →   + 1010 0001
    $a2                          1010 0010
```

The only flag of interest at this point is the *C* flag. In this case the *C* flag is set, indicating that a borrow is required. We can see this by looking at the original arguments. The subtrahend, $5f, is larger than the minuend, $01. The next subtract-with-borrow instruction will then use the *C* flag to subtract one from the result of that subtraction. Following is the CPU operation for the subtract-with-borrow instruction:

```
    $80        1000 0000         H  N  Z  V  C
 -  $00     -  0000 0000         – │0 │0 │1 │0
    $80        1000 0000
 -  (C)     -          1
    $7f        0111 1111
```

Now let's look at the signed and unsigned interpretation of the results.

<table>
<tr><td colspan="3" align="center">**Unsigned Numbers**</td><td colspan="3" align="center">**Signed Numbers**</td></tr>
<tr><td>$8001</td><td>→</td><td>32769</td><td>$8001</td><td>→</td><td>(-32767)</td></tr>
<tr><td>- $005f</td><td>→</td><td>- 95</td><td>+ $005f</td><td>→</td><td>- (+95)</td></tr>
<tr><td>$7fa2</td><td>→</td><td>32674</td><td>$7fa2</td><td>→</td><td>(+32674)</td></tr>
</table>

The unsigned 16-bit result is correct. We can detect this by checking the *C* flag, which in this case is zero, indicating a correct result. The signed 16-bit result, however, is incorrect. Again, this is detected by checking the overflow flag. In this case *V* is set to indicate a two's-complement overflow. The desired result is –32862, but this is beyond the 16-bit two's-complement range, –32768 to +32767. We can extend this same process for larger word sizes by using more subtract-with-borrow instructions on the next higher-order bytes.

Read-Modify-Write Instructions. Before we look at compare instructions, let's look at the memory negate instruction *neg* in more detail. The *neg* instruction is the first instruction we have covered so far that is a read-modify-write instruction. A *read-modify-write* instruction changes the contents of a memory location by first reading the location, then modifying the read value, and then writing the modified data back into the location. In this case the *neg* instruction reads (M), performs a two's-complement conversion, and then writes the result back to (M). Besides the fact that two bus access cycles are required, there is an important implication for any read-modify-write instruction—the location must be readable and writable. Normally the data is a RAM location or an MCU control register location. For these cases the *neg* instruction works as expected. If the location is a write-only output port, however, a read-modify-write instruction cannot be used. In this case the instruction will read garbage, modify it, and write the modified garbage to the output.

Any time you see an operation of the form

```
(M) op -> (M)
```

where *op* is any operation, you know it is a read-modify-write instruction because *(M)* is both a source and destination.

5.3.9 BCD Subtraction

Performing BCD subtraction on the CPU11 or CPU12 is not as straightforward as BCD addition. When we needed a BCD addition, we performed a binary addition followed by a *daa* instruction. The *daa* instruction, however, only works for converting a binary addition. We cannot use it to convert a binary subtraction. This is not true for all processors, however. The Z80, for example, uses an additional flag that indicates an addition or subtraction. The Z80's *daa* instruction then uses that flag to determine whether it is to make an adjustment for addition or for subtraction.

To perform BCD subtraction on a CPU11 or CPU12-based MCU, we have to use ten's-complement arithmetic. This is similar to the two's-complement process that is used by the CPU to perform binary subtraction. We find the ten's-complement of the subtrahend and use a BCD addition.

The ten's-complement of an n digit number, N, is

$$C_{10, N} = 10^n - N$$

Using the ten's-complement of a number allows you to convert a BCD subtraction to a BCD addition:

$$Y = X - N = X + C_{10, N}$$

For a two-digit number

$$C_{10, N} = 100 - N$$

EXAMPLE 5.1 _____

Ten's-Complement Subtraction

For $X = 81$ and $N = 9$

$$Y = X - N = 81 - 9 = 72$$

Using ten's-complement arithmetic

$$C_{10,\,9} = 100 - 9 = 91$$
$$Y = X + C_{10,\,N} = 81 + 91 = 172 \Rightarrow 72_{\text{mod}100}$$

Example 5.1 shows that the ten's-complement process does result in the correct solution. If we look at the operations involved, however, we just replaced one BCD subtraction, $X - N$, with another, $100 - N$. In addition, if we are working with two-digit numbers, we cannot even represent 100. The solution to this is to adjust the ten's-complement calculation to

$$C_{10,\,N} = 100 - N = 99 - N + 1$$

We can represent 99 with two digits, so we have solved that problem. Now if we look at the remaining subtraction, $99 - N$, notice that we can use binary subtraction. N ranges from 00 to 99, so for all N, a simple binary subtraction gives the correct BCD result.

So to perform a BCD subtraction we use the following process:

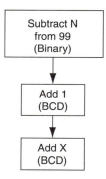

To implement Example 5.1 we could use the following program:

```
ldaa #$99               ;ten's-comp of N
suba #$09
inca
daa
adda #$81               ;Add X
daa
```

One additional case to consider is when this subtraction results in a negative number. We can detect a negative result by looking at the C flag after performing the last addition and decimal adjust. If the C flag is set, then the result is positive and can be used as is. If the C flag is clear, then the result is negative. Of course we cannot represent negative numbers with BCD. So after detecting a negative number, you either have to borrow one from the next most-significant digit, if there is one, or set an additional sign flag to be used for

Chapter 5

displaying the data. To use the result, however, you also first must calculate the complement. Following is the process for detecting and converting negative results:

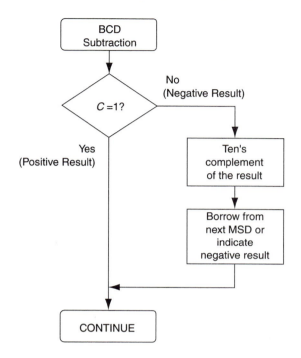

5.3.10 Compares and Tests

Compare instructions are used to compare two arguments and set the appropriate flags for a conditional branch instruction. All the CPU12 and CPU11 compare instructions are shown in Table 5.10. Notice in the operation description that the compare instructions perform

TABLE 5.10 COMPARE AND TEST INSTRUCTIONS

Mnemonic	Function	CPU11	CPU12 Addressing Modes					
			INH	IMM	DIR	EXT	IDX	[IDX]
CBA	ACCA–ACCB	X	X					
CMPA	ACCA–(M)	X		X	X	X	X	X
CMPB	ACCB–(M)	X		X	X	X	X	X
CPD	ACCD–(M:M+1)	X		X	X	X	X	X
CPS	SP–(M:M+1)			X	X	X	X	X
CPX	IX–(M:M+1)	X		X	X	X	X	X
CPY	IY–(M:M+1)	X		X	X	X	X	X
TST	(M)–0	X				X	X	X
TSTA	ACCA–0	X	X					
TSTB	ACCB–0	X	X					

the same operation as the subtraction instructions except that the result of the subtraction is not saved. This is required to preserve the two arguments that were compared. The test instructions are simply compare to zero. The test instructions always clear the *C* and *V* flags, however. Therefore they are not useful for all unsigned branches. We will use the compare and test instructions later when conditional branches are covered.

5.3.11 Decrement and Increment Instructions

The increment and decrement instructions shown in Table 5.11 are used to add or subtract one. At first glance we might think they are the same as the corresponding add or subtract instructions. Be careful. There are some important differences. It is a good idea to always check the instruction description when using these instructions with a conditional branch. There are three different ways the flags are affected. The 8-bit increments and decrements (*inc, inca, incb, dec, deca,* and *decb*) do not affect the carry flag. This means that you cannot perform an unsigned conditional branch based only on the flags set by these instructions. The index register increments and decrements (*dex, dey, inx,* and *iny*) only change the *Z* flag, so you can only use *beq* or *bne* instructions. The stack pointer increment and decrement instructions (*des* and *ins*) do not affect any flags, so conditional branches cannot be used without first executing a compare.

Notice both *dec* and *inc* are read-modify-write instructions, so they will only work on a bidirectional location.

Multiple-Byte Increments and Decrements.   Notice in Table 5.11 that the only multiple-byte increments or decrements are those applied to the index registers and the stack pointer. There are many times when a 16-bit value in ACCD or a 16-bit value in memory must be incremented or decremented. The best way to do this is to load the data into ACCD and use

```
addd #1
```

TABLE 5.11 INCREMENT AND DECREMENT INSTRUCTIONS

Mnemonic	Function	CPU11	CPU12 Addressing Modes					
			INH	IMM	DIR	EXT	IDX	[IDX]
DEC	(M)–1 → (M)	X				X	X	X
DECA	ACCA–1 → ACCA	X	X					
DECB	ACCB–1 → ACCB	X	X					
DES	SP–1 → SP	X	X					
DEX	IX–1 → IX	X	X					
DEY	IY–1 → IY	X	X					
INC	(M)+1 → (M)	X				X	X	X
INCA	ACCA+1 → ACCA	X	X					
INCB	ACCB+1 → ACCB	X	X					
INS	SP+1 → SP	X	X					
INX	IX+1 → IX	X	X					
INY	IY+1 → IY	X	X					

104

or

```
subd #1
```

We cannot use an increment instruction followed by an add-with-carry or a decrement followed by a subtract-with-borrow because the increment and decrement instructions do not affect the C flag.

▶ 5.4 SHIFTING AND ROTATING

Shifting or rotating data is often required for arithmetic, parity, bit alignment, and I/O applications. Table 5.12 shows all the CPU11 and CPU12 shift and rotate instructions. All shift and rotates shift the contents of a word by one bit position. The rotate instructions are normally used for multiple-byte shifts. Source 5.7 shows two examples.

TABLE 5.12 SHIFT AND ROTATE INSTRUCTIONS

Mnemonic	Operation	CPU11	CPU12 Addressing Modes			
			INH	EXT	IDX	[IDX]
LSLA		X	X			
LSLB	(C ← b7 ... b0 ← 0)	X	X			
LSL		X		X	X	X
LSRA		X	X			
LSRB	(0 → b7 ... b0 → C)	X	X			
LSR		X		X	X	X
LSLD	(C ← b7 ACCA b0 ← b7 ACCB b0 ← 0)	X	X			
LSRD	(0 → b7 ACCA b0 → b7 ACCB b0 → C)	X	X			
ASLA		X	X			
ASLB	(C ← b7 ... b0 ← 0)	X	X			
ASL		X		X	X	X
ASLD	(C ← b7 ACCA b0 ← b7 ACCB b0 ← 0)	X	X			
ASRA		X	X			
ASRB	(b7 ... b0 → C)	X	X			
ASR		X		X	X	X
ROLA		X	X			
ROLB	(C ← b7 ... b0)	X	X			
ROL		X		X	X	X
RORA		X	X			
RORB	(b7 ... b0 → C)	X	X			
ROR		X		X	X	X

Source 5.7 Multiple-Byte Shifts

```
*******************************************
* 16-bit logical shift left

        lsl var16+1
        rol var16


*******************************************
* 16-bit arithmetic shift right

        asr var16
        ror var16+1

var16   rmb 2
```

A common application for the shift and rotate instructions is multiplication and division. Shifts can be used to multiply or divide by constant powers of 2. We will look at this in more detail in Chapter 7.

▶ 5.5 BOOLEAN LOGIC, BIT TESTING, AND BIT MANIPULATION

In this section we focus on bit-oriented operations used to test and manipulate individual bits in a word. These operations are common in embedded systems. Applications include reading an input switch, activating an LED, generating a pulse, or communicating over a serial interface.

Bit testing and manipulation is fundamentally accomplished using Boolean logic operations on the words containing the bits. Since bit-oriented operations are so common in embedded designs, however, some microcontrollers have additional instructions that are specialized for bit-oriented operations.

The CPU11 and CPU12 do have bit-oriented instructions; however, there are times when it is more appropriate to use Boolean logic instructions. Also, when using a CPU that does not have bit-oriented instructions or when using a high-level language, all bit operations must be done using Boolean logic. Therefore we first look at the fundamentals of bit-oriented operations using Boolean logic. We then look at the CPU11 and CPU12 bit-oriented instructions. Later in Chapter 13 we cover bit-oriented operations using the C language.

Throughout this section two sets of operator conventions are used when describing Boolean operations. These conventions are shown in Table 5.13. In program text or any

TABLE 5.13   BOOLEAN OPERATOR CONVENTIONS

Operator	C-Type Operator	Equation Operator
AND	&	·
OR	|	+
Ex-OR	^	⊕
NOT	~	′

text-only file, the C-type operators are used. Otherwise the Boolean equation-type operators are used.

5.5.1 Boolean Logic Instructions

Boolean logic functions are an essential part of most embedded microcontroller programs. Boolean logic is used for implementing outputs that depend on a Boolean combination of inputs and for bit testing and manipulation. Table 5.14 shows the CPU12 logic instructions. They include bitwise AND, bitwise Ex-OR, bitwise OR, and one's-complement (bitwise NOT) instructions.

It is important to understand the difference between a bitwise operator and a logical operator. All of these instructions are bitwise operators. A bitwise operator performs the operation on each pair of aligned bits in the two arguments. A logical operator on the other hand performs the operation on the complete word and returns a value of $01 (TRUE) or $00 (FALSE). To illustrate the difference, let's compare the results of a bitwise AND, &, and a logical AND, &&.

The bitwise AND, $81 & $80, returns $80 after the following operation:

$$
\begin{array}{c}
1000\ 0001 \\
\cdot\ \underline{1000\ 0000} \\
1000\ 0000
\end{array}
$$

The logical AND, $81 && $80, returns a $01 (TRUE) because both $81 and $80 are nonzero (TRUE).

5.5.2 Bit Manipulation

Bit manipulation involves setting, clearing, or inverting individual bits in a word without affecting the other bits in the word. In this section we learn how to use Boolean logic instructions to manipulate bits. Source 5.8 shows three bit manipulation code snippets.

TABLE 5.14 LOGIC INSTRUCTIONS

Mnemonic	Operation	CPU11	CPU12 Addressing Modes					
			INH	IMM	DIR	EXT	IDX	[IDX]
ANDA	ACCA · (M) → ACCA	X		X	X	X	X	X
ANDB	ACCB · (M) → ACCB	X		X	X	X	X	X
ANDCC	CCR · (M) → CCR							
EORA	ACCA ⊕ (M) → ACCA	X		X	X	X	X	X
EORB	ACCB ⊕ (M) → ACCB	X		X	X	X	X	X
ORAA	ACCA + (M) → ACCA	X		X	X	X	X	X
ORAB	ACCB + (M) → ACCB	X		X	X	X	X	X
ORCC	CCR + (M) → CCR			X				
COMA	ACCA′ → ACCA	X	X					
COMB	ACCB′ → ACCB	X	X					
COM	(M)′ → (M)	X				X	X	X

Source 5.8 Bit Manipulation Using Logic Instructions

```
***************************************************************
* Clear bits 4-7 of ACCA

        anda #%00001111              ;ACCA & %00001111 -> ACCA


***************************************************************
* Set bits 0-3 of out_reg

        ldaa #%00001111             ; %00001111 -> ACCA
        oraa out_reg                ; ACCA | out_reg -> ACCA
        staa out_reg                ; ACCA -> out_reg


***************************************************************
* Invert bit0 of ACCB

        eorb #%00000001             ; ACCB ^ %00000001 -> ACCB
```

Clearing Bits Using the AND Instructions.   The bitwise AND instructions perform an AND operation on each pair of aligned bits. A typical application of the AND instructions is to clear some of the bits of a word without affecting the other bits in the word. This is based on the two Boolean identities

$$X \cdot 1 = X \qquad\qquad X \cdot 0 = 0$$

So to clear a bit, we AND it with a zero. We AND all the other bits in the word with a one, which results in no change. The first code snippet in Source 5.8 is an example that clears bits 4 through 7 of ACCA. The resulting CPU operation for this instruction is

$$
\begin{array}{r}
A_7 A_6 A_5 A_4\ A_3 A_2 A_1 A_0 \\
\cdot\quad \underline{0\ 0\ 0\ 0\ \ 1\ 1\ 1\ 1} \\
0\ 0\ 0\ 0\ \ A_3 A_2 A_1 A_0
\end{array}
$$

Every bit in ACCA corresponding to a zero in the argument is cleared.

Setting Bits Using the OR Instructions.   The bitwise OR instructions perform an OR operation on each pair of aligned bits. OR instructions are normally used to set bits in a word without affecting the other bits. This is based on the identities

$$X + 1 = 1 \qquad\qquad X + 0 = X$$

The second code snippet in Source 5.8 uses the OR instruction *oraa* to set bits 0 through 3 of the memory location *out_reg*. The CPU operation for the *oraa* instruction is

$$
\begin{array}{r}
A_7 A_6 A_5 A_4\ A_3 A_2 A_1 A_0 \\
+\quad \underline{0\ 0\ 0\ 0\ \ 1\ 1\ 1\ 1} \\
A_7 A_6 A_5 A_4\ \ 1\ 1\ 1\ 1
\end{array}
$$

In this example every bit in the memory location *out_reg* corresponding to a one in ACCA is set.

Inverting Bits Using the Ex-OR Instructions. The bitwise EX-OR instructions perform an exclusive-OR operation on each pair of aligned bits. EX-OR instructions can be used to invert bits. This is based on the identities

$$X \oplus 1 = X' \qquad X \oplus 0 = X$$

The last code snippet in Source 5.8 inverts bit 0 of ACCB using the EX-OR instruction *eorb*. The CPU operation for the *eorb* instruction is

$$
\begin{array}{c}
B_7B_6B_5B_4 \ B_3B_2B_1B_0 \\
\oplus \ \ 0\ 0\ 0\ 0\ \ \ 0\ 0\ 0\ 1 \\
\hline
B_7B_6B_5B_4 \ B_3B_2B_1B_0'
\end{array}
$$

Every bit of ACCB corresponding to a one in the argument is inverted. In this case only B_0 is inverted.

5.5.3 Bit Testing

Bit testing involves determining the status of an individual bit or bit field in a word. The basic strategy is to mask out all other bits using the AND instruction and then check the status of the Z flag to determine whether the bits are all zeros. Source 5.9 shows three code snippets that branch based on the status of individual bits.

Source 5.9 Bit Testing

```
*****************************************************************
* branch to foo if bit 3 of ACCA is set
*****************************************************************
BIT3     equ %00001000
BIT7     equ %10000000

         anda #BIT3          ;ACCA & %00001000 -> ACCA
         bne foo


*****************************************************************
* branch to bar if bits 7 and 3 of out_reg are clear
*****************************************************************
         ldaa out_reg
         anda #(BIT3 | BIT7) ;ACCA & %10001000 -> ACCA
         beq bar


*****************************************************************
* branch to bar if bits 7 and 3 of out_reg are set
*****************************************************************
         ldaa out_reg
         coma                ;~ACCA -> ACCA
         anda #(BIT3 | BIT7) ;ACCA & %10001000 -> ACCA
         beq bar
```

In the first example all the bits in ACCA except bit 3 are set to zero. This is referred to as masking. The value of bit 3 alone determines the state of the *Z* flag. In the second example both bit 3 and bit 7 of the memory location *out_reg* must be zero to set the *Z* flag. Therefore, the branch will only occur if both bit 3 and bit 7 are cleared. In the last example we want to branch to *bar* if both bit 3 and bit 7 are set. This can only be done by first inverting the argument. At first glance you might try

```
anda #(BIT3 | BIT7)
bne bar
```

This is not quite right, however, because it would branch to *bar* if either bit 3 or bit 7 is set. We only want to branch if both are set.

Bit testing using logic instructions is more efficient when the bits to be tested are already in an accumulator. When the bits to be tested are in memory, it is better to use the CPU12 bit-conditional branches described in Section 5.6.

CCR Flag Manipulation. The CPU12 also has two logic instructions specifically designed to manipulate the CCR flags, *andcc* and *orcc*. These instructions allow you to set or reset a flag bit manually. Source 5.10 shows a code snippet to set the *C* flag and clear the *N* flag.

Source 5.10 Manual CCR Flag Manipulation

```
*****************************************************************
* Manually set C and clear N.
*****************************************************************
C_FLAG      equ %00000001
N_FLAG      equ %00001000
               ...
               ...
               ...
            orcc #C_FLAG              ;CCR | %00000001 -> CCR
            andcc #(~N_FLAG)          ;CCR & %11110111 -> CCR
```

In this example symbols were used to make the program more readable. The constant *C_FLAG* has a one in the bit position of the *C* flag and *N_FLAG* has a one in the bit position of the *N* flag. To set the *C* flag, we use the *orcc* instruction shown, which is equivalent to

```
CCR | %00000001
```

To clear the *N* flag, we cannot simply use *andcc* with *N_FLAG*. That would clear all bits except the *N* flag. The *N_FLAG* constant must first be inverted at assembly time. This is done by using the assembler NOT operator "~".

The CPU11 does not have the *andcc* or *orcc* instructions. Instead there are individual instructions for setting and clearing CCR flags, as shown in Table 5.15.

If it is necessary to change flags other than *C, V,* or *I* on the CPU11, you must use the *tap* and *tpa* instructions.

TABLE 5.15   CCR FLAG MANIPULATION INSTRUCTIONS

Mnemonic	Operation	CPU11	CPU12 Addressing Modes					
			INH	IMM	DIR	EXT	IDX	[IDX]
CLC	$0 \rightarrow C$	X	X					
CLI	$0 \rightarrow I$	X	X					
CLV	$0 \rightarrow V$	X	X					
SEC	$1 \rightarrow C$	X	X					
SEI	$1 \rightarrow I$	X	X					
SEV	$1 \rightarrow V$	X	X					

5.5.4 Bit Test and Manipulation Instructions

Table 5.16 shows the CPU11 and CPU12 bit test and manipulation instructions. These include bit test instructions (*bita* and *bitb*), bit manipulation instructions (*bclr* and *bset*), and bit-conditional branches (*brclr* and *brset*). Here we only look at the bit test and manipulation instructions. The bit-conditional branches will be covered in Section 5.6.

As you can see, the bit-oriented instructions are actually microinstructions involving the same Boolean logic operations described in the previous section. *bita* and *bitb* are bit testing instructions. These instructions perform the same Boolean operations as the bit testing examples shown in Source 5.9.

The *bclr* and *bset* instructions are used to set or clear bits in a word in memory. The syntax for these instructions is

```
bclr M,msk            ;(M) & ~msk -> (M)
bset M,msk            ;(M) | msk -> (M)
```

where

M = the memory location affected using direct, extended, or indexed addressing

msk = a mask word that contains a one in each bit position to be set or cleared

Notice that both *bset* and *bclr* are read-modify-write instructions. Like all read-modify-write instructions, this requires the location to be both readable and writable.

TABLE 5.16   BIT TEST AND MANIPULATION INSTRUCTIONS

Mnemonic	Operation	CPU11	CPU12 Addressing Modes					
			INH	IMM	DIR	EXT	IDX	[IDX]
BITA	$ACCA \cdot (M)$	X	X					
BITB	$ACCB \cdot (M)$	X	X					
BCLR	$(M) \cdot msk' \rightarrow (M)$	X			X	X	X	
BSET	$(M) \mid msk \rightarrow (M)$	X			X	X	X	

Syntax Note

Notice the arguments for the *bset* and *bclr* instructions are separated by commas. For CPU12 assemblers, this should always be the case. However, for some CPU11 assemblers, the arguments were separated by spaces. The most notable examples are Motorola's AS11 and all of the code described in the *M68HC11 Reference Manual*. For the CPU11, some assemblers like the Introl-CODE assembler use comma-separated arguments, and some like the AS11 use space-separated arguments. This can cause some minor headaches when porting CPU11 assembly code between assemblers. This is also true for the bit-conditional branch instructions described later.

Source 5.11 shows some code snippets using the *bset* and *bclr* instructions. These snippets are functionally equivalent to the first two snippets in Source 5.8.

Source 5.11 Bit Manipulation Using Bit-Manipulation Instructions

```
*****************************************************************
* Clear bits 4-7 of ACCA

        psha                    ;SP-1->SP;ACCA->(SP)
        bclr 0,sp,%11110000     ;(SP) & %00001111->(SP)
        pula                    ;(SP)->ACCA;SP+1->SP


*****************************************************************
* Set bits 0-3 of out_reg

        bset out_reg,%00001111  ;(out_reg) | %00001111->(out_reg)
```

The first example uses the *bclr* instruction to clear bits 4 through 7 of ACCA. In this case, since the bit-manipulation instructions only operate on a memory location, we first have to move the contents of ACCA to memory by pushing it onto the stack. Then we use indexed addressing to clear the bits using the *bclr* instruction. The result is then pulled off the stack back into ACCA. If we compare this code to the equivalent code in Source 5.8, we can see that it is more efficient to use the AND instruction. In general the logic instructions are more efficient when the data are contained in a CPU register, and the *bset* or *bclr* instructions are more efficient when manipulating bits in memory.

The second code snippet in Source 5.11 sets bits 0 through 3 of the memory location *out_reg*. In this case direct or extended addressing is used to access *out_reg*. If we compare this code with the equivalent code in Source 5.8, we can see it is more efficient to use the *bset* instruction. This is because the data are already located in memory.

One thing to remember is that *bset* and *bclr* only change the bits corresponding to the ones in the mask. A very common error is to assume that

```
bset DDRP,%11110000
```

sets bits 4 through 7 and also clears bits 0 through 3. That is not the case. It only sets bits 4 through 7. Bits 0 through 3 remain unchanged. In order to clear the lower four bits also, we would have to use

```
bset DDRP,%11110000
bclr DDRP,%00001111
```

▶ 5.6 BRANCHES AND JUMPS

So far the instructions we have examined are data transfer or arithmetic instructions. We now look at two types of program control instructions: branches and jumps. Program control instructions change the normal execution sequence of the CPU by changing the contents of the program counter. Without program control instructions, we could only write programs made up of straight-line sequences. Every instruction in the sequence would have to be executed, and the program counter would always continue to the next instruction in program memory. To say the least, a CPU without program control instructions would not be very useful.

Both branches and jumps change the program flow by changing the program counter to a destination address. To jump to a different part of the program or conditionally branch past a block of code, a branch or jump instruction is used.

5.6.1 Jump Instruction

CPU11 and CPU12 both have a single unconditional jump instruction, *jmp*, as shown in Table 5.17. In all cases the destination of the jump is implemented by the program counter being loaded with the effective address of the argument. Either extended or indexed addressing modes can be used to determine the jump destination. Source 5.12 shows a simple example of using the *jmp* instruction to jump to the destination *next*.

Source 5.12 Simple Jump Example

```
               CPU12
Addresses      Opcodes      Label      Instruction      Description
00000800       060900                  jmp next         ;$0900 -> PC
00000803                                 ...

00000900                    next         ...
```

In this example the program flow will change when the *jmp* instruction is reached. It uses extended addressing to jump to *next* by loading $0900 into the program counter. Normally without a jump instruction, the program counter would point to $0803 and the instruction (not shown) at that location would be executed. Notice the CPU12 opcode. The instruction opcode for the *jmp* instruction is $06. It is followed by the effective address $0900, which is the value that gets loaded into the program counter.

The *jmp* instruction can address any location in the 64K-byte memory map, but it cannot be used for relocatable code unless the destination address is fixed.

TABLE 5.17 JUMP INSTRUCTION

Mnemonic	Function	CPU11	CPU12 Addressing Modes					
			INH	IMM	DIR	EXT	IDX	[IDX]
JMP	EA → PC	X				X	X	X

5.6.2 Branches

Branch instructions also change the program flow by changing the program counter. They use PC relative addressing, however. This means that instead of loading the destination address into the program counter, they add a relative offset, *rel*, to the program counter. The instruction operation is

$$PC + rel \rightarrow PC$$

where

$$rel = A_D - PC$$

$$A_D = \text{the destination address}$$

$$PC = \text{the current value of the program counter}$$

The CPU12 has both short branches and long branches. The relative offset for the short branches is limited to a signed 8-bit number. This means the short branch destination address must be within −128 bytes and +127 bytes of the current PC. Long branches, however, have signed 16-bit offsets. They require two extra bytes of program space, but they can reach any location in the 64K-byte map. Source 5.13 shows an example of a short branch, *bra*, and a long branch, *lbra*.

Source 5.13 Simple Branch Examples

Addresses	CPU12 Opcodes	Label	Instruction	Description
00000800	182000fc		lbra far	;PC+$00fc -> PC
00000804			...	
00000900	20fe	far	bra far	;PC+$fe -> PC
00000902				

For the programmer, the branch instructions are used in the same way as a *jmp* instruction. In both cases the label used to identify the destination address is given as the argument. The difference between branches and jumps is the way the argument is stored and the way the instruction calculates the new value of the program counter.

Let's first look at the short branch, *bra far*. This is a branch instruction that makes the program loop back to the same branch instruction forever. In C this is equivalent to

```
while(TRUE){
}
```

It is often referred to as a *trap*.

At assembly time, the assembler calculates the 8-bit relative offset to store in the program code. Using the equation given previously

$$rel = A_D - PC = \$0900 - \$0902 = \$FFFE$$

Remember that the program counter always points to the next instruction, so for short branch instructions the program counter is

$$PC = A_S + 2$$

where

$$A_S = \text{the address of the branch instruction itself}$$

Since the offset for a short branch can only be 8 bits, the 16-bit value, $FFFE, must be truncated to $FE. You can see from Source 5.13 that this is the value stored after the opcode at address $0901.

At run time the CPU sign extends the relative offset and adds it to the program counter

$$PC + rel = \$0902 + \$FFFE = \$0900$$

which results in the correct destination address, $0900.

The long branch instruction in Source 5.13 is implemented the same way, only the offset is 16 bits. In this case

$$rel = A_D - PC = \$0900 - \$0804 = \$00FC$$

Notice for long branches

$$PC = A_S + 4$$

because there are four program bytes for the long branch instruction.

Let's look at what would happen if a short branch were used in place of the long branch. The offset would be

$$rel = A_D - PC = \$0900 - \$0802 = \$00FE$$

If $00FE were then truncated to 8 bits, the result would be $FE, the branch would subtract two from the PC, and we would end up with an unintended trap. This is an example of an *out-of-range* error. Your assembler should detect this and give you an error message.

As programmers, we luckily do not have to calculate the relative offsets while writing our source code. That would be very inefficient. This does not mean we can just forget about all of this stuff, however. We do have to debug the code. While debugging it is very important that we know how to calculate the offset values.

5.6.3 Conditional Branches

The branch and jump instructions given in the previous examples are all unconditional. That is, they always perform the jump or the branch when they are executed. Conditional branch instructions branch only if certain conditions are satisfied. If the conditions are not satisfied, the program flow continues to the next instructions and the branch instruction has no effect. The conditions are always based on the contents of the condition code register (except for the bit-conditional branches covered later). Conditional branches

TABLE 5.18 SIMPLE CONDITIONAL BRANCHES

Branch			Complementary Branch		
Test	Mnemonic	Condition	Test	Mnemonic	Condition
R = M	*BEQ*	Z = 1	R ≠ M	*BNE*	Z = 0
Carry	*BCS*	C = 1	No Carry	*BCC*	C = 0
Positive	*BPL*	N = 0	Negative	*BMI*	N = 1
Overflow	*BVS*	V = 1	No Overflow	*BVC*	V = 0
Always	*BRA*	—	Never	*BRN*	—

are used to implement conditional and loop constructs, which will be covered in more detail in Chapter 6.

The CPU11 and CPU12 have many different conditional branches—simple conditional branches, conditional branches for signed numbers, and conditional branches for unsigned numbers. The simple conditional branches are all based on the value of a single CCR flag. Table 5.18 shows all the simple conditional branches. For many CPUs, these are the only branch instructions available. You can implement any of the other branches using a combination of simple branches.

Notice there is a branch-never, *brn*. It traditionally was used for debugging machine code back in the days when the code was entered with a hex keypad. To bypass branches you could simply replace the actual branch opcode with a *brn* opcode. On the CPU11 you could also use a *brn* as a three-cycle no-op instruction. On the CPU12 the *brn* only takes a single cycle, so it has no advantage over the *nop* instruction.

The *beq* and *bne* instructions are shown here because they are only based on the zero flag. Because the condition logic is the same, they are used for both signed and unsigned numbers.

Unsigned Conditional Branches. The conditional branches for unsigned numbers are shown in Table 5.19. They are based on a specific combination of flag bits including at least the carry flag.

Since these instructions depend on the carry flag, we have to be careful about using them. Some instructions do not affect the *C* flag, which could result in a branch taken or a branch missed in error. Source 5.14 shows a typical error of this type. The example shows a loop that is based on a count that is decremented using the *deca* instruction. The count is an unsigned number and each time through the loop it is decremented. The decrement does not affect the *C* flag, however, so the *bhi* may or may not branch when expected. Also

TABLE 5.19 CONDITIONAL BRANCHES FOR UNSIGNED NUMBERS

Branch			Complementary Branch		
Test	Mnemonic	Condition	Test	Mnemonic	Condition
R > M	*BHI*	C + Z = 0	R ≤ M	*BLS*	C + Z = 1
R ≥ M	*BHS*	C = 0	R < M	*BLO*	C = 1

shown in Source 5.14 is a way to correct this error by using a *cmpa* instruction. Of course a *bne* would also work, because it only depends on the *Z* flag.

Source 5.14 Unsigned Conditional Branch Error

```
**************************************************
* Loop with potential unsigned branch error.
**************************************************
            ldaa #COUNT
bad_loop

            ...
            deca
            bhi bad_loop


**************************************************
* Replace with
**************************************************
            ldaa #COUNT
good_loop

            ...
            deca
            cmpa #0
            bhi good_loop
```

It is important to understand the *R* and *M* listed in the test column of Table 5.19. The *R* refers to a register and the *M* refers to memory. For example, in the instruction sequence

```
cmpa #$80               ;compare ACCA with $80.
bhi foo                 ;if ACCA > $80 branch to foo.
```

R refers to ACCA and M refers to the immediate data, #$80.

Notice that *bhs* has the same condition as *bcc,* and *blo* has the same condition as *bcs.* In fact *bhs* and *bcc* have the same opcode, and *blo* and *bcs* have the same opcode. To improve code readability, always use the correct mnemonic for the application. If you are comparing two unsigned numbers, use *bhs* or *blo*. If the *C* flag is used for some other function like an error flag, however, use *bcc* or *bcs*. The only problem this can cause is during disassembly. A disassembler cannot know the intent of the program, so it always uses only one of the mnemonics. For example if D-Bug 12 or BUFFALO is used to trace through a *bhs* instruction, the display will actually show a *bcc* instruction.

Signed Conditional Branches.   The conditional branches for signed numbers are shown in Table 5.20. They are based on a specific combination of flag bits including at least the overflow flag.

Notice the complexity of the Boolean conditions for these instructions. Although we could perform the same branches using simple branch instructions, these extra instructions are convenient. Source 5.15 shows an example of using simple branch instructions to implement the *bgt*. Thank Motorola for providing these instructions!

TABLE 5.20 CONDITIONAL BRANCHES FOR SIGNED NUMBERS

Branch			Complementary Branch		
Test	Mnemonic	Condition	Test	Mnemonic	Condition
R > M	*BGT*	Z+(N $\oplus$ V)=0	R $\leq$ M	*BLE*	Z+(N $\oplus$ V)=1
R $\geq$ M	*BGE*	(N $\oplus$ V)=0	R < M	*BLT*	(N $\oplus$ V)=1

Source 5.15 A Branch-If-Greater-Than Using Simple Branches

```
* Branch to foo if ACCA is greater than NUM.
*
                    cmpa #NUM
                    beq skip_br
                    bmi chk_v
                    bvc foo
                    bra skip_br
chk_v               bvs foo
skip_br
```

Be Careful to Use the Correct Branch. Carefully choosing the branch instruction can never be overemphasized. For beginning programmers, one of the most common errors is to use the incorrect branch. Most common is the use of the complementary branch in error, such as using a *bne* instead of a *beq*. Close behind is the use of a branch instruction intended for the wrong data type, such as using a *bgt* instead of a *bhi* with unsigned numbers. This can be a subtle bug. The program may work with the incorrect branch because the range of the data that is actually compared does not result in an incorrect branch. Thanks to Murphy's law, it is only at the worst possible moment that new data exposes the bug.

5.6.4 Long Conditional Branches

The CPU12 also has a long branch instruction for every short branch instruction, as shown in Table 5.21. To make a long conditional branch, we simply use the long branch

TABLE 5.21 LONG BRANCHES

Branch			Complementary Branch		
Test	Mnemonic	Condition	Test	Mnemonic	Condition
R = M	*LBEQ*	Z = 1	R $\neq$ M	*LBNE*	Z = 0
Carry	*LBCS*	C = 1	No Carry	*LBCC*	C = 0
Positive	*LBPL*	N = 0	Negative	*LBMI*	N = 1
Overflow	*LBVS*	V = 1	No Overflow	*LBVC*	V = 0
Always	*LBRA*	—	Never	*LBRN*	—
R > M	*LBHI*	C + Z = 0	R $\leq$ M	*LBLS*	C + Z = 1
R $\geq$ M	*LBHS*	C = 0	R < M	*LBLO*	C = 1
R > M	*LBGT*	Z+(N $\oplus$ V)=0	R $\leq$ M	*LBLE*	Z+(N $\oplus$ V)=1
R $\geq$ M	*LBGE*	(N $\oplus$ V)=0	R < M	*LBLT*	(N $\oplus$ V)=1

instruction in place of the short branch instruction. It will cost a couple of bytes of program code, so you should use long branches only if it is required.

There are no conditional jump instructions for either CPU11 or CPU12. For the CPU12, this is not a problem because the long branch instructions can be used. The CPU11 does not have long branches, however. How can we make a condition branch, or jump, to an address that goes beyond the limits of the signed 8-bit offset? Source 5.16 shows the normal method for implementing a long conditional branch using a 68HC11. In this example we want to jump to *far* if a result is equal.

Source 5.16 68HC11 Solution to Conditional Long Branch

Addresses	CPU12 Opcodes	Label	Instruction	Description
			...	
00001000	2603		bne skip_bra	;Skip jump to far
00001002	7e8000		jmp far	;$8000 -> PC
00001005		skip_bra		
			...	
00008000	20fe	far	bra far	

We would normally use the *beq* instruction, but since *far* is out of range, we have to use a *jmp* instruction. To do this, we find the complementary branch instruction, in this case *bne,* and branch around a *jmp* instruction. This solution requires three extra program bytes, and for the 68HC11, three extra clock cycles. This solution is also not relocatable, but since it is not practical to create relocatable code on the 68HC11 anyway, this is not a big deal.

Source 5.17 shows the CPU12 solution to the same conditional long branch. In this case the cost is two extra program bytes and one extra clock cycle if the branch is made, two extra clock cycles if the branch is not made.

Source 5.17 CPU12 Solution to Conditional Long Branch

Addresses	CPU12 Opcodes	Label	Instruction	Description
00001000	18276ffc		lbeq far	;PC+$6ffc -> PC
			...	
00008000	20fe	far	bra far	

5.6.5 Bit-Conditional Branches

The CPU11 and the CPU12 also have branch instructions that are conditional on individual bits or a bit field within a word. Table 5.22 shows the bit-conditional branch instructions for the CPU12 and CPU11.

The syntax for these instructions is

```
brclr M,msk,A_D
brset M,msk,A_D
```

TABLE 5.22 BIT-CONDITIONAL BRANCHES

Mnemonic	Condition	CPU11	CPU12 Addressing Modes					
			INH	IMM	DIR	EXT	IDX	[IDX]
BRCLR	$(M) \cdot msk = 0$	X			X	X	X	
BRSET	$(M)' \cdot msk = 0$	X			X	X	X	

where

M = the memory location containing the bits to be tested

msk = the bit mask, which contains a one for each bit to be tested

A_D = the destination address if the branch is performed

The operation of these two instructions is the same as the code using logic instructions shown in Source 5.9, except the bits to be tested must be in a memory location, not a CPU register. Source 5.18 shows the equivalent code for the last two snippets in Source 5.9. You can see that the *brset* and *brclr* instructions are very efficient when testing bits in memory.

Source 5.18 Bit-Conditional Branch Examples

```
*************************************************************
* branch to bar if bits 7 and 3 of out_reg are clear
*************************************************************

        brclr out_reg,%10001000,bar


*************************************************************
* branch to bar if bits 7 and 3 of out_reg are set
*************************************************************

        ldx #out_reg
        brset 0,x,%10001000,bar
```

There is no good way to use a *brset* or *brclr* instruction on a byte in a CPU register. The data must first be moved into memory. If we try using the stack as we did in Source 5.11, we may end up branching with no way to balance the stack. Another method would be to put the data in a global variable location used for temporary data. This technique should be avoided, however, so the best bet is to use logic instructions when the bits are in a register and *brset* or *brclr* when the bits are in memory.

▶ 5.7 SUBROUTINES

In this section, we learn how to use the program control instructions for implementing subroutines. *Subroutines* are the fundamental building block for assembly language programs. They are important for at least three reasons:

- They improve a program's structure.
- They improve a program's portability.
- They help reduce code space by avoiding repeated code.

TABLE 5.23 SUBROUTINE INSTRUCTIONS

Mnemonic	Function	CPU11	CPU12 Addressing Modes					
			INH	IMM	DIR	EXT	IDX	[IDX]
BSR	SP–2 → SP	X						
	PC → (SP:SP+1)							
	PC+rel → PC							
JSR	SP–2 → SP	X			X	X	X	X
	PC → (SP:SP+1)							
	A_D → PC							
RTS	(SP:SP+1) → PC	X	X					
	SP+2 → SP							
CALL	SP–2 → SP					X	X	X
	PC → (SP:SP+1)							
	SP–1 → SP							
	PPAGE → (SP)							
	Page → PPAGE							
	A_D → PC							
RTC	(SP) → PPAGE		X					
	SP+1 → SP							
	(SP:SP+1) → PC							
	SP+2 → SP							

In today's environment the first two reasons, program structure and portability, have become far more important.

In this section, we learn how to use the subroutine-related instructions, examine the program flow, and see how the CPU implements subroutines. Table 5.23 shows all the subroutine-related instructions on the CPU12. The traditional subroutine instructions used for 64K-byte systems are *bsr, jsr,* and *rts*. The new CPU12 instructions for subroutines that may fall on different memory pages are *call* and *rtc*. In this section we focus on the traditional subroutine instructions and assume all code is contained in the 64K-byte map.

5.7.1 Subroutine Program Flow

So far we have looked at branch and jump instructions to change the program flow. Now let's say we have a snippet of code that performs a function that we would like to reuse. We will first try to implement this with branch instructions, as shown in Figure 5.3. The program starts at the top of Figure 5.3(a). It reaches the first branch to *func,* and the branch goes to the function code. Once the function is finished, we need to go back to the instruction following the branch instruction. In this case we use a *bra* instruction with the label *back*. The program branches back and continues until we need to call the function again. Again, as shown in Figure 5.3(b), we use a branch instruction to get to the function. Only this time, when the function is finished, the branch takes program control back to the instruction following the first branch instead of the one following the second branch.

A creative programmer can come up with many ways to fix this problem. Do not even try. The result will invariably be classic spaghetti code. Subroutine instructions were designed for this very purpose.

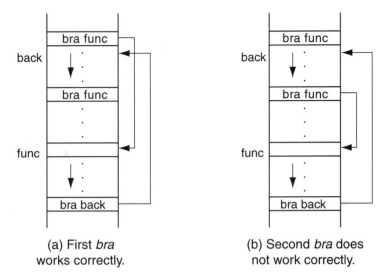

(a) First *bra*
works correctly.

(b) Second *bra* does
not work correctly.

Figure 5.3   An Attempt to Use Branches to Implement a Subroutine

Figure 5.4 shows the program flow when subroutine instructions are used. All subroutines are called with either branch to subroutine, *bsr,* or jump to subroutine, *jsr,* and they all end with a return from subroutine, *rts.* The return from subroutine always returns program control to the instruction following the last *jsr* or *bsr* that has not been serviced. As you can see from Figure 5.4, the program works as desired. Now instead of putting the function code everywhere, we only need to include it one time, add an *rts,* and call it using *bsr* or *jsr.*

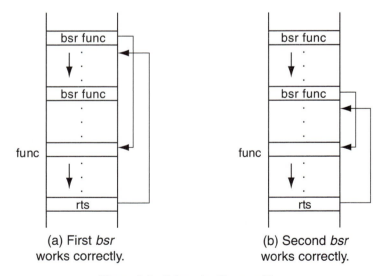

(a) First *bsr*
works correctly.

(b) Second *bsr*
works correctly.

Figure 5.4   Subroutine Program Flow

Chapter 5

Beyond reusing code, there are many advantages to this program flow. Because we know the program flow always returns to the next instruction, we can treat the branch instruction itself as an instruction that performs the function. This means at a higher level that the program flow can be seen as a simple sequence. The contents of the subroutine, all the complications of its implementation, are hidden. This is extremely important for top-down design and for debugging. If you were debugging the program code using simple branch instructions shown in Figure 5.3, you would have to follow all the branches, because you have no way to know where the program flow will go after a branch. With a subroutine, however, you know it will return to the next instruction, so you can ignore the branch to subroutine altogether. This of course assumes the subroutine was written correctly. We will look at some rules for writing subroutines later.

How It Works.   We saw that the *rts* instruction returns program flow to the instruction following the last subroutine call. To do this, the *rts* instruction must return to a different address for each call, and when subroutines are nested, it means the correct return address must go to the instruction following the correct subroutine call. How can this be done? Well, recall that the program counter always points to the next instruction. The *bsr* and *jsr* instructions save the program counter on the stack before going to the subroutine. The *rts* instruction can then use it to return to the correct address by pulling the PC off the stack. The stack is the ideal place to store the PC, as we see in Source 5.19. Source 5.19 is a program that uses two subroutines, *sub1* and *sub2*. To simplify the code, only the instructions related to the subroutines are shown.

Source 5.19 Subroutine Implementation

```
        CPU12
Addr    Opcodes    Label    Instruction    Description
0000               main
                            ...
0010    0720                bsr sub1       ;SP-2-> SP,PC -> (SP:SP+1),PC+rel -> PC
0012                        ...

0030    20ce                bra main       ;PC+rel=main -> PC
0032               sub1
                            ...
0040    160051              jsr sub2       ;SP -2 -> SP,PC ->(SP:SP+1),sub2-> PC
0043                        ...

0050    3d                  rts            ;(SP:SP+1)-> PC,SP+2 -> SP
0051               sub2
                            ...
0060    3d                  rts            ;(SP:SP+1)-> PC,SP+2-> SP
0061
```

First let's look at the program flow of this program. At the highest level the program is an endless loop between *main* and the *bra main* instruction. This part of the code is often referred to as the main code. Because *sub1* is implemented as a subroutine, the *bsr sub1* instruction can be treated like any other instruction, and we can ignore the detailed program flow into the subroutine code.

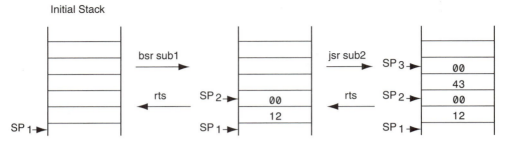

Figure 5.5 Stack Contents for Source 5.19

The detailed flow of this program is a bit more complex. The program starts out at *main,* and continues until it reaches the *bsr sub1* instruction. It then goes to *sub1* and continues executing the code in *sub1* until it reaches the *jsr sub2* instruction. At this point it goes to *sub2* and executes the code in *sub2* until the *rts* instruction is reached. The *rts* instruction passes control back to the instruction following the last subroutine call, which in this case is the *jsr sub2* instruction. The program flow then continues with the rest of *sub1* until its *rts* instruction is reached. It then goes back to the next instruction following the *bsr sub1* instruction and continues with the rest of the main program. When it reaches the *bra main* instruction, it returns to *main* and repeats.

In this example the subroutines are said to be *nested subroutines,* because *sub2* was called from within *sub1.* As the last paragraph illustrated, the detail program flow was complex. Because we are using subroutines, however, at the highest level the main program is simply an endless loop, and the complexity of the subroutine nesting is hidden. Now let's look at the CPU operations required to implement the subroutines. Figure 5.5 shows the stack contents and stack pointer at various places in the program flow.

At the start of the main program, the initial stack pointer is as shown on the left in Figure 5.5. When the *bsr sub1* instruction is executed, the address of the instruction following the *bsr* is pushed onto the stack. In this case the program counter is pointing to address $0012, so the stack now looks like the center figure in Figure 5.5. The program then jumps to *sub1* and executes until it reaches the *jsr sub2* instruction. Again the program counter is pushed onto the stack. In this case $0043 is on the stack, and the stack looks like the figure on the right in Figure 5.5. When *sub2* is finished, the *rts* instruction pulls the top two bytes off the stack into the program counter. So $0043 is loaded into PC, and program execution continues through *sub1.* The stack is now back to the center figure in Figure 5.5. When *sub1* is finished, the *rts* instruction again pulls the top two bytes off the stack into the program counter. This time $0012 is loaded into the PC, and the program returns to the main program following the *bsr sub1* instruction.

In this example we can see how the LIFO stack enables nested subroutines. The LIFO structure of the stack is perfect for nested routines or for nested instances of the same routine. We will later use the stack not only for return addresses of routines but also for parameters and local variables in routines. Using the stack enables both nesting and reentrant routines.

5.7.2 Basic Subroutine Techniques

There are some important rules and basic techniques to follow when designing subroutines. In Chapter 6 techniques that are more complex are introduced for implementing local variables and passing parameters.

Never Branch out of or into a Subroutine. There is never a good reason to branch out of a subroutine with a branch, a conditional branch, or a jump instruction, even if the code branches back into the subroutine. A subroutine should always have a single entry point and a single exit. The entry point is at the label that defines the subroutine name. To enter a subroutine, a *bsr* or *jsr* must be used. The exit should always be a single *rts* at the end of the subroutine. If there are other points in the code that should return from the subroutine, a branch to the single *rts* should be made instead of adding another *rts*. Having a single *rts* can simplify debugging and reduce the number of bugs introduced later in the development process.

Branching out of or into a subroutine with a simple branch or jump is a common mistake for unknowing programmers. It can cause a disastrous bug. If you branch into a subroutine without using a *bsr* or *jsr*, the program counter is not pushed onto the stack. When the *rts* is reached, garbage is loaded into the PC and the program crashes. Branching out of a subroutine creates a subtler bug. Each time you call a subroutine, the program counter is pushed onto the stack. If you branch out of the subroutine, however, the PC is not pulled off by the *rts*. Eventually the stack grows beyond its bounds, and the program crashes.

The source code of a subroutine should be a self-contained block starting at the subroutine header and ending at the *rts*. Of course this does not include another subroutine call. It is perfectly acceptable to call one subroutine from another as long as a *bsr, jsr,* or *call* instruction is used.

Subroutine Documentation. One of the advantages of using subroutines is their portability. When writing a subroutine, we must keep this in mind. How should the subroutine be designed and documented so it provides a general function that easily can be used elsewhere? We talk about design issues throughout this text. All subroutines must be documented using comments in the source code. A good practice is to create a subroutine template that includes a comment header. This template can be used every time you create a new subroutine. Items to include in the header include the following:

- Subroutine name and functional description
- MCU and monitor dependencies (if any)
- Parameters and variables, especially arguments, return values, and global variables used
- Stack requirements
- Registers destroyed or preserved
- Name, date, and revision history
- Pseudo-C representation of the routine

Source 5.20 shows a sample of a subroutine header. Notice the subroutine name is defined as a C prototype. Although this is a subroutine written in assembly code, the C prototype helps show the parameters involved with the routine. It is also a good idea to design any subroutine to use the parameter passing conventions of your C compiler. This enables subroutines written in assembly to be called from a C program without modification.

Source 5.20 Subroutine Header Sample

```
**************************************************************************
* char *OutByte(char *) - This subroutine outputs one hex byte to the sci.
*
* Entry Args: ACCD contains a pointer to the output byte
* Returns: ACCD points to the byte following output byte in memory
* MCU: 68HC912B32
* Monitor: D-Bug12
* Stack Usage: 12 bytes
* Registers: All registers but ACCD and CCR are preserved
* Notes: Uses D-Bug12 PUTCHAR()
* Todd Morton 2/17/98
**************************************************************************
```

Not all headers will look the same. For example many subroutines will not depend on a monitor being present. If not, there is no need to indicate a monitor. Make sure that the MCU description has the correct level of specificity. In the example the MCU listed is the 68HC912B32. This means that this routine would not necessarily work without modification on the 68HC812A4 or any other CPU12-based microcontroller. It may work, but it was not considered as part of the design, and it was not tested.

Preserve Registers. Consider the code shown in Source 5.21. This code is supposed to output a string of 10 asterisks. However, its operation relies on the contents of both ACCA and ACCB. ACCA is used as a counter for the number of times the character is sent, and the subroutine *putchar* uses ACCB to access the character to be sent. If *putchar* uses ACCA and ACCB without preserving the contents first, the program will not work correctly. For example, if we are using the D-Bug12's version of *putchar,* a carriage return is always returned in ACCD. This means after a single asterisk is displayed, the program will output carriage returns forever.

Source 5.21 Subroutine Call without Preserving Registers

```
                ldaa #10
                ldab #'*'
starloop        jsr putchar
                deca
                bne starloop
```

We have two choices for solving this problem. We can preserve the registers in the calling routine, or we can preserve the registers in the subroutine. The method used depends on the situation. It is preferable to preserve registers in the subroutine, that way the code is in the subroutine a single time instead of adding it every time the subroutine is called. However, if the subroutine is provided by a third party or a C compiler, and it is not practical to edit the source, the registers must be preserved in the calling routine before the subroutine call. Source 5.22 shows revised code in which the registers are preserved in the calling routine.

Source 5.22 Registers Preserved in the Calling Routine

```
                    ldaa #10
                    ldab #'*'
starloop            pshd
                    jsr putchar
                    puld
                    deca
                    bne starloop
```

In this case ACCA and ACCB are pushed onto the stack before the subroutine call. They are then pulled off the stack after the subroutine call. This is a reasonable use of the stack because the push and pull instructions are close together. In Chapter 6 we look at a better but more complex solution, creating local variable storage on the stack for the count and the output character.

If you are writing a new subroutine, the registers can be preserved in the subroutine itself. For example, if we have access to the *putchar* source used above, we could add code in the subroutine to preserve all registers used in the subroutine. To do this we add a push instruction for every register used in the subroutine. If *putchar* uses ACCA, ACCB, and IX, we could add the following code to the subroutine:

```
putchar             pshd                ;preserve IX and ACCD
                    pshx
                    ...                 ;putchar code
                    pulx                ;pull preserved registers
                    puld
                    rts
```

This method is appropriate only for simple subroutines. We are assuming here that these registers are not being used to pass parameters and we are not already implementing the register contents as local variables. These topics are covered in Chapter 6.

Notice that there has been no attempt to preserve the CCR. This is normally the case. If a subroutine is called, it is normal to assume that the CCR contents are destroyed. However, we can preserve the CCR if needed. In fact the CPU12 has made this simple with the addition of the *pshc* and *pulc* instructions. For the CPU11 we would have to use the following instructions to preserve the CCR:

```
                    tpa
                    psha
                    ...
                    pula
                    tap
```

We have just covered the basic rules and techniques to follow when designing sub-routines. Again it cannot be overemphasized that it is always best to design a subroutine for the long term. Think of it being used in many different instances and many different projects. Design for reusability, follow your C compiler parameter passing conventions, and document the routine carefully.

▶ 5.8 POSITION INDEPENDENCE

Position independence refers to the ability of a program to function in more than one block of memory. In this text we cover three levels of position independence: position dependent, source relocatable, and object relocatable. In most computer literature the terms *relocatable* or *position independent* refer to object-relocatable code.

5.8.1 Position-Dependent Code

Position-dependent code cannot be moved to another location in memory without rewriting and reassembling the source. There is no reason for a program to be position de-pendent. This type of code is unacceptable and easily avoided. The use of labels is all that is required to keep code from being position dependent. Source 5.23 shows a code snippet that is position dependent.

Source 5.23   Position-Dependent Code

Address	Instruction
	org $0000
	ldaa $0025
	...
00000025	rmb 1

In this example we need to load a variable into ACCA. The variable, however, is ac-cessed by referring to its address instead of a label. If the program is moved to another lo-cation by changing the origin address, it will not work. The variable will no longer be at $0025, but the load instruction will still load the contents $0025. To make this program work, we would have to rewrite the source by changing the address in the load instruction to the new location of the variable.

Unfortunately many programmers get their start by writing programs like this. This is because many programmers start out hand assembling. In addition they may load the code into a trainer with a hex keypad, or their textbook may be filled with code like this in its examples. Hand assembly code examples can introduce some bad habits, so it is impor-tant to use labels for all references as soon as an assembler is introduced.

5.8.2 Source-Relocatable Code

Source-relocatable code is code that can be moved to a new location just by changing the address in the origin directive and reassembling. If a linker is used, only the linker command file needs to be changed. No change is required in the source, but the source does have to be reassembled. Source 5.24 shows the code from Source 5.23 revised to be source relocatable. The only change was to use a label to reference the variable.

Source 5.24 Source-Relocatable Code

```
Address                                    Instruction
                                           org $0000
                                           ldaa var
                                            ...
          var                              rmb 1
```

Now if we want to move this code to a new location, we can simply change the address in the origin directive and reassemble. The assembler will calculate a new value for *var* and use it for the *ldaa* instruction.

Most embedded programs fall into this category. It is rare for object-relocatable code to be required. The object code is typically contained in a ROM device at a fixed address and remains there throughout the life of the product.

5.8.3 Object-Relocatable Code

Object-relocatable code is what most people refer to as position-independent or relocatable code. Object-relocatable code can be moved by changing the location of the object code. The program does not have to be rewritten or reassembled. It is not always practical to generate object relocatable code. It requires a CPU that has a PC relative addressing mode for most of its instructions. Source 5.25 shows three different lines that jump to a subroutine. The first line uses extended addressing and the last two lines use PC relative addressing.

Source 5.25 Object-Relocatable Examples

```
          CPU12
Address   Object        Instruction
00008000  169000        jsr sub1
00008000  15fa0ffc      jsr sub1,pcr
00008000  15fa0ffc      jsr sub1-(*+4),pc
```

Figure 5.6 shows program memory containing the object code for the first line. The figure on the left shows the object code at $8000. It was set to this location when the code was assembled. Extended addressing is used to jump to the subroutine *sub1* located at $9000. On the right the figure shows the same object code after it has been moved to $a000. When the jump to subroutine is reached, it again jumps to $9000, but *sub1* is now at address $b000, so the program will not work.

The second and third lines of Source 5.25 call the same subroutine, but they use PC relative addressing. If *sub1* had been within range, a *bsr* could also have been used. The CPU12 allows the program counter to be used as an indexing register, resulting in PC relative addressing. The second line in Source 5.25 uses the symbol *pcr* where the indexing

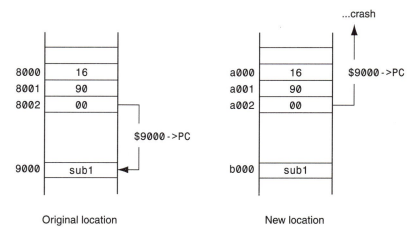

Figure 5.6 Extended Addressing Code

register is normally indicated. This special assembler operator tells the assembler to set the offset to the difference between the destination address and the program counter. In effect it creates PC relative addressing. The third line in the example shows the equivalent instruction using the normal indexed addressing mode.

Note that the *pcr* is not a new addressing mode. It simply tells the assembler to use indexed addressing with PC as the indexing register and to set the offset, *n*, equal to

$$n = A_D - \mathrm{PC}$$

Remember this is happening at build time, so the only way the assembler knows where the program counter would be is by taking the address of the current instruction, *, and adding the number of bytes contained in the instruction. The result will be the location of the next instruction, which is the correct value for the current PC at run time. In this case the *jsr* instruction is four bytes, so the PC would be

$$\mathrm{PC} = * + 4$$

and

$$n = A_D - (* + 4)$$

The *pcr* operator is quite handy. It not only makes the code more readable, but it also saves the programmer from having to look up the number of bytes for each instruction. In fact the number of bytes in some instructions depends on the destination location. Remember indexed addressing may take two, three, or four bytes. The actual number is not known without hand calculating the offset first. Unfortunately not all assemblers support the *pcr* operator.

Now let's go back to looking at position independence. Figure 5.7 shows the object code for either one of the last two lines in Source 5.25. The object code is made up of an opcode ($15), a postbyte ($FA), and a 16-bit offset ($0FFC). The figure on the left shows

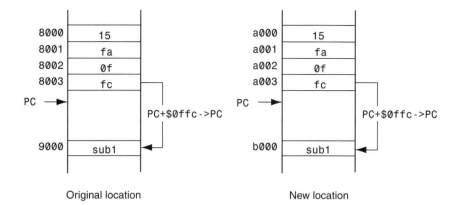

Figure 5.7 Relative Addressing Code

the object code at $8000. The opcode and postbyte tell the CPU to jump to the subroutine *sub1* by using PC as the indexing register and $0FFC as the offset. Since the PC is pointing to the next instruction, the instruction jumps to

$$\$8004 + \$0FFC = \$9000$$

The figure on the right shows the same object code moved to $A000. This time the program counter is pointing to $A004, so the instruction jumps to

$$\$A004 + \$0FFC = \$B000$$

which is the new location of *sub1*.

Of course this assumes that *sub1* is part of the code that was moved. PC relative addressing is only object relocatable if all the code is moved as a unit. We cannot move some of the code to one location and then move other parts of the code somewhere else.

Absolute Locations. There are times when we want to create relocatable code, but we are referencing objects that have a fixed or absolute location. These locations do not move with the code. Examples include MCU control registers, peripheral registers, or monitor subroutines. To access these fixed locations, we cannot use PC relative addressing. We have to use one of the non-PC relative addressing modes like extended or indexed using an index register.

Most 8-bit processors do not have PC relative addressing available to all of their instructions. There is simply not enough demand for object-relocatable code to justify the die space required to implement PC relative addressing. The most notable exception is Motorola's M6809, a popular but unsupported processor that has practically become a cult icon.

Object-relocatable code, although rarely needed for embedded systems, is the most common type of object code for most programmers, whether they know it or not. All programs that run on your desktop computer must have object-relocatable code because they are disk-based systems. To run a program, the object code is first loaded from the disk to RAM. It then executes the code out of RAM. It would certainly not be practical to reserve specific RAM blocks for specific programs. So when the object is loaded, it can be loaded into any available RAM block. This requires object-relocatable code. Similarly, a disk-based embedded system would also require object-relocatable code. This is becoming more common at the high end of embedded systems, but for small systems it will remain rare.

▶ SUMMARY

In this chapter we covered a lot of material. We learned how to use most of the instructions in the CPU12, how to use stacks, and how to perform basic arithmetic operations, Boolean logic, and bit-oriented operations. We also looked at most of the program control instructions including jumps, branches, and subroutines.

All the program examples in the chapter have been small snippets of code. These snippets illustrate how the instructions can be used, and they represent the ingredients of a complete program. If we use a building project analogy, the instructions represent the materials, the lumber, the nails, and the concrete. The development software is our tool, our hammer and saw. In this chapter we looked at the basic skills required to apply our tools and materials. To continue the building project analogy, learning the skills presented in this chapter is like learning to flatten a nail's point when nailing close to the end or to set the circular saw blade to the shallowest setting that will still make the cut.

You cannot expect to pick all the skills up by reading through this chapter. You must start programming. After reading through this chapter, you will still need to use it as a reference. Eventually, with experience this material (and much more) will become second nature. You will become a very good carpenter; however, you cannot yet build a house. In the next chapter, we look at program design, how to look at and design the big picture. So far we have only covered the details.

EXERCISES

1. Given the following definitions, write a source code snippet that performs the following sequence:
   - Load ACCA with CONST1
   - Load ACCB with the contents of CONST2
   - Load IX with the contents of Ptr1
   - Load SP with INITSP

   ```
   CONST1      equ $80
   INITSP      equ $0a00
               org $2000
   CONST2      fcb $08
   Ptr1        rmb 2
   ```

2. Show the machine code for the program in Exercise 1.

3. Using a load-effective address instruction, add 8 to IX.

4. Repeat Exercise 3 using the *abx* instruction. Compare execution times for both programs.

5. Use a load-effective address instruction to balance the stack after the following instructions:

```
pshc
pshx
pshd
```

6. Using a single *sty* instruction, push IY onto the stack.

7. The CPU11 does not have a *pshcc* instruction. Write a program compatible with the CPU11 using other instructions that push the CCR onto the stack without changing the CCR.

8. Write a program using the *sex* instruction to convert the 8-bit signed variable *Var8* to a 16-bit signed variable, *Var16*.

9. Repeat Exercise 8 using any instructions but without using a CPU register (CCR can change).

10. The CPU11 does not have a *pshd* instruction. Implement *pshd* using CPU11 instructions.

11. Use stack instructions to exchange ACCD and IY.

12. Give the contents of ACCA, ACCB, and the CCR bits *H, N, Z, V,* and *C* after each instruction in the instruction sequence shown. Use a dash if the flag is not affected and a question mark if a result is unknown.
   (a) `ldaa #$80`
   (b) `adda #$78`
   (c) `tab`
   (d) `suba #-112`
   (e) `mul`

13. Show the contents of ACCA and the CCR bits *H, N, Z, V,* and *C* after each instruction. The instructions are executed individually with the initial conditions shown. If the instruction does not affect the flag use a dash or if the contents are unknown show a question mark.
   (a) `adda #$78` with initial conditions
```
   A:B    CCR = HNZVC
   00:00        00000
```
   (b) `adda #$80` with initial conditions
```
   A:B    CCR = HNZVC
   80:00        00000
```
   (c) `suba #-1` with initial conditions
```
   A:B    CCR = HNZVC
   90:00        00000
```
   (d) `asra` with initial conditions
```
   A:B    CCR = HNZVC
   81:00        00000
```

14. Write a program to add the unsigned 8-bit variable *Var8* to the unsigned 16-bit variable *Var16*. Flags must be set in case a carry is required.

15. Write a program that increments the unsigned 24-bit variable *LongVar*. Include directives to define the variable.

16. Write a program that adds the packed BCD variables *bcd1* and *bcd2*.

17. Write a program that subtracts the packed BCD variable *bcd1* from *bcd2*. Include code to detect a negative result. If a negative result occurs, set bit 0 of the variable *Sign* and convert the result so it has the correct magnitude.

18. Write a program snippet to implement the following 16-bit rotation. Each time the program runs, the bits are shifted one time.

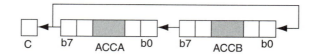

19. Write a subroutine that calculates the even parity bit of the contents of ACCB. The parity bit should be contained in ACCA so ACCD contains a 9-bit word (parity + 8 bits).

20. Write a subroutine that implements an 8-bit LFSR pseudo-random number generator. The LFSR is a single shift-left operation where

$$B_0^* = B_3 \oplus B_4 \oplus B_5 \oplus B_7$$

where B_0 is the least-significant bit of ACCB and B_0^* is the next B_0 after the shift.

21. Given two unsigned numbers in accumulators A and B, write a routine that will put the greater number in accumulator A and the lesser in B.

22. Write a snippet that will branch to *next* if the bits in the 8-bit variable *Status* are as follows (do not change *Status*):
    (a) bit3 $\oplus$ bit1 = 0
    (b) bit2 | bit0 = 0
    (c) bit2 · (bit3 $\oplus$ bit1) = 1

23. Set bit 0 and bit 2 and clear bit 3 of the variable *Status*. All other bits must remain the same.

24. Write a program that fills memory locations $E000 to $FFFF with $AA.

25. How many times does *Process* get executed when the instruction *BRN* is replaced with the following instructions?
    (a) *bne*
    (b) *bpl*
    (c) *blt*

```
              ldaa #200
    loop      bsr Process
              deca
              BRN loop
```
    Assume *Process* preserves all registers.

26. The variable *Status* initially contains the value $A5. Will a branch occur after these instructions?

(a) `brclr Status,$82,next`

(b) `brclr Status,64,next`

(c) `brclr Status,$01,next`

(d) `brset Status,$84,next`

27. Calculate the required relative offsets for the following branch instructions and destination addresses. If the destination is too far, indicate and show the equivalent branch using a long branch instruction.

Instruction	Branch Address	Destination Address	Offset
a. `bra last`	$8010	$7FFF	
b. `bra next`	$0800	$0938	
c. `brset var,$01,next`	$0880	$0938	
d. `brclr 2,x,$02,wait`	$0810	$0810	

28. Show the machine code for the branch instructions in Exercise 27. Assume *Var* is at $08A0.

29. A division program that requires 25 bytes of code space is needed at 10 different places in a large program. How many bytes are saved if the division program is implemented as a subroutine rather than directly inserted into the program when the division is required? Assume that only *jsr* (EXT) and *rts* instructions are needed to convert the original division program into a subroutine.

30. The following program is intended as a routine that adds a block of memory bytes:

```
****************************************************************
* Subroutine SumBlock - Sum a block of memory bytes.
*      Arguments: Block start address passed in IX.
*                 Block end address passed in IY.
*                 Sum returned in ACCD
****************************************************************
SumBlock    pshy
            ldd #$0000
sumlp       addb ,x
            adca #$00
            cpx 0,sp
            beq sumend
            inx
            bra sumlp
sumend      puly
            rts
```

(a) Given the following memory and register contents before *SumBlock* is called, find the resulting sum generated by the routine:

```
PC     SP     X      Y      D = A:B    CCR = SXHI NZVC
08A0   0C02   0890   089f     00:10           1101 0100

0890   01 F1 01 81 - 01 37 01 01 - 01 40 00 00 - 01 01 FF 00
```

(b) Find the time it takes to execute the routine as a function of the number of locations that are summed, assuming an 8MHz E-clock.

(c) What is the maximum number of locations that can be summed by this routine before there is a possible error in the sum?

31. Is the program shown in Exercise 30 object relocatable? (Assume that the stack is not moved.)

32. Write a snippet that uses the routine in Exercise 30 to find the sum of the block $8000–$8040 and then stores the result in *Sum*. Also show the definition of *Sum*.

6

Assembly Program Design and Structure

In this chapter we cover some techniques and tools for designing assembly language programs. In the last chapter we covered assembly language coding techniques, and now it is time to learn how to put these pieces together to form reliable and reusable routines and programs.

It used to be adequate for an engineer to pick up the manual for a CPU and piece together a program. In today's environment, a working program is not enough. Programs not only have to work reliably, but they also have to be reusable. This is because programs are becoming larger and more complex with shorter time-to-market expectations. Because of this, most of the programming that is done today is actually revising existing code instead of creating new code. So our code must be reusable. This means it must be readable, revisable, and portable. It is important to avoid unstructured, tricky solutions. Any experienced programmer would rather work with a clear, simple program than an obscure, tricky one. In fact many times it is faster to start over from scratch than to revise a poorly structured program.

While programming it is important to keep in mind that the program will probably have a long life span. Programmers in the 1970s generally had no idea that their code would need to pass Y2K compliance testing. Not only did they not foresee the Y2K problem, but they probably did not even think their code still would be used in the year 2000. Embedded system programs, in general, have longer life spans than computer applications. A well-written program module can live multiple lives by being reused in new products. This means we need to place much more emphasis on program design, structure, and style. We also need to start to build a collection of well-designed, reusable routines. Most importantly, we should try to avoid this self-fulfilling prophecy:

> *We'd better start coding right away because we're going to have a lot of debugging to do.*[1]

Let's look at some tools, common design structures, and conventions we can use to design and document reliable and reusable assembly routines.

▶ 6.1 DESIGN AND DOCUMENTATION TOOLS

In this section we look at two tools for designing and documenting assembly language programs: flow diagrams and pseudo-C. These tools provide a way to describe a program at a higher level than the actual assembly code. Each one has its own strengths and weaknesses. Choosing a method to use depends on the application, company conventions, and personal preferences. There are many other techniques for program design, such as statecharts, data flow diagrams, and structure charts. These are used for documenting larger real-time and multitasking programs.

6.1.1 Flow Diagrams

Flow diagrams are a widely used tool for graphically designing and documenting programs. In general they provide a good way to visualize the program structure. Table 6.1 shows some of the advantages and disadvantages associated with flow diagrams.

The main problem with flow diagrams is that they can be time consuming to create and they cannot be included in the program source file because they are not text based. This can be a real problem for documentation. Programmers tend not to go back and change the flow diagram after changes are made to the source code. Consequently the flow diagram is not representative of the actual code in the end.

One of the advantages of flow diagrams is their wide use and simplicity. We can design flow diagrams with three basic components: process blocks, decision blocks, and connectors.

Process Blocks. A process block represents a structured block of code. To be structured, it always has a single entrance and a single exit or no exit. Figure 6.1 shows three types of process blocks. Actually only one is required, but if you want to distinguish processes that are implemented as routines, you can add the double lines at the sides of the block in the center. In addition many programs in embedded systems are endless loops. The process block shown on the right in the figure indicates that the process is repeated forever. This endless loop is a higher-level representation of a looping construct, which is shown later in this section.

The code represented by a process block may consist of anything from a single line of code to several large routines. You should be able to represent all flow diagrams by a single higher-level, process block.

---

[1]Steve McConnell, *Code Complete* (Redmond, WA: Microsoft Press, 1993).

TABLE 6.1   ADVANTAGES AND DISADVANTAGES OF FLOW DIAGRAMS

Advantages	Disadvantages
Good graphical representation for visualization	Not text; need drawing tools and cannot be included in the program source file
Close correlation with code	Overly complex with some data structures
Simple and widely used	Poor representation of asynchronous events and concurrency
Easy to detect nonstructured code	

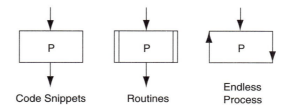

Figure 6.1   Process Blocks

Decision Blocks.   Decision blocks are used for conditionally executing processes. They represent a high-level language *if* and an assembly language conditional branch instruction. They should always have a single entrance and only two exits. The condition that the decision is based on is always binary, with each state corresponding to an exit path. Figure 6.2 shows three decision blocks. Functionally they are all the same, that is, if the condition *C* is true, the path *T* is taken, and if the condition is false, the path *F* is taken. The three fig-· ures simply show that the two exit paths can emerge from any two corners. The entrance also can come in from a side corner.

Connectors.   There are two types of connectors used in this text, as shown in Figure 6.3. Start and termination connectors are shown on the left. They represent the entrance and exit of a flow diagram. For snippets of code, *START* and *CONTINUE* are typical labels for these connectors. If the flow diagram represents a routine, the entrance connector should be labeled with the name of the routine, and the exit label should be *RETURN*.

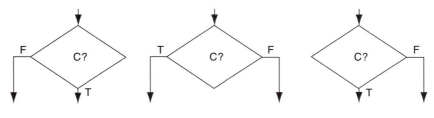

Figure 6.2   Decision Blocks

Figure 6.3   Flow Diagram Connectors

Shown on the right are continuation connectors. They should be used if a flow diagram is very complex or if it spans more than a single page. But remember that flow diagrams are valuable because they can help the programmer visualize the operation of the code. If the flow diagram becomes too complex, and there are so many crossed lines that it looks more like a bowl of spaghetti, it no longer helps the programmer visualize the code. In fact if your flow diagram does look like that, there is a good chance that your code will, too. Use a continuation connector to avoid crossing lines, but before you use it, see if you can create a simpler structure. The same goes for connecting multiple pages.

If your flow diagram spans more than a single page, it has lost some of its value as a visualization tool. Flow diagrams should be limited to a single page. This helps with visualization and forces you to design the program hierarchically. Following are some general rules for using flow diagrams.

Always Design the System Hierarchically.   Start with the top-level program flow and then design the flow diagrams for each process. There may be several levels in the hierarchy depending on the size and complexity of the program. The top level is the most abstract and includes the complete program. The bottom level is the closest to the actual code. If you feel as if you are simply writing the code in the flow diagram, however, you have probably gone too far. It should be straightforward to go from a flow diagram to code, but it should not be redundant.

Flow Diagrams Should Be Implementation Independent.   The flow diagram text, or labels, should be microprocessor and language independent. This means you should not be referring to actual register names or instructions. Remember that your design tools should always be at a higher level of abstraction than the code and should be usable for any implementation.

Limit All Flow Diagrams to a Single Page.   Keep your diagrams as short and as simple as possible. Remember that flow diagrams are valuable because they help visualize the code. If they look like spaghetti or span multiple pages, they no longer add value.

Always Update the Flow Diagram to Reflect Your Actual Program.   Program design and coding is an iterative process. You may design a flow diagram and while writing the code, stumble on a better design. Do not forget to go back and change your flow diagram. Remember that your flow diagram is also important for documenting the design.

Always Use Structured Constructs.   We will cover the structured constructs in Section 6.2, but the most important characteristic is that every flow diagram must have a single

TABLE 6.2   ADVANTAGES AND DISADVANTAGES OF PSEUDO-C

Advantages	Disadvantages
Lowest level design tool; the transition to code should be very simple	Poor representation of asynchronous and concurrent events
Text-based, therefore easier to change and more likely to change with software revisions; provides prewritten comments	Not as good for visualization
Starting point for interfacing the code with C or rewriting it in C	Not as good for large system descriptions

entrance and a single exit or no exit. This makes every flow diagram structured, and it can be represented by a higher-level process block. This is so important that it should be seen as a requirement.

6.1.2 Pseudo-C

In this text we will also be using pseudo-C as a programming design language or PDL. A programming design language, or pseudocode, is a higher-level description of the code in a form understandable by most readers. Most PDLs use the English language to describe the program structure and flow, normally in the form of program comments. The advantage to using English is that it is widely understood and there is not a strict syntax. The disadvantage is that PDLs written in English tend not to be as precise. Another option for a PDL is to use a high-level language like C. It is more precise and can easily be converted to a C program. The disadvantage of using C is the clutter and complexity caused by the language's syntax. Also, not all the syntax is understood by programmers that do not know C.

A very practical solution is to use a combination of English language comments and pseudo-C. Pseudo-C includes the programming constructs of C without including all the syntax details. Most programmers will understand the basic C constructs, and if you are required to rewrite the code in C, you already have the basic program design in your pseudo-C comments. Table 6.2 lists the advantages and disadvantages of using pseudo-C.

The primary advantage of pseudo-C versus flow diagrams is that it is text-based, so it can be included in the source file as comments. This means it is more likely to be changed when the program source code is changed. It also provides excellent comments from the start, instead of tacking on comments after the program is complete. The process of designing an assembly program using pseudo-C is shown in Example 6.1. From this example you can see that the process has commenting built into the design.

EXAMPLE 6.1 _____

Design Process Using Pseudo-C

Write a Pseudo-C Function Prototype and Description.   This serves the purpose of defining the routine name and arguments and describing the routine's function. It is not necessary to include such details as the *void* data type.

```
*******************************************************************
* AbsValue(var)
* Description: Returns the absolute value of a signed byte.
```

Write the Program in Pseudo-C.   Use only the basic constructs. Do not include implementation details.

```
AbsValue(var) {
    if(var < 0){
        var = -var;
    }
    return var;
}
```

Convert to Assembly.   Write the assembly code to implement the pseudo-C constructs, including implementation details.

```
AbsValue     tstb
             bpl absrtn
             negb
absrtn       rts
```

Create Commenting.   Convert the pseudo-C to comments and add implementation comments to complete the program.

```
*******************************************************************
* AbsValue(var)
* Description: Calculates the absolute value of a signed byte.
* Arguments: ACCB contains the input byte.
*            ACCB contains the returned absolute value.
* Pseudo-C:
* AbsValue(var) {
*     if(var < 0){
*         var = -var;
*     }
*     return(var);
* }
*******************************************************************

AbsValue     tstb                    ;Test if positive
             bpl absrtn
             negb                    ;If not, negate
absrtn       rts


*******************************************************************
```

In summary, both flow diagrams and pseudo-C are useful tools for designing and documenting an assembly program. They help keep track of the overall structure while you are programming the details. Because of their visual qualities, flow diagrams are especially helpful for beginning programmers. With experience a programmer may switch to pseudo-C for efficiency. Pseudo-C is especially useful for experienced programmers who are working with both C programs and assembly code. Both flow diagrams and pseudo-C are used throughout this text.

▶ 6.2 STRUCTURED CONTROL CONSTRUCTS

Programming in assembly can be described as a free-for-all. You may have noticed that the only "rules" in assembly language programming are syntax rules. There are no rules enforcing program structure or data types. This means that it is easy to write poor programs in assembly. To avoid this mistake, all programs should be structured and therefore have a single entrance and a single exit.

In this section we cover the basic structured control constructs. These constructs are all based on the program control instructions for the C programming language. For each construct we will look at the flow diagram and the pseudo-C code, and then see how to translate the construct to assembly code. Once you know how to translate each construct to assembly, designing your program becomes a process of putting these structured constructs together. This is a process equivalent to programming in a high-level language.

6.2.1 Sequence

The first and most obvious construct is a simple sequence of processes. Of course all programs involve a sequence. Figure 6.4 shows the flow diagram and pseudo-C code for a sequence. In the figure *P2* is shown as a routine. In the flow diagram this is indicated by double lines on the sides of the process block. In pseudo-C we show that process as a C

Flow Diagram	Pseudo-C
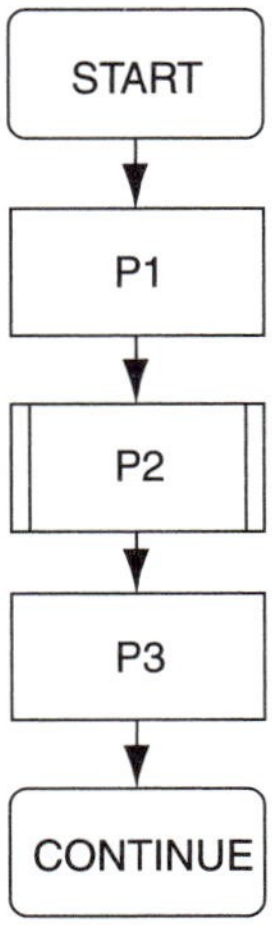 START P1 P2 P3 CONTINUE	/*P1*/ P2() /*P3*/

Figure 6.4 The Sequence Construct

function. The other two processes are snippets of code. In pseudo-C these are shown as comments or as the actual pseudo-C code to implement the processes.

6.2.2 Conditional Constructs

Conditional constructs are used to conditionally execute processes. The *if, if-else, if–else-if–else,* and *case* constructs are covered here. The fundamental construct is the *if.* All other conditional constructs can be created by using multiple *if* constructs.

The *If* Construct. The *if* construct is used to conditionally execute a process. Figure 6.5 shows the flow diagram and the pseudo-C implementations. The *if* construct translates to a decision block in the flow diagram and a conditional branch in assembly code. When using the *if* construct, a process, *P,* is executed only if the condition, *C,* is true. This may or may not be what you really need. A good defensive programming practice is always to consider what needs to be done if the condition is not true. It may be that an *if-else* is required.

Figure 6.6 shows an example of an absolute value program to illustrate the *if* construct. The condition tests to see if *var* is less than zero. If it is less than zero, it negates *var;* otherwise it does nothing.

The *If-Else* Construct. The *if-else* construct is used to conditionally execute one of two mutually exclusive processes. Figure 6.7 shows the flow diagram and pseudo-C code for the *if-else* construct. Notice that the flow diagram paths join before exiting. This seems like a minor detail, but it is important for creating a structured program. If they did not join, the *if-else* construct would have two exit points and would violate the requirements for a structured program.

Figure 6.8 shows a simple thermostat program that uses an *if-else* construct. If *Temp* is greater than *TSET,* the *HeatOff* subroutine is executed. If *Temp* is not greater than *TSET,* the *HeatOn* routine is executed.

Flow Diagram	Pseudo-C

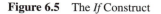	`if(C){`
	`    /*P*/`
	`}`

Figure 6.5 The *If* Construct

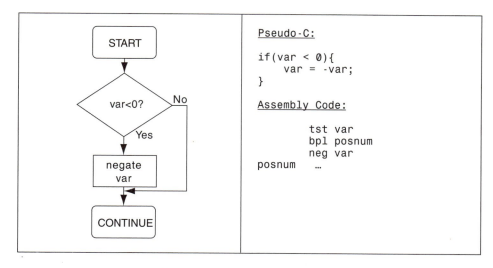

Figure 6.6 Absolute Value Program

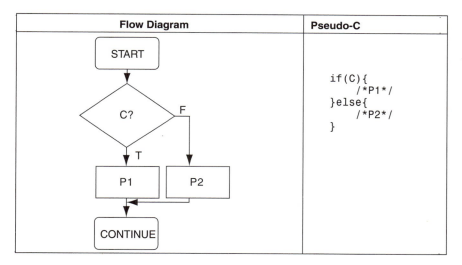

Figure 6.7 *If-Else* Construct

Compare the assembly code to the flow diagram. The complication in converting the flow diagram to assembly is because the flow diagram is a two-dimensional diagram, whereas the assembly code has in effect only one dimension. We handle this by adding extra *bra* instructions to branch over processes.

Notice that all assembly code used to implement the function falls between *start* and *cont*. This should always be the case if the program is to be structured. A common mistake occurs because of the required extra branch. Source 6.1 is an example of the code written incorrectly. The tendency is to branch somewhere, anywhere out of the way, to call

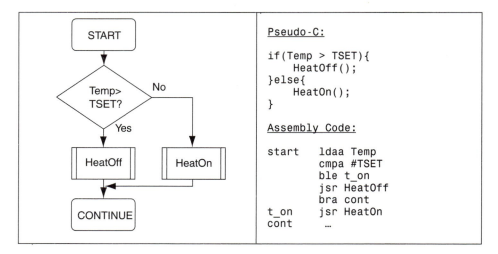

Figure 6.8   A Simple Thermostat Using an *If-Else* Construct

HeatOn. This is a classic example of spaghetti code. Not only are extra branches required, but the program is also no longer structured, because there may be code unrelated to the thermostat function after the label *cont*.

Source 6.1 Spaghetti Code Version of the Simple Thermostat

```
start     ldaa Temp
          cmpa #TSET
          ble t_on
          jsr HeatOff
          bra cont
cont      ...
          bra around
t_on      jsr HeatOn
          bra cont
around    ...
```

This is probably the most difficult part of converting a flow diagram to assembly code. Look at the example code and make sure you understand how to convert it correctly.

The *If–Else-If–Else* Construct.   The *if–else-if–else* construct is used to conditionally execute one of several mutually exclusive processes. Figure 6.9 shows the flow diagram and the pseudo-C code. The *if–else-if–else* construct is an extension to the *if-else* construct that includes multiple conditional tests. The last *else* is optional, but a good programming practice is to always at least consider the final *else*.

To illustrate the *if–else-if–else* construct let's look at a better thermostat than the one shown in Figure 6.8. The problem with the simple thermostat is that there is no hysteresis built into the controller. The heat will immediately turn on when the temperature goes

Flow Diagram	Pseudo-C

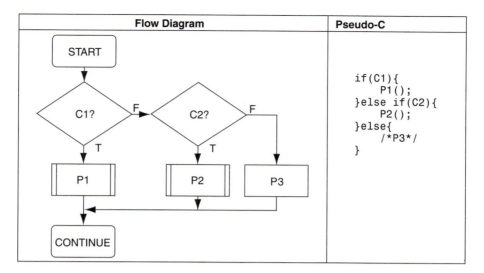

```
if(C1){
     P1();
}else if(C2){
     P2();
}else{
     /*P3*/
}
```

Figure 6.9 *If–Else-If–Else* Construct

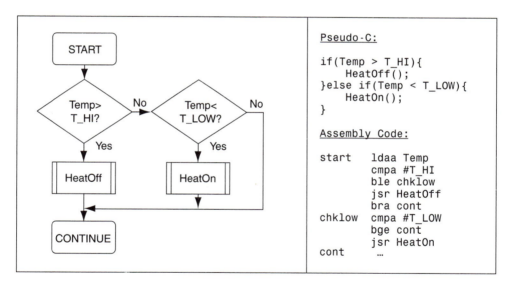

Pseudo-C:

```
if(Temp > T_HI){
    HeatOff();
}else if(Temp < T_LOW){
    HeatOn();
}
```

Assembly Code:

```
start    ldaa Temp
         cmpa #T_HI
         ble chklow
         jsr HeatOff
         bra cont
chklow   cmpa #T_LOW
         bge cont
         jsr HeatOn
cont     ...
```

Figure 6.10 A Better Thermostat Program Using *If–Else-If* Construct

below *TSET*. As soon as the heat turns on, the temperature will rise and the program will immediately turn the heat off again. Consequently the heater will cycle on and off much too quickly. The solution to this problem is shown in Figure 6.10. This thermostat will not turn the heater off until the temperature is greater than *T_HI*. It will not turn the heater on until the temperature goes below *T_LOW*. By setting the difference between *T_HI* and *T_LOW*, we can control the cycle rate of the heater.

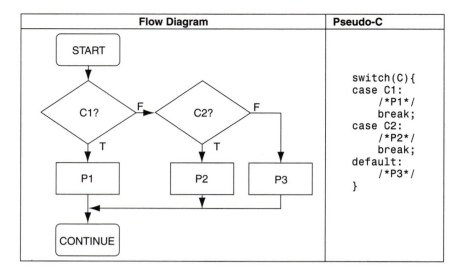

Figure 6.11 The *Case* Construct

The program in Figure 6.10 uses an *if–else-if–else* construct, but notice that the last *else* is not used. This is because if *Temp* is not less than *T_LOW* or greater than *T_HIGH*, the heater setting is not changed. The conversion from the flow diagram to assembly code is very similar to the *if-else* example.

The *Case* Construct.   The *case* construct is essentially the same as the *if–else-if–else* construct. There are some subtle differences in the implementation when using C, but we will not address those differences here. The *case* construct simply will be an alternate notation to the *if–else-if–else* in the pseudo-C code. Figure 6.11 shows the flow diagram and pseudo-C code for the case construct.

Notice that in a C *case* construct, a *break* is required to execute only one process. There is also a default process defined. This is equivalent to the last *else* in the *if–else-if–else* construct. Figure 6.12 shows a simple command parser that uses the *case* construct. Again you can see that we also could have implemented this program using an *if–else-if–else* construct.

6.2.3 Looping Constructs

A looping construct is a special case of a conditional construct that conditionally repeats one or more processes. Let's look at the basic looping constructs used in C, the *do-while, while,* and *for* constructs. Minor variations to these three constructs are also discussed.

The *Do-While* Loop.   The *do-while* loop runs a process, and if a condition is true, it repeats the process. Figure 6.13 shows the flow diagram and pseudo-C code for a *do-while* loop. The distinguishing characteristic of a *do-while* loop is that it always runs the process at least one time. Consequently it is not as versatile as the *while* loop; however, it can be more efficient than the *while* loop.

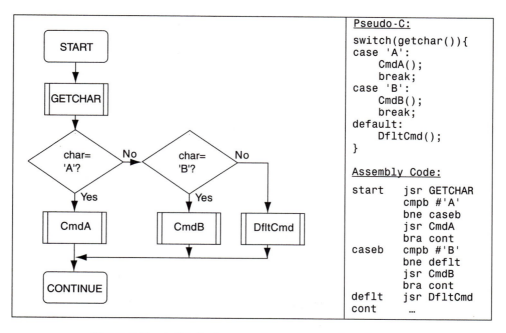

Figure 6.12   A Simple Command Parser Using a *Case* Construct

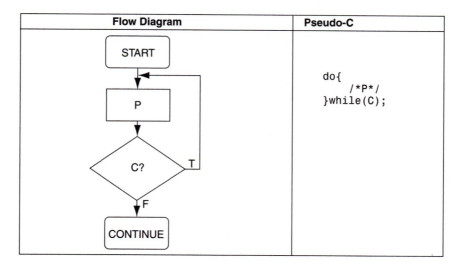

Figure 6.13   The *Do-While* Construct

Figure 6.14 shows an example of a *do-while* loop that inputs and stores data until a carriage return is received. Since the *do-while* loop was used, the carriage return will be stored in memory, even if it was the first character received. This may not be the desired result, but as you can see, the assembly code for this loop is very efficient.

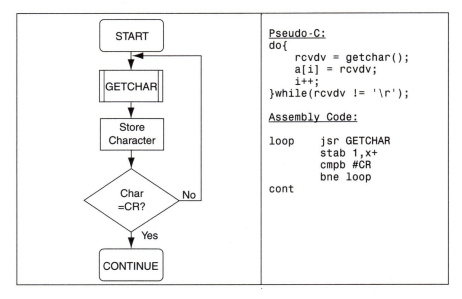

Figure 6.14 Input Line Using a *Do-While* Loop

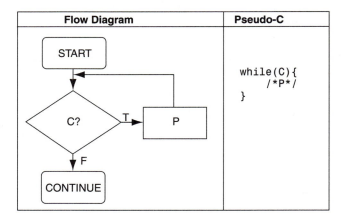

Figure 6.15 The *While* Loop

The *While* Loop. As shown in Figure 6.15, the *while* loop tests a condition, and if the condition is true, it repeats one or more processes. The *while* loop is more versatile than the *do-while* because it can run a process zero or more times. We can also write a *while* loop to run a process at least one time if that is what is required, but it will not be as efficient as a *do-while*.

Figure 6.16 shows the input line example using a *while* loop. This time if a carriage return is entered, the store process is not executed, so it is not stored. If the carriage return is the first character received, nothing will be stored. You can see that the assembly code is not as efficient as the *do-while* implementation.

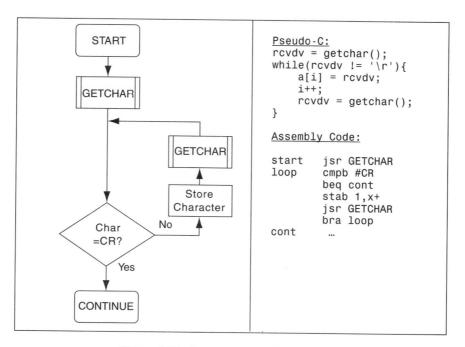

Figure 6.16 Input Line Using a *While* Loop

Actually the example in Figure 6.16 is interesting because it has two processes that have different requirements. *GETCHAR* must be run at least one time, whereas the character storage may run zero or more times. This means the *do-while* loop is most appropriate for the *GETCHAR* process, but the *while* loop is required for the character storage process. By running *GETCHAR* one time before the loop, then again at the end of each loop, the program did implement a *do-while* with a *while* loop.

By making a minor variation to the *while* loop, we can make the assembly code more efficient and we can create a generalized loop construct. Notice that in Figure 6.16 the *GETCHAR* process is called in two different places. If we move *GETCHAR* inside the loop before the decision block, we only need it one time, as shown in Figure 6.17. This is now a combination of a *while* loop and a *do-while* loop. To show this in pseudo-C, we have to use the rather obscure comma operator to put the *do-while* type processes in the *while* expression. With the comma operator, expressions are evaluated from left to right every time through the loop, so *GETCHAR* is called before the received character is checked. We could also put the assignment in a single expression, but that is considered poor programming practice.

Figure 6.18 is a generalized combination *do-while-while* loop. *P1* represents all the *do-while*-type processes, and *P2* are all the *while*-type processes. The pseudo-C notation is a bit odd, but the resulting assembly code is efficient and structured.

The *For* Loop. A *for* loop is a *while* loop with loop parameters and operations defined in an argument list. We can implement the *for* loop with a *while,* but the *for* improves the program structure by placing the loop parameters in the argument list. The person reading the

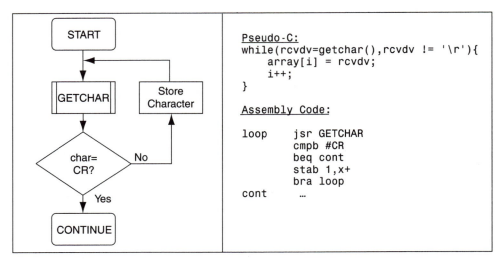

```
Pseudo-C:
while(rcvdv=getchar(),rcvdv != '\r'){
    array[i] = rcvdv;
    i++;
}

Assembly Code:

loop    jsr GETCHAR
        cmpb #CR
        beq cont
        stab 1,x+
        bra loop
cont    ...
```

Figure 6.17   A Hybrid *While* Loop

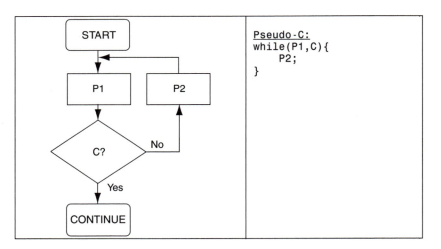

```
Pseudo-C:
while(P1,C){
    P2;
}
```

Figure 6.18   Generalized Combination *Do-While-While* Loop

code only has to read the *for* argument list to understand exactly how the loop will operate. Figure 6.19 shows the block diagram and pseudo-C code for the *for* loop.

The *for* loop executes the first statement in the argument list, *S1*. This initializes the looping parameter. Next it goes into a *while* loop. It checks the condition, and if the condition is true, it runs the process. Then it executes the last statement in the *for* argument list, *S2*. This is the operation performed on the looping parameter every time through the loop. Typically it is an increment or decrement operation.

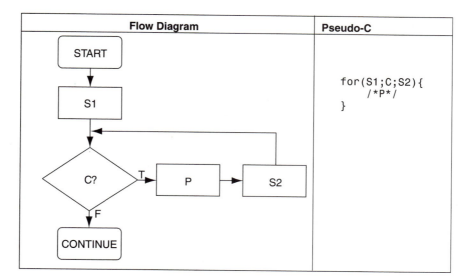

Figure 6.19 The *For* Loop

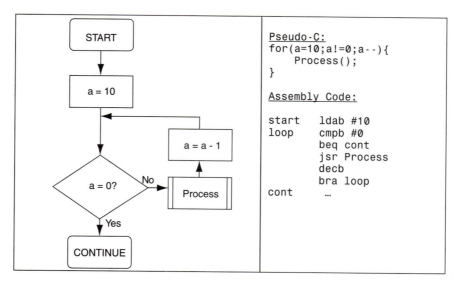

Figure 6.20 A Programmed Loop Using a *For* Construct

For loops are commonly used for programmed loops, loops that are programmed to execute a process a desired number of times. Figure 6.20 shows a programmed loop that executes *Process* 10 times. It is important to remember the order of events, especially if the loop parameter is used in the process.

Terminal Count Loop Bug. Before we leave looping constructs, there is a common and serious bug that needs to be addressed. This bug is encountered when the looping condition tests for a terminal count. Examples of terminal counts are as follows:

Data Size	Count Direction	Terminal Count
8-bit	Count up	$FF
16-bit	Count up	$FFFF
Any size	Count down	0

For example if an 8-bit variable is used as a loop counter, it cannot be greater than $FF. If a loop condition is based on the variable being *less than or equal to* $FF, the loop will never end.

Figure 6.21 shows a memory-fill program that contains the terminal count bug. The program was designed to fill memory locations $E000 through $FFFF with a fill character, *fillc*. The problem is that the decision block tests with a lower-or-same instruction, *bls*. Logically this design is correct. The problem is that when the pointer is $FFFF and is incremented, it rolls over to $0000, which is still less than $FFFF. Consequently the loop will never end and the program will attempt to fill all 64K-bytes of the memory map. The same result will also occur if the program is decremented down to $0000 and a *higher-or-same* test is made. Note that this bug also can be created when programming in C unless the compiler can detect it and generate a warning.

To fix this program, we need to change the test to *equal-to* or *not-equal-to* and use a *do-while* type loop. Figure 6.22 shows one way to fix the memory fill program so it will work correctly.

To summarize, any program can be broken down into these structured constructs. By using these constructs, your program will be structured, and you will be able to design

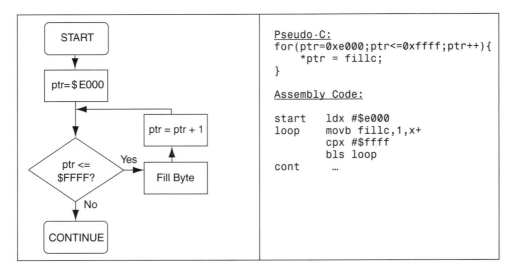

Figure 6.21   Example of a Terminal Count Bug

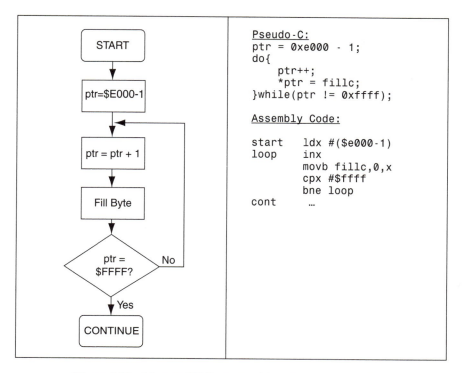

Figure 6.22 Memory Fill Program without Terminal Count Bug

hierarchically. It also makes programming assembly language more like programming in a high-level language.

▶ 6.3 DATA STORAGE

In this section we look at ways to store and access data in assembly language programs. The storage methods we cover are stored constants, register variables, local variables, and global variables. We analyze these methods by looking at the scope of the data, the resulting program efficiency including RAM space and execution speed, program modularity, and program reentrancy.

6.3.1 A Data Object

A *data object* is a memory location that contains data. Figure 6.23 shows the parts of a data object. The name of the object corresponds to its address, and the value of the object corresponds to its contents. In assembly programs the label is used as the data object's name. The data objects we have looked at in previous chapters are stored constants created with the *fcb, fdb,* or *fcc* directives and global variables created with the *rmb* directive. Note that equated symbols are not considered data objects because they do not have a storage location.

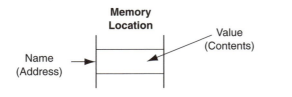

Figure 6.23 A Data Object

We can access the data object by name or by reference. To access data by name, we use direct or extended addressing and a label. To access an object by reference, we use indexed addressing, or with the CPU12, the pointer may be stored in memory and indirect addressing can be used. Source 6.2 shows several ways to access data objects.

Source 6.2 Examples of Accessing Data Objects

```
*****************************************************************
* Access by name
*****************************************************************
          ldab Const             ;(Const)=$55 -> ACCB
          ldd Var16              ;(Var16:Var16+1) -> ACCD


*****************************************************************
* Access by reference using IX
*****************************************************************
          ldx #Var8              ;Var8 -> IX
          ldab 0,x               ;(IX)=(Var8) -> ACCB


*****************************************************************
* Two ways to access by reference using a pointer and IX
*****************************************************************
          ldx Ptr                ;(Ptr:Ptr+1) -> IX
          ldab 0,x               ;(IX)=((Ptr:Ptr+1)) -> ACCB

          ldx #Ptr               ;Ptr -> IX
          ldab [0,x]             ;((IX))=((Ptr:Ptr+1)) -> ACCB


*****************************************************************
Const     fcb $55      ;Stored constant

Var8      rmb 1        ;8-bit global variable
Var16     rmb 2        ;16-bit global variable
Ptr       rmb 2        ;16-bit global variable used as a pointer
*****************************************************************
```

A data object has three main attributes: its persistence, its scope, and its data type. In assembly language programs, it is up to the programmer to keep track of these. For example in Source 6.2 the variable *Ptr* is used as a pointer. However, there are no assembler rules that say you cannot use those two bytes for a data value.

TABLE 6.3 DATA OBJECT PERSISTENCE

Data Object	Persistence
Stored Constants	Forever
Global Variables	As long as power is applied
Local Variables	Within the defining routine
Volatile Variables	None

Persistence. The persistence or volatility of a variable refers to the length of time the contents of the variable are valid. The persistence of a stored constant, for example, is forever. Table 6.3 shows the four levels of persistence normally encountered: volatile variables, local variables, global variables, and stored constants.

Generally the longer the persistence of a variable, the longer that location must be reserved for the variable, which results in more memory being used.

Scope. The scope of a variable refers to the region of the program in which the variable is accessible. Scope is controlled by the assembler or compiler. For assembly programs contained in a single module (file), the scope is equivalent to the persistence. For modular code, the scope of an object may be limited to one file. Chapter 13 covers data in C and for modular programs.

6.3.2 Register Variables

A *register variable* is a variable stored in a CPU register. Because inherent addressing can be used, register variables result in the most efficient program code. Using register variables comes at a cost, however. As long as a register is holding a variable, it cannot be used for anything else. Since the CPU11 and CPU12 have very few registers, we must restrict the use of register variables to those variables that require a limited persistence. It would be like parking your car in middle of the street—it is okay if you are dropping someone off, but if you leave the car, all the other traffic would have to be rerouted to another street, or more likely, your car would not be there when you return. Register variables are reasonable for small, simple subroutines as long as the data do not need to persist beyond the routine. Because register variables are so efficient, they are often used in assembly programs. Source 6.3 shows a small routine that uses a register variable. In this case IX contains a variable pointer *ptr*, which is incremented through a block of memory. This routine is so simple that using a register for *ptr* is appropriate.

Source 6.3 Example Routine That Uses Register Variables

```
*******************************************************************
* FillMem(){
*      ptr = StartAddr - 1;
*      do{
*          ptr++;
*          *ptr = Fillc;
```

```
*       }while(ptr != EndAddr);
*******************************************************************
FillMem    pshx
           ldx StartAddr
           dex
loop       inx
           movb Fillc,0,x
           cpx EndAddr
           bne loop
           pulx
           rts


*******************************************************************
Fillc      fcb $55
StartAddr  fdb $8000
EndAddr    fdb $FFFF


*******************************************************************
```

6.3.3 Global Variables

When RAM locations are permanently allocated with the *rmb* or *ds* directives, they are *global variables.* Global variables are easy to create and can result in efficient code. For this reason global variables tend to be the variable of choice by beginning programmers. Global variables are also appropriate for even the most complex systems, such as multitasking systems in which direct parameter passing is not possible. If used indiscriminately, however, global variables can be problematic. The following paragraphs discuss several reasons to avoid using global variables.

Note _____

These variables are more appropriately called *static* variables, but *global* is the more traditional term in assembly language literature. Strictly, *global* refers to a variable that has a scope that includes all program modules. The term *static* refers to a variable that persists until power is removed. These distinctions are covered in more detail in Chapter 13.

Increased RAM Requirements. If you use global variables every time temporary storage is required, you will use more RAM space. Once defined, the RAM location is permanently allocated and cannot be used for anything else.

One misguided attempt at reducing the RAM requirements while using global variables is to use shared variables. A *shared global variable* is a global variable that is used temporarily for more than one unrelated variable. The idea is to use the same location for different data at different times. The main problem with shared global variables is the potential for inadvertent changes caused by some seemingly unrelated routine. Every time you use a shared global, you must make sure that it is not being used by another routine at that time. This can be a considerable problem, especially if the variable must persist for a long time. In effect a shared global variable has an unknown persistence, which practically equates to no persistence. It is like parking in the vice president's parking spot. As long as you know her schedule, you can get away with it. If she arrives unexpectedly when your car is there, however, you will most likely return to find a different car in the parking space.

Decreased Portability.   It is important to try to keep a program portable so it can be reused in other products. When a routine uses a global variable, it is no longer portable by itself. The global variable is now part of the routine, so if we want to use the same routine in another program, we have to remember to include the global variable. When a global variable is required, it is important that the routine's documentation clearly define its global memory requirements.

Reentrant Code.   A routine is reentrant if it can be used by two programming threads at the same time. This means that if a routine is interrupted, it can be reentered and both instances of the routine will operate correctly. To be concerned about reentrancy, you must have two different programming threads. Programs have multiple threads if they use either a preemptive multitasking kernel or interrupts. Since most embedded systems are interrupt driven, routines will be more portable if they are reentrant, so it is a good practice to always write reentrant routines. If a routine is not reentrant, you can still use it in a multitasking system, but you have to provide a mutual exclusion mechanism, such as semaphores, to control access to the routine. We cover semaphores in Chapter 16. Recursive routines also must be reentrant. A function is recursive if it calls itself. Few embedded applications benefit by using recursion, and since recursion can be problematic, it is best avoided.

When a routine uses a global variable, it may no longer be reentrant. To illustrate this, Source 6.4 shows a very simple subroutine, *MultBy8,* that uses a global variable. Note this routine is for illustrating a reentrancy error only. It should not be used in a real program, especially when using a CPU with a fast multiply instruction like the 68HC12.

Source 6.4 A Routine That Operates on a Global Variable

```
MultBy8              asl Var
                     asl Var
                     asl Var
                     rts
```

Figure 6.24 shows the program flow and the variable contents when the routine is reentered. First the main routine wants to multiply N by 8, so it loads *Var* with N. Then it branches to the *MultBy8* subroutine. The *MultBy8* subroutine does two shift-lefts so *var* is equal to $4N$. Between the second and third shift, an interrupt request occurs, which switches context to an interrupt service routine, ISR.

The ISR then wants to multiply M by 8, so it loads *Var* with M. At this point the original $4N$ in *Var* is lost. The ISR branches to reenter *MultBy8,* which completes its task and returns to the ISR with *Var* equal to $8M$. This is what the ISR requested, so there is no error yet.

Now the ISR finishes and returns to the interrupted instance of *MultBy8*. This instance of *MultBy8* still has one more shift to go, so it shifts *Var* to the left for the last time. But instead of returning with *Var* equal to $8N$, it returns with *Var* equal to $16M$. With respect to *main,* the *MultBy8* routine failed to function correctly.

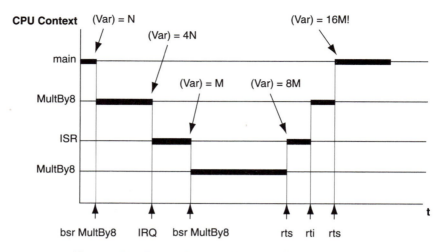

Figure 6.24 Context Diagram Showing a Reentrancy Error

Global variables and register variables are easy to use, but they both have their disadvantages. Register variables must be restricted to simple routines and variables with limited persistence. Global variables permanently use a memory location even if it is only required temporarily and can result in nonreentrant routines.

6.3.4 Local Variables

The two problems with global variables, permanent allocation and reentrancy, can be solved by using the stack to temporarily allocate memory space for variables in a routine. These are called *local variables*. Since the space is allocated during the routine, new space is allocated for each instance of the routine, which solves the reentrancy problem. However, it also means the persistence and scope of a local variable is limited to the routine. Once the routine is exited, the local variable no longer exists. Compared to using global variables, local variables require less RAM space and result in reentrant routines, but they are also more complex to program. To create reliable routines using local variables, these steps should be followed:

1. The stack space required for all local variables is allocated immediately upon entering the routine.
2. The local variables are accessed by using indexed addressing with constant offsets into the allocated stack space. If the stack pointer is used as the indexing register, it should not be changed throughout the routine unless it is to call another routine.
3. The stack space is deallocated immediately before exiting the routine.

Figure 6.25 shows the typical stack contents after the space for local variables has been allocated. This is commonly called a *stack frame*. Once the stack frame has been generated, the variables in the stack frame can be accessed by using indexed addressing with a

160   Chapter 6

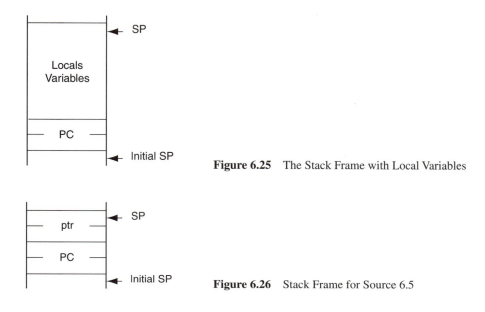

Figure 6.25   The Stack Frame with Local Variables

Figure 6.26   Stack Frame for Source 6.5

constant offset. Each time you use the stack, the offsets to the parameters and variables change. Because of this, it is not a good idea to use the stack after the space is allocated. When the routine is complete, the space for the local variables must be *deallocated* so the stack pointer will point to the return PC.

Source 6.5 shows the same memory-fill routine shown in Source 6.3, except it uses a local variable for *ptr* instead of a register variable. Immediately upon entering, the routine allocates two bytes of stack space by subtracting two from the stack pointer. This space will be used throughout the routine to store the current value of *ptr*.

Figure 6.26 shows the resulting stack frame during the routine. To access *ptr,* we use indexed addressing with the stack pointer and a zero offset. As long as the stack is not used within the routine, it can always be accessed in this way.

Source 6.5 Memory-Fill Routine Using a Local Variable

```
*****************************************************************
* A memory-fill routine to demo local variables.
*
* FillMem(){
*     UBYTE *ptr;
*     ptr = StartAddr - 1;
*     do{
*         ptr++;
*         *ptr = Fillc;
*     }while(ptr != EndAddr);
*****************************************************************

FillMem     leas -2,sp                    ;Allocate local space
            movw StartAddr,0,sp           ;Initialize pointer
```

```
                ldd 0,sp                          ;Decrement ptr
                subd #1
                std 0,sp
loop            inc 1,sp                          ;Fill loop
                bne no_c                          ;Increment ptr
                inc 0,sp
no_c            ldx 0,sp
                movb Fillc,0,x
                cpx EndAddr
                bne loop
                leas 2,sp                         ;Deallocate local space
                rts

****************************************************************
Fillc           fcb $55
StartAddr       fdb $8000
EndAddr         fdb $FFFF
****************************************************************
```

Before reaching the *rts* instruction to return from the routine, the local variable's stack space is deallocated by adding two to the stack pointer.

Compare Source 6.3 with Source 6.4 and you can see the increased complexity added by using a local variable. For this routine the obvious choice would be to use a register variable for *ptr*. This is often not the case, however. Many routines cannot use a register variable because there are no spare registers to allocate to the variable. In fact this is one of the primary reasons that C code is less efficient than assembly code. The C compiler is often not smart enough to know that a register variable can be used, so it will create a local variable instead. Local variables will be addressed more in Section 6.5 when we look at passing parameters on the stack and in Chapter 13, *Creating and Accessing Data in C*.

▶ 6.4 PROGRAM STRUCTURE

In this section, we only cover program structure as applied to assembly programs in a single module. Modules are covered in more detail in Chapter 12 and Chapter 14. As we saw in Chapter 3, an assembly program is made up of a main program and subroutines. If the system is interrupt driven, there will also be interrupt service routines. Designing a program hierarchically creates a more reliable and reusable program. In order to design hierarchically, you must use structured programming constructs and carefully designed routines. We have already covered designing structured programs using flow diagrams and pseudo-C constructs. Let's now concentrate on routines.

6.4.1 Routines

Well-designed routines can have a major impact on the reliability and reusability of your program. There are two primary concerns when designing a routine: cohesion and coupling. *Cohesion* refers to the contents or function of the routine. A routine with strong

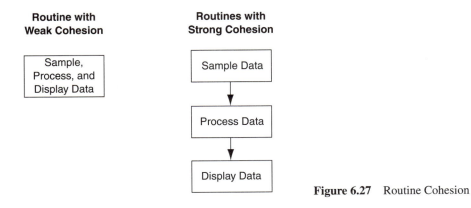

Figure 6.27 Routine Cohesion

cohesion performs one clearly defined function. *Coupling* refers to the way that the parameters are passed to and from the calling routine. We will cover parameter passing in Section 6.5.

Routines that have strong cohesion are more versatile and reusable. For example, Figure 6.27 shows two different ways to design routines for a data acquisition device. The first design has a single routine that does everything. This routine is not cohesive because it performs more than one function. A function like this is not very reusable. If you want to change the display type, for example, you would have to rewrite the routine.

The second design breaks the routines down by function. To change the display type, you would only have to replace the *Display Data* routine and reuse the other routines.

Designing routines for reuse becomes easier with experience. After writing the same routine several different ways, a programmer understands what the basic requirements are and how to design routines so they can be reused.

Creating a library of well-designed reusable routines is probably the best way to decrease time-to-market and the cost of product development. Keep this in mind when designing your routines.

▶ 6.5 PASSING PARAMETERS

Passing parameters is a form of communication between two routines. The process of passing parameters creates temporary data storage that is only used to hold the parameters while they are passed from one routine to another. The process should be as efficient as possible. Think of parameter passing like a courier business. You can have couriers deliver with a bicycle or with a truck. Of course the truck may be more versatile and can carry larger loads, but it is also more expensive and can be slowed by traffic problems more easily. In this section we analyze two different methods for passing parameters, using registers and using the stack. Using global variables will also be covered as an alternative to passing parameters.

6.5.1 Parameter Types

When passing parameters between two routines, we can pass the data, the name of the data, or a reference to the data. These are often referred to as call-by-value, call-by-name, and call-by-reference. Source 6.6 shows some code snippets that illustrate these three methods. All three snippets use a register to pass the parameter.

Source 6.6 Different Parameter Types

```
***********************************************************************
* Call-by-value.
***********************************************************************
        ldab Var            ;The value is passed in ACCB.
        bsr sub


***********************************************************************
*Call-by-name.
***********************************************************************
        ldx #Var            ;The name (label) is passed in IX.
        bsr sub2


***********************************************************************
*Call-by-reference.
***********************************************************************
        movw #Var,Varptr          ;A pointer to Var is passed in IX.
        ...
        ldx Varptr
        bsr sub3


***********************************************************************
```

Call-by-value works well when the parameter is small enough to conveniently fit into a CPU register. In the first example, *Var* is an 8-bit value that is passed in ACCB. The call-by-name example passes the name, or label, of *Var* in IX. The call-by-reference example passes the value of a pointer to *Var*. Since the pointer contains the address of *Var*, the same value is passed as in the second example. The difference is that the pointer is a variable, whereas the name of *Var* is a constant. In this case we would probably use call-by-value because *Var* can easily fit into ACCB. Call-by-name and call-by-reference work well when the parameter consists of a larger data structure or a string that is too large to fit in a register.

6.5.2 Using CPU Registers

The simplest and most efficient method for passing parameters is to use a register. It is the bicycle in our courier analogy. Parameter storage only needs to persist during the context change from one routine to the other; therefore using registers is ideal. To pass a parameter to a subroutine, the calling routine simply places the parameter in a register before jumping to the subroutine. To return a parameter to the calling routine, the called routine places the result in a register before returning. Source 6.6 showed three different ways to pass parameters using registers.

The *MultBy8* routine in Source 6.7 is a trivial subroutine that is used throughout this section to illustrate the different methods of passing parameters. Here the variable to be multiplied is passed in ACCB and the result is returned in ACCB. For a simple routine like this, passing the parameter in a register is the obvious choice. It is efficient and simple.

Source 6.7 A Multiply-by-8 Routine That Uses ACCB for Input and Output Parameters

```
MultBy8       aslb
              aslb
              aslb
              rts
```

Limitations to Passing Parameters in Registers.   As we saw, this type of parameter passing is efficient and straightforward. It is limited by the number of available registers, however. If there are more parameters than registers, the stack must be used. Before using the stack, we should see if call-by-name or call-by-reference can be used.

Care must also be taken when using registers to pass parameters in a multitasking or interrupt-driven system. In order for the routine to be reentrant or even interruptible, the registers must be saved when the context is switched. This is normally the case, but not always. The CPU11 and CPU12 microcontrollers automatically preserve all registers when an interrupt occurs, so if the context is switched by an interrupt, the routine is interruptible and reentrant. It should be noted, however, that not all CPUs automatically preserve the registers when an interrupt occurs. For these CPUs, it is up to the programmer to include the register preservation in the interrupt service routine. If this task is neglected, the routine will not be interruptible or reentrant.

Using a CCR Bit for Error Recovery.   When a single bit needs to be passed, it is possible to pass the bit in a CCR flag instead of using a whole register.

For example when handling error conditions in a routine, it is preferable to return through the *rts* and inform the calling routine of the error rather than jumping directly to an error handler from the routine. This preserves the structure of the routine and results in a more flexible error-handling process. In the *MultBy8* routine in Source 6.7, there is a possibility that the result is too large for ACCB. Source 6.8 shows a revision that passes the error condition through the *C* flag.

Source 6.8 Multiply by 8 with Error Condition Passed in the C Flag

```
MultBy8                 aslb
                        bcs m8_exit
                        aslb
                        bcs m8_exit
                        aslb
m8_exit                 rts
```

In this case the *C* flag will indicate an error. The calling routine can check the *C* flag after calling *MultBy8* and choose the appropriate action. If the error were handled inside the routine, there would be no choice of what action to take.

Care must be taken when choosing a CCR bit to pass the status. You do not want to use a flag that may be needed for normal operation. In the previous example the *C* flag was an obvious choice because of the way the shift instruction works. Passing status flags in CCR bits is only possible when writing code in assembly. When programming in the C language, the error status must be passed through one of the CPU registers.

6.5.3 Using the Stack

The most versatile method to pass parameters is to use the stack. As long as there is enough stack space, any number of parameters can be passed and the resulting routine will always be reentrant. When writing code in assembly language, however, we must consider that using the stack is not as efficient and is the most complex to design. Using the stack for passing parameters is analogous to using a truck in our courier analogy.

In this section we cover conventions for using the stack for both parameter passing and local variables. These conventions can help guide us through this relatively complex process so our code is reliable and understandable to other programmers.

These conventions are also similar to but not necessarily the same as the process used by a C compiler. By learning these conventions, you are not only learning how to design assembly routines but also learning how to debug code generated by a C compiler. It will also help you design routines that can be called by or can call a C program. For more information on writing assembly routines for C code, see Chapter 14.

Stack Usage Conventions.   If the stack is used in an ad hoc way for passing parameters or local variables, it is very easy to create stack-related bugs. With this in mind, if we follow standard conventions for using the stack we can create more reliable and understandable code. Figure 6.28 shows a general process for passing parameters on the stack, and creating local variables. This process also provides for passing parameters in the registers.

The calling process first pushes the parameters that will be passed on the stack. If there are also parameters that will be passed in registers, those registers are loaded. The subroutine is then called using the normal *bsr, jsr,* or *call* instructions. Once execution returns from the subroutine, the stack space that was used for the parameters is deallocated.

By keeping the allocation and deallocation of the stack space in the same routine, it is clearer for the programmer to see that the stack is balanced. This is a good programming practice when writing this type of code in assembly. Since compilers are better at keeping track of stack space than we are, they may deallocate the stack space in the called routine.

In the subroutine the register parameters are pushed onto the stack. In effect this creates a local variable with an initial value equal to the parameter value. This step is only required if the parameter value must be preserved throughout the whole routine. Stack space is then allocated for local variables. When the routine is complete, the stack space used for the local variables and the register parameters must be deallocated to return the stack pointer to the return PC. Figure 6.29 shows the typical stack contents after the space for local variables has been allocated. It is often referred to as a *stack frame*.

Once the stack frame has been generated, the variables in the stack frame can be accessed by using indexed addressing with constant offsets. For this reason the stack pointer should not be changed throughout the rest of the routine. If you need to temporarily save

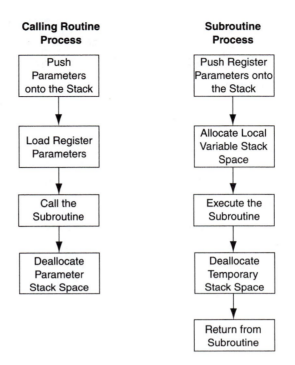

Figure 6.28 Stack Usage Process

data on the stack, create a new local variable by allocating more stack space at the beginning of the routine.

Source 6.9 shows another version of the *MultBy8* subroutine. In this example the stack is used to pass the 8-bit parameter to *MultBy8*. The result is returned in ACCB. In this routine there are no local variables.

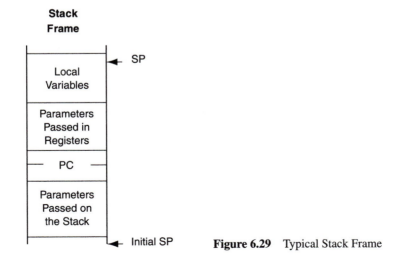

Figure 6.29 Typical Stack Frame

Source 6.9 MultBy8 Routine with the Parameter Passed on the Stack

Calling Routine	Subroutine	
ldab Number	MultBy8	ldab 2, sp
pshb		aslb
bsr MultBy8		aslb
leas 1,sp		aslb
stab Result		rts

Source 6.10 shows the memory block-fill program that was used in Source 6.3 and Source 6.5. In this version the routine is much more versatile because the parameters are passed to the routine instead of being constants or global variables. This routine can be used to fill any block with any fill character by simply designating the parameters.

Source 6.10 A Memory-Fill Program That Passes Parameters on the Stack

```
*****************************************************************
*Calling routine
*Pseudo-C:
* FillMem(ubyte fillc, ubyte *startaddr, ubyte *endaddr);
*****************************************************************
            ...
            ldd endaddr             ;push parameters on stack
            pshd
            ldd startaddr
            pshd
            ldab fillc              ;fillc passed in ACCB
            jsr FillMem             ;Call subroutine
            leas 4,sp               ;Deallocate parameter stack space
            ...
*****************************************************************
* Called Routine
* Pseudo-C:
* FillMem(ubyte fillc, ubyte *startaddr, ubyte *endaddr){
*      ptr = startaddr - 1;
*      do{
*          ptr++;
*          *ptr = fillc;
*      }while(ptr != endaddr);
*****************************************************************
FillMem     pshb                    ;push fillc
            leas -2,sp              ;Allocate local space for ptr
            movw 5,sp,0,sp          ;Initialize ptr
            ldd 0,sp
            subd #1
            std 0,sp
loop        inc 1,sp                ;Fill loop
            bne no_c
            inc 0,sp
no_c        ldx 0,sp
```

```
            movb 2,sp,0,x
            cpx 7,sp
            bne loop
            leas 3,sp              ;Deallocate local space
            rts
```

* *

The calling routine first pushes the block addresses onto the stack. It then loads the fill character into ACCB and jumps to the subroutine. Once in the subroutine, the fill character *fillc* is pushed onto the stack. This preserves *fillc* so it is available throughout the whole routine. Then two bytes are allocated on the stack for the local variable *ptr*.

Figure 6.30 shows the resulting stack frame. Once this stack frame has been created, all of the parameters and variables are accessed using indexed addressing with a constant offset. Following are the offsets that correspond to the different variables and parameters:

```
endaddr:    7,sp
startaddr:  5,sp
fillc:      2,sp
ptr:        0,sp
```

Once the stack frame is allocated, the subroutine executes the fill loop. Then the stack space for *ptr* and *fillc* is deallocated. This moves the stack pointer back to the return PC. Program control returns to the calling routine, and the rest of the stack frame, *startaddr* and *endaddr,* is deallocated.

Writing assembly language programs using the stack may seem overly complex at first. Nevertheless it is required for reliable, reentrant code when the parameters do not fit in registers. Luckily this type of programming is needed only when our programs become more complex. And if a program is complex, it is usually best to use C, in which case the compiler will generate this code.

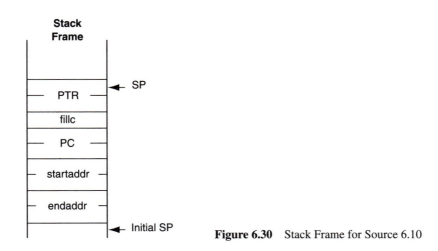

Figure 6.30   Stack Frame for Source 6.10

Using the Stack with the 68HC11. In the example given previously, we used the stack for passing parameters and local variables. This type of code depends heavily on stack-oriented instructions. Therefore, in order for the code to be efficient, the stack-oriented instructions for the given CPU must be efficient. Since this is the type of code generated by a C compiler, stack-oriented instructions play a large role in the overall efficiency of code generated by a compiler. The CPU12 introduced new stack-oriented addressing modes and instructions for this very reason. To illustrate the difference in code efficiency, the equivalent program is shown in Source 6.11 using only CPU11 instructions. This program requires 67 bytes of program space, whereas the program in Source 6.10 only required 54 bytes.

Source 6.11 The Block-Fill Routine Written for the 68HC11

```
******************************************************************
*Calling Routine
*Pseudo-C:
* FillMem(ubyte fillc, ubyte *startaddr, ubyte *endaddr);
******************************************************************
                ...
                ldd startaddr
                pshb
                psha
                ldd endaddr
                pshb
                psha
                ldab fillc
                jsr FillMem
                ins
                ins
                ins
                ins
                ...
******************************************************************
*Pseudo-C:
* FillMem(ubyte fillc, ubyte *startaddr, ubyte *endaddr){
*       ptr = startaddr - 1;
*       do{
*           ptr++;
*           *ptr = fillc;
*       }while(ptr != endaddr);
******************************************************************
FillMem         pshb                    ;push fillc
                des                     ;Allocate local space
                des
                tsy
                ldd 5,y                 ;Initialize ptr
                std 0,y
                subd #1
                std 0,y
loop            inc 1,y                 ;Fill loop
```

```
              bne no_c
              inc 0,y
no_c          ldx 0,y
              ldab 2,y
              stab 0,x
              cpx 7,y
              bne loop
              ins                        ;Deallocate local space
              ins
              ins
              rts
```

In this case the CPU12 instructions that help the most to increase efficiency are the *leas* instructions to allocate and deallocate stack space and the ability to use the stack pointer as an indexing register. In the CPU11 code it is necessary to transfer the stack pointer to IX or IY to access the local variables and parameters. This can present a real problem if two pointers are required for the function of the program. At least one of the pointers will have to be a local variable on the stack.

In this discussion we have not looked at returning parameters on the stack. In most situations return values will fit into a CPU register so the stack does not have to be used. If there are parameters that must be returned on the stack, space is allocated before the routine is called. This space is then part of the stack frame between the parameters passed to the routine and the PC. Since this is rarely used, it is not covered in any more detail.

6.5.4 Communicating with Global Variables

The previous two sections described two methods for passing parameters between two routines. Passing parameters is a form of interroutine communication. By using parameters for this communication, the routines can be made more portable and reentrant. Global variables can also be used for interroutine communication. Remember that global variables persist as long as power is applied, so data can be placed in a global variable by one routine and read by another routine. In fact for interrupt service routines, global variables are the only way to communicate, because the programmer does not know when the routine will be executed.

Global Variables versus Parameter Passing. The *MultBy8* program shown in Source 6.4 does not have its parameters passed to and from the routine. Instead it uses the global variable *Var* for both the input parameter and the result. We saw that this resulted in a non-reentrant routine, but there is another reason for avoiding this practice. When a routine affects the value of a global variable, it is being sneaky. Instead of acting on the parameters passed to the routine, it is sneaking out the "back door" and changing the data stored in *var* as a side effect.

It is good programming practice to always pass the parameters directly to and from the routine to make it more versatile. This makes it easier to see what the routine does and what data are affected. Source 6.12 shows another *MultBy8* routine that changes a global variable, but the variable to be changed is passed to the routine.

Source 6.12 A Routine That Uses Parameters to Affect a Global Variable

```
*************************************************************************
* Calling routine
*************************************************************************
              …
              ldx #Var               ; point to global
              jsr MultBy8            ; multiply by 8
              …
*************************************************************************
* Subroutine MultBy8
*    A routine that multiplies a memory location by 8.
*       Input requirements: IX points to the global.
*       Output characteristics: IX points to the result.
*************************************************************************
MultBy8       asl ,x                 ;Mult (IX) by 8
              asl ,x
              asl ,x
              rts
*************************************************************************
```

Here the *MultBy8* routine has been changed so the name of the global variable is passed to the routine instead of having the variable changed as a side effect. Not only is the routine better because it is easier to read, because parameters are used, the routine is also much more flexible. It can now be used to multiply any variable by 8, not just *Var*.

It must be noted that global variables are an efficient way to get data from one task to another in interrupt-driven and multitasking systems. In order to do this reliably, semaphores are used to control access and avoid reentrancy errors. Mutual exclusion by using semaphores is covered later in Chapter 16.

▶ SUMMARY

In this chapter, we have covered design tools, techniques, and conventions for designing and documenting reliable and reusable assembly programs. Some of the most important concepts we covered are the following:

- *Structured programming.* Always use structured constructs to create a structured program. A structured program has a single entrance and a single exit. It allows you to design the system hierarchically and avoid spaghetti code.
- *Document code with design tools.* Remember that the design tools are not only to assist in the design phase. They are also important as documentation. This means you must make sure to go back and change a flow diagram to reflect your new program. For pseudo-C, it means to use the pseudo-C code as part of your program's comments.
- *Structured constructs.* Use the structured constructs to design your program. Make sure you can convert the construct to assembly. This will make programming in assembly more like programming in a high-level language and the resulting program will be structured.

172

- *Carefully design your data objects.* Make sure you are using the appropriate data object for the application. The tendency at first is to use global variables. Global variables are appropriate for some but not most situations.
- *Design routines for reuse.* Always design your routines for reuse and a long life span. This means passing parameters instead of using global variables and designing routines with strong cohesion.
- *Follow parameter passing conventions.* If the parameters cannot be passed in registers, make sure to use the standard parameter passing and local variable conventions. This will result in more reliable code, more understandable code, and code that can interface with C code with little or no changes. Avoid ad hoc stack operations.

EXERCISES

1. Based on the assembly program in Figure 6.8, are the parameters *Temp* and *TSET* signed or unsigned numbers?
2. Suppose the thermostat program in Figure 6.10 must be repeated as long as the variable *ControlOn* is nonzero. Show the flow diagram and pseudo-C code for the loop. The thermostat code should be represented as the process *Control Temperature*.
3. Add a case to the command parser in Figure 6.12 that executes *CmdC* if a "C" is received. Show the flow diagram, pseudo-C code, and assembly code.
4. Show the flow diagram, the pseudo-C code, and the assembly code for a *while* loop that implements the same function as the *do-while* example in Figure 6.14.
5. Show the flow diagram, the pseudo-C code, and the assembly code to implement the memory-fill program in Figure 6.22 using a *while* loop construct.
6. A subroutine requires three local variables. *ptr1* and *ptr2* are 16 bits each, and *lvar* is eight bits. Write the stack allocation code and deallocation code for these variables. Show where the code is relative to the subroutine beginning and end. Also show the constant offset values for each variable.
7. A subroutine called *PUTSTRG* outputs the ASCII string that starts at the location passed in ACCD. Given the following definition, write the code to output *HaltMsg*.
   ```
   HaltMsg fcc 'System Down'
   ```
8. Design a memory-block move routine. The parameters are the starting address of the source block, *sblock,* the starting address of the destination block, *dblock,* and the block size, *bsize*. All parameters are 16 bits wide and must be passed on the stack. Show the flow diagram, the pseudo-C code, and the assembly source.

Assembly Applications

In this chapter we cover some common applications and techniques used when programming in assembly language. The routines and code snippets are not only presented as possible solutions for their applications, they are also presented to illustrate some of the program design concepts covered in Chapter 6. Some of the applications should be written in assembly, like the software delay routines. But many could also be written in C. For those programs the pseudo-C code is shown in the program headers. This not only helps document the program but also can be used to see how an assembly program would have been written in C.

▶ 7.1 SOFTWARE DELAY ROUTINES

So far all of our programs have run at full processor speed with no concern for synchronization with the outside world. Software delay routines are used to synchronize the execution of a program with respect to time. This might be required, for example, to temporarily display a message, to control the timing of a software serial transceiver, or to wait for an EEPROM to be erased. There are many applications in which the execution of a routine must be timed to meet external requirements. In general this is called *real-time programming,* which is covered in detail in the next chapter. In this section we look at the design of software delay routines to time the execution of a program.

At first it appears that software delays are important routines because most embedded designs require real-time programming. However, as we see in Chapter 8, there are better ways to implement real-time software, and once these other techniques are used, software delays do not work. The main problem with software delays is that their timing characteristics change if interrupts are used. The fact is that most embedded systems use interrupts, so for these systems software delay routines are useless unless they are used while interrupts are disabled. This usually means they are restricted to system initialization code. Despite these limitations, software delays are commonly found in simple embedded designs. By covering them at this point, we can also develop techniques for calculating the execution time of a program. This can be useful for other programs in which execution time is critical.

7.1.1 Instruction Timing

In order to design a software delay routine, we must know how to calculate the time it takes to execute the instructions that make up the routine. Instruction timing is covered in Section 4.3. To be able to use software delays, you must be using a CPU that has a deterministic instruction execution time. This excludes most CPUs that have a pipeline in which more than one instruction can be executed at a time. The problem with these systems is that the execution time depends on the state of the pipeline during run time. Fortunately the CPU12 uses an instruction queue, not a true pipeline. In the instruction queue, only one instruction is executed at a time so the execution times are deterministic. In Section 4.3, we found that the execution time for the CPU12 is

$$T_{ex} = N \times T_E$$

where

$$T_{ex} = \text{total execution time}$$

$$N = \text{the total number of clock cycles}$$

$$T_E = \text{the E-clock period, } 1/f_E$$

The frequency of the E-clock is one-half the crystal frequency found on the XTAL and EXTAL inputs. For the 68EVB912B32 EVB, the crystal frequency is 16MHz, so f_E is equal to 8MHz.

The assembly code determines the total number of clock cycles in the routine. Therefore knowing how long it takes to execute a routine also implies that you know what the assembly code is that makes up that routine. This means that it is not practical to create software delays routines in a high-level language. For example if we write a delay routine in C, we do not know what the compiler will generate without looking at the resulting assembly code. It is likely that one compiler will generate different assembly code than another compiler for the same C source. This would mean different delay times for different compilers. And if you have to count assembly instructions anyway, you might as well write the code in assembly in the first place.

7.1.2 Delay Routine Design

The overall structure of a program that uses a software delay is shown in Figure 7.1. The software delay routine is simply a routine that does nothing for a desired amount of

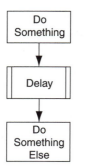

Figure 7.1 Structure of a Program That Uses a Software Delay

time. In normal operation the CPU is always running, so we need to have the CPU execute code that in effect does nothing but waste time.

To waste time, the delay routine executes a programmed loop that is executed the number of times required to make the total execution time equal to the required time delay. To design a delay routine, you must know the required delay time and accuracy. It is also wise to look at the requirements for the whole system so you can write one routine or a small set of related routines that meet all the requirements for the system.

Delay routines can be characterized by their range and resolution. The range is the maximum delay that can be achieved by the routine. The resolution is the time it takes for a single pass through the delay loop. We start with high-resolution short delays and then move on to longer delays. Source 7.1 shows a 100μs delay routine. Remember that a delay routine should in effect do nothing, so it should preserve any registers that it uses. This will also make the routine more versatile for use in long delay routines that use nested routines.

Source 7.1 A 100μs-Delay Subroutine

```
********************************************************************
* Subroutine call
********************************************************************
          jsr Dly100us        ;[4]


********************************************************************
* Subroutine Dly100us - 100us delay loop.
* Assuming 68HC12, E=8MHz,no clock stretching, 16-bit bus
*    -preserves all registers except CCR.
*    -requires 1 byte stack space
********************************************************************
TC100U    equ 197             ;Delay count for 100us delay
********************************************************************
Dly100us  psha                ;[2]
          ldaa #TC100U        ;[1]
d100ulp   deca                ;[1]
          bne d100ulp         ;[3/1]
          pula                ;[3]
          rts                 ;[5]


********************************************************************
```

To design this program so it has an execution time of 100μs, we need to calculate the total number of clock cycles as a function of $TC100U$, the number of times through the loop. The instructions outside the loop run one time, and the instructions inside the loop run $TC100U$ times. The last time through the loop, however, the *bne* instruction takes only one cycle because it does not branch. Therefore the *bne* instruction takes three cycles $TC100U - 1$ times and one cycle one time. The resulting equation for the total number of cycles is

$$N = 4 + 2 + 1 + (1 + 3)(TC100U - 1) + (1 + 1) + 3 + 5$$

The first set of parentheses represent the instructions in the loop when a branch is made. They will run $TC100U - 1$ times. The third set of parentheses shows the last cycle through the loop when the *bne* instruction takes only one cycle. The *jsr* instruction is also included because this is a high-resolution routine, and we want it to be as accurate as possible. This equation can be simplified to

$$N = 17 + 4(TC100U) - 4 = 13 + 4 \, (TC100U)$$

For a delay of 100μs we want

$$N = \frac{T_{ex}}{T_E} = \frac{100\mu s}{125 ns} = 800 \text{ cycles}$$

We can rearrange the equation and solve for $TC100U$:

$$TC100U = \frac{800 - 13}{4} = 196.75$$

We cannot have a noninteger count, so this must be rounded to 197 for the smallest error. With $TC100U$ equal to 197, the resulting delay is 100.125μs for an error of +0.125%. If we need more accuracy, we could adjust the subroutine using *nop* instructions to add one-cycle delays as required. In this case if we use three *nop* instructions outside the loop, the equation becomes

$$TC100U = \frac{800 - 16}{4} = 196$$

This time no rounding is required, so the delay will be as accurate as the CPU crystal frequency.

The routine in Source 7.1 has a small loop with a resolution of 500ns. The maximum delay is only 129.6μs, however, which can be realized by setting $TC100U$ to zero (256 times through the loop).

The following examples show how we can generate longer delays. The routine shown in Source 7.2 generates a 1ms delay. It uses the same structure as Source 7.1, but the range has increased by making use of a 16-bit register for the loop counter.

Source 7.2 A 1ms-Delay Subroutine

```
********************************************************************
* Subroutine call
********************************************************************
```

```
              jsr Dly1ms              ;[4]
****************************************************************
* Subroutine Dly1ms - 1ms delay loop.
* Assuming 68HC12, E=8MHz,no clock stretching, 16-bit bus
*      -preserves all registers except CCR.
*      -requires 2 bytes stack space
****************************************************************
TC1MS         equ 1996              ;Delay count for 1ms
****************************************************************
Dly1ms        pshx                  ;[2]
              ldx #TC1MS            ;[3]
d1mslp        dex                   ;[1]
              bne d1mslp            ;[3/1]
              pulx                  ;[3]
              rts                   ;[5]


****************************************************************
```

Because the instructions changed, we have to recalculate the total number of cycles as a function of the loop count *TC1MS*:

$$N = 15 + 4(TC1MS)$$

For a delay of 1ms, we need

$$N = \frac{T_{ex}}{T_E} = \frac{1\text{ms}}{125\text{ns}} = 8,000 \text{ cycles}$$

This requires a loop count *TC1MS* of

$$TC1MS = 1,996.25 \rightarrow 1,996$$

The rounding error is –0.0125%. This routine actually has the same resolution, 500µs, but it has a higher range. The maximum delay for this program is 32.7699ms, which occurs if the loop is executed 65,536 (*TC1MS* = 0) times.

For longer delays nested subroutines should be used. Source 7.3 shows a 1-second-delay subroutine that uses the 1ms-delay routine from Source 7.2.

Source 7.3 A 1-Second-Delay Subroutine Using Nested Subroutines

```
****************************************************************
* Subroutine Dly1sec - 1s delay loop.
* Assuming 68HC12, E=8MHz,no clock stretching, 16-bit bus
*      -preserves all registers except CCR.
*      -requires 6 bytes stack space
****************************************************************
TC1S          equ 1000     ;Delay count for 1ms
****************************************************************
Dly1sec       pshx         ;[2]
              ldx #TC1S    ;[2]
```

```
d1slp        jsr Dly1ms   ;[7999]
             dex          ;[1]
             bne d1slp    ;[3/1]
             pulx         ;[3]
             rts          ;[5]
```

****

As the delay gets longer, it is normally less important to count every instruction. For a 1-second delay, the total number of cycles should be 8,000,000. When the total is this large, each instruction represents a very small percentage of the total delay. So typically some instructions are ignored to simplify the process. In the previous example it is obvious that you can make a 1-second delay by calling the 1ms delay 1,000 times. The actual number of clock cycles in this program, including a subroutine call, is 8,003,014. This results in an actual delay of

$$T = 125\text{ns} \ (8,003,014) = 1.000377\text{s}$$

which results in an error of only +0.0377%. This is more than adequate for most applications. Applications that result in this error accumulating for a long time, however, such as a real-time clock, could not use this routine. With this error, a real-time clock could be off by more than two hours in six months.

Another method for creating long delays is to use nested loops. Nested loops have no advantage over nested subroutines, however, and are more difficult to design and use. Nested subroutines also result in a better design because there are multiple delay routines that can be used. A system that uses the 1-second-delay example would have both a 1-second-delay routine and a 1ms-delay routine available.

It turns out that the 1ms-delay routine in Source 7.2 is a useful routine. Many applications require time delays that are a multiple of 1ms. Source 7.4 shows a more flexible version of the routine that has the delay time in milliseconds passed in ACCD.

Source 7.4 A Programmable Delay Subroutine

```
**********************************************************************
* Subroutine WaitDms - A programmable delay in ms.
* Assuming 68HC12, E=8MHz,no clock stretching, 16-bit bus
*     -The number of ms is passed in ACCD.
*     -preserves all registers except CCR.
*     -requires 6 bytes stack space
**********************************************************************
WaitDms      pshd
msdlp        jsr Dly1ms
             subd #1
             bne msdlp
             puld
             rts


**********************************************************************
```

As you can see, by using nested subroutines it is easy to create flexible delay routines. Again, it is not often that we have the opportunity to use these routines, because most of our systems will have interrupts.

▶ 7.2 I/O DATA CONVERSIONS

In this section we look at the algorithms and code examples for detecting and converting data types normally used in small embedded systems. These data types include ASCII, BCD, and binary. ASCII is the standard for sending text characters from one computer or peripheral to another. BCD is used to represent decimal data, which is normally required in a human interface. And of course binary is the normal data type for the processor.

Most of the routines in this section can be found in the *Basic I/O* module in Appendix B. Most of these routines also rely on the serial port driver routines *sci_open, sci_write,* and *sci_read.* These routines will be covered in Chapter 9.

7.2.1 ASCII Conversions

ASCII.   The ASCII character set is a standard for representing text characters with binary codes. In our case we focus on the 7-bit ASCII character set, so each character is represented by a 7-bit binary number. The 7-bit ASCII table can be found in Appendix B.

ASCII character codes are standard for virtually all computer text. If you need your system to interface with a PC, an LCD, or a terminal, you will normally transmit the characters in ASCII. The ASCII character set not only includes printable characters, it also includes control characters used for cursor positioning, handshaking, graphics, and beeps.

Character Detection.   Let's start with some simple routines to detect different types of ASCII characters. The first routine, shown in Figure 7.2, emulates the standard C function

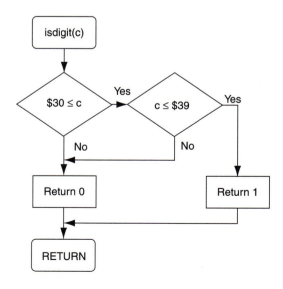

Figure 7.2   Flow Diagram for *isdigit(c)*

isdigit(c). It returns a nonzero value if the ASCII character is a digit (0..9). Referring to the ASCII table in Appendix B, we can see that an ASCII character is a digit if its value falls between $30 and $39.

The assembly code shown in Source 7.5 implements this routine. In the code, notice that instead of using the hexadecimal number for the character, we use the character itself in single quotation marks. This is the standard assembler notation for ASCII. For example, instead of using $30 for zero, we use '0'. We will use this notation from now on, because it makes the flow diagrams and source code easier to read and understand. We do not need to look it up in the ASCII table to understand what the code is doing.

Source 7.5 Assembly Source for isdigit(c)

```
;**********************************************************************
;  isdigit(c)
;
;  Description: A routine to check if an ASCII character is a
;               digit (0-9).
;  Arguments: Character is passed in ACCB.
;             Return value is passed in ACCB and Z bit of CCR.
;  Pseudo-C:
;  UBYTE isdigit(c){
;     if('0'<= c <= '9'){
;        return 1;
;     else
;        return 0;
;  }
;**********************************************************************
isdigit       cmpb #'0'
              blo isd_not
isd_9         cmpb #'9'
              bhi isd_not
              ldab #1
              bra isd_rtn
isd_not       clrb
isd_rtn       rts


;**********************************************************************
```

Another similar routine that is useful is *ishex(c)*. This routine checks to see if an ASCII character is a hexadecimal character. Its flow diagram is shown in Figure 7.3 and its assembly code is shown in Source 7.6. Notice that there are two sets of alphabetic characters, uppercase and lowercase. The flow diagram and source code shown check for both uppercase and lowercase characters. We could also use the *toupper(c)* routine shown later to change all alpha characters to uppercase before using them. This would simplify routines like *ishex(c)*, because they would only have to check for uppercase characters. However, it would also result in a system that is not case sensitive.

Because there are several decision blocks in this program, it takes careful thought and planning to avoid extra branches.

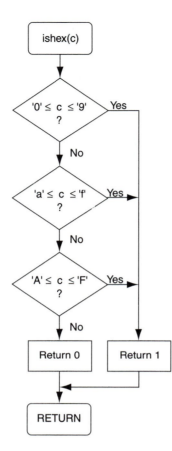

Figure 7.3 Flow Diagram for *ishex(c)*

Source 7.6 Assembly Code for ishex(c)

```
;****************************************************************************
; ishex(c)
;
; Description: A routine to check if an ASCII character is a
;              hexidecimal character (0..9),(a..f), or (A..F).
; Arguments: Character is passed in ACCB.
;            Return value is passed in ACCB and Z bit of CCR.
;
; Pseudo-C:
; UBYTE ishex(UBYTE c){
;     if(('0' <= c <= '9') || ('A' <= c <= 'f') || ('a' <= c <= 'F')){
;         return 1;
;     }else{
;         return 0;
;     }
; }
;
;****************************************************************************
```

```
ishex           cmpb #'0'              ;Check if between '0' and '9'
                blo isnoth
                cmpb #'9'
                bls ish
                cmpb #'A'              ;check if between 'A' and 'F'
                blo isnoth
                cmpb #'F'
                bls ish
                cmpb #'a'              ;check if between 'a' and 'f'
                blo isnoth
                cmpb #'f'
                bls ish
isnoth          clrb                  ;Not hex, return 0
                bra ish_rtn
ish             ldab #1               ;is hex, return 1
ish_rtn         rts
```

;***

ASCII conversion is accomplished by a simple mapping process that involves addition or subtraction. We can convert a lowercase character to uppercase, convert a character to BCD, and convert a character to hexadecimal. We also can see how to convert BCD and hexadecimal characters to ASCII.

Converting to Uppercase. The first conversion is an ASCII-to-ASCII conversion that changes the case of a lowercase character to uppercase. This routine emulates the C function *toupper(c)*. The flow diagram is shown in Figure 7.4, and the assembly code is shown in Source 7.7.

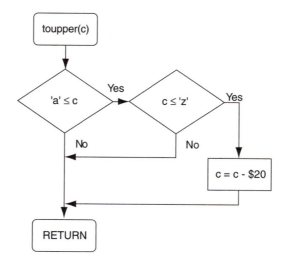

Figure 7.4 Flow Diagram for *toupper(c)*

The program first determines whether the character is a lowercase alphabetic character. If it is, it subtracts $20 from the character. Referring to the ASCII table, we can see that the difference between all lowercase characters and their equivalent uppercase character is $20. Sometimes this is expressed in a program as

'a' – 'A'

Source 7.7 Assembly Code for toupper(c)

```
;**********************************************************************
; toupper(c)
;
; Description: A routine that converts lowercase alpha characters to
;              uppercase. All other characters remain unchanged.
; Arguments: Character is passed in ACCB.
;            Return value is passed in ACCB.
; Pseudo-C:
; UBYTE toupper(UBYTE c){
;    if('a' <= c <= 'z'){
;        c = c - 0x20;
;    }
; }
;**********************************************************************
toupper        cmpb #'a'
               blo tou_rtn
               cmpb #'z'
               bhi tou_rtn
               subb #$20
tou_rtn        rts

;**********************************************************************
```

7.2.2 BCD and Hex Conversions

Binary coded decimal, BCD, is used when decimal data, usually from a human interface, is used. Each BCD digit is represented by a 4-bit binary number ranging from zero to nine. The values 10 through 15 are considered illegal. We can store BCD numbers in two ways, packed and unpacked. Figure 7.5 shows how the number 36 is stored as packed and unpacked BCD. Packed BCD numbers require less storage space and result in programs

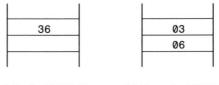

(a) Packed BCD 36   (b) Unpacked BCD 36

Figure 7.5 Packed and Unpacked BCD Digits

that are more efficient. Unpacked BCD numbers are the direct result of an ACSII-to-BCD conversion, because each ASCII digit corresponds to one BCD digit.

To pack or unpack BCD digits, a sequence of masking and shifting is used. Source 7.8 shows example code snippets for both packing and unpacking. In these two examples the unpacked BCD digits are pointed to by IX, and the packed BCD digit is in ACCB. Certainly the stack could also be used to hold the unpacked digits and the code would be essentially the same. It is best to perform the shifting in ACCA or ACCB because it requires only one clock cycle. It takes at least three cycles to shift a memory location.

Source 7.8 Assembly Code for Packing and Unpacking BCD Digits

```
**********************************************************************
* packbcd
*
* Description: Two unpacked digits pointed to by IX (ubcdu:ubcdl) are
*              converted to a packed bcd digit (bcd) in ACCB.
* Pseudo-C:
* bcd = *ubcdl | (*ubcdu<<4);
**********************************************************************
            ldab 0,x
            lslb
            lslb
            lslb
            lslb
            orab 1,x


**********************************************************************
* unpackbcd
*
* Description: One packed digit in ACCB is unpacked into two bytes
* pointed to by IX (ubcdu:ubcdl).
* Pseudo-C:
* *ubcdl = bcd & 0x0f;
* *ubcdu = bcd >> 4;
*
**********************************************************************
            tfr b,a
            anda #$0f
            staa 1,x
            lsrb
            lsrb
            lsrb
            lsrb
            stab 0,x
**********************************************************************
```

The packing and unpacking process is also required for hex-to-binary conversions. Source 7.16 and Source 7.18 contain code examples that require hexadecimal packing or unpacking.

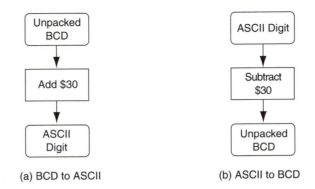

(a) BCD to ASCII   (b) ASCII to BCD

Figure 7.6   Basic Unpacked BCD ↔ ASCII Conversion Processes

ASCII↔BCD Conversions.   Recall that ASCII characters are a standard for data received from a computer system. If a user is entering decimal numbers, the system will receive the ASCII characters for the digits, not the decimal digits themselves. So to use the digits, we must convert ASCII to BCD. Figure 7.6 shows the basic process involved in the conversion. To go from ASCII to BCD, we subtract $30; to go from BCD to ASCII, we add $30. These are very straightforward and simple conversions.

Hex↔ASCII Conversions.   Most of the time it is preferable to use decimal numbers for a human interface. If computer-related parameters are being transferred, however, it may be more appropriate to use hexadecimal numbers. For example if a memory address needs to be entered, then you would use hex numbers. Hexadecimal numbers are also easier to convert to binary because each hex digit corresponds directly to a binary nibble.

Figure 7.7 shows the processes and Source 7.9 shows the code required for converting between an ASCII character and a hexadecimal nibble. The hex-to-ASCII conversion process

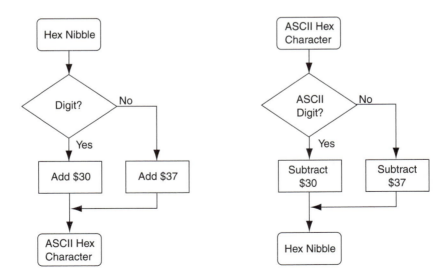

Figure 7.7   Basic Hex ↔ ASCII Conversion Processes

186   Chapter 7

generates an uppercase hex character. To generate a lowercase character, we would add $57 to the alphabetic nibbles instead of $37. The ASCII-to-hex process shown in Figure 7.7 assumes that the ASCII characters are all uppercase. The *htob()* routine in Source 7.9 also accepts lowercase characters. If the hex character is *'a'* through *'f'*, the program subtracts $57 instead of $37. Another way to handle this is to convert all characters to uppercase with *toupper(c)* and then use the simplified process shown in Figure 7.7.

The *htoa()* routine in Source 7.9 assumes that ACCB contains a zero in the high-order nibble, so the binary word must be unpacked before it is converted.

Source 7.9 Routines That Convert Hex-to-ASCII and ASCII-to-Hex

```
;*********************************************************************
; htoa()
; Description: Converts one binary nibble to uppercase ASCII.
; Arguments: Nibble passed in ACCB (LCN)
;            ASCII character returned in ACCB
; Pseudo-C:
; UBYTE htoa(UBYTE bin){
;    if(bin <= 9){
;        return(bin + 0x30);
;    }else{
;        return(bin + 0x37);
;    }
; }
;*********************************************************************
htoa          cmpb #9
              bhi ha_alpha
              addb #$30
              bra ha_rtn
ha_alpha      addb #$37
ha_rtn        rts


;*********************************************************************
; htob(c)
;
; Description: A routine converts an ASCII hex character to binary.
;              Assumes a legal hex character (run ishex() first).
; Arguments: The ASCII digit is passed in ACCB.
;            Returns the binary nibble in ACCB.
; Pseudo-C:
; UBYTE htob(UBYTE c){
;    if('0' <= c <= '9') {
;        bin = c - '0';
;    }else if('a' <= c <= 'f'){
;        bin = c - 'a';
;    }else if('A' <= c <= 'F'){
;        bin = c - 'A';
;    }
;    return bin;
; }
```

```
;*********************************************************************
htob        cmpb #'0'        ;is digit?
            blo hb_rtn
            cmpb #'9'
            bhi hb_lw
            subb #'0'        ;subtract $30
            bra hb_rtn
hb_lw       cmpb #'A'        ;is uppercase hex?
            blo hb_rtn
            cmpb #'F'
            bhi hb_up
            subb #$37        ;subtract $37
            bra hb_rtn
hb_up       cmpb #'a'        ;is lowercase hex?
            blo hb_rtn
            cmpb #'f'
            bhi hb_rtn
            subb #$57        ;subtract $57
hb_rtn      rts

;*********************************************************************
```

7.2.3 Binary Conversions

So far our conversions have involved a relatively simple mapping process. To convert between decimal digits and binary numbers, however, we need a more complex process. Typically if data are passed between a user interface and a computer, the conversions required are

$$\text{ASCII} \leftrightarrow \text{BCD} \leftrightarrow \text{Binary}$$

The ASCII characters are first converted to BCD, and BCD is then converted to binary. To convert BCD to binary, we need to use the following equation:

$$N_2 = BCD_0 10^0 + BCD_1 10^1 + \ldots + BCD_{N-1} 10^{N-1}$$

where $BCD_0 \ldots BCD_{N-1}$ are the digits for an N-digit BCD number and N_2 is the binary equivalent.

For example a three-digit BCD number can be described by

$$N_2 = BCD_0 10^0 + BCD_1 10^1 + BCD_2 10^2$$

Or to fit a simple computer algorithm, this equation can be manipulated as follows:

$$N_2 = BCD_0 + (BCD_1 + BCD_2 \cdot 10)10$$

BCD-to-Binary. From this equation we can see that converting BCD to binary involves an iterative process of multiplying by 10 and adding each BCD digit from the most-significant digit to the least. Figure 7.8 shows a flow diagram to implement the equation to convert an N-digit BCD number to binary.

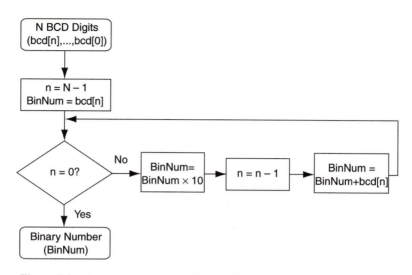

Figure 7.8 General Process for *N*-Digit BCD Number to Binary Conversion

TABLE 7.1 NUMBER OF BCD DIGITS THAT
CAN BE HELD BY A BINARY WORD

Binary Word Size	BCD Digits
8 bits	1,2
16 bits	3,4
24 bits	5,6,7

In Figure 7.8 the number of BCD digits, *N*, is an unknown. For a real routine we must at least know the maximum number of digits to be converted in order to know the required data size of the binary result. Table 7.1 shows the number of BCD digits that can be held in a given binary word size. This is assuming that all the digits can range from 0 to 9. There are applications in which this is not the case. For example a parameter may range from 0 to 100, which would require three BCD digits but only an 8-bit binary word.

Source 7.10 shows an assembly program that converts three ASCII digits to one binary byte. The routine also has some extra functionality. It is designed to be used with any string, so it must check for nondigit characters and check for strings that result in a binary value greater than 255.

Source 7.10 Three-Digit Decimal String to 1-Byte Binary Conversion

```
;********************************************************************
; decstob()
;
; Description: A routine that converts a string of decimal
;              characters to a binary byte until a CR, white space,
;              or a NULL is encountered.
```

```
; Arguments: Pointer to the string is passed in ACCD
;            Pointer to binary byte is passed on the stack
;            Error code is returned in ACCB
;                0 -> No Error
;                1 -> Nondecimal character
;                2 -> Too large
;
; Pseudo-C:
; UBYTE decstob(UBYTE *strg,UBYTE *bin){
;     if(isdigit(*strg)){
;         *bin = (*strg-0x30);
;     }else if(*strg == NULL){
;         return 0;
;     }else{
;         return 1;
;     }
;     strg++;
;     while(*strg != NULL || *strg != ' ' || *strg != '\t'){
;         if(isdigit(*strg)){
;             *bin = *bin * 10 + (*strg-0x30);
;             if(*bin > 255){
;                 return 2;
;             }
;         }else{
;             return 1;
;         }
;         strg++;
;     }
;     return 0;
; }
;************************************************************************
decstob     pshx            ;preserve IX
            tfr d,x         ;strg pointer->IX
            clr [4,sp]      ;clear bin
            ldab 0,x        ;get first digit
            cmpb #NULL       ;if NULL done
            beq dsb_finlp
            jsr isdigit     ;if not digit error 1
            cmpb #0
            beq dsb_err1
            ldab 0,x
            subb #$30       ;convert char to binary
            stab [4,sp]
            inx             ;point to next char
dsb_lp      ldab 0,x
            cmpb #NULL       ;finished if NULL or white space
            beq dsb_finlp
            cmpb #' '
            beq dsb_finlp
            cmpb #TAB
```

```
            beq dsb_finlp
            jsr isdigit        ;error if not digit
            beq dsb_err1
            ldab [4,sp]        ;bin*10+bcdx
            ldaa #10
            mul
            tsta               ;check for overflow
            bne dsb_err2
            ldaa 0,x
            suba #$30          ;convert net digit to bcd
            aba
            bcs dsb_err2       ;check for overflow
            staa [4,sp]
            inx                ;point to next digit
            bra dsb_lp
dsb_finlp   clrb               ;done no error
            bra dsb_rtn
dsb_err1    ldab #1            ;nondigit error
            bra dsb_rtn
dsb_err2    ldab #2            ;too-large error
dsb_rtn     pulx               ;recover IX
            rts


;**********************************************************************
```

This routine starts by setting the binary byte equal to the most-significant BCD digit after it has been converted from ASCII. Then it enters a loop that multiplies the current binary byte by 10 and then adds the next BCD digit. This loop implements an iterative execution of the equation

$$bin = bin \times 10 + bcd_n$$

Notice this is another way to implement the equation for a BCD to binary conversion. To multiply the binary byte by 10, the CPU12's *mul* instruction is used. Between each multiplication and addition, the result is checked to see if it is too large. Since the *mul* instruction's result is placed in ACCD, and the result must be small enough to fit in a single byte, ACCA is checked to verify that it is zero. If not, the result was greater than 255. The result is also checked for a result greater than 255 after the addition.

Binary-to-BCD. To convert a binary number to BCD, we use an iterative process involving a divide by 10. Figure 7.9 shows the general process for converting a binary number to *N* BCD digits. The binary number *BinNum* is divided by 10. The quotient becomes the new *BinNum* and the remainder is the next most-significant BCD digit.

Source 7.11 is an example of a routine to convert a 1-byte binary word to a three-digit BCD array. This example uses the CPU12 *idiv* instruction to perform the division. The *idiv* instruction places the quotient in IX and the remainder in ACCD. The pseudo-C code is a bit cryptic and inaccurate because it shows both a division operation and a modulus operation. Actually, since the *idiv* instruction produces both the quotient and the remainder, only a single division operation is required.

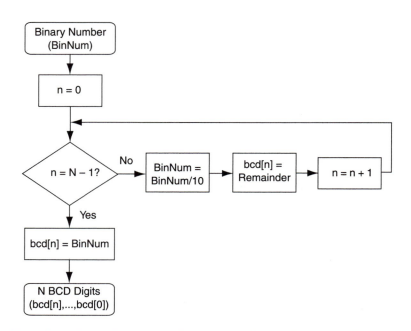

Figure 7.9 General Process for a Binary to an *N*-Digit BCD Number Conversion

Source 7.11 One-Byte Binary to Three-Digit BCD Array Conversion

```
;************************************************************************
; btod()
;
; Description: One binary byte is converted to three unpacked digits
;              in bcd[] (bcd2:bcd1:bcd0).
; Arguments: Binary byte is passed in ACCB.
;            Address of bcd[0] is passed on the stack.
; Pseudo-C:
; void btod(UBYTE bin, UBYTE *bcd){
;     bcd[2] = bin%10;
;     bin = bin/10;
;     bcd[1] = bin%10;
;     bcd[0] = bin/10;
; }
;************************************************************************
btod          pshy          ;preserve IY, IX, and ACCA
              pshx
              psha
              ldy 7,sp      ;Array pointer -> IY
              clra
              ldx #10       ;divide bin by 10
              idiv
              stab 2,y      ;remainder = bcd0
              tfr x,d       ;divide quotient by 10
```

```
                ldx #10
                idiv
                stab 1,y      ;remainder = bcd1
                tfr x,d
                stab 0,y      ;quotient = bcd2
                pula          ;recover IY, IX, and ACCA
                pulx
                puly
                rts
```

;***

If the CPU does not have a division instruction, we have to use a different technique to do the divide by 10. Source 7.12 shows another version of the *btod()* program that does not use a division instruction. In this case division is performed by multiple subtraction for each digit. Each time 100 or 10 is subtracted, the corresponding BCD digit is incremented.

Source 7.12 Binary Byte to BCD Converter Code
without Using a Division Instruction

```
*********************************************************************
* btod()
*
* Description: One binary byte is converted to three unpacked digits
*              in bcd[](bcd2:bcd1:bcd0). This version does not
*              use a division instruction.
* Arguments: Binary byte is passed in ACCB.
*            Address of bcd[0] is passed on the stack.
* Pseudo-C:
* void btod(UBYTE bin,UBYTE *bcd){
*      bcd2 = 0;
*      bcd1 = 0;
*      bcd0 = 0;
*      while(bin >= 100){
*          bin -= 100;
*          bcd2++;
*      }
*      while(bin >= 10){
*          bin -= 10;
*          bcd1++;
*      }
*      bcd0 = bin;
*}
*********************************************************************
btod        pshx                ;preserve IX
            ldx 4,sp            ;bcd pointer -> IX
            clr 0,x             ;clear BCD digits
            clr 1,x
```

```
                clr 2,x
DoHnds          cmpb #100              ;Count hundreds
                blo DoTens
                subb #100
                inc 0,x
                bra DoHnds
DoTens          cmpb #10               ;Count tens
                blo DoOnes
                subb #10
                inc 1,x
                bra DoTens
DoOnes          stab 2,x               ;store ones
                pulx                   ;recover IX
                rts
************************************************************************
```

One problem with this routine is the wide range of execution times. If the binary number is between zero and nine, the routine is fast because there is no looping required. If the binary value is 190 through 199, however, the program ends up looping 10 times and is relatively slow. There are also other methods for converting binary to BCD, but they all result in code that is not as efficient as the code in Source 7.11 or Source 7.12.

▶ 7.3 BASIC I/O ROUTINES

The first subroutine library used by most programmers is a library to perform basic I/O tasks via the serial port. Most of the time these libraries are provided as part of a debug monitor program or as a high-level language's I/O library. The routines described here are similar to those provided by a monitor and the standard C I/O library. A complete listing for this library is shown in Appendix B.

We are all probably familiar with the "Hello world" example. It is often the first programming example in compiler manuals or textbooks. Here is the typical "Hello world" example written in C.

```
#include <stdio.h>
int main(int argc, char **argv){
    printf("Hello world\n");
}
```

This example uses the *printf()* function to output the string *Hello world*. As you can imagine, outputting a string to a user display is a common task. The problem with this program when using embedded microcontrollers is that the resulting code is extremely inefficient. The code above results in more than 6,000 bytes of code space! This is because *printf()* was designed to do a lot more than output constant strings. Using *printf()* to output a constant string is an example of overdesign. It is like pounding in a furniture tack with a sledgehammer. Using *printf()* is commonplace when writing computer code, and it can be appropriate for some embedded systems. But for small embedded systems that only require basic I/O functionality, it is inappropriate.

In this section, we look at the design of a module of basic I/O routines that will be more practical for embedded microcontrollers. Some of these routines are similar to the routines provided by the BUFFALO monitor, whereas others are more similar to some of the low-level routines found in the C *stdio* library. The D-Bug12 monitor also provides access to some I/O routines, but they are inefficient. In fact the basic I/O module shown in Appendix B is designed to be loaded into the byte-erasable EEPROM in an M68EVB912B32 EVB as an alternative to the D-Bug12 routines. Since these routines use the 68HC12's serial port, SCI, they directly or indirectly use *sci_open, sci_read(), and sci_write()*. These routines as well as the 68HC12's SCI are described in Chapter 9.

7.3.1 Character I/O

To receive or transmit characters through the serial port, we can use the SCI routines *sci_read()* and *sci_write()*, which are described in Chapter 9. In this section we also look at another character input routine, *getchar()*.

getchar(). *sci_read()* checks the serial port one time and returns either the character received or zero if no character had been received. Most of the time our program must wait for a character to be received. Source 7.13 shows the assembly routine for *getchar()*. It is a polling loop that calls *sci_read()* until a character is received. Note it is a blocking routine because it does not return until a character is received. This can present a problem when multiple tasks need to run concurrently in the system.

Source 7.13 Assembly Code for getchar()

```
;**********************************************************************
; getchar() - Read input until character is received.
;             Returns character in ACCB.
;Pseudo-C:
; UBYTE getchar(void){
;     while(1){
;         c = sci_read();
;         if(c != 0){
;             return c;
;         }
;     }
; }
;**********************************************************************
getchar      jsr sci_read          ;wait for character
             cmpb #0
             beq getchar
             rts


;**********************************************************************
```

We could also define the function *putchar()* to send a character out the serial port. For the applications in this chapter, however, the function of *putchar()* is identical to *sci_write()*, so it will not be shown. In all the I/O routines that follow, *putchar()* is used instead of *sci_write()*.

7.3.2 String I/O

A *string* is an array of ASCII characters. A terminating character is placed at the end of the array to indicate the end of the string. In this text we follow the C string standard and place a NULL or zero at the end of a string. By using a NULL terminator, the string can be interpreted by the C string functions. Unfortunately the BUFFALO monitor uses an EOT character ($04), so if you are working with some old 68HC11 programs, you may have to change the terminator.

putstrg(). The first routine, *putstrg()*, outputs a NULL terminated string. Source 7.14 shows the code for *putstrg()*. The program is a loop that outputs each character until a NULL is reached.

Source 7.14 Assembly Code for putstrg()

```
;**********************************************************************
; putstrg() - Output string to sci.
;    ACCD contains pointer to NULL terminated string
;    Differs from standard C puts() because it does not send '\n'
;    Returns with ACCD pointing to the byte following the NULL
;    All other registers but CCR preserved
; Pseudo-C:
; UBYTE *putstrg(UBYTE *strg){
;     while(*strg != NULL){
;         putchar(*strg);
;         strg++;
;     }
; }
;**********************************************************************
putstrg      pshx               ;preserve IX
             tfr d,x            ;strg pointer->IX
ps_nxt       ldab 1,x+
             cmpb #NULL         ;put until NULL
             beq puts_rtn
             jsr putchar
             bra ps_nxt
puts_rtn     tfr x,d
             pulx               ;recover IX
             rts

;**********************************************************************
```

To use *putstrg()*, we need to load the pointer to the string in ACCD. Here is a program to output "Hello world."

```
             ldd #HelloMsg
             jsr PUTSTRG
             ...
HelloMsg     fcc 'Hello world'
             fcb 0
```

This example shows the importance of choosing the appropriate routine to perform a function. The *printf()* function took about 6,000 bytes of code. *putstrg()* took about 20! It is not the fact that *printf()* is written in C and *putstrg()* is assembly that causes the difference in the amount of code. The answer lies in selecting the correct function. The C function *puts()* gives almost the same result as *putstrg()*.

getstrg(). In order to input a string of characters, we need a more complex routine. Source 7.15 shows the code for a *getstrg()* routine that reads the characters received and saves them into an array. It will continue to input characters until a carriage return is received or the string length parameter is exceeded. *getstrg()* uses the parameter *strglen* to limit the size of the string, which determines the amount of memory to allocate for the array. The number of characters indicated in *strglen* should include the NULL character that will be placed at the end of the string. The other argument to *getstrg()* is a pointer to the array. *getstrg()* also handles the backspace key so the user can replace mistyped entries. The last thing *getstrg()* does is replace the carriage return at the end of the string with a NULL (zero). Now the saved string is a NULL terminated string and can be used with any string function in the Basic I/O module or any standard C string function.

Source 7.15 Assembly Code for getstrg()

```
;*****************************************************************************
; getstrg()
;
; Description: A routine that inputs a character string to an array
;              until a carriage return is received or strglen is exceeded.
;              Only printable characters are recognized except carriage
;              return and backspace.
;              Backspace erases displayed character and array character.
;              A NULL is always placed at the end of the string.
;              All printable characters are echoed.
;              Return vaule:
;                   0 -> ended with CR
;                   1 -> if strglen exceeded.
; Arguments: Pointer to array is passed in ACCD.
;            strglen is passed on the stack. strglen includes CR/NULL.
;            Return value is passed in ACCB.
; Pseudo-C:
; UBYTE getstrg(UBYTE *strg, UBYTE strglen){
;     charnum = 0;
;     while(c = getchar(),(c != '\r') && (charnum < (strlen-1))){
;         if(' ' <= c <= '~'){
;             putchar(c);
;             *strg = c;
;             strg++;
;             charnum++;
;         }else if(c = '\b' && charnum > 0){
;             putchar('\b');
;             putchar(' ');
;             putchar('\b');
;             strg--;
```

```
;            charnum--;
;          }
;        }
;      outcrlf();
;      *strg = NULL;
;      if(c == '\r'){
;          return 0;
;      }else{
;          return 1;
;      }
; }
;*********************************************************************
getstrg     pshx              ;preserve IX
            tfr d,x           ;strlen -> 4,sp;  strg -> IX
            clra              ;charnum -> ACCA;  c -> ACCB
gs_nxt      jsr getchar       ;Character input
            cmpb #CR          ;done if CR
            beq gs_finish
            cmpb #' '         ;Check if printable
            blo gs_chkbs
            cmpb #'~'
            bhi gs_nxt
            inca              ;increment charnum
            cmpa 4,sp         ;break out if too many chars
            beq gs_finish
            jsr putchar       ;echo char
            stab 1,x+         ;store char and increment strg
            bra gs_nxt        ;get next character
gs_chkbs    cmpb #BS          ;Check for backspace
            bne gs_nxt
            tsta              ;ignore if no chars yet
            beq gs_nxt
            jsr putchar       ;erase displayed char
            ldab #SPACE
            jsr putchar
            ldab #BS
            jsr putchar
            dex               ;decrement strg and charnum
            deca
            bra gs_nxt        ;get next character
gs_finish   jsr outcrlf       ;echo CR
            clr 0,x           ;place NULL at end of array
            cmpb #CR          ;determine return value
            bne gs_ovrfl
            clrb              ;return 0 if CR
            bra gs_rtn
gs_ovrfl    ldab #1           ;return 1 if overflow
gs_rtn      pulx              ;recover IX
            rts

;*********************************************************************
```

To call *getstrg()* we must allocate memory for the array and set up the parameters. The following code will input and store a string that has as many as four characters:

```
STRGLEN          equ 5
                 ...
                 ldab #STRGLEN
                 pshb
                 ldd #strgbuf
                 jsr GETSTRG
                 leas 1,sp
                 ...
strgbuf          rmb STRGLEN
```

Notice that the value of *STRGLEN* must also include the NULL that ends the string. After *getstrg()* is called, the stack space used for passing *STRGLEN* must be deallocated. This is important to remember. If the *leas* instruction is removed, the stack will become un-balanced.

This *getstrg()* function is useful for user responses that involve a single argument. If the user must enter multiple arguments, however, the function *slicestr()* can be used to split the string into tokens that can then be evaluated individually. *slicestr()* is found in the Basic I/O module in Appendix B.

7.3.3 Data Input and Output

In this section we look at some routines used for hexadecimal and decimal data I/O. These routines combine the character conversion functions with character or string I/O functions.

Hexadecimal Output.   Source 7.16 shows two routines, *outhexb()* and *outhexw(),* that output hex digits for a single byte or a 16-bit word. *outhexb()* unpacks the byte, converts each nibble to ASCII, and outputs the characters, most-significant digit first.

Source 7.16 Hexadecimal Output Routines

```
;************************************************************************
;
; outhexb() - Output one hex byte.
;    ACCD contains pointer to byte to be sent.
;    Returns with ACCD pointing to the next byte in memory.
;    All other registers but CCR preserved.
; Pseudo-C:
; UBYTE *outhexb(UBYTE *bin){
;     putchar(htoa(*bin >> 4));
;     putchar(htoa(*bin & 0x0f));
; }
;************************************************************************
;
outhexb     pshx            ;preserve IX
            tfr d,x         ;pointer to bin -> IX
```

```
            ldab  0,x          ;put most sig. nibble
            lsrb
            lsrb
            lsrb
            lsrb
            jsr htoa           ;convert to ASCII
            jsr putchar
            ldab  0,x          ;put least sig. nibble
            andb  #$0f
            jsr htoa
            jsr putchar
            inx                ;point to next byte
            tfr x,d
            pulx               ;recover IX
            rts
```

```
;*********************************************************************
; outhexw() - Outputs two hex bytes.
;    ACCD contains pointer to displayed word.
;    Returns with ACCD pointing to the next byte in memory.
;    All other registers but CCR preserved.
;Pseudo-C:
; UWDYE *outhexw(UWYDE *bin){
;    outhexb(bin[0]);
;    outhexb(bin[1]);
; }
;*********************************************************************
outhexw     jsr outhexb
            jsr outhexb
            rts
```

```
;*********************************************************************
```

To output a 16-bit word, *outhexw()* simply calls *outhexb()* two times. This can be done because *outhexb()* returns the next address in ACCD, which is then passed to the second *outhexb()*.

Decimal Output.   To output a binary value in decimal is a bit more complex because we have to make a binary-to-decimal conversion. Source 7.17 shows two routines, *outdecb()* and *outdecw()*, to output a single byte or a 16-bit word in decimal.

outdecb() must first convert the binary byte into an array of three BCD digits. Three bytes of stack space are allocated for this array so we do not have to use global variables. The function *btod()* then is used to make the binary-to-decimal conversion. When displaying data in decimal, it is normally desired to remove leading zeros. *outdecb()* skips past any leading zeros and then converts the BCD digits to ASCII and outputs them. When writing code to delete leading zeros, be careful it does not delete all zeros.

200   Chapter 7

Source 7.17 Decimal Output Routines

```
;*********************************************************************
; outdecb()
;
; Description: A routine that outputs the decimal value of one byte.
; Arguments: Pointer to byte is passed in ACCD.
;
; Pseudo-C:
; void outdecb(UBYTE *bin){
;     bcd[] = btod(*bin);
;     i = 0;
;     while(bcd[i] == 0 && i < 2){
;          i++;
;     }
;     while(i < 3){
;          putchar(bcd[i] + 0x30);
;          i++;
;     }
; }
;*********************************************************************
outdecb      pshx                ;preserve IX
             leas -3,sp          ;allocate space for BCD digits
             tfr d,x             ;get binary byte
             ldab 0,x
             tfr sp,x            ;convert bin to BCD
             pshx
             jsr btod
             leas 2,sp
             ldaa #0
odb_lp1      cmpa #2             ;ignore leading zeros
             bhs odb_lp2
             ldab A,sp
             bne odb_lp2
             inca
             bra odb_lp1
odb_lp2      cmpa #3             ;convert bcd[i] to ASCII and put
             bhs odb_finlp
             ldab A,sp
             addb #$30
             jsr putchar
             inca
             bra odb_lp2
odb_finlp    leas 3,sp           ;deallocate BCD space
             pulx                ;recover IX
             rts


;*********************************************************************
; outdecw()
;
; Description: A routine that outputs the decimal value of a 16-bit word.
; Arguments: Pointer to word is passed in ACCD.
```

```
;
; Pseudo-C:
; void outdecw(UWYDE *bin){
;     bcd[] = bwtod(*bin);
;     i = 0;
;     while(bcd[i] == 0 && i < 4){
;         i++;
;     }
;     while(i<5){
;         putchar(bcd[i] + 0x30);
;         i++;
;     }
; }
;****************************************************************************
outdecw        pshx                    ;preserve IX
               leas -5,sp              ;allocate space for BCD digits
               tfr d,x                 ;get binary word
               ldd 0,x
               tfr sp,x                ;convert bin to BCD
               pshx
               jsr bwtod
               leas 2,sp
               ldaa #0
odw_lp1        cmpa #4                 ;ignore leading zeros
               bhs odw_lp2
               ldab A,sp
               bne odw_lp2
               inca
               bra odw_lp1
odw_lp2        cmpa #5                 ;convert bcd[i] to ASCII and put
               bhs odw_finlp
               ldab A,sp
               addb #$30
               jsr putchar
               inca
               bra odw_lp2
odw_finlp      leas 5,sp               ;dealocate BCD space
               pulx                    ;recover IX
               rts


;****************************************************************************
```

Unlike the *outhexw()* routine, we cannot output a 16-bit word in decimal by simply calling *outdecb()* twice. *outdecw()* is very similar to *outdecb()*, however. The only difference is that five bytes must be allocated for the BCD characters, and the routine *bwtod()* is used to make the binary-to-decimal conversion. *bwtod()* is a 16-bit version of *btod()*.

These two routines convert and immediately output the converted characters. We could also write a routine that converts the binary data to a NULL terminated string stored in memory. The output string then could be sent using *outstrg()*. It would be less efficient but would allow us to use the same binary-to-string conversion routine regardless of the output device. For example we would use the same function for a terminal display or an LCD display.

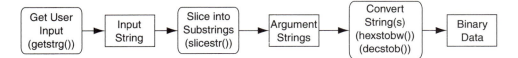

Figure 7.10 Data Input Process

String Conversions. To input data from a user, we have two choices. We can write routines that read and convert an input string, or we can use *getstrg()* to input the string and then write routines to convert the stored string. If we choose to write routines that handle both the string input and conversion, these routines would have to handle character input with all of its complications. Therefore it is more flexible and efficient to use *getstrg()* to input the string and then convert the string. This way the conversion routines can make the assumption that it is working on a standard NULL terminated string.

This is an example of good routine design that results in a more reusable set of routines. To test this concept, try to write a routine that inputs and converts a four-character hex string to binary. You will find the routine is just as complex as the combination of *getstrg()* and the string conversion program shown in Source 7.18, but it can only be used for one thing. If you want to input a hex byte or decimal, you would have to completely rewrite the routine.

The normal data flow for the process to input binary data is shown in Figure 7.10. First *getstrg()* handles the actual user input. It does not care what the content of the string is. It simply stores the data in memory as a standard NULL terminated string.

The next step is to slice the input string into a substring or *token* for each argument. This step is only required if there are more than one arguments. It can be handled by *slicestr()*, which detects white space, adds a NULL after the next token, and sets a pointer to the start of that token. You have to run *slicestr()* one time for each argument in the string.

Now the tokens are converted into binary data using a string-to-binary conversion program. The conversion process depends on the type of data entered. Was it a hexadecimal string or was it a decimal string? Source 7.18 shows an example of a hex string-to-binary conversion program, *hexstobw()*. It will convert a string of hexadecimal characters to a 16-bit binary word. It also ignores white space, counts characters for length errors, and detects nonhex-character errors. This routine is relatively complex. Notice the use of local variables. See if you can follow the code and understand not only the conversion process but also the use of the stack for parameters and local variables.

Source 7.18 Hex String to 16-Bit Binary Conversion

```
;*************************************************************************
; hexstobw()
;
; Description: A routine that converts a string to a hex 16-bit word
;              until white space or NULL is encountered.
; Arguments: Pointer to the string is passed in ACCD
;            Pointer to binary word is passed on the stack
;            Error code is returned in ACCB
;              0 -> No Error
```

```
;                1 -> Too large
;                2 -> Nonhex character
; Pseudo-C:
; UBYTE hexstobw(UBYTE *strg,UWYDE *bin){
;     cnt = 0;
;     *bin = 0;
;     while(*strg != NULL || *strg != ' ' || *strg != '\t'){
;         if(ishex(*strg)){
;             *bin = (*bin << 4) | htob(*strg);
;         }else{
;             return 2;
;         }
;         strg++;
;         cnt++;
;         if(cnt > 4){
;             return 1;
;         }
;     }
;     return 0;
; }
;*********************************************************************************
hexstobw        pshx                ;preserve IX
                tfr d,x             ;string pointer -> IX
                leas -2,sp          ;cnt -> 1,sp; tmpc -> 0,sp
                clr 1,sp            ;clear cnt
                ldd #0
                std [6,sp]          ;clr binary number
sh_lp           ldab 1,x+           ;get next character
                cmpb #NULL          ;finish if NULL, space, or tab.
                beq sh_finlp
                cmpb #' '
                beq sh_finlp
                cmpb #TAB
                beq sh_finlp
                jsr ishex           ;check for legal hex digit
                beq err2            ;if not return error 2
                jsr htob            ;convert ASCII hex to binary
                stab 0,sp
                ldd [6,sp]          ;pack binary nibble into bin
                lsld
                lsld
                lsld
                lsld
                orab 0,sp
                std [6,sp]
                inc 1,sp            ;check if four digits done
                ldab 1,sp
                cmpb #4
                bhi err1            ;error 1 if too many chars
                bra sh_lp
```

```
sh_finlp      clrb              ;return 0
              bra sh_rtn
err1          ldab #1
              bra sh_rtn
err2          ldab #2
sh_rtn        leas 2,sp         ;reallocate stack
              pulx              ;recover IX
              rts
```

;**

If the string was entered as a decimal string, a decimal string-to-binary conversion must be used. Source 7.19 is the assembly code for *decstob()*. *decstob()* converts a string of up to three decimal characters to one binary byte. It also checks for nondecimal-character errors and a value greater than 255.

Source 7.19 Decimal String-to-Binary Conversion Routine

```
;**********************************************************************
; decstob()
;
; Description: A routine that converts a string of decimal
;              characters to a binary byte until a CR, white space,
;              or a NULL is encountered.
; Arguments: Pointer to the string is passed in ACCD
;            Pointer to binary byte is passed on the stack
;            Error code is returned in ACCB
;               0 -> No Error
;               1 -> Nondecimal character
;               2 -> Too large
;
; Pseudo-C:
; UBYTE decstob(UBYTE *strg,UBYTE *bin){
;     if(isdigit(*strg)){
;         *bin = (*strg - 0x30);
;     }else if(*strg == NULL){
;         return 0;
;     }else{
;         return 1;
;     }
;     strg++;
;     while(*strg != NULL || *strg != ' ' || *strg != '\t'){
;         if(isdigit(*strg)){
;             *bin = *bin * 10 + (*strg - 0x30);
;             if(*bin > 255){
;                 return 2;
;             }
;         }else{
```

```
;              return 1;
;          }
;         strg++;
;      }
;    return 0;
; }
;**********************************************************************
decstob        pshx                ;preserve IX
               tfr d,x             ;strg pointer -> IX
               clr [4,sp]          ;clear bin
               ldab 0,x            ;get first digit
               cmpb #NULL          ;if NULL done
               beq dsb_finlp
               jsr isdigit         ;if not digit error 1
               cmpb #0
               beq dsb_err1
               ldab 0,x
               subb #$30           ;convert char to binary
               stab [4,sp]
               inx                 ;point to next char
dsb_lp         ldab 0,x
               cmpb #NULL          ;finished if NULL or white space
               beq dsb_finlp
               cmpb #' '
               beq dsb_finlp
               cmpb #TAB
               beq dsb_finlp
               jsr isdigit         ;error if not digit
               beq dsb_err1
               ldab [4,sp]         ;bin*10+bcdx
               ldaa #10
               mul
               tsta                ;check for overflow
               bne dsb_err2
               ldaa 0,x
               suba #$30           ;convert net digit to bcd
               aba
               bcs dsb_err2        ;check for overflow
               staa [4,sp]
               inx                 ;point to next digit
               bra dsb_lp
dsb_finlp      clrb                ;done no error
               bra dsb_rtn
dsb_err1       ldab #1             ;nondigit error
               bra dsb_rtn
dsb_err2       ldab #2             ;too-large error
dsb_rtn        pulx                ;recover IX
               rts


;**********************************************************************
```

The routine first converts the ASCII character to BCD and then implements the equation

$$bin = BCD_0 + (BCD_1 + BCD_2 \cdot 10)10$$

The binary variable *bin* is first set to the most-significant BCD digit, and then the routine enters a loop that does the following:

$$bin = bin \times 10 + BCD_x$$

where BCD_x is the next BCD digit. The loop runs until it adds BCD_0 or until the binary result is greater than 255. This routine could easily be expanded for 16-bit data. The extended multiply would have to be used, but the algorithm would be the same.

EXAMPLE 7.1

A Decimal Input Example

In this example we follow a user input that contains two decimal arguments through the process required to input the string, slice the string, and convert the tokens to two binary numbers. The user input is

```
SP 1 0 TAB 1 1 CR
```

We can then call *getstrg()* input and save the user input as the following NULL terminated string:

```
20 31 30 09 31 31 0D 00
```

To split the string into tokens corresponding to the two arguments, we can call *slicestr()* two times, which would affect the string as follows:

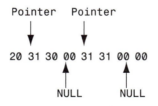

The first call to *slicestr()* replaced the TAB with a NULL and set a pointer to point to the first character of the first token. The second call to *slicestr()* replaced the carriage return with a NULL and set a pointer to point to the first character of the second token. Now the two pointers can be used to access the individual tokens, and we can use *destob()* to convert the two tokens to binary.

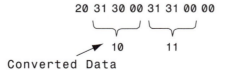

7.3.4 Section Summary

Most basic I/O requirements can be met using the programs described in the last two sections. The code examples shown not only illustrate the techniques for data and string conversions and I/O but also show several program design and assembly programming techniques described in the last two chapters.

▶ 7.4 FIXED-POINT ARITHMETIC

In this section we expand on the arithmetic techniques covered in Chapter 5 by looking at fixed-point data representations and techniques for multiplication and division of fixed-point numbers.

Fixed-point numbers are binary representations in which an imaginary radix point is at a fixed location in the word. This is in contrast to floating-point numbers. Floating-point numbers have the form

$$M \times 2^E$$

where M is the mantissa, which controls the number's precision, and E is the exponent, which controls the number's range. This representation allows for a higher range of number values. There are many potential ways to represent a floating-point number in a binary system. The IEEE single-precision floating-point number is represented by a 32-bit word, as shown in Figure 7.11.

When working with small microcontrollers, especially when writing assembly code, it is rarely practical to use floating-point numbers. This is especially true if you are trying to design the arithmetic functions themselves. It may be more practical to use floating-point numbers if a floating-point library is available. And it is easier to work with floating-point numbers if you are writing code in C. Libraries and a high-level language only make the programming task easier; however, they do not help improve the execution time or the code size. Floating-point routines in general are large and slow when written for an 8-bit microcontroller. The CPU12's 16-bit bus and additional extended arithmetic instructions help, but for fast routines it is usually best to use fixed-point arithmetic.

Of course if you were writing programs for a PC, where the CPU has a larger data bus and an arithmetic unit that supports floating-point arithmetic, it is perfectly reasonable to use floating-point arithmetic. Unfortunately this means most standard math libraries use floating-point arithmetic because they were originally designed for computer systems.

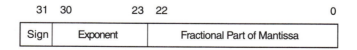

Figure 7.11　IEEE Single-Precision Floating-Point Number Representation

7.4.1 Fractional and Mixed Binary Numbers

We have seen that binary numbers can be interpreted in different ways. We can interpret binary numbers as signed or unsigned integers or, as we just learned, we can also interpret binary numbers as floating-point numbers.

Binary numbers also can be used to represent fractions or mixed numbers with a fixed radix point. These are called *fixed-point* numbers. So far we have only worked with binary integers. A binary integer's radix point is always interpreted as being to the right of bit 0, as shown in Figure 7.12. To create fractional or mixed numbers, we simply interpret the binary integer as if the radix point has moved to another location in the word.

Unsigned Binary Fractions.   To interpret a binary number as an unsigned fraction, we moved the radix point to the left of the most-significant bit of the word. Figure 7.12 shows the interpretation for an 8-bit binary integer and 8-bit binary fraction.

From the figure you can see that b_7 has a weight of $1/2$ instead of 128, b_6 has a weight of $1/4$ instead of 64, and so on. The range of a binary fraction is

$$0 \le N_F \le \frac{2^n - 1}{2^n}$$

and the resolution is

$$\frac{1}{2^n}$$

Another Way of Looking at It.   When we interpret a binary number as a fraction, we are in effect shifting the integer to the right eight times. As we see later in this section, shifting a binary number right one time is the same as a dividing that number by two. So in Figure 7.12 the binary number is effectively divided by 256 when the fractional interpretation is made. In general, to convert a binary number to its fractional equivalent we use

$$N_F = \frac{N_I}{2^n}$$

where

$$N_F = \text{the fraction represented by the binary number}$$

$$N_I = \text{the binary integer}$$

$$n = \text{the number of bits in the word}$$

b_7	b_6	b_5	b_4	b_3	b_2	b_1	b_0

Integer Interpretation: 2^7 2^6 2^5 2^4 2^3 2^2 2^1 2^0 .

Fractional Interpretation: $\cdot 2^{-1}$ 2^{-2} 2^{-3} 2^{-4} 2^{-5} 2^{-6} 2^{-7} 2^{-8}

Figure 7.12   Fractional and Integer Binary Number Interpretations

For an 8-bit number such as the one shown in Figure 7.12

$$N_F = \frac{N_I}{256}$$

This equation is the best way to quickly calculate binary fractions.

EXAMPLE 7.2 _____

Decimal-to-Fractional Binary Conversions

Find the binary fraction for 7/32.

Solution 1

We can add a one in each bit position in a way that makes the sum equal to 7/32.

$$\frac{7}{32} = \frac{1}{8} + \frac{1}{16} + \frac{1}{32} = 2^{-3} + 2^{-4} + 2^{-5} = \%00111000 = \$38$$

This method is a bit cumbersome because you have to break down the fraction into a sum of inverse powers of 2.

Solution 2

A better way to make the conversion is to use

$$N_F = \frac{N_I}{256} \Rightarrow N_I = N_F \times 256 = \frac{7}{32} \times 256 = 56 = \$38$$

_____

Unsigned Mixed Numbers. If we move the radix point to any other position in the binary word, the result will be a mixed number. A mixed number is a number that has an integer part and a fractional part. Following is an 8-bit mixed number with a 3-bit integer and a 5-bit fraction:

b$_7$	b$_6$	b$_5$	b$_4$	b$_3$	b$_2$	b$_1$	b$_0$

Mixed Number
Interpretation: 2^2 2^1 2^0 . 2^{-1} 2^{-2} 2^{-3} 2^{-4} 2^{-5}

$$m = 5$$

To convert a mixed number to decimal we use

$$N_M = \frac{N_I}{2^m}$$

where

$$N_M = \text{the unsigned mixed number represented by the number}$$

$$N_I = \text{the unsigned binary integer represented by the number}$$

$$m = \text{the number of bits in the fractional part of the number}$$

By moving the radix point, we can control the range and resolution of the number. The range of an unsigned mixed number is

$$0 \le N_M \le \frac{2^n - 1}{2^m}$$

and the resolution is

$$\frac{1}{2^m}$$

where m is the number of bits in the fractional part of the number. In the 8-bit example shown previously, m is 5. Increasing m improves the resolution, but it also decreases the range. This makes sense because the dynamic range for a given binary number is always the same.

Signed Mixed Numbers.   To represent a signed mixed number, we interrupt the most-significant bit as the sign bit and its weight is a negative number. The representation for an 8-bit signed mixed number with a 4-bit integer and 4-bit fraction is

Signed Mixed Number	b_7	b_6	b_5	b_4	b_3	b_2	b_1	b_0
Interpretation	-2^3	2^2	2^1	2^0 .	2^{-1}	2^{-2}	2^{-3}	2^{-4}

$$m = 4$$

To convert a signed mixed number to decimal, we use

$$N_{SM} = \frac{N_{SI}}{2^m}$$

where

$$N_{SM} = \text{the signed mixed number represented by the number}$$

$$N_{SI} = \text{the signed binary integer represented by the number}$$

In general the range of a signed mixed number is

$$-2^{n-m-1} \le N_{SM} \le 2^{n-m-1}\left(1 - 2^{-(n-1)}\right)$$

and the resolution is

$$\frac{1}{2^m}$$

Again the farther we move the radix point to the right, the greater the range, and the farther left we place the radix point, the better the resolution. For signed numbers the radix point cannot be moved to the left of the most-significant bit. When there is only one integer bit to the left of the radix point, the number is sometimes called a *signed fraction.*

To negate signed mixed numbers, including signed fractions, we use the same process as with binary integers. We take the two's-complement of the number. In fact we can use any of the arithmetic operators on signed or unsigned mixed numbers in the same way we use them for integers. The actual values only depend on our interpretation.

EXAMPLE 7.3 _____

Finding Signed Mixed Numbers

Using the mixed number representation of an 8-bit signed mixed number with m equal to four, find the binary representation for $1\,^1/4$.

Solution

We can find the binary representation either by adding weights for each bit or by using the conversion equation. Adding weights, we get

$$1\frac{1}{4} = 1 + \frac{1}{4} = \%00010100 = \$14$$

Using the conversion equation, we get

$$N_{SM} = \frac{N_{SI}}{2^m} \Rightarrow N_{SI} = N_{SM} \times 2^m = 1\frac{1}{4} \times 16 = 20 = \$14$$

Normally using the conversion is easier because you don't have to break the number down into powers of 2. You can then add weights to verify your answer.

EXAMPLE 7.4 _____

Negating Signed Mixed Numbers

Find the signed mixed number representation for $-1\,^1/4$ using the solutions given in Example 7.3.

Solution

We can find the representation of $-1\,^1/4$ by negating the solution for $1\,^1/4$. To negate, we just find the two's-complement as follows:

$$-(\%00010100) = \%11101100 = \$EC$$

We can verify this by adding the weights for each one bit:

$$\%11101100 = -2^3 + 2^2 + 2^1 + 2^{-1} + 2^{-2} = -8 + 6\frac{3}{4} = -1\frac{1}{4}$$

7.4.2 Errors Due to Mixed Numbers

We do not need to be concerned with errors while working with binary integers because every integer can be represented exactly. This is not true for mixed numbers, which have a fractional part. There are an infinite number of possible fractions within any given range, but a binary fraction can only represent 2^m values, where m is the number of fractional bits. In Example 7.2, the fraction was an exact sum of fractions what are inverse powers of 2, so there was no error. Most fractions cannot be represented exactly by a binary fraction, however. To represent these numbers, we must approximate by using the closest binary fraction to the desired fraction.

The error introduced by a binary fraction is

$$\varepsilon = \frac{Rnd(N \cdot 2^m) - N \cdot 2^m}{2^m}$$

where

$$Rnd() = \text{a rounding function}$$

$$N = \text{desired mixed number}$$

$$m = \text{the number of fractional bits in the word}$$

The peak error occurs when the numerator of the error equation is equal to $\pm^1/2$, which corresponds to a peak error of

$$\varepsilon_{pk} = \pm\frac{1}{2^{m+1}} = \pm\frac{Resolution}{2}$$

The error associated with fractional binary numbers can be problematic because the size of the error may be a large percentage of the desired fraction. This results in a large percentage error that can accumulate during further arithmetic processes. The actual percentage error involved in a binary mixed number is

$$\varepsilon\% = \left(1 - \frac{Rnd(N \cdot 2^m)}{N \cdot 2^m}\right) \times 100\%$$

where

$$N = \text{the desired mixed number}$$

$$m = \text{the number of fractional bits in the word}$$

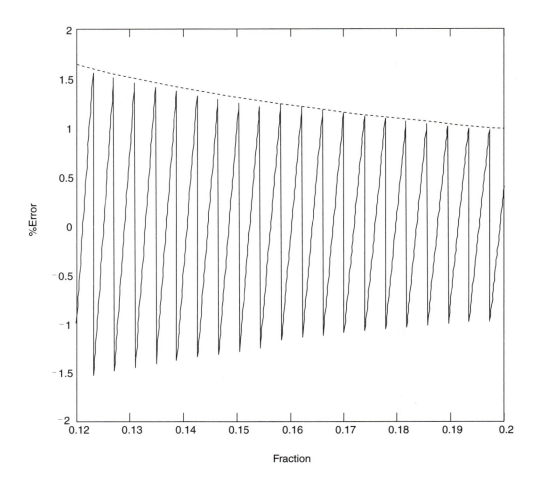

Figure 7.13   Actual and Peak Error for an 8-Bit Binary Fraction

Figure 7.13 shows the error for 8-bit fractions ($m = 8$) between 0.12 and 0.2. The actual error ranges from zero to plus or minus the peak values that follow a curve described by

$$\varepsilon_{pk}\% = \left(\frac{1}{N_F \cdot 2^{m+1}} \right) \times 100\%$$

which is also shown in Figure 7.13.

As shown in the figure, any fraction smaller than 0.2 has the potential to have an error greater than ±1%. On the other hand, the error can be zero down to 0.003906 if the fraction is an exact combination of inverse powers of 2.

When designing a program we must decide which error function to use—the actual error or the peak error. If the fraction is a variable, its actual value is unknown so the worst-case error is the peak error. If, however, the fraction is a constant, we know what the exact

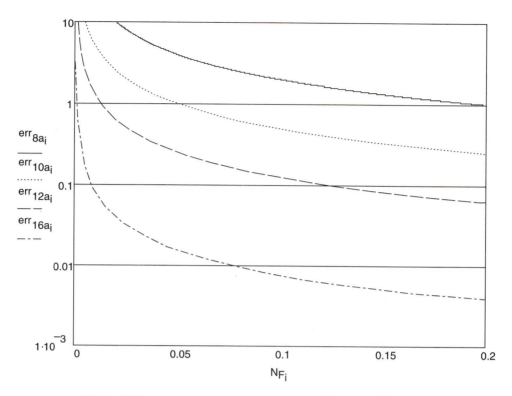

Figure 7.14   Peak Errors for 8-, 10-, 12-, and 16-Bit Binary Fractions

error will be. While designing a program, if we try to make the constant fractions equal to sums of inverse powers of two, the program will have no error.

We can also decrease the error dramatically by using a larger word size. Figure 7.14 shows the peak error for 8-bit, 10-bit, 12-bit, and 16-bit fractions. For example a 16-bit binary number can represent any fraction greater than 1/1310 (0.000763) with an error less than 1% and any fraction greater than 1/131 with an error less than 0.1%.

EXAMPLE 7.5

Decimal-to-Fractional Binary Conversion with Error

In Example 7.2, the fraction was an exact sum of inverse powers of 2. Now let's look at one that is not. Find the binary representation of 0.1.

Solution 1

Again, we will use

$$N_I = 256 \times N_F = 256 \times \frac{1}{10} = 25.6$$

Since the binary fraction is not an integer, we cannot represent 1/10 exactly using an 8-bit fraction. The best we can do is to round the number and use the closest 8-bit binary fraction to 0.1.

$$25.6 \Rightarrow 26 = \$1A = \%00011010$$

This results in an error of +1.56%, which is consistent with Figure 7.14.

Solution 2

We can decrease the error by using a larger binary word size. Let's find the 16-bit fraction to represent 0.1:

$$N_I = 2^m \times N_F = 65536 \times \frac{1}{10} = 6553.6 \Rightarrow 6554 = \$199A$$

Now the error is only +0.006%.

Obviously it is important to understand fixed-point mixed numbers, so we are not restricted to using only integers. Taking advantage of their properties can also significantly improve the efficiency of our programs.

In Chapter 5 we covered the basic arithmetic operations including addition, subtraction, and negation of binary integers. In the next two subsections we build on that material by covering multiplication and division of signed and unsigned integers and mixed numbers.

7.4.3 Multiplication

The CPU12 multiplication instructions in Table 7.2 represent a significant improvement over the 68HC11 microcontrollers, along with most other 8-bit or 16-bit microcontrollers. For example the CPU11 *mul* instruction executes in 10 clock cycles, whereas the CPU12's *mul* takes only three clock cycles. These speed improvements, along with the addition of signed and extended multiplies, make more efficient signal processing with the CPU12 possible.

Many 8-bit CPUs and MCUs have no multiply or divide instructions at all. When using these processors, the programmer must write routines for multiplication and division. In this subsection we will cover some of these routines along with techniques for using the CPU12 instructions.

TABLE 7.2 CPU12 MULTIPLICATION INSTRUCTIONS

Mnemonic	Function	CPU11	CPU12 Addressing Modes					
			INH	IMM	DIR	EXT	IDX	[IDX]
EMUL	16×16 Multiply, Unsigned		X					
EMULS	16×16 Multiply, Signed		X					
MUL	8×8 Multiply, Unsigned	X	X					

Result Size. When multiplying two numbers, we need to be aware of the data size required for the result. In general when multiplying an n_1-bit number to an n_2-bit number we need an $(n_1 + n_2)$-bit result. This is both a necessary and sufficient condition unless one or both of the numbers do not span the full range possible for its data size.

Multiply by Repeated Addition. This "brute force" method of multiplication uses a programmed loop to repeatedly add one of the arguments. The number of times it is added is equal to the other argument. Because the routine execution time depends on the argument that is used for the adding loop, this method is only practical for multiplying by a constant. If it is used to multiply variables, the addition loop may run zero times or 255 times, depending on the value of the variable. This wide range of execution time is not desirable. Of course when making a worst-case timing analysis, the case of running through the loop 255 times must be used, so this method is also slow for multiplying variables. It may be reasonable in some CPUs to use repeated addition for multiplication by a constant. For example if we needed to multiply a number by three, this method would be relatively fast and efficient. However, even this best case scenario is not as efficient as using the CPU12 *mul* instruction.

Shift-and-Add Multiplication. This method implements the multiplication method we use when multiplying two numbers with pen and paper. The least-significant digit of one argument is multiplied to the other argument, the result is shifted, and then the next most-significant digit is multiplied and the result is added to the last result. This continues until all digits are multiplied, shifted, and added. This type of routine is slow, but it is faster than using repeated addition for multiplying variables. It may be the only alternative if we are multiplying two variables and the CPU does not have a multiply instruction.

Multiplying by Shifting. An efficient method of multiplying a number by a constant that is a power of 2 is by shifting left. This only works for constant powers of 2. To understand how this method works, note that a binary number can be described as

$$N^2 = 2^0 + 2^1 + \ldots + 2^n$$

and if we multiply it by two

$$2 \times N^2 = 2(2^0 + 2^1 + \ldots + 2^n) = 2^1 + 2^2 + \ldots 2^{n+1}$$

This is the same as shifting the binary number to the left one bit. Because shift instructions are fast, this has long been a popular method. To use this method, we need to plan ahead to only multiply by a constant power of 2. For example an analog circuit that amplifies a signal for an ADC may be designed so the scaling required in the CPU is always by a constant power of 2.

When multiplying by a constant that is not a power of 2, we can combine the shift-left method with the repeated-adding method. For example in Section 7.2 we needed to multiply a number by 10 for the BCD-to-binary conversion programs. Source 7.20 shows two code snippets for multiplying a BCD digit by 10. The first snippet uses a combination of shifting and adding. It takes advantage of the efficiency of shifts by breaking down the equation into multiplies by two and adding.

Source 7.20 Two Ways to Multiply a BCD Digit by 10

```
*******************************************************************
* An example of multiplication of ACCA by 10 using shifting and
* adding.
*
*Pseudo-C:
*     bin = ((bin <<1 ) + (bin << 3));
*******************************************************************
mult10       lsla        ; 2x
             tfr a,b
             lslb        ; 4x
             lslb        ; 8x
             aba         ; 2x + 8x
*******************************************************************
* An example of multiplication of ACCB by 10 using mul.
*
*Pseudo-C:
*     bin = 10 * bin;
*******************************************************************
             ldaa #10
             mul
```

The shift-and-add code is fast. It only takes six clock cycles. But even this is two cycles slower than the snippet that uses the CPU12's *mul* instruction. Because of this, when using the CPU12 even the shift-left method is not often required.

Unsigned Multiply Instructions.   The CPU12 has an 8-bit × 8-bit unsigned multiply instruction, *mul,* and a 16-bit × 16-bit unsigned multiply instruction, *emul.* These multiply instructions can be used for integers or mixed numbers. Following is a simple example using the *mul* instruction:

```
mult         ldaa #$20        ;$20 -> ACCA
             ldab #$35        ;$35 -> ACCB
             mul              ;$20 × $35 = $06A0 -> ACCD
```

The actual values of the arguments and the results depend on our interpretation. First we interpret the arguments as 8-bit integers:

$$
\begin{array}{rcr}
\$20 & \rightarrow & 32 \\
\times \ \underline{\$35} & \rightarrow & \times \ \underline{53} \\
\$06A0 & \rightarrow & 1696
\end{array}
$$

For integers the multiply instruction works as expected.

Now we interpret both arguments as 8-bit unsigned mixed numbers. When multiplying mixed numbers, the radix point of the result is placed so that the number of fractional bits in the result is equal to the sum of the fractional bits in each argument.

$$
m_r = m_1 + m_2
$$

Following is an example in which the first argument has five fractional bits and the second argument has four:

$$
\begin{array}{rcl}
\texttt{iii.fffff} & \rightarrow & m_1 = 5 \\
\times \underline{\texttt{\quad iiii.ffff}} & \rightarrow & m_2 = 4 \\
\texttt{iiiiiii.fffffffff} & \rightarrow & m_1 + m_2 = 9
\end{array}
$$

The result of the multiplication has nine fractional bits, so the interpretation of the multiplication operation is

$$
\begin{array}{rcl}
\texttt{\$20} & \rightarrow & 1.0000 \\
\times \underline{\texttt{\$35}} & \rightarrow & \times \underline{3.3125} \\
\texttt{\$06A0} & \rightarrow & 3.3125
\end{array}
$$

Again the same operation produced the correct binary result. Make sure you understand the process used for this interpretation.

If we interpret these numbers as binary fractions, m_1 and m_2 are both eight. So the result is a 16-bit fraction and the interpretation is as follows:

$$
\begin{array}{rcl}
\texttt{\$20} & \rightarrow & 0.12500000 \\
\times \underline{\texttt{\$35}} & \rightarrow & \times \underline{0.20703125} \\
\texttt{\$06A0} & \rightarrow & 0.02587890625
\end{array}
$$

Again the same operation produced the correct results.

As shown in the example, the 16-bit fractional result is an exact solution. But what if our application required that the result be an 8-bit word? If we truncate the least-significant byte, the result then would be $06. This is not as accurate as rounding the result to eight bits. To round the result, we look at bit 7 of the least-significant byte. If it is one, we round up. If it is zero, we round down. In this case bit 7 of ACCB is one, so we need to round up to $07.

To implement this in a program, we do not need to test bit 7. We can take advantage of the *mul* instruction, which sets the C flag if bit 7 of the result is set. So to round off the result to eight bits, we can use an add-with-carry instruction after the multiply as follows:

```
mult        ldaa #32        ;$20 -> ACCA
            ldab #53        ;$35 -> ACCB
            mul             ;$20 x $30 = $06A0 -> ACCD
            adca #0         ;round fraction to 8 bits, $07 -> ACCA
```

Signed Multiply Instructions.   Unlike addition and subtraction, we have to use a different operation for signed multiplication. We cannot use the *mul* instruction because it does not include sign extension. The CPU12 has one signed multiply instruction, *emuls*. Following is a simple example that uses *emuls* to multiply two signed numbers:

```
mults       ldd #1000       ;$03e8 -> ACCD
            ldy #-13        ;$fff3 -> IY
            emuls           ;$ffff -> IY,$cd38 -> ACCD
```

Again we can interpret the results in several ways. First let's look at the signed integer interpretation

$$
\begin{array}{rcr}
\$03E8 & \rightarrow & 1000 \\
\times \quad \$FFF3 & \rightarrow \times & (-13) \\
\hline
\$FFFFCD38 & \rightarrow & -13000
\end{array}
$$

The integer result turned out as expected.

Now let's make a different interpretation. Let's say that the first argument is an integer and the second argument is a signed binary fraction. The number of fractional bits in the first argument, m_1, is zero. The number of fractional bits in the second argument, m_2, is 15. Therefore the result has

$$
m_r = m_1 + m_2 = 15 \text{ fractional bits}
$$

The resulting value can then be calculated by

$$
N_{SM} = \frac{N_{SI}}{2^m} \frac{-13{,}000}{2^{15}} = -0.3967285156
$$

So the interpretation is

$$
\begin{array}{rcr}
\$03E8 & \rightarrow & 1000.0000000000 \\
\times \quad \$FFF3 & \rightarrow \times & (-0.0003967285) \\
\hline
\$FFFFCD38 & \rightarrow & (-0.3967285156)
\end{array}
$$

Again the multiply instruction works correctly for both signed integers and signed mixed numbers.

7.4.4 Division

Just like multiplication, there are several ways to divide two numbers. We will look at four methods: repeated subtraction, shift-rights, shift-and-subtract division, and division using the CPU12 division instructions. First let's take a closer look at the required result size for division of mixed numbers.

Result Size. Unlike multiplication, we cannot define a data size that will guarantee an exact result. When dividing, you may end up with a result with a fractional part that is not a combination of inverse powers of 2. All we can do is design the result's data size so it meets the accuracy requirements for the full range of expected values. We can, however, define a minimum result size that covers the full range of the integer part of the result. In general the result size must be

$$
n_q \geq n_n - m_n + m_d + m_q
$$

where n_q is the total number of bits in the result, n_n is the total number of bits in the numerator, and m_n, m_d, and m_q are the number of fractional bits in the numerator, denominator, and quotient. If the numerator, denominator, and quotient are all integers, the result size must be at least the numerator size. This result size does not include a fractional part. The size of the fractional part can be determined by the required accuracy and the peak error shown in Figure 7.14.

Division by Repeated Subtraction.   We can perform division by repeated subtraction, but this is only practical when the denominator is a constant. We actually saw an example of division by repeated subtraction in Source 7.12 earlier in this chapter.

Division by Shifting Right.   As with multiplication, we can take advantage of the properties of a binary number to divide a number by a constant power of 2. In the case of division, we use a shift-right operation to divide a number by two. There is, however, a slight complication. A different operation must be used for signed numbers than unsigned numbers. Source 7.21 shows two programs that use shift-right instructions to perform an integer division of $CA by 16. The first example is an unsigned division. To divide an unsigned number by two using a shift instruction, a zero must be shifted into the most-significant bit. Therefore we have to use the logical shift-right instruction, *lsra*. Because this is an integer division, the fractional part of the result is truncated.

Source 7.21 Integer Divide-by-16 Using Shift-Rights

```
******************************************************************
* Integer divide-by-16 using shift-rights
******************************************************************
uidiv16     ldaa #$ca          ;%11001010 = 202  -> ACCA
            lsra               ;%01100101 = 101
            lsra               ;%00110010 = 50
            lsra               ;%00011001 = 25
            lsra               ;%00001100 = 12
******************************************************************
sidiv16     ldaa #$ca          ;%11001010 = -54
            asra               ;%11100101 = -27
            asra               ;%11110010 = -14
            asra               ;%11111001 = -7
            asra               ;%11111100 = -4
******************************************************************
```

Let's look at what would happen if we were to use the same program for signed numbers. The original argument would be interpreted as –54. After the first shift, the result would be +101, which is obviously incorrect. The reason for this error is that a signed number division requires sign extension. The second example in Source 7.21 uses an arithmetic shift-right instruction, *asra*. This instruction copies the most-significant bit into bit 7, as shown in Section 5.4. This implements an automatic sign extension. Using *asra,* the signed division works as required. Notice for negative integers that when a fractional part is truncated, the magnitude of the result is rounded up, not down.

Because we are dividing by two, we can have a result with full accuracy as long as the result is a mixed number with at least $n_1 + n_2$ bits, where n_1 is the number of bits in the numerator and n_2 is the number of bits in the denominator. Source 7.22 shows signed and unsigned divide-by-16 programs in which the result is a mixed number with an 8-bit integer part in ACCA and an 8-bit fractional part in ACCB.

Source 7.22 Mixed Number Divide-by-16 Using Shift-Rights

```
******************************************************************
* Mixed number divide-by-16 using Shift-Rights
* ACCA - Integer, ACCB - Fraction
******************************************************************
umdiv16     ldaa #$ca           ;%11001010 = 202 -> ACCA
            clrb                ;0 -> ACCB
            lsra
            rorb                ;%01100101 00000000 = 101
            lsra                ;
            rorb                ;%00110010 10000000 = 50.5
            lsra                ;
            rorb                ;%00011001 01000000 = 25.25
            lsra                ;
            rorb                ;%00001100 10100000 = 12.625
******************************************************************
smdiv16     ldaa #$ca           ;%11001010 = -54
            clrb                ; 0 -> ACCB
            asra
            rorb                ;%11100101 00000000 = -27
            asra                ;
            rorb                ;%11110010 10000000 = -13.5
            asra                ;
            rorb                ;%11111001 01000000 = -6.75
            asra                ;
            rorb                ;%11111100 10100000 = -3.375
******************************************************************
```

Using the shift instructions, we can create a relatively fast division program. The programs in Source 7.21 require only four clock cycles, excluding the load instruction. The speed of the mixed number programs in Source 7.22 take eight clock cycles. So if you can design your system for division operations that are by powers of 2, you may be able to save clock cycles by using the shift instructions. If the denominator is not a power of 2, or it is a variable, then the CPU12 division instructions should be used.

General Division Algorithm.   For general integer division on processors that do not have division instructions, we can use a shift-and-subtract algorithm that implements long division. Source 7.23 shows an 8-bit example written in assembly.

Source 7.23 A General Unsigned 8-Bit-by-8-Bit Division Routine

```
******************************************************************
* Div8x8 - An unsigned 8-bit-by-8-bit division routine
*
* Arguments: Numerator is passed in ACCB
*            Denominator is passed in ACCA
* Return Values: Quotient is returned in ACCA
*                Remainder is returned in ACCB
```

```
**************************************************************************
Div8x8          psha                      ;denominator -> 2,sp
                leas -2,sp                ;count -> 0,sp; quotient -> 1,sp
                clra
                movb #8,0,sp              ;count = 8
                clr 1,sp                  ;quotient = 0
divlp           lsl 1,sp                  ;Shift left quotient
                lsld                      ;Shift left extended numerator
                bcs divsub                ;Subtract if carry
                cmpa 2,sp                 ;Subtract if extnum >= denom
                blo divnxt
divsub          suba 2,sp                 ;Subtract denominator
                inc 1,sp                  ;Increment quotient
divnxt          dec 0,sp                  ; done?
                bne divlp
                tfr a,b                   ;remainder -> ACCB
                ldaa 1,sp                 ;quotient -> ACCA
                leas 3,sp                 ;recover stack space
                rts


**************************************************************************
```

This type of routine will work when the denominator is a variable. The general algo-rithm for an N_n-bit numerator and an N_d-bit denominator is

```
Extend Numerator to Nn+Nd+1 bits (Numerator = extNum:Num)
Quotient = 0;
for(i = 0; i < Nn; i++){
    Quotient << 1;
    Numerator << 1;
    if(extNum >= Denominator){
        extNum -= Denominator;
        Quotient++;
    }
}
```

If you are programming in assembly, you can extend the numerator to only $N_n + N_d$ bits because you have direct access to the carry flag. This is what we did in Source 7.23. Luckily, when using the 68HC12, we have division instructions, which are much easier to use, require less code space, and are much faster.

Division Instructions. Table 7.3 shows the CPU12 division instructions. These instruc-tions represent a significant improvement over the 68HC11 microcontrollers along with most other 8-bit or 16-bit microcontrollers. The CPU11 *idiv* and *fdiv* instructions take 41 clock cycles, whereas the CPU12 versions only take 12 clock cycles. These speed im-provements, along with the addition of signed and extended division instructions, make possible more efficient signal processing with the CPU12.

TABLE 7.3   CPU12 DIVISION INSTRUCTIONS

Mnemonic	Function	CPU11	CPU12 Addressing Modes					
			INH	IMM	DIR	EXT	IDX	[IDX]
EDIV	32-by-16 Divide, unsigned		X					
EDIVS	32-by-16 Divide, signed		X					
FDIV	16-by-16 Fractional divide, unsigned	X	X					
IDIV	16-by-16 Integer divide, unsigned	X	X					
IDIVS	16-by-16 Integer divide, signed		X					

Integer Division.   If we only need an integer result, using a division instruction is a straightforward task. We can use any division instruction except *fdiv* and only use the integer part of the result. For example the first part of the program in Source 7.24 divides 48 by 10. Since it is only an integer division, the result is truncated to four. Notice the *bcs* instruction after the *idiv*. This is used to detect an attempted divide-by-zero. Normally we only need this when the denominator is a variable. In this example it would take quite a brain glitch to explicitly load IX with zero and then perform a division.

Source 7.24 16-Bit Integer Division

```
*****************************************************************
* Integer division using idiv
*****************************************************************
uidiv       ldd  #48           ;Numerator, 48 -> ACCD
            ldx  #10           ;Denominator, 10 -> IX
            idiv              ;Quot,$0004->IX, Rem,$0008 -> ACCD
            bcs  divby0_err
*****************************************************************
* Rounding
*****************************************************************
            cpd  #10/2         ;Remainder < denominator/2?
            blo  uid_done      ;Yes, done
            inx               ;No, increment Quotient
uid_done
*****************************************************************
```

If we need more accuracy than the integer division result, we may be able to round the result instead of truncating it. The second part of the code in Source 7.24 shows a method for rounding when the denominator is the constant 10. If the denominator is a constant, we can simply compare the remainder in ACCD to one-half the denominator. If it were greater than one-half, we would round up; if not, we would round down.

Note

Before we go on, notice that the examples we are using here have been dividing one constant by another. If this were really the case, the CPU would not have to perform the division at all. Since constants are known at assembly time, the assembler could perform the division itself. In a real program one or both of the arguments actually would be variables. We are using constants here to simplify the illustration.

If the numerator or denominator is a signed number, the *idiv* instruction is replaced with the *idivs* instruction.

Mixed Number Division. When using mixed numbers, division is a bit more complex. First we need to determine the radix point location of the result. In general

$$m_q = m_n - m_d$$

where

$$m_q = \text{the number of fractional bits of the quotient}$$
$$m_n = \text{the number of fractional bits of the numerator}$$
$$m_d = \text{the number of fractional bits of the denominator}$$

So if the radix point is in the same location in both the numerator and the denominator, the result is an integer. In the example above both m_n and m_d are zero, so m_q is zero.

First let's look at a case in which we want a result that is a mixed number. By using a mixed number, the result will be more accurate. Remember that since we are dealing with binary fractions, it may not be possible to have an exact result no matter how large the word size.

Source 7.25 and Source 7.26 show two ways to produce a result that has an 8-bit integer part and an 8-bit fractional part. In these programs the numerator and denominator are both 8-bit words. Let's first interpret the numerator and denominator as integers. This results in

$$m_n = m_d = 0$$

The numerator is equal to 48 and the denominator is equal to 10. The expected result is 4.8. Since we want m_q to be eight and $m_n - m_d$ is zero, we need to shift the result to the left eight times. This is the same as multiplying the result by 2^8.

The program in Source 7.25 first performs the *idiv* like the previous example. The integer part of the result is in IX, but we need to shift this to the left eight times. This is done by transferring the quotient to ACCD and then transferring ACCB to the most-significant byte of the result instead of the least-significant byte. This in effect shifted the result eight times to the left. The next step requires converting the remainder produced by *idiv* to an 8-bit binary fraction. To convert a remainder to the equivalent fraction, it must be divided by the original denominator. This will always produce a result less than one, so the program uses *fdiv,* which produces a fractional result with m_{qf} equal to 16. The solution we want has only an 8-bit fractional part, so the 16-bit fraction is rounded and placed into the least-significant byte of the result.

Source 7.25 An 8-Bit-by-8-Bit to 16-Bit (8-Bit Quotient: 8-Bit Fraction) Division

```
************************************************************
* Without scaling
* 8-bit/8-bit -> 8-bit Quotient:8-bit Fraction
************************************************************
divide1
            ldd #$30
            ldx #$0a
            idiv                ;$0004 -> IX,$0008 -> ACCD
            bcs divby0_err
            xgdx
            stab result         ;Store Quotient
            xgdx
            ldx #$0a            ;Convert remainder to fraction
            fdiv                ;$CCCD -> IX,$0008 -> ACCD
            bcs divby0_err
            xgdx
            bitb #%10000000     ;Round off fraction to 8 bits
            bpl save_frac
            adda #1
save_frac   staa result+1
************************************************************
```

This program produces a result equal to $04CD. For a mixed number interpretation with m equal to eight the value is

$$N_{MR} = \frac{N_{IR}}{2^m} = \frac{\$04CD}{256} = 4.800078125$$

In this example the result is within 0.02%. This is an accurate result, but if the numerator or denominator is an 8-bit variable, we have to assume the accuracy would follow the 8-bit curve in Figure 7.14. If the range is limited to results greater than 0.2, we can assume the error will be less than 1%. We also can make the result data size larger and repeat the *fdiv* operation to produce a result with any desired accuracy.

The code in Source 7.25 is the traditional way of using the *fdiv* instruction to find the fractional part of the result. We may be able to write a more efficient program by scaling the numerator so the quotient of the *idiv* instruction is already a mixed number. Remember

$$m_q = m_n - m_d$$

If we want m_q to be eight, then we can make m_n equal to eight before performing the division. In effect we are multiplying the numerator by 256 before dividing, instead of multiplying the result by 256 after dividing. Source 7.26 shows a program that implements this. It first multiplies the numerator by 256 by loading the integer into ACCA instead of ACCB. Now the numerator is a 16-bit word with m_n equal to eight. After the denominator is loaded, the *idiv* instruction is executed. The result in this case is again $04CD, which is the same as the result of Source 7.25. The difference is that this program is much more efficient.

226

Source 7.26 Effects of Scaling on 7.25

```
*******************************************************************
* With scaling
* 8-bit/8-bit -> 8-bit Quotient:8-bit Fraction
*******************************************************************
            ldaa #$30               ;48x256 first
            clrb
            ldx #$0a
            idiv                     ;$04CD -> IX,$0008 -> ACCD
            bcs divby0_err
            stx result
*******************************************************************
```

This program was not only more efficient, but it also demonstrated a way of calculating a mixed number result without using a fractional division. This method could be used if a fractional division routine was not available. The requirements to use this method are that the division instruction can accept the data size of the shifted numerator and that the integer part of the result is large enough. In this example, since

$$n_q = 16 \geq n_n - m_n + m_d + m_q = 16 - 8 + 0 + 8 = 16$$

the requirements for result size are met.

Now let's look at a case in which the denominator is a mixed number, so m_d is greater than m_n. When the denominator is a mixed number, special care must be taken, because it may be less than one. If the denominator is less than one, the result may be larger than the numerator. Again we need a result size as follows:

$$n_q \geq n_n - m_n + m_d + m_q$$

If we divide a 16-bit integer by a 16-bit fraction, we would need at least a 32-bit result just for the integer part. Notice that there are no division instructions that produce a 32-bit result. This means we have to use a combination of an integer division and a fractional division to produce the full integer part of the result.

We again use the program in Source 7.24, only we interpret the numerator as an 8-bit integer and the denominator as a 16-bit mixed number with an 8-bit fraction. In this case m_n is equal to zero and m_d is eight. The fraction in the denominator is now

$$N_{FD} = \frac{10}{2^8} = 0.0390625$$

The expected result from the division would be

$$N_{IR} = \frac{48}{0.0390625} = 1228.8$$

The actual result was $0004. To make the correct interpretation, we have to move the radix point of the result eight bits to the right. This is equivalent to multiplying the result by 256. Now we have

$$N_{IR} = \$0400 = 1,024$$

This is not very close to the expected 1228.8! The reason for this large error is that the result of the division was in effect only eight bits, which resulted in a significant portion of the result being truncated before shifting left.

We can solve this problem by first scaling the numerator and then performing the division. This is what we did in Source 7.26. By shifting the numerator to the right eight times, we made m_n equal to eight. The number of fractional bits in the result is now

$$m_q = m_n - m_d = 8 - 8 = 0$$

The quotient from Source 7.26 was $04CD, which is 1,229. This result is much closer to the expected result of 1,228.8.

Reciprocals. A common task that requires division is to find a reciprocal of a number. For example if we have the period of a waveform, we can find the frequency by calculating the reciprocal.

Source 7.27 shows a basic 16-bit reciprocal routine. It uses the *fdiv* instruction to divide one by a 16-bit variable. You must be careful with this routine, however. The fractional result has an error that follows the 16-bit curve in Figure 7.14. So if the variable is greater than 1,310, the accuracy of the result may be worse than ±1%. If we restrict the variable to eight bits, the result will always be within ±0.2%.

Source 7.27 A 16-Bit Reciprocal Routine

```
*********************************************************************
* recip(c)
*
* Description: A routine that calculates the reciprocal of a
*              16-bit variable.
* Arguments: Variable is passed in ACCD.
*            Return value is passed in ACCD.
* Pseudo-C:
* UWYDE recip16(UWYDE c){
*     return(1/c);
*
*********************************************************************
recip16     pshx
            tfr d,x
            ldd #1
            fdiv
            tfr x,d
            pulx
            rts
*********************************************************************
```

Note that this routine also works for mixed numbers. Let's look at an interpretation in which the denominator is a 16-bit binary fraction. If the variable equals $0510, it represents the fraction 0.019989. The reciprocal of this fraction should be 50.02748. The actual result is 50. The radix point in the result is

$$m_q = m_{qf} - m_d = 16 - 16 = 0$$

where m_{qf} is the number of fractional bits resulting from the *fdiv* instruction, which always gives the result as a binary fraction. Therefore the result is an integer with a value of 50. This result represents an error of –0.055%.

► SUMMARY

In this chapter we looked at several applications that are typically implemented using assembly code. Most of them can also be written in C, so pseudo-C code is shown in most of the program headers.

These routines do not necessarily serve as *the* solution. You should look at each routine as a possible solution. Embedded systems by their very nature are application specific. It is rare that a routine simply can be dropped into a project without some changes. When designing the routines shown in this chapter, however, we did try to keep reusability in mind.

EXERCISES

1. Write a subroutine that implements a 10μs delay with no error beyond that of the MCU clock.
2. Write a program that uses the *WaitDms* delay routine in Source 7.14 to wait 100ms (nominal).
3. Design the flow diagram, pseudo-C code, and assembly code for *isalpha(c)*. This routine detects an alphabetic character (a...z, A...Z) and is implemented the same as *isdigit(c)* and *ishex(c)*.
4. Design the flow diagram, pseudo-C code, and assembly code for *isprint(c)*. This routine detects any printable character and is implemented the same as *isdigit(c)* and *ishex(c)*. A printable character is defined as a character that can be displayed on a terminal. These include ASCII characters within the range $21 to $7E.
5. Design the flow diagram, pseudo-C code, and assembly code for *tolower(c)*. This routine converts uppercase characters to lowercase characters. All other ASCII characters are returned unchanged.
6. Calculate and compare the execution times required for Source 7.11 and Source 7.12.
7. The *getchar()* routine in the Basic I/O module does not echo the received character automatically. Rewrite *getchar()* so it echoes the received character if the 8-bit global variable *autoecho* is nonzero.
8. Design a program that implements the user-input portion of the D-Bug12 Block Fill, BF, command. Use only the I/O functions included in the Basic I/O module in Appendix B. Place the start address in the variable named *startaddr*, the ending address in the variable *endaddr*, and the fill character in the variable *fillc*.

9. A thermometer project needs to display the temperature on a terminal in decimal. The temperature range is 0 to 120 degrees and is contained in the 8-bit variable *temp*. Write a program that displays the current temperature every second. Use the delay routine in Source 7.4 for the delay and use the Basic I/O module to display the temperature. The new temperature should be displayed on the same line so it overwrites the last temperature.

10. Give the decimal values of the following unsigned fractional numbers:
    (a) %10100101
    (b) %00010000
    (c) $8400
    (d) $0123

11. Repeat Exercise 10 interpreting the numbers as signed fractional numbers.

12. Repeat Exercise 10 interpreting the numbers as unsigned mixed numbers with a 4-bit fractional part.

13. Repeat Exercise 10 interpreting the numbers as signed mixed numbers with a 4-bit fractional part.

14. For the following specifications, give the minimum word size and minimum number of fractional bits to represent the parameter. The actual range may exceed or be less than the specified range by the given accuracy.
    (a) Range: −1 to +1; Accuracy: ±2%
    (b) Range: 0.1 to +2; Accuracy: ±0.1%
    (c) Range: −10,000 to +10,000; Accuracy: ±10%
    (d) Range: −50 to +150; Accuracy: ±0.1

15. Write a routine that multiplies an 8-bit unsigned number in *num* by five. Assume the result can be held in an 8-bit word and write the routine using each of the following methods:
    (a) Using the *mul* instruction
    (b) Using a combination of shifting and adding

16. Compare execution times for the two programs written for Exercise 15.

17. A system has a variable that contains the temperature value in Celsius. The Celsius variable is an 8-bit signed mixed number with one fractional bit.
    (a) What is the range and resolution of the Celsius variable?
    (b) Write a routine that converts the Celsius variable to Fahrenheit. The Fahrenheit value is a 16-bit signed integer. The required accuracy and resolution is ±1.0 degree Fahrenheit.

18. What is the data size required to hold a 12-bit variable multiplied by 10?

19. In Source 7.27 we calculated the reciprocal of a 16-bit integer using *fdiv*. Write a program that calculates the reciprocal of a 16-bit integer. The reciprocal must be a 16-bit fractional number. Write the program using each of the following methods:
    (a) Using *ediv* only
    (b) Using *edivs* for the reciprocal of a signed number

20. To divide a 16-bit mixed number with eight fractional bits by an 8-bit fractional number, what is the required quotient size and number of fractional bits for 1% accuracy or better?

Introduction to Real-Time I/O and Multitasking

This chapter is an introduction to the analysis and design of real-time and multitasking systems. These are important topics because virtually all embedded systems require real-time response while running more than one task.

We start with some general terms, models, and CPU load analysis and then go over event detection and response for a single task using event loops and interrupts. This discussion is followed by some basic multitasking concepts, including a simple, cooperative multitasking kernel that is relatively simple to write in assembly. Finally we cover the use of the CPU12 interrupts in more detail.

This chapter also introduces information about how a CPU's load and responsiveness can be dramatically improved by using hardware. Because of its on-chip resources, this is one of the fundamental strengths of a microcontroller.

▶ 8.1 REAL-TIME SYSTEMS

A *real-time system* is a system that must respond to signals within explicit and bounded time requirements. A real-time system's response time must be somewhat deterministic so we can know if it will meet the requirements. The timing requirements may be specified with a maximum value, a minimum value, or both. Note that real-time does not only mean "faster."

Most embedded systems also are required to perform more than a single task. This complicates the design because CPU-based systems are sequential systems. They can only run one task at a time, so the time spent executing one task will affect the timing of the other tasks. This is in contrast to parallel systems like digital logic, which can perform multiple tasks in parallel.

The complexity of a real-time system can range from a simple software delayed sequencer to a complex multitasking system with a full-featured real-time operating system. In this chapter we cover simple systems that are practical to design ourselves. They are practical to write in assembly language or C. More complex systems will be covered in Chapters 15 and 16.

A CPU-Based I/O Model. First let's look at a model for a microprocessor-based real-time system. Figure 8.1 shows a CPU that is running two tasks. These two tasks cannot run in parallel, but they can run concurrently by sharing the CPU. The difference between these two tasks and two normal routines is that their execution is asynchronous to each other. This means that one task does not control the other by calling it directly.

Input signals are sampled by the tasks, which in turn respond by generating output signals. In addition to input signals and output signals, the system may have signals between the tasks. These are called *intertask signals* and are realized in software. For example an alarm variable set by a sensor-scanning task and detected by a siren-output task would be considered an intertask signal as long as the two tasks run asynchronously.

The sampling of signals and the generation of output signals are all controlled by the program flow of the CPU. We call this *program time*. Events on the signals are asynchronous to the program flow, so they are called *real-time* events. This means that within the program, we cannot determine where a real-time event will occur so the input signals must be sampled periodically to detect the event. As shown in Figure 8.2, the sampling process of a task can be modeled as a synchronizer that is clocked when the program reads the inputs. In effect this synchronizes the real-time events to program time so we know when the data are valid within the program flow.

When designing a real-time system, we must consider both the output response function and the output response times.

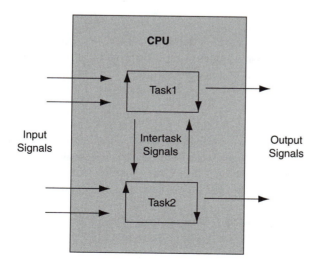

Figure 8.1   CPU I/O Model

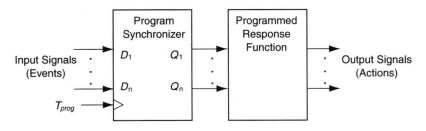

Figure 8.2 Logic Model for a Task

▶ 8.2 CPU LOADS

Before we look at the event detection process in more detail, we need to define a way to determine the load that a task (or a set of tasks) places on the CPU. By calculating the CPU load, we will be able to determine whether a system is realizable.

As we just saw, our real-time tasks are endless loops that sample the signals periodically. Therefore virtually all the programs we write for embedded systems are periodic. For a system with a single periodic task, the CPU load is

$$L = \frac{T_T}{T_p}$$

where

$$T_T = \text{the task's execution time}$$

$$T_p = \text{the task's execution period}$$

Most systems require more than a single task, so in general the load for n tasks is

$$L = \frac{T_{T1}}{T_{p1}} + \frac{T_{T2}}{T_{p2}} + \cdots + \frac{T_{Tn}}{T_{pn}}$$

The value of the load parameter L can help determine whether a design is realizable. A system cannot possibly work if the load is greater than one. After all a system with a single task and a load greater than one means that the task takes longer to execute than the time between executions. Obviously this cannot work. So a necessary condition for a design to be realizable is

$$L \leq 1$$

This is not a sufficient condition for a design to be realizable, however. A design may not be possible even if the load is less than one. This is especially true as the number of tasks increase. A general rule of thumb is to keep the CPU load less than 0.7. This allows for the task switching requirements and some minimal future expansion.

It is important to note that this is a system-level parameter, so the CPU load does not guarantee that the specific real-time requirements for a design will be met. We still need to analyze timing parameters such as the response time for the completed program.

A task's response to a real-time input signal can be classified as unconditional or event driven. A task with an unconditional response executes the same program code to generate the response regardless of the contents of the input signal. An event-driven response means the program response is conditional on the detection of an event. In an event-driven task, the program flow is changed to respond to the detected event.

8.3.1 Unconditional I/O

When using unconditional I/O, there are no requirements to synchronize the program flow with an event. Source 8.1 shows two simple examples that perform unconditional input and output. These examples show how simple unconditional I/O programs can be.

Source 8.1 Simple Unconditional I/O Examples

```
*******************************************************************
* Save the state of a configuration jumper tied to PP0 in the
* global variable, Mode.
* Pseudo-C:
*      Mode = PORTP & BIT0;
*******************************************************************
            ldaa  #BIT0
            anda  PORTP
            staa  Mode


*******************************************************************
* Turn on an active-low LED connected to PP1.
* Pseudo-C:
*      PORTP = PORTP & ~BIT1;
*******************************************************************
            bclr  PORTP,BIT1


*******************************************************************
```

The first example reads the value of an input signal that is tied to a configuration jumper. The second example turns on an active-low LED. In both of these examples, the program made a direct data transfer. There was no need to synchronize the read or the write with an external event.

Periodic Unconditional I/O.   In the last example the program performed a single read or write. Many systems, however, may require a continuous response. For example we may be able to replace some digital logic hardware with a CPU task. We can do this by periodically sampling the input and updating the output. There is a difference, however. A logic gate's output is a continuous function of its inputs, but as shown in Figure 8.2, a CPU-based system is a sampled system.

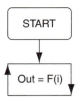

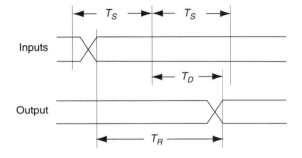

Figure 8.3 Continuous Unconditional I/O

Figure 8.4 Unconditional I/O Timing

Figure 8.3 shows the flow diagram for a periodic unconditional response program. As you can see, the program is an endless loop that sets the output equal to a function of the inputs each time through the loop. Notice that in this case, there is no attempt to synchronize the response with an event contained in the signal.

The timing diagram for this system is shown in Figure 8.4. The sample period, T_S, is the time it takes the program to make one pass through the sampling loop. T_R is the response time; it is the delay from an input change to the responding output change. T_D is the system delay time, which includes the software and hardware delays once the input is sampled.

The worst-case response time for this system is

$$T_R \leq T_S + T_D \cong 2 \times T_S$$

The worst-case response occurs when the input changes immediately after a sample is taken and it is not detected until the next sample. After the input change is detected, the program executes the response function and then returns to the start of the loop to take a new sample.

In this type of program, T_D is typically only slightly less than T_S because the output action normally occurs immediately before the program branches back to the beginning of the sample loop. Therefore a reasonable and conservative approximation for the worst-case response time is two times the sample period.

The sample period T_S also determines the minimum valid input time. If the input is valid for a time less than T_S, it could be missed. For example if a pulse occurs on the input, the output may not respond to the pulse if the pulse-width is less than T_S.

The program in Source 8.2 uses unconditional I/O to sample the two SCI serial lines *Txd* and *Rxd*. If either signal is active, it turns on an LED connected to *PORTP*, bit 0. This serves a similar function to the LEDs on the front panel of a modem. Because *Txd, Rxd,* and the LED are active-low, the response function for this system is

$$PP0' = Txd' + Rxd'$$

To simplify the program code, we can use DeMorgan's theorem to convert this equation to

$$PP0 = Txd \cdot Rxd$$

Source 8.2 A Serial Activity Monitor

```
***********************************************************************
* This program outputs a low-level on PP0, to indicate serial
* activity on either the SCI Rxd pin or Txd pin.
*
* Output Port: PORTP,bit0.
* MCU: 68HC912B32
*
* Pseudo-C:
* while(1){
*     if((PORTS & RXD_BIT) && (PORTS & TXD_BIT)){
*         PORTP |= BIT0;
*     }else{
*         PORTP &= ~BIT0;
*     }
* }
***********************************************************************
                                                    ;[~]
main          bset PORTP,BIT0
              bset DDRP,BIT0
smon_lp       brset PORTS,RXD_BIT¦TXD_BIT,sm_idle   ;[4]
              bclr PORTP,BIT0                       ;[4]
              bra smon_lp                           ;[3]
sm_idle       bset PORTP,BIT0                       ;[4]
              bra smon_lp                           ;[3]


***********************************************************************
```

The sample time for this loop can be calculated by counting clock cycles for the instructions in the loop. In this case there are two possible paths in the loop, one that turns the LED on and the other that turns the LED off. When there is more than one path, the longest path should be used for these calculations. Assuming an 8MHz E-clock frequency on the M68HC912B32, the cycle time through the loop is

$$T_S = 125\text{ns} \cdot (4 + 4 + 3) = 1.375\mu\text{s}$$

This means that the minimum pulse-width on *Txd* or *Rxd* that can be detected is $1.375\mu s$. In this case we are sampling a serial bit stream, so the bit rate would have to be greater than

$$R_{max} = \frac{1}{T_S} = 727.27 \text{bps}$$

to not be detected. This is far greater than the normal bit rates used on the SCI port. The response time T_R is

$$T_R = T_S + T_D = 1.375 + 1.04 = 2.379\mu s$$

where

$$T_D = T_{TD} + t_{PWD} = 125 \text{ns}(4 + 4) + 40 \text{ns} = 1.04\mu s$$

T_{TD} is the task delay and t_{PWD} is the M68HC912B32 peripheral port circuit delay time. For most cases the approximation

$$T_R = 2 \times T_S = 2 \times 1.375 = 2.75\mu s$$

would be adequate. Compared to logic-gate standards, the response delay for this program is slow, but for this application it is far more than adequate. A person watching this LED would never notice a $2.75\mu s$ delay!

8.3.2 Event Driven I/O

Event-driven I/O requires a task to detect and respond to an event. Once the event is detected, the program flow changes to a service routine or to code that sets an event flag. We can implement the detection process and the change in program flow with software only, a combination of software and hardware, or with hardware only.

An *event* is a predetermined characteristic of a signal. If the signal is an electrical signal, the event may be defined as the level or a change in the signal. For a software signal, the event may be defined as a value or a change in value of a variable. Events may also be much more complex. For example the detection of a specific word on a serial port can be considered an event. In this section we concentrate on simple events.

Periodic Events. Events that have a deterministic period are periodic. By knowing the event period, we can predict when the next event will occur. If we know when the next event will occur, we can determine the required sample rate to detect the event. This allows us to reduce the CPU load by reducing the sample rate. Examples of applications that use periodic events include flashing LEDs, timers, and ADCs.

Sporadic Events. Sporadic events are nonperiodic events with a period that is not deterministic. Because we use a sampling process to detect the event, however, the minimum time between sporadic events must be known. If not, it would be impossible to determine whether the system would capture every event. Some designs may also require that the maximum time and the average time between events be known. The minimum time between events can be used to determine the peak load on the CPU and the average time between events can be used to determine the average load on the CPU. Examples of sporadic events include switch inputs and words received on a serial port.

Event Detection and Response. There are two ways to implement the event-driven I/O process—event loops and interrupts. Event loops can be implemented with software only, or they may include a hardware detection circuit. Interrupts are a hardware-only solution for both event detection and the change in program flow to the event service routine.

To detect an event with software, we use a free-running event loop as shown in Figure 8.5. Event loops are also called *polling loops* or *gadfly loops*. The event loop is an endless loop that samples, or polls, the signal, waiting for the event to occur. Once it occurs the program responds by servicing the event or setting a flag. In this text we also refer to this as a *free-running event loop* to distinguish it from a timed-event loop, which is covered later in this section.

Some things must be considered when using an event loop. First, an event loop is a blocking routine. That is, the program cannot do anything else but poll the signal until the event occurs. Depending on your application this may not be acceptable. For example if there are other tasks to be performed, they cannot be executed while the event loop is polling. This problem is especially serious if the source of the event fails, which would cause the program to be stuck in the event loop indefinitely. We look at different techniques to poll for an event while also servicing other tasks in Section 8.4. In this section we treat each task as if it were the only task running on the CPU.

We also have to make sure that the sample rate is high enough to detect the event. For example if the event is a signal pulse, we must be sure that the period of the polling loop is shorter than the pulse-width. If this is not possible, a hardware detector must be used.

Level Detection. Level detection means that the signal event is a predefined value. Two types of applications use level detection, *as-long-as* applications and event flags.

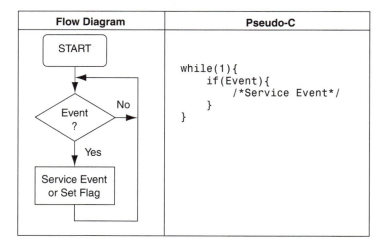

Figure 8.5 Basic Event Loop

As-long-as applications service the event "as long as" the event is present. An example of an application that must respond as long as an event is active is a pulse-width counter. The event is defined as the active level for the pulse and as long as it is active, a counter is incremented. This technique only works if the sample rate of the event loop is constant, because the number of times the counter is incremented depends on the sample rate and the length of the pulse. For a constant sampling time, we need to use a timed event loop.

An application that uses an event flag clears the event signal after the event is detected, so it is serviced only one time. Most applications require that an event be serviced one time and only one time. Using level detection alone, this means that the event time must be less than the service time so it is not active when the signal is sampled again.

It is not always possible to do this without clearing the event before the next sample. To clear the event signal we set up an event flag or status flag. Figure 8.6 shows the functional model of an event flag. Event flags are usually associated with detection hardware, but we can use the same model for software signals.

When the signal transitions to an active event, the event flag goes high. After the event is serviced, the flag is cleared. The event flag will not go high again unless another event occurs on the signal.

To clear the event, we must be able to access and change the signal. If the signal is an electrical signal external to the MCU, the clearing function may require extra hardware, which would result in a recurring hardware cost.

The event loop to detect an event flag is shown in Figure 8.7. At this point we do not care if the source of the event flag is a hardware signal or a software signal.

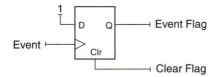

Figure 8.6   Event Flag Model

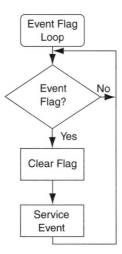

Figure 8.7   Event Flag Detection Loop

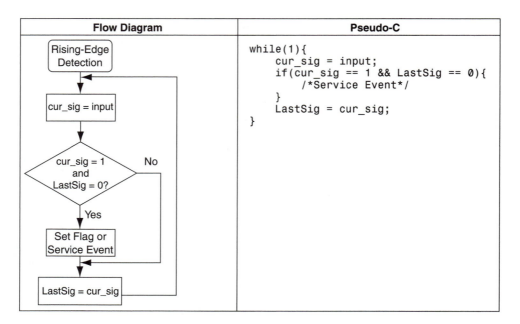

Flow Diagram	Pseudo-C

```
while(1){
    cur_sig = input;
    if(cur_sig == 1 && LastSig == 0){
        /*Service Event*/
    }
    LastSig = cur_sig;
}
```

Figure 8.8   Rising-Edge Detection Flow Diagram

Edge Detection.   Edge detection is used to respond to an event one time and only one time, without the added cost of the event flag hardware. It does this by detecting a change in the signal instead of the signal level. For binary signals, these changes are edges. We can detect rising edges, falling edges, or any edge. For multivalued signals, edge detection corresponds to the detection of a change of word values.

Figure 8.8 shows a flow diagram to detect a rising edge on a binary input signal. Edge detection uses an event loop that compares the last input sample to the current input sample. In Figure 8.8, the last sample is *LastSig* and the current sample is *cur_sig*. The last sample must use a static variable for persistent storage.

The rising-edge detector shown will service the event if the last sample is a zero and the current sample is a one. To convert this program to a falling-edge detector, we would service the event if the last sample is a one and the current sample is zero. To change the program to detect any edge, we would wait until the two samples are not equal.

Edge detection using hardware or software does not work well if the event signal contains unexpected edge transitions. Unexpected edges can occur if the signal source is noisy or generated with a bouncy mechanical device such as a switch or relay. Switch debouncing and noise immunity are covered in the subsection on timed event loops.

The program in Source 8.3 is an example of a rising-edge detector in CPU12 assembly code. This program is used to count pulses on *PORTT,* bit 0. When a rising edge is detected, the service routine *UpdateCnt* is called to increment a pulse counter. Note that there is no overflow code, so if more than 65,535 pulses occur, the counter will restart at zero.

Source 8.3 A Rising-Edge Detection Program

```
**************************************************************************
* RECntr()
* Description: A demonstration of rising-edge detection. This
*              program counts pulses on PORTT,bit0.
* Globals: LastSig, PulseCnt
* MCU: 68HC912B32
*
* Pseudo-C:
* RECntr(){
*    LastSig = 1;
*    PulseCnt = 0;
*    while(1){
*        cur_sig = PORTT & BIT0;
*        if(cur_sig == 1 &&  LastSig == 0){
*            CountEdge();
*        }
*        LastSig = cur_sig;
*    }
* }
*    CountEdge(){
*        PulseCnt++;
*    }
*
**************************************************************************
RECntr      movb #1,LastSig     ;Initialize variables
            movw #0,PulseCnt

redet_lp    ldab PORTT          ;Get new sample
            andb #BIT0
            beq no_redge        ;skip service if new sample != 1
            tst LastSig         ;skip service is last sample != 0
            bne no_redge
            bsr UpdateCnt       ;rising edge detected, service
no_redge    stab LastSig        ;update last sample for next time
            bra redet_lp

**************************************************************************
UpdateCnt   pshd                ;increment pulse counter
            ldd PulseCnt
            addd #1
            std PulseCnt
            puld
            rts

**************************************************************************
```

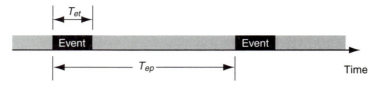

Figure 8.9 Event-Timing Parameters

Free-Running Event Loop Timing Parameters. Let's look at some of the timing requirements and response times for a free-running event loop. The timing characteristics of the event signal determine the requirements for the event-detection loop. Figure 8.9 shows the general timing characteristics for an event signal.

To detect every event, the polling period T_{poll} of the loop must be less than the event time T_{et}, therefore

$$T_{poll} < T_{et}$$

However, the polling period depends on whether the service is executed or not. When the service is not executed, the polling time for the event loop is

$$T_{poll} = T_{det}$$

where T_{det} is the event detection time. When the service is executed, the polling time for the event loop is

$$T_{poll} = T_{det} + T_{srv}$$

where T_{srv} is the service time.

The worst-case timing occurs after an event has been detected. In order not to miss the next event, the timing depends on characteristics of the event signal and the type of detection being used. For level detection the polling period must be

$$T_{poll} = T_{det} + T_{srv} < T_{ep} + T_{et2} - T_{et1}$$

where T_{et1} is the event valid time for the first event and T_{et2} is the event valid time for the second event. If the event valid times are the same, or if the signal is an event flag, this becomes

$$T_{poll} = T_{det} + T_{srv} < T_{ep}$$

For edge detection, the requirement is

$$2T_{det} + T_{srv} < T_{ep}$$

because we must have at least one sample that falls between the two events to reset the edge detector.

Event Response Time. The response time is the time from an event becoming active to the time a response action is made. In general the event response time is

$$T_R = T_{poll} + T_{det} + T_{srv} + T_{CIR}$$

where T_{CIR} includes the circuit delays for both the event signal path and the response signal path. (This response time equation assumes that there are no interrupts used in the system.) The worst-case polling period, T_{poll}, occurs when the event is serviced. So the worst-case response time is

$$T_R = 2\,(T_{det} + T_{srv}) + T_{CIR}$$

Typically the circuit delays are negligible but may play a role if the response time requirement is very short.

EXAMPLE 8.1

Timing Requirements for Counting Pulses with Software Edge Detection

This is the first of three examples that use different ways to detect and count pulses on *PORTT*, bit 0. For each example, we look at the CPU load, verify the event timing requirements, and calculate the event response time. Following are the timing parameters for the pulse train:

$$\text{Pulse width: } T_{et,\ min} = 10\mu s$$
$$\text{Pulse period: } T_{ep,\ min} = 500\mu s,\ T_{ep,\ ave} = 10ms$$

This first example uses the program in Source 8.3, which uses software for both event detection and response. The timing parameters for this example assume an M68HC912B32 with an E-clock frequency of 8MHz. The detection and service times are

$$T_{det} \cong 2\mu s,\ T_{srv} \cong 2.75\mu s$$

These times are found by adding instruction times in the detection loop and the service routine. For large routines this can be tedious, in which case we can make a conservative approximation or find them empirically with a logic analyzer or an oscilloscope.

Required CPU Load. Of course the program in Source 8.3 as shown has a CPU load of one. The CPU is always polling to detect the event or servicing the event. However, we are polling at a higher rate than required. We could decrease the polling rate to $T_{et,\ min}$ to allow other tasks to be performed. This gives a system-level CPU load requirement for this method of detecting and counting the pulses. If we do this the peak load is

$$L_{pk} = \frac{T_{det,\ max}}{T_{et,\ min}} + \frac{T_{srv,\ max}}{T_{ep,\ min}} = \frac{2\mu s}{10\mu s} + \frac{2.75\mu s}{500\mu s} = 0.2055$$

This is a sporadic pulse train, so we can also calculate the average CPU load:

$$L_{ave} = \frac{T_{det,\ max}}{T_{et,\ min}} + \frac{T_{srv,\ max}}{T_{ep,\ ave}} = \frac{2\mu s}{10\mu s} + \frac{2.75\mu s}{10ms} = 0.2003$$

Because most of the load is due to the event polling, the average load is about the same as the peak load. With a design goal of keeping the CPU load less than 0.6, this pulse counter uses almost one-third of the available load. This is acceptable if there are little or no other tasks to be performed by the CPU. Also note that if an equivalent routine is executed on an M68HC11F1 with a crystal frequency of 8MHz, the peak load is 1.29, which means that this system would not be realizable.

Polling Time Requirements.   To not miss an event when using edge detection

$$T_{poll} = T_{det} < T_{et} \text{ and } 2T_{det} + T_{srv} < T_{ep}$$

For the program in Source 8.3

$$T_{det} = 2\mu s < 10\mu s$$

and

$$2T_{det} + T_{srv} = 2 \times 2\mu s + 2.75\mu s = 6.75\mu s < 500\mu s$$

So the sampling time requirements for the M68HC912B32 system are met. For the M68HC11F1 system, T_{det} is 12.5μs, so events could be missed using software event detection.

Event Response Time.   In this example the event response time is probably not a critical requirement. The response action is that the pulse counter is incremented. It does not really matter when it is incremented. As an exercise, however, we can calculate the response time:

$$T_R = T_{CIR} + 2 (T_{det} + T_{srv}) \cong 0 + 2 (2\mu s + 2.75\mu s) = 9.5\mu s$$

The circuit delays are listed as zero because they are much smaller than 9.5μs. Again this time is not critical, so there is no requirement to which we can compare it.

In this example the M68HC912B32 system was able to perform the pulse counting with a load of ~0.2. The M68HC11F1 system would not work, however. Its load was greater than one, and its sample time was greater than the event time.

8.3.3 Hardware Event Detection

The process of detecting an event can place a high load on the CPU. In Example 8.1 we saw that virtually all the CPU load was due to the event detection loop. To decrease the load on the CPU or to meet event detection requirements, we can use a hardware circuit to detect the event. This is the primary role of on-chip I/O resources such as timers and serial ports. These circuits detect an event and then signal the CPU that the event occurred with either an event flag or an interrupt.

The event loop shown in Figure 8.5 performs both event detection and event service as a single routine. To use a different method for event detection, we need to split the processes into two separate tasks. Figure 8.10 shows the flow diagram for the event loop separated into two tasks, one for event detection and the other for event service.

Another good reason for separating the event detection and the event servicing tasks is that it results in routines that have stronger cohesion. This results in a more portable and maintainable program.

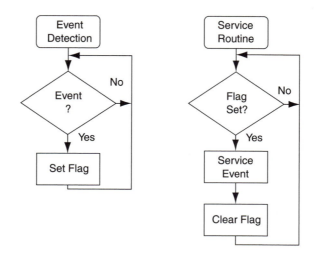

Figure 8.10 Event Routines Coupled by Using an Event Flag

The program to implement these two routines in software is shown in Source 8.4. An event flag, *PulseFlg,* is used to couple the two routines. The event detection loop will set the flag to indicate an event occurred, and the service routine will detect the flag and service the event if the flag is set. Notice that both tasks are event loops. One is detecting the actual event from the signal, and the other routine is detecting the event flag.

Source 8.4 Software Pulse Detection and Flag Detection

```
**********************************************************************
* SoftRE()
* Description: A demonstration of rising-edge detection. This
*              program sets a flag at each rising edge on PT0.
* Globals: LastSig, PulseFlg
* MCU: 68HC912B32
* Pseudo-C:
* SoftRE(){
*     LastSig = 1;
*     PulseFlg = 0;
*     while(1){
*         cur_sig = PORTT & BIT0;
*         if(cur_sig == 1 && LastSig == 0){
*             PulseFlg = 1;
*         }
*         LastSig = cur_sig;
*     }
* }
**********************************************************************
SoftRE      movb #1,LastSig     ;initialize last signal
            clr PulseFlg        ; and pulse flag
```

```
redet_lp     ldab PORTT          ;get current signal
             andb #BIT0
             beq no_redge        ;no redge if zero
             tst LastSig
             bne no_redge        ;no redge if last one
             inc PulseFlg        ;set pulse flag
no_redge     stab LastSig        ;update last signal
             bra redet_lp
```

```
**********************************************************************
* FlgDet()
* Description: A demonstration of event-flag detection. This
*              program increments the pulse count for each flag.
* Globals: PulseCnt, PulseFlg
* Pseudo-C:
* FlgDet(){
*     PulseCnt = 0;
*     while(1){
*         if(PulseFlg != 0){
*             PulseFlg = 0;
*             PulseCnt++;
*         }
*     }
* }
**********************************************************************
FlgDet       movw #0,PulseCnt    ;initialize pulse count
fdet_lp      ldab PulseFlg       ;wait for flag to be set
             beq fdet_lp
             clr PulseFlg        ;clear flag
             ldd PulseCnt        ;increment pulse count
             addd #1
             std PulseCnt
             bra fdet_lp

**********************************************************************
```

Splitting the code into two tasks has actually made the code more complex and would result in a higher CPU load. In this case we have traded program efficiency for better routine cohesion. Because the routines are more cohesive, however, we can change the implementation of either task without affecting the other. So we can now easily implement the event detection with hardware.

On the M68HC912B32, we can use an on-chip hardware resource called an *input capture* to detect edges on a signal. An input capture is part of the MCU's timer resource, which we cover in detail in Chapter 9. The program shown in Source 8.5 uses an input capture to detect rising edges and set an event flag. Since the detection is done in hardware, the software consists only of the event flag detection and service routine.

Source 8.5 Pulse Counting Program Using a Hardware Event Flag

```
******************************************************************
* A demonstration of a pulse counter that uses an Input Capture
* to detect each rising edge. Uses the flag only.
* Globals: PulseCnt
* MCU: 68HC912B32
* Pseudo-C:
* ICFlg(){
*     PulseCnt = 0;
*     TSCR = 0x80;
*     TCTL4 = 0x01;
*     TFLG1 = COF;
*     while(1){
*         if((TFLG1 & COF) != 0){
*             TFLG1 = COF;
*             PulseCnt++;
*         }
*     }
* }
******************************************************************
ICFlg        movw #0,PulseCnt          ;initialize count
             movb #$80,TSCR            ;turn on timer
             movb #$01,TCTL4           ;capture rising edges
             movb #C0F,TFLG1           ;clear CH1 flag
icflg_lp     brclr TFLG1,C0F,icflg_lp  ;wait for flag
             movb #C0F,TFLG1           ;clear CH1 flag
             ldd PulseCnt              ;increment pulse count
             addd #1
             std PulseCnt
             bra icflg_lp
******************************************************************
```

The first part of the program initializes the timer circuit to detect rising edges. When-ever an edge on *Timer Channel 0* is detected, the *COF* flag bit in the *TFLG1* register will be set by the timer hardware. The program then enters the flag detection and event service loop.

The program in Source 8.5 is much simpler than the programs in Source 8.3 or Source 8.4 because we no longer need the rising-edge detection code. This also signifi-cantly decreases the CPU load, because we are detecting the event flag, not the event itself. When detecting the event flag, our polling period only has to be less than the pulse period, not the pulse-width.

EXAMPLE 8.2 _____

Timing Requirements Using a Hardware Event Detector and Event Flag

Calculate the CPU load, the timing requirements, and the response time for the program in Source 8.5. Use the same pulse train described in Example 8.1 and compare the results with the software-only system.

Solution

The timing parameters for the event flag detection routine in Source 8.5 are

$$T_{det} = 625\text{ns}, \ T_{srv} = 1.875\mu\text{s}$$

Again these times were calculated by adding instruction times.

CPU Load. The program in Source 8.5, as it is shown, has a CPU load of one. The CPU is always polling the event flag or servicing the event. We could decrease the polling rate to $T_{ep, \, min}$ to allow other tasks to be performed, however. Notice that in this case we only have to poll the flag at the event rate, not at the event time. If we do this, the peak load is

$$L_{pk} = \frac{T_{det}}{T_{ep, \, min}} + \frac{T_{srv}}{T_{ep, \, min}} = \frac{625\text{ns}}{500\mu\text{s}} + \frac{1.875\mu\text{s}}{500\mu\text{s}} = 0.005$$

$$L_{ave} = \frac{T_{det}}{T_{ep, \, min}} + \frac{T_{srv}}{T_{ep, \, ave}} = \frac{625\text{ns}}{500\mu\text{s}} + \frac{1.875\mu\text{s}}{10\text{ms}} = 0.0014$$

Because we no longer had to poll to detect the pulse's edge, the CPU load dropped dramatically. Even for an equivalent program for the 68HC11F1 system described in Example 8.1, the peak load is only 0.03.

Event Polling Time Requirements. Since the event is actually sampled by the input capture circuit, the timing requirements are based on the electrical specifications for the M68HC912B32. The critical parameter is the input capture pulse-width, PW_{TIM}, which must be less than the minimum pulse width.

$$PW_{TIM} = 270\text{ns} < T_{et, \, min} = 10\mu\text{s}$$

This parameter is easily met and is far better than our software detection timing. For the 68HC11F1 system, PW_{TIM} is 520ns, which is also more than adequate for this example.

Event Flag Polling Time Requirements. To not miss an event, the event flag must be sampled at a higher rate than the pulse rate. The requirements are

$$T_{det} < T_{ep} \ \text{and} \ T_{det} + T_{srv} < T_{ep}$$

The worst case is when the event service is executed:

$$T_{det} + T_{srv} = 625\text{ns} + 1.875\mu\text{s} = 2.5\mu\text{s} < 500\mu\text{s}$$

so the sampling time requirements for the M68HC912 system are met. For the M68HC11F1 system

$$T_{det} + T_{srv} = 3.5\mu\text{s} + 15\mu\text{s} = 18.5\mu\text{s} < 500\mu\text{s}$$

so it will also work.

There would be no doubt about getting this system to work on either the 68HC12 or the 68HC11. This example illustrates the dramatic effect that hardware peripheral circuits can have on the CPU load and timing specifications.

Event Response Time.   In this example the event response time depends on the time it takes for the input capture to set the event flag, and then the time it takes to detect and service the event. The time it takes the input capture to set the input capture flag adds additional time to the circuit delays. Although this time is not specified, it is reasonable to use the *TCNT* clock period. With an E-clock frequency of 8MHz and the timer module's prescaler set to divide by one, the delay would be 125ns. The resulting response time is

$$T_R = T_{cir} + 2\,(T_{det} + T_{srv}) \cong 125\text{ns} + 2\,(625\text{ns} + 1.875\mu\text{s}) = 5.125\mu\text{s}$$

There may be additional circuit delays depending on the external circuit. Typically the external circuit delays are much smaller than 5.125μs.

8.3.4 Interrupt-Based Detection and Response

An *interrupt* is a hardware solution for both event detection and changing the program flow to a service routine. When using an interrupt, the service routine is called the *interrupt service routine* or *ISR*. The ISR may service the event itself or may set a flag that is detected by another task. If the ISR services the event itself, there is no longer a need for an event loop, which allows the CPU to perform other tasks and still respond to an asynchronous event.

In this section we look at the overall program flow for an interrupt. Section 8.3 covers using the CPU12 interrupts in detail.

Interrupt Program Flow.   Figure 8.11 shows the overall program flow in an interrupt driven system. The main program, or *background,* is shown running an endless task loop. Once an interrupt event is detected, the background will be interrupted and the CPU will automatically jump to and execute the interrupt service routine. This is like a subroutine call instruction that is initiated by a hardware event.

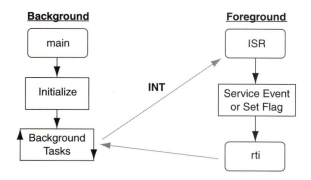

Figure 8.11   Interrupt Program Flow

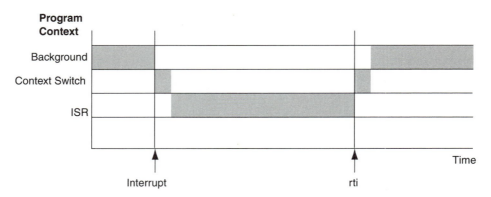

Figure 8.12 Context Diagram for an Interrupt

The interrupt service routine ends with an *rti* instruction, which returns the CPU to the background to resume execution at the instruction that would have run if the interrupt had not occurred.

The interrupt occurs asynchronously to the background program, so we do not know when it will occur; therefore, the complete CPU context must be saved. This process is called a *context switch*. Figure 8.12 shows the context diagram for the interrupt process.

The context switch is made automatically by the CPU. In most context diagrams the context switch is assumed and is not shown; this leaves the background context and the ISR context.

There are two ways to use interrupts to service events.

Periodic Interrupts.　These interrupts are generated by a periodic timer event. A periodic interrupt is used as a timing source for the system. Typically periodic interrupts are used for timed event loops that allow the polling rate to be reduced so other tasks can run. It also allows polling at a constant rate, which allows precise timing and the generation of periodic signals.

Event-Generated Interrupts.　These interrupts are used to detect and service the event directly. When the event occurs, the CPU control is passed to the interrupt service routine for that interrupt source.

The method used for a specific system depends on the characteristics of the event and the overall structure of the program.

Source 8.6 is a program that counts pulses by using an event-generated interrupt. In this case the input capture is configured to generate an interrupt when a rising edge occurs. At this point it is not important to understand the details for setting up the input capture or the interrupt. If we look at the program flow, we can see that the program initializes the input capture, enables interrupts with the *cli* instruction, and enters an endless task loop.

Source 8.6　Pulse Counting Using an Input Capture Interrupt

```
****************************************************************
* A demonstration of a pulse counter that uses an Input Capture
* to detect each rising edge. This version uses the IC interrupt.
* Globals: PulseCnt
```

```
* MCU: 68HC912B32
* Pseudo-C:
* ICInt(){
*     PulseCnt = 0;
*     TSCR = 0x80;
*     TCTL4 = 0x01;
*     TFLG1 = COF;
*     TMSK1 = COF;
*     while(1){
*         /* Other Tasks */
*     }
* }
* Ic0Isr(){
*     TFLG1 = COF;
*     PulseCnt++;
* }
******************************************************************
ICInt       movw #0,PulseCnt              ;initialize count
            movb #$80,TSCR                ;turn on timer
            movb #$01,TCTL4               ;capture rising edges
            movb #C0F,TMSK1               ;enable CH1 interrupt
            movb #C0F,TFLG1               ;clear CH1 flag
            cli                           ;Enable interrupts

icflg_lp    jsr Tasks                     ; Do other tasks
            bra icflg_lp


******************************************************************
Ic0Isr      movb #C0F,TFLG1               ;clear CH1 flag
            ldd PulseCnt                  ;increment pulse count
            addd #1
            std PulseCnt
            rti
******************************************************************
```

The program will continue to execute the tasks in the background task loop until a rising edge is detected by the input capture. Once the rising edge is detected, the input capture generates an interrupt, and the context is switched to the interrupt service routine, *Ic0Isr*. The interrupt service routine clears the input capture's flag and increments the pulse count. Notice that the main program does not require an event detection loop or an event flag loop to count pulses.

EXAMPLE 8.3 _____

Timing Requirements Using an Event Generated Interrupt

Calculate the CPU load, the timing requirements, and the response time for the program in Source 8.6. Use the same pulse train described in Example 8.1 and compare the results with Example 8.1 and Example 8.2.

Solution

The timing parameters for the interrupt-based design in Source 8.6 are

$$T_{srv} = 3.625\mu s$$

This time was calculated by adding the context switch time, the ISR execution time, and the context switch time to return. There is no need for a T_{det} time because there is no event detection loop.

CPU Load. The resulting CPU loads for Source 8.6 are

$$L_{pk} = \frac{T_{srv}}{T_{ep,\,min}} = \frac{3.625\mu s}{500\mu s} = 0.0073$$

$$L_{ave} = \frac{T_{srv}}{T_{ep,\,ave}} = \frac{3.625\mu s}{10ms} = 0.0004$$

Because the interrupt service time is about the same as the event flag detection and service time in Example 8.2, the peak load is about the same as the peak load when using a hardware event flag. Because polling for an event flag is not required, however, the average load was reduced.

In general interrupts have the biggest effect on the CPU load when the events are sporadic, with a large average event period and a small service time. This is because when we are using an event loop we are still polling for the event even if it does not occur. With an interrupt the program does nothing until an event occurs.

Event Sampling Time Requirements. Since we are using the input capture hardware to detect the rising edge, the sampling time is identical to the program that uses a hardware event flag shown in Example 8.2.

Event Rate Requirements. The rate at which an interrupt-based system can detect events can get complex. The calculations will be simplified if we do not allow nested interrupts. Without nested interrupts, the service must be complete before another event can be detected. So the service time must be less than the event period. In this case

$$T_{instr} + T_{srv} = 5.25\mu s < 500\mu s = T_{ep,\,min}$$

so the requirement is easily met. For the M68HC11F1 system

$$T_{srv} = 17.5\mu s = 17.5\mu s < 500\mu s = T_{ep,\,min}$$

so it also easily meets the requirement.

Event Response Time. In this example the event response time depends on the event service time and the interrupt latency. There is no detection time or polling time involved because there is no polling loop. The time it takes the input capture to initiate an interrupt is the same as the time it takes to set the input capture flag given in Example 8.2. With an

E-clock frequency of 8MHz and the timer module's prescaler set to divide by one, the delay would be 125ns. The resulting response time is

$$T_R = T_{cir} + T_{srv} \cong 125\text{ns} + 6.375\mu\text{s} = 6.5\text{ms}$$

There may be additional circuit delays depending on the external circuit. As you can see, the interrupt-based response time is significantly lower than the other two examples.

This example along with Example 8.1 and Example 8.2 have shown the effects of using different methods for event detection and event service. Because the example pulse train's pulse-width was much smaller than the minimum pulse period, the hardware event flag program in Example 8.2 resulted in a much smaller CPU load than the software event detection in Example 8.1. In addition, since the pulse train was sporadic, with a relatively long average event period, the interrupt-based system resulted in an extremely small average load. Some signals would not result in much of a difference among these three methods. For example if the signal were a periodic square wave with a longer service time, the CPU load for all three methods would be about the same.

8.3.5 Timed Event Loops

So far, the event detection loops we have covered are free-running event loops, which have polling times that are determined by the code that makes up the loop and the service. In general these times are nondeterministic, especially once interrupts are introduced into the system. In the pulse counting examples, we only had to be concerned with the minimum polling time. There are many applications, however, that require a constant polling rate. For example if we are converting an analog signal with an ADC, we must sample the analog signal at a fixed rate to have meaningful samples. Another example is a sequencer that generates a sequence of output actions separated by specific time delays.

Timed Loops Using Software Delays.   One method for implementing timed event loops is to use software time delays. Figure 8.13 shows the flow diagram for a timed event loop

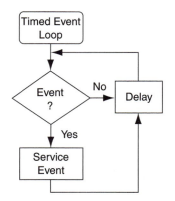

Figure 8.13   Timed Loop and Sequence Using Software Delays

using a software delay. If the delay time is much larger than the detection and event service time, the loop will sample at approximately a fixed rate.

If the delays are generated using a software delay routine, the event loop requires a CPU load of one. The CPU cannot be performing other tasks because it is always either detecting an event, servicing an event, or stuck in the delay routine. Also, as we have seen, once interrupts are used, software delays are no longer accurate. Timed event loops using software delays must be restricted to simple systems. Once it is necessary to perform multiple tasks or use an interrupt, it is no longer practical to use software delays.

Timed Event Loops Using a Periodic Timer. Timed event loops are normally implemented by making a pass through the loop conditional on a periodic timer event. This in effect synchronizes the execution of the loop to a periodic timer event. As we will see a periodic timer is also very important as a timing source for most multitasking kernels. Once we base execution on periodic timer events, it is easier to start thinking about scheduling and executing multiple tasks.

Figure 8.14 shows the basic timed event loop. As you can see the event service is now conditional on two events, the timer event and the actual event. Many unconditional I/O applications also use timed loops. For these systems the timed loop is the same as Figure 8.14 only without the event detection.

The source of the timer event is a hardware circuit that can generate the periodic timer events. When using a 68HC12 or 68HC11, we can use the free-running counter, the real-time interrupt system, or an output compare.

To ensure consistent and accurate timing of an event loop, it is important that the timer event is a periodic event that is generated from a source that is independent of the pro-

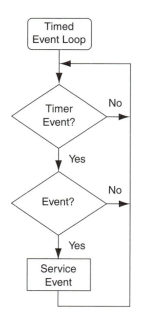

Figure 8.14 Basic Timed Event Loop Based on a Timer Event

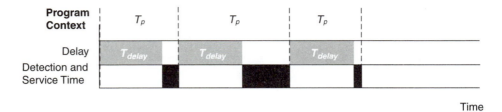

Figure 8.15 Inconsistent Loop Times Due to Using a Timer Delay

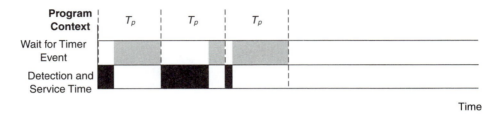

Figure 8.16 Consistent Loop Times Due to Using a Periodic Timer Event

gram. This means that a simple delay cannot be used for the timer event, even if the delay is generated by an external source. This is because the overall loop period would be the sum of the time delay and the execution time for event detection and service. The loop periods for a system that uses a timer delay are shown in Figure 8.15. As you can see the loop period, T_p, is inconsistent due to the variations in event detection and service times. If the detection and service times are much smaller than the delay time T_{delay}, the errors may be negligible and a delay can be used.

Figure 8.16 shows a system that uses periodic timer events. Notice that the loop periods are always the same because the loop timing is not a function of the detection and service times.

The loop times are consistent only if the timer event period, T_p, is greater than the worst-case detection and service time. Otherwise a timer event may be missed. Therefore

$$T_p > T_{T, max}$$

Timing Parameters for Timed Event Loops. Let's look at the timing requirements and response times for a timed event loop. We will compare these with the timing for a free-running event loop. Figure 8.9 on page 242 showed the general timing characteristics for an event signal.

Again, in order not to miss an event, the event signal must be sampled with a period that is less than the event time. For a timed event loop this means

$$T_p < T_{et}$$

In addition, for level detection

$$T_p < T_{ep} + T_{et2} - T_{et1}$$

Introduction to Real-Time I/O and Multitasking

255

if the event is realized as an event flag

$$T_p < T_{ep}$$

and for edge detection

$$2T_p < T_{ep}$$

Notice that these timing equations are much simpler than the free-running event loops. Not only are they simpler, but T_p is deterministic and easily found. We no longer have to count instructions to determine the loop time or the service time, as long as T_p is greater than the detection time and service time. This does come at a cost, however. The cost is a longer event response time for some systems.

The event response time for a timed event loop is

$$T_R = T_p + T_{det} + T_{srv} + T_{CIR}$$

Since T_p must be greater than $T_{det} + T_{srv}$, this response time is always larger than the response time for a free-running event loop. This makes sense because a free-running event loop does not have to wait for the timer event. Timed event loops, however, are easier to work with, especially when precise timing is required.

8.3.6 Switch Debouncing and Noise Immunity

Simple Switch Debouncing.   One application that requires a timed event loop is switch debouncing. As mentioned earlier in this section, mechanical push buttons will bounce when they are pressed or released. The bouncing causes multiple edges to occur for each keypress and release. This can cause a problem when using edge-detection software. Figure 8.17 shows the timing parameters for an event with signal bounce.

In order to avoid detecting multiple edges due to switch bounce, only one sample can occur during either bounce time. So the polling period, T_p, must be greater than the key activation bounce time, T_{ba}, and the key release bounce time, T_{br}.

$$T_p > T_b$$

where

$$T_b = max[T_{br}, T_{ba}]$$

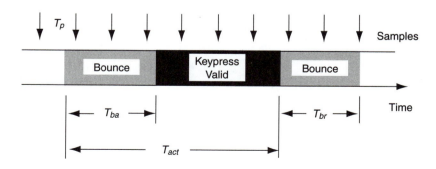

Figure 8.17   An Event with Signal Bounce

This defines the minimum polling period. The maximum polling period is set to make sure that at least one sample is made during the key activation time. So

$$T_p < T_{act} - T_{ba}$$

Therefore the range of acceptable sample times is

$$T_b < T_p < T_{act} - T_{ba}$$

Notice that this places the following requirement on the signal itself:

$$T_{act} > T_{ba} + T_b$$

If the key is not pressed longer than the sum of the activation bounce time and the worst-case bounce time, it is not possible to reliably detect the keypress.

For push-button switches or keypads, the typical activation time, T_{act}, is greater than ~30ms, depending on the switch. If we design the sampling time to be greater than 30ms, it may be possible that a quick keypress will be missed. This can be a serious problem because it is very irritating for the user. Bounce times for normal push buttons and keypads range from 4ms to 12ms. Source 8.7 shows a program that counts switch presses on *PORTP,* bit 0.

Source 8.7   A Basic Switch Debouncing Routine

```
*****************************************************************
* SwCntr()
* A switch debounce program. SwCnt is incremented every
* time the active-low switch on PORTP, bit-0 is pressed.
*    - Debounced for switch bounce time <16ms.
* Controller: 68HC912B32
* Monitor: D-Bug12
* Pseudo-C:
* SwCntr(){
*    RTICTL = 0x05;
*    RTIFLG = RTIF;
*    LastSw = 0;
*    SwCnt = 0;
*    while(1){
*        while(RTIF & RTIFLG == 0){
*        }
*        RTIFLG = RTIF;
*        cur_sw = PORTP & BIT0;
*        if(cur_sw == 0 && LastSw == 1){
*            SwCnt++;
*        }
*        LastSw = cur_sw;
*    }
* }
*****************************************************************
SwCntr      movb #%00000101,RTICTL      ;Init. RTI to 16.384ms
```

```
            movb #RTIF,RTIFLG              ;clear RTI flag
            clr SwCnt                      ;Init. Globals
            clr LastSw

swc_wait    brclr RTIFLG,#RTIF,swc_wait    ;wait for rti event
            movb #RTIF,RTIFLG              ;clear RTI flag
            ldaa PORTP                     ;detect falling edge
            anda #BIT0
            bne swc_last
            ldab LastSw
            beq swc_last
            inc SwCnt                      ;increment switch count
swc_last    staa LastSw
            bra swc_wait
```

****

This example uses the 68HC12's real-time interrupt to generate timer events every 16.384ms. The timer events are detected by polling the RTI flag. With this timer rate, the routine will be good for switches with bounce times less than 16.384ms. If the key activation bounce time is 4ms, this routine would detect key activation times greater than 20.384ms.

EXAMPLE 8.4 _____

A Switch Debounce Example

A push-button switch has a key activation bounce time of 4ms and a key release bounce time of 10ms. For a key activation time of 25ms minimum, find the range of acceptable sample times.

Solution

The key release time is the worst-case bounce time. So the range is

$$T_b = 10\text{ms} < T_s < T_{act} - T_{ba} = 25\text{ms} - 4\text{ms} = 21\text{ms}$$

The program shown in Source 8.7, with a sample time of 16.384ms, would be appropriate for this switch.

_____

Switch Debouncing with Noise Immunity.   The problem with the simple switch debouncing routine is that a small glitch caused by noise may be detected as a keypress. For most switch applications, there would be little noise on the signal, but if the switch is not directly attached to the PCB with the microcontroller, there may be noise coupled into the switch signal.

To make the system immune to noise, we require that the switch is at its active level for two consecutive samples. So the program will detect the keypress edge and then at the next sample time will confirm that the switch is at its active level.

258   Chapter 8

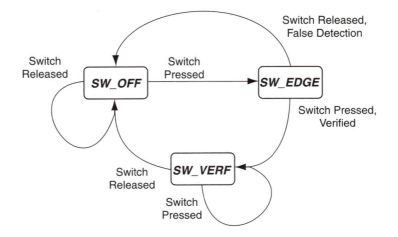

Figure 8.18 Switch Debouncing with Noise Immunity State Machine

To do this our program implements the state machine shown in Figure 8.18. The *SW_OFF* state is reached whenever the switch is sampled in the inactive state. The *SW_EDGE* state is reached if a switch activation is detected. In effect this transition represents a detected edge. Now if the next sample shows that the switch is still active, the keypress is verified, the *SW_VERF* state is reached, and the appropriate action will be made. If an inactive switch is detected while in the *SW_EDGE* state, the machine transitions to the *SW_OFF* state. This transition represents a false switch detection—probably caused by noise. The *SW_VERF* state is required to wait for the switch to be released. Once an inactive switch is detected, the machine transitions to the *SW_OFF* state. We also could add a counter to the *SW_VERF* state to implement an auto-repeat feature.

Source 8.8 shows a routine that implements this state machine for switch debouncing. Like the program in Source 8.7, this routine will increment a counter every time a valid keypress is detected.

Source 8.8 A Switch Debounce Routine with Noise Immunity

```
**************************************************************************
* SwCntr()
* A switch debounce program. SwCnt is incremented every
* time the active-low switch on PORTP, bit-0 is pressed.
*     - Debounced for switch bounce time <16ms.
*     - With noise immunity
*
* Controller: 68HC912B32
* Monitor: D-Bug12
* Pseudo-C:
* SwCntrN(){
*     RTICTL = 0x04;      /* RTI rate = 8.196ms */
*     RTIFLG = RTIF;
*     SwState = RELEASED;
*     SwCnt = 0;
```

```
*      while(1){
*          while(RTIF & RTIFLG == 0){
*          }
*          RTIFLG = RTIF;
*          cur_sw = PORTP & BIT0;
*          if(SwState == SW_OFF){          /* wait for switch edge */
*              if(cur_sw == PRESSED){
*                  SwState = SW_EDGE;
*              }
*          }else if(SwState == SW_EDGE){ /*Verify switch press    */
*              if(cur_sw == PRESSED){
*                  SwState = SW_VERF;     /*switch press verified */
*                  SwCnt++;
*              }else{                     /*false switch press     */
*                  SwState = SW_OFF
*              }
*          }else if(SwState == SW_VERF){ /* Wait for release       */
*              if(cur_sw == RELEASED){
*                  SwState = SW_OFF
*              }
*          }
*      }
* }
******************************************************************************
* Program
******************************************************************************
SwCntrN     movb #%00000101,RTICTL          ;Init. RTI to 8.192ms
            movb #RTIF,RTIFLG               ;clear RTI flag
            clr SwCnt                       ;Init. Globals
            movb #RELEASED,SwState

swcn_wait   brclr RTIFLG,#RTIF,swcn_wait    ;wait for rti event
            movb #RTIF,RTIFLG               ;clear RTI flag
            ldaa PORTP                      ;cur_sw -> ACCA
            anda #BIT0
            ldab SwState

            cmpb #SW_OFF                    ;SW_OFF state
            bne swcn_edge
            cmpa #PRESSED                   ;RELEASED state
            bne swcn_wait                   ;check for edge
            movb #SW_EDGE,SwState           ;if edge change state
            bra swcn_wait

swcn_edge   cmpb #SW_EDGE                   ;SW_EDGE state
            bne swcn_verf
            cmpa #PRESSED
            bne swcn_false
            movb #SW_VERF,SwState
            inc SwCnt                       ;increment switch count
```

```
            bra swcn_wait

swcn_false  movb #SW_OFF,SwState
            bra swcn_wait

swcn_verf   cmpb #SW_VERF                           ;SW_VERF state
            bne swcn_wait
            cmpa #RELEASED
            bne swcn_wait
            movb #SW_OFF,SwState
            bra swcn_wait
```

****

The timing requirements for this switch debouncing routine are different from the requirements for the simple switch debouncer. The transitions shown in the state diagram are made based on a sample of the switch signal. The debouncing only works if the polling period is within the required range. First, to avoid multiple key detection due to switch bouncing

$$2T_p < T_b$$

where

$$T_b = max[T_{br}, T_{ba}]$$

In addition, in order not to miss a keypress

$$2T_p < T_{act} - T_{ba}$$

This results in a polling period range of

$$\frac{T_b}{2} < T_p < \frac{T_{act} - T_{ba}}{2}$$

Compared to the simple debouncing routine, we must sample the switch twice as often. You may have noticed that the RTI rate for the routine in Source 8.8 is only 8.192ms compared to 16.384ms for the routine in Source 8.7. Signal glitches caused by noise will be rejected by this routine as long as they are less than T_p wide.

▶ 8.4 BASIC COOPERATIVE MULTITASKING

So far the programs we have looked at perform only a single task. Most embedded microcontroller programs, however, must perform several tasks concurrently. For example, if the two event loops shown in Figure 8.10 on page 245 are both implemented in software, they would have to be executing at the same time. The CPU, however, is a sequential system. It can only execute one program at a time. To solve this problem, we can take advantage of the time spent waiting in one task to execute another task.

To coordinate this process, we need a program to schedule and run tasks, and we need tasks that will give up control of the CPU so other tasks can run. The program to schedule and run the tasks is called a *kernel,* or sometimes an *executive.* The kernel is the part of an operating system that schedules and dispatches tasks. There are many different types of kernels, ranging from simple task loops to kernels designed for full-featured operating systems.

In this section we look at one basic multitasking technique that can be used for most simple systems. It is simple enough to be written in assembly language and designed from scratch. Other kernel designs will be covered in more detail in Chapter 15.

8.4.1 Tasks and Kernels

So far we have been using the term *task* as an abstraction used to help break a system down into functional blocks. Now we need to define a task in more detail. A task is normally implemented as a routine. The distinction between a task and a regular routine is that a task can stand alone, and the actions of one task occur asynchronously to another task. A regular routine is always called by another routine, so they are synchronized in program time.

The process of designing a complete multitasking system involves breaking down the system into tasks, designing each task, and using a kernel to run the tasks. Most embedded systems require deterministic response times, so we need a real-time, multitasking kernel.

As we have seen most tasks are endless loops, either to detect and respond to an event or to generate unconditional outputs. So when there are multiple tasks, we need a kernel that can preempt or interrupt a task, or we need a task that is not stuck in an endless loop. Therefore the design of each task will depend on the type of kernel we are using.

Preemptive Kernels.   In all of our code examples, the tasks are endless loops that apparently have the CPU to themselves. For example the two tasks in Figure 8.10 each appear to have full control of the CPU. Defining tasks this way is the best way for conceptualizing and designing a multitasking program. To design this type of task, however, we need to use a more sophisticated preemptive multitasking kernel. Figure 8.19 shows the flow diagram for a preemptive multitasking kernel with three tasks. Each task is an individual event loop that is designed as if it were the only task running on the system. The kernel determines which task should run at any one time and may preempt one task to execute another higher-priority task.

Because of their complexity, it is usually not a good use of time to design your own preemptive multitasking kernel. There are many good kernels available off the shelf. Preemptive

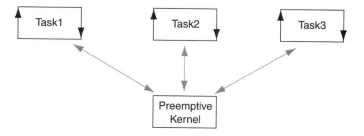

Figure 8.19   Preemptive Multitasking Program Flows

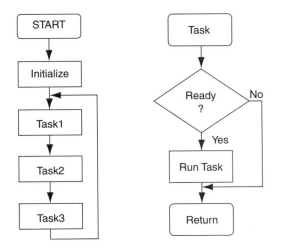

Figure 8.20 Cooperative Multitasking Loop and Cooperative Task

kernels are covered in more detail in Chapters 15 and 16, and in that discussion we use an off-the-shelf kernel, µC/OS-II.

Cooperative Kernels.   In this section we look at one type of cooperative multitasking kernel. The kernel and task structures are shown in Figure 8.20. The kernel shown is often called a *cyclic scheduler,* which is a loop that sequentially executes each task. We can think of this as a big event loop that services multiple events.

In order to use this type of kernel, the tasks can no longer be endless loops. You can imagine if *Task1* in Figure 8.20 was implemented as an endless loop, the scheduler would never get to *Task2* or *Task3*. Each task must cooperate by running only one time and then returning to the scheduler loop. We will call these tasks *cooperative tasks.*

8.4.2 Time-Slice Cyclic Scheduler

We saw in the previous section that most tasks must run at a periodic rate. If it is an event-driven task we call this the polling rate. In general we refer to the rate at which a task runs as the task period. If the system uses a free-running event loop, we can sometimes find the period by adding instruction times. When we have a multitasking loop like the one shown in Figure 8.20, however, it is generally impossible to determine the loop time. The scheduler loop time is dependent on the event-detection time and service time for every task. Usually these times are not deterministic, and we do not even know if a service is executed during a given pass through the loop. The best we can do is to determine the worst-case task time, assuming every service is executed.

To make the tasks run periodically, we can design the cyclic-scheduler loop as a timed loop. A timed scheduling loop is often called a *time-slice* scheduler because it slices time into known periods. During each time slice, each task has an opportunity to run. The flow diagram for a scheduler that uses a timed loop is shown in Figure 8.21. Each time through the loop, the scheduler starts by waiting for a timer event. The timer event may be a timer flag or a flag set by a timer's interrupt service routine.

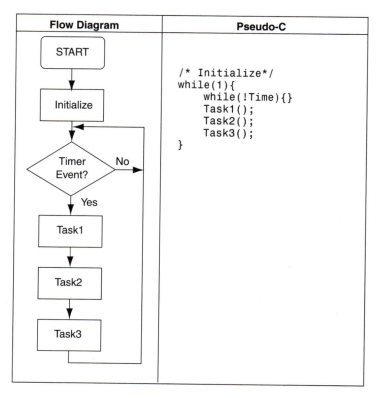

Figure 8.21 Background Time-Slice Scheduler

Figure 8.22 shows the program context for a time-slice cyclic scheduler. Once the timer event occurs, each task has the opportunity to be executed. If the task has to run every time slice, then it is run unconditionally. If a task is event driven, it checks to see if the event has occurred, and if it has occurred, it services the event; if not, it immediately returns.

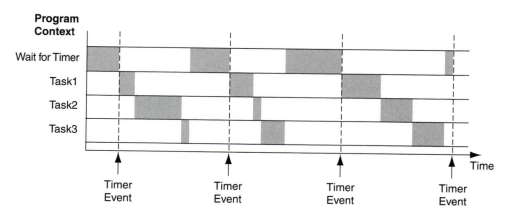

Figure 8.22 Program Context for the Time-Slice Cyclic Scheduler

Since each task can only run as often as one time per time slice, the time slice period, T_{slice}, must be less than or equal to the greatest common divisor of all the task periods.

$$T_{slice} \leq gcd\left[T_{p1},\ldots,T_{pn}\right]$$

where gcd is the greatest common divisor.

When the timer event occurs, the tasks are run sequentially. After all tasks are executed, the time remaining until the next timer event is made up in the timer event loop. Notice that in order for this system to work as expected, all tasks must be finished before the next timer event. So

$$T_{slice} > T_{T1,\,max} + T_{T2,\,max} + T_{T3,\,max} + \Sigma T_{int,\,T_{slice}}$$

where $T_{Tn,max}$ is the maximum possible execution time for *Taskn*, and $\Sigma T_{int,\,T_{slice}}$ is the time spent servicing interrupts during this time slice. So the time-slice period must be greater than the sum of all the task times but smaller than the greatest common divisor of the task periods.

Event Response Time. The response time for each task in a time-slice cyclical scheduler is a function of the time-slice period. For the nth task in the scheduler, the response time is

$$T_{Rn} \leq T_{pn} + \sum_{i=1}^{i=n-1} T_{Ti} + T_{TDn} + T_{CIR} + \Sigma T_{int,\,T_{Rn}}$$

where

$$T_{pn} = \text{the task period for the } n^{th} \text{ task}$$

$$\sum_{i=1}^{i=n-1} T_{Ti} = \text{the sum of all task times for the tasks that are executed before the } n^{th} \text{ task}$$

$$T_{TDn} = \text{the task delay time for the } n^{th} \text{ task}$$

$$T_{CIR} = \text{the circuit delays}$$

$$\Sigma T_{int,T_{Rn}} \text{ is the sum of all interrupt service times that are executed during } T_{Rn}$$

One of the assumptions made for this response time is that T_{pn} is synchronous to T_{slice}. This is normally the case because each task period is normally set to a multiple of the time-slice period.

$$T_{pn} = x_n \cdot T_{slice}$$

If T_{pn} is not synchronous to the time-slice period, we need to add both periods to the total response time.

As you can see the response time for this type of scheduler may be relatively large. It is not a responsive system, but it is simple to design. Therefore this type of scheduler is best suited for simple designs. For example, if there are no interrupts, the last term can be removed. If a task requires a fast response time, it should be the first to execute. This would remove the second term.

Finally the largest part of the response time is the task period. To lower the task response, we need to lower the task period. To lower the task period, we may need to lower the time-slice period. So lowering the time-slice period can be the most important way to lower the response time.

One Way to Lower the Time-Slice Period. Sometimes it is difficult to find a time-slice period that meets both timing requirements. Let's look at one of several techniques to reduce the time-slice period while still meeting the requirement that all tasks are finished before the next timer event.

In general not all tasks need to execute every time slice. Some tasks may need to run only every *nth* slice. One way to reduce the time-slice period is to try to spread these tasks out to mutually exclusive time slices. For example, if we needed *Task2* and *Task3* to execute every second slice, we could add a counting variable, execute *Task2* during even-numbered slices, and run *Task3* during odd-numbered slices. In this way *Task2* and *Task3* would never execute during the same slice. For this situation

$$T_{slice} > T_{T1,\,max} + max[T_{T2,\,max},\, T_{T3,\,max}]$$

The overall task time for a given time slice has been reduced because we know that *Task2* and *Task3* will never run during the same slice.

The other techniques used to reduce the scheduler period involve breaking the tasks down into smaller pieces. We will look at some of these techniques in Chapter 15.

EXAMPLE 8.5

Time-Slice Scheduler Timing Design

Imagine a system that has four tasks that need to run concurrently using the time-slice cyclic scheduler. The four tasks have the following requirements.

Task	Execution Time	Task Period
Task1	40µs	1ms
Task2	200µs	2ms
Task3	400µs	2ms
Task4	500µs	10ms

Find the required time-slice period, the CPU load, and a task schedule that will meet all the given timing requirements.

Solution

Time Slice. The time-slice period must be less than or equal to the greatest common divisor of the task polling periods. In this case, we will set the time-slice period to

$$T_{slice} = gcd[T_{p1},\, T_{p2},\, T_{p3},\, T_{p4}] = 1ms$$

CPU Load. For load calculations, assume that the task detection time for each task is negligible and that the task execution time is run at the task period. So, the CPU load is

$$L_{pk} = \frac{T_{T1}}{T_{p1}} + \frac{T_{T2}}{T_{p2}} + \frac{T_{T3}}{T_{p3}} + \frac{T_{T4}}{T_{p4}} = \frac{40\mu s}{1ms} + \frac{200\mu s}{2ms} + \frac{400\mu s}{2ms} + \frac{500\mu s}{10ms} = 0.39$$

With a load of 0.39, it should be easy to design a program that can run all four tasks.

Task Schedule. Now we have to design a task schedule that will meet the required task periods and meet the requirement that all tasks be finished within a given time slice. *Task1* runs every time slice, *Task2* and *Task3* must run every other time slice, and *Task4* runs every 10 time slices. If we simply sum the four task times, we get

$$T_{T1, max} + T_{T2, max} + T_{T3, max} + T_{T4, max} = 40\mu s + 200\mu s + 400\mu s + 500\mu s = 1.140ms$$

This is greater than the time-slice period so this design will not work—even with a CPU load of only 0.399.

We have to design the system so some tasks are never executed together. In this case, to make this system work, we can make sure that *Task3* and *Task4* never run during the same time slice. We can do this by making *Task3* run every odd time slice and *Task4* run every tenth (even) time slice. Now the worst-case total task time is

$$T_{T1, max} + T_{T2, max} + max[T_{T3, max}, T_{T4, max}] = 40\mu s + 200\mu s + 500\mu s = 740\mu s$$

This is less than the scheduler period, so the system will work.

Response Times. We can calculate the response time for each task if we assume that there are no interrupts, the task delay is equal to the task time, and we will disregard the circuit delays. We will also assume that the tasks are executed in order from *Task1* to *Task4*.

For Task1:

$$T_{R1} \leq T_{p1} + T_{TD1} = 1ms + 40\mu s = 1.04ms$$

For Task2:

$$T_{R2} \leq T_{p2} + T_{T1} + T_{TD2} = 2ms + 40\mu s + 200\mu s = 2.24ms$$

For Task3:

$$T_{R3} \leq T_{p3} + T_{T1} + T_{T2} + T_{TD3} = 2ms + 40\mu s + 200\mu s + 400\mu s = 2.64ms$$

For Task4:

$$T_{R4} \leq T_{p4} + T_{T1} + T_{T2} + T_{TD4} = 10ms + 40\mu s + 200\mu s + 500\mu s = 10.74ms$$

Notice that the response time for *Task4* does not include the task time for *Task3* because they will never be executed during the same time slice.

8.4.3 A Simple Stopwatch Example

To illustrate the program code for a system that uses a time slice scheduler, we will look at a basic stopwatch program. This program is a simplistic one-button stopwatch that counts in seconds from 0 to 99.

The main program including the initialization and scheduling loop is shown in Source 8.9. The routine *WaitSlice* controls the timing of the scheduler loop. It creates the time slices and handles a timer counter, which is used to derive a one-second count. There are three tasks: *ScanSw, UpdateSecs,* and *OutSecs. ScanSw* debounces a switch signal and based on a switch press changes the stopwatch mode by calling the *ChangeTimMode* function. The stopwatch modes are *CLEAR, COUNT,* and *STOP. UpdateSecs* updates a variable, *Seconds,* so it contains the number of seconds since the *Count* mode was entered. *OutSecs* outputs the value of *Seconds* to the serial port. It outputs a new value every time a change is detected in *Seconds.*

Source 8.9 A One-Second Stopwatch Using a Cyclic Scheduler

```
;****************************************************************
; A 1-button stopwatch based on the 68HC12 RTI System.
; This version has a debounced switch to stop/reset/start timer.
; Controller: 68HC912B32
; Monitor: D-Bug12
; Uses BasicIO module
; Pseudo-C:
; main(){
;     RTICTL = 0x04;
;     RTIFLG = RTIF;
;     SCIOPEN();
;     /*initialize globals*/
;     while(1){
;         WaitSlice();
;         ScanSw();
;         UpdateSecs();
;         OutSecs();
;     }
; }
;****************************************************************
main:       movb #%00000100,RTICTL    ;Enable RTI for 8.192mS
            movb #RTIF,RTIFLG
            jsr SCIOPEN               ;Initialize SCI port
            clr Seconds               ;Initialize Globals
            clr RtiCnt
            movb #CLEAR,TimMode
            movb #RELEASED,SwState
            movb #$ff,LastDisp        ;Force display update

sched_lp    jsr WaitSlice             ;Wait for time slice
            jsr ScanSw                ;Run Tasks
```

```
            jsr UpdateSecs
            jsr OutSecs
            bra sched_lp
;*********************************************************************
```

One thing we should notice about the main scheduling loop is how simple it is. We can look at this loop and know exactly what the system does because all the tasks are separate subroutines. In this case we can see that we wait for the timer event, check the mode switch, update the seconds value, and output the number of seconds. This program gives a good high-level view of our system.

Source 8.10 shows the kernel's timing routine, *WaitSlice*. The timing is based on the RTI flag *RTIF*. *WaitSlice* waits for the flag and then increments a counter, *RtiCnt,* for the *UpdateSecs* task. The RTI is initialized in the main program to generate an event flag every 8.192ms, which is the time-slice period. This period was selected out of the available RTI rates to meet the switch debouncing requirements of *ScanSw*. Of all the tasks, *ScanSw* has the lowest required task period. Remember that the total of the task execution times must be less than this time-slice period. In this case if the task times exceed the time-slice period, we may not be able to accurately detect a switch press.

Source 8.10 The Stopwatch Scheduling Loop Timer

```
;*********************************************************************
;
;WaitSlice - The time-slice timer routine. It uses the RTI system flag
;            RTIF. The RTI is set for an 8.192ms for the switch debounce.
;
;
; Pseudo-C:
; WaitSlice(){
;     while(RTICTL&RTIF == 0){}
;     RTIFLG = RTIF;
;     RtiCnt++;
; }
;*********************************************************************
;
WaitSlice   brclr RTIFLG,RTIF,WaitSlice
            movb #RTIF,RTIFLG
            inc RtiCnt
            rts
;*********************************************************************
```

Source 8.11 shows the *ScanSw* task. It detects a keypress, and based on the keypress it changes the timer mode variable *TimMode*.

Source 8.11 Switch Debounce Task and Timer Mode Function

```
;*********************************************************************
; ScanSw() - A switch debounce program. ScanSw detects and debounces an
;            active-low switch on PORTP, bit-0.
;
;     - Debounced for switch bounce time <16ms.
;     - With noise immunity
```

```
;      - Assumes that is executed every 8.192ms.
;
; Pseudo-C:
; ScanSw(){
;     cur_sw = PORTP & BIT0;
;     if(SwState == SW_OFF){          /* wait for switch edge */
;         if(cur_sw == PRESSED){
;             SwState = SW_EDGE;
;         }
;     }else if(SwState == SW_EDGE){ /*Verify switch press    */
;         if(cur_sw == PRESSED){
;             SwState = SW_VERF;     /*switch press verified */
;             ChangeTimMode();       /*change stopwatch mode */
;         }else{                     /*false switch press     */
;             SwState = SW_OFF;
;         }
;     }else if(SwState == SW_VERF){ /* Wait for release      */
;         if(cur_sw == RELEASED){
;             SwState = SW_OFF;
;         }
;     }
; }
;********************************************************************************
ScanSw        pshd
              ldaa PORTP                        ;cur_sw -> ACCA
              anda #BIT0
              ldab SwState

              cmpb #SW_OFF                       ;SW_OFF State
              bne ssw_edge
              cmpa #PRESSED                      ;RELEASED state
              bne ssw_exit                       ;check for edge
              movb #SW_EDGE,SwState              ;if edge state -> SW_EDGE
              bra ssw_exit

ssw_edge      cmpb #SW_EDGE                      ;SW_EDGE state
              bne ssw_verf
              cmpa #PRESSED                      ;verify switch press
              bne ssw_false
              movb #SW_VERF,SwState
              jsr ChangeTimMode                 ;change stopwatch mode
              bra ssw_exit

ssw_false     movb #SW_OFF,SwState              ;False keypress
              bra ssw_exit

ssw_verf      cmpb #SW_VERF                      ;SW_VERF state
              bne ssw_exit
              cmpa #RELEASED                     ;check for release
              bne ssw_exit
```

```
                movb #SW_OFF,SwState
ssw_exit        puld
                rts

;*********************************************************************
; ChangeTimMode() - Changes the stopwatch timer mode.
;         Timer Modes: CLEAR, COUNT, STOP
;
; Pseudo-C:
; ChangeTimMode() {
;     if(TimMode == CLEAR){
;         TimMode = COUNT;
;         RtiCnt = 0;
;     }else if(TimMode == COUNT){
;         TimMode = STOP;
;     }else{
;         TimMode = CLEAR;
;         Seconds = 0;
;         RtiCnt = 0;
;     }
; }
;*********************************************************************
ChangeTimMode   pshb                    ;preserve registers
                ldab TimMode            ;change mode
                cmpb #CLEAR
                bne chm_cnt
                movb #COUNT,TimMode ;to count mode
                clr RtiCnt              ;reinitialize RtiCnt
                bra chm_rtn
chm_cnt         cmpb #COUNT
                bne chm_stp
                movb #STOP,TimMode  ;to stop mode
                bra chm_rtn
chm_stp         movb #CLEAR,TimMode ;to clear mode
                clr RtiCnt              ;reinitialize RtiCnt and clear Seconds
                clr Seconds
chm_rtn         pulb                    ;restore registers
                rts
;*********************************************************************
```

To accurately detect switch presses, *ScanSw* must run during every pass through the scheduler loop. When a switch press is detected and verified, *ScanSw* calls *ChangeTimMode* to control the stopwatch mode.

ChangeTimMode implements the state machine in Figure 8.23. *SW_VERF* indicates that the switch debounce task reached the *SW_VERF* state, which indicates a verified switch press. Some state transitions require actions. For example, when it transitions from the *STOP* mode to the *CLEAR* mode, the *Seconds* variable is cleared. This action should be performed here and not in the *UpdateSecs* routine because it should occur only at the mode transition. If it were done in *UpdateSecs, Seconds* would be cleared every time *UpdateSecs*

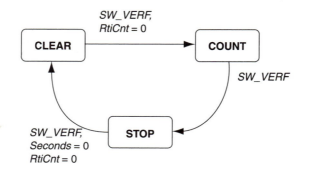

Figure 8.23   Timer Mode State Machine

runs in the *CLEAR* mode. *RtiCnt* is cleared in the *STOP* to *CLEAR* transition and the *CLEAR* to *COUNT* transition for an accurate first second.

This routine illustrates the use of a global variable to communicate with another asynchronous task. The *TimMode* variable is used by the *UpdateSecs* task to determine whether it should count or not.

Source 8.12 shows the task that updates the *Seconds* variable every second if the stopwatch is in the *COUNT* mode. It uses two global variables that are set by other tasks, *RtiCnt* and *TimMode*. It only increments the *Seconds* variable if *RtiCnt* is equal to 122 and *TimMode* is equal to one. *RtiCnt* is incremented by the *WaitSlice* task every time slice. Therefore it takes approximately one second for *RtiCnt* to reach 122. After *RtiCnt* reaches 122, this task clears it for the next one-second count.

Source 8.12 Stopwatch Seconds Counter Task

```
;**************************************************************************
; UpdateSecs - This routine increments Seconds every second.
;    - Seconds is a two-digit decimal number.
;    - RtiCnt is used to determine one second.
;    - Actual rate: 122 x 8.192ms = 0.999424S (0.058% error)
;    - Count is enabled when Mode is COUNT.
; Pseudo-C:
; UpdateSecs(){
;     if((TimMode == COUNT) && (RtiCnt == 122)){
;            RtiCnt = 0;
;            AddOneSecBCD();
;     }
; }
;**************************************************************************
UpdateSecs  psha                ;preserve ACCA
            ldaa TimMode         ;check switch state
            cmpa #COUNT          ;1: increment seconds
            bne upsec_rtn
            ldaa RtiCnt          ;one second?
            cmpa #122
            bne upsec_rtn        ;no: return
            clr RtiCnt
```

```
                ldaa Seconds
                inca
                daa
                staa Seconds
upsec_rtn       pula                ;recover ACCA
                rts
*********************************************************************
```

Seconds actually contains a BCD count, so when it is incremented, the *daa* instruction is used to convert the binary addition to a BCD addition. Since *Seconds* is only one byte, the range of this stopwatch is 0 to 99 seconds.

Finally Source 8.13 shows the routine to display the number of seconds on a terminal. For this example the display is updated every time a change is detected in *Seconds*. This design results in the lowest average CPU load and is more responsive than executing it at a fixed rate. It not only updates the display when *Seconds* is incremented by *UpdateSecs*, it also updates it when *Seconds* is cleared by the *TimMode* task.

Source 8.13 Stopwatch Output Routine

```
;*********************************************************************
; OutSecs - Displays Seconds on the terminal if Seconds has
;           changed.
; Pseudo-C:
; OutSecs() {
;     if(Seconds != LastDisp){
;         LastDisp =  Seconds;
;         OUTHEXB(&Seconds);
;     }
; }
;*********************************************************************
OutSecs         pshb                ;preserv ACCB
                ldab Seconds        ;Seconds changed?
                cmpb LastDisp
                beq outs_rtn        ;no: return
                stab LastDisp
                ldab #BS
                jsr PUTCHAR
                jsr PUTCHAR
                ldd #Seconds        ;yes: update display
                jsr OUTHEXB
outs_rtn        pulb
                rts
*********************************************************************
```

Another issue that comes up in the *OutSecs* routine is the potential for a long task time because of the display routines. When the bit rate of the SCI is set to 9600bps, it takes about 1ms for each character to be transmitted. This means the execution time for this routine is more than 2ms. This is less than the scheduler loop time, but you can imagine that it would not take a very long display message to reach the 8ms limit.

The primary limitation to this type of kernel is when there is a task that takes longer than one time slice to execute. When this occurs we have to determine the seriousness of missing some timer events. For some systems there will be an unnoticeable pause, while for other systems it would have catastrophic results. For our stopwatch example, if the output routine were long enough to miss an RTI event, our accuracy would be reduced and the switch might not be detected accurately. These would be considered serious problems. The most practical way to solve problems such as these is to use a preemptive kernel with an output task that can be interrupted. This requires using a more complex kernel, but that may be a better alternative than to try to get it to work with this scheduler. Another solution may be to use the task decomposition techniques covered in Chapter 15.

Section Summary

In this section we looked at one type of multitasking kernel. The time-slice cyclic kernel was chosen because of its simplicity and flexibility. It is appropriate for many applications and is very simple to design, debug, and understand. It is not very responsive, however. It is only one of many variations for a cooperative multitasking kernel. Chapters 15 and 16 cover more techniques for designing and using multitasking kernels.

▶ 8.5 USING CPU12 INTERRUPTS

In Section 8.3, we saw that an interrupt is a hardware solution that detects an event and automatically jumps to a service routine. As we saw in Example 8.3, interrupts are especially beneficial when detecting and servicing sporadic events. This is because an event loop is no longer required to detect the event and jump to the service routine.

In this section we look at the CPU12 interrupts in enough detail to illustrate the concepts and processes involved in using interrupts. For complete detail on the CPU12 interrupts, refer to the *CPU12 Reference Manual* and the *Technical Summary* for the specific microcontroller being used.

8.5.1 CPU12 Interrupt Sources

Table 8.1 shows the interrupt sources for the M68HC912B32 microcontroller. Most of these interrupt sources also can be found on the other CPU12-based microcontrollers. Compared to a microprocessor-based system, there is a large number of interrupt sources. This is one of the advantages to using a microcontroller. The internal hardware on a microcontroller allows for a large number of sources that can each make the CPU jump directly to an appropriate service routine.

We can divide the interrupt sources into four groups: interrupts from on-chip resources, external interrupts, software interrupts, and reset exceptions. In this chapter we cover the external interrupts and the software interrupt. The on-chip interrupt sources are covered in Chapter 9 and the reset exceptions (Opcode Trap, COP Failure, RESET, and Clock Monitor Failure) are covered in Chapter 10.

Table 8.1 also shows the interrupt mask bit and the local enable bits associated with each interrupt source. All on-chip interrupt sources and the *IRQ* can be enabled or disabled

TABLE 8.1 68HC912B32 INTERRUPT SOURCES

Interrupt Source	Vector Address	CCR Mask	Local Enables
Reserved	`$FF80-$FFCF`	I	—
BDLC	`$FFD0:$FFD1`	I	`IE`
ADC	`$FFD2:$FFD3`	I	—
Reserved	`$FFD4:$FFD5`	I	—
SCI 0 Serial System	`$FFD6:$FFD7`	I	`TIE, TCIE, RIE, ILIE`
SPI Serial Transfer	`$FFD8:$FFD9`	I	`SPIE`
Pulse Accumulator Input Edge	`$FFDA:$FFDB`	I	`PAI`
Pulse Accumulator Overflow	`$FFDC:$FFDD`	I	`PAOVI`
Timer Overflow	`$FFDE:$FFEF`	I	`TOI`
Timer Channel 7	`$FFE0:$FFE1`	I	`C7I`
Timer Channel 6	`$FFE2:$FFE3`	I	`C6I`
Timer Channel 5	`$FFE4:$FFE5`	I	`C5I`
Timer Channel 4	`$FFE6:$FFE7`	I	`C4I`
Timer Channel 3	`$FFE8:$FFE9`	I	`C3I`
Timer Channel 2	`$FFEA:$FFEB`	I	`C2I`
Timer Channel 1	`$FFEC:$FFED`	I	`C1I`
Timer Channel 0	`$FFEE:$FFEF`	I	`C0I`
Real-Time Interrupt	`$FFF0:$FFF1`	I	`RTIE`
/IRQ	`$FFF2:$FFF3`	I	`IRQEN`
/XIRQ	`$FFF4:$FFF5`	X	—
SWI	`$FFF6:$FFF7`	—	—
Unimplemented Instruction Trap	`$FFF8:$FFF9`	—	—
COP Failure Reset	`$FFFA:$FFFB`	—	`COP Rate Select`
Clock Monitor Failure Reset	`$FFFC:$FFFD`	—	`CME, FCME`
RESET	`$FFFE:$FFFF`	—	—

by the *I* interrupt mask bit in the condition code register and a local enable bit. To enable one of these sources, the mask bit must be cleared, and the local enable bit must be set. For example if we want to enable the RTI interrupt, we need to set the *RTIE* bit in the *RTICTL* register and clear *I* in the condition-code register.

8.5.2 CPU12 Interrupt Processing

The interrupt process consists of hardware that detects an event and signals the CPU and CPU exception processing that sample the interrupt signals and automatically changes the program flow to the interrupt service routine. At the end of the service routine, the return-from-interrupt instruction *rti* changes context back to the interrupted program. We saw the high-level program flow in Figure 8.11 on page 249. Now let's look at the interrupt process in more detail.

The detailed CPU12 interrupt process is shown in Figure 8.24. As we can see all the interrupt processing occurs between the execution of two instructions. This process is often called *exception processing* and interrupts are often called *exceptions* because the CPU will execute the next instruction *except* when one of the interrupts is pending. Because some of the special purpose CPU12 instructions are long, they are interruptible. This is not a common practice but it is necessary to keep the interrupt latency down to a reasonable level.

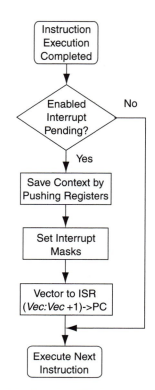

Figure 8.24   Detailed CPU Interrupt Processing

An interrupt is said to be pending when it is enabled, its event has been detected, and it has not been serviced. If the CPU detects a pending interrupt, it saves the context by pushing all the CPU registers, sets the appropriate interrupt mask bits, and vectors to the interrupt service routine.

Pending Interrupt Detection.   The first part of the interrupt process is to detect the interrupt event and signal the CPU that the event occurred. Most interrupt events are detected by the on-chip resources. Only IRQ and XIRQ are detected directly by the CPU. The CPU checks the interrupt signals from all sources after it has finished executing each instruction. In effect the sample time is equal to the longest instruction or the longest time that the interrupts are disabled during normal operation.

The interrupt signals from the on-chip resources are implemented either as event flags that must be cleared explicitly or as automatic event flags that are cleared automatically when a related register is accessed. Because they are flags, it is easy to service an interrupt one time and only one time. This is not true, however, for the IRQ and XIRQ interrupts, which are external active-low signals. The CPU uses level detection, so these sources are serviced as long as the interrupt signal is low.

Saving Context.   Once a pending interrupt is detected, the CPU12 saves the context by pushing all the CPU registers onto the stack. This is similar to a subroutine call except

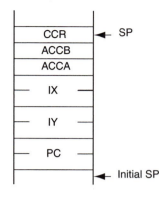

Figure 8.25 The CPU12 Interrupt Stack Frame

that it saves all the registers instead of only the program counter. All registers must be saved because the interrupt is asynchronous to the program flow and we cannot predict when it will occur.

Figure 8.25 shows the order that the registers are pushed. Since the CPU12 saves all registers automatically, there is no need to preserve registers by pushing them onto the stack in the ISR as we do when writing a subroutine. One anomaly of the CPU12 interrupt stack frame is the order of ACCB and ACCA. Normally ACCB would be pushed first to be consistent with a big-endian processor. This anomaly originated in the 6800 CPU when 16-bit operations were not foreseen. It was not changed in the 68HC12 MCUs to preserve code compatibility with the 68HC11.

Not all CPUs push all the registers when an interrupt occurs. For the CPUs that do not, the registers must be pushed by the ISR before they are used. On one hand this allows us to save only the registers that are used in the ISR, which may save time and stack space. We have to use the push and pull instructions, however, which require more program space and are slower per register than the automatic method. It also forces the programmer to remember to save the registers, which if forgotten can produce some serious bugs.

At the end of the interrupt service routine, the return-from-interrupt instruction *rti* pulls the interrupt stack frame from the stack, which switches the CPU context back to the interrupted program.

Set Interrupt Masks. After the context is saved on the stack, the CPU sets the appropriate interrupt masks. If the interrupt source is one controlled by the *I* mask, then *I* will be set automatically. If the interrupt source is the XIRQ, then both *X* and *I* masks are set. This is done to prevent nested interrupts before the system is ready. For example, imagine if an on-chip interrupt source interrupted the CPU and the *I* mask was not set automatically. All on-chip interrupt sources are seen as pending by the CPU as long as their flag bit is set. Normally the flag bit is cleared in the ISR. But if the mask were not set, the CPU would see the same interrupt as still pending and initiate the interrupt process again. This would occur before the first ISR instruction was executed, so there would be no chance to manually set the flag. The CPU would continue to be interrupted without running any instructions until it crashed.

To implement nested interrupts, we can clear the mask bit after entering the ISR, which would allow the ISR to be interrupted by another pending interrupt. Before clearing the mask, however, we need to prepare the system for another interrupt. This means that at least the source's flag must be cleared before the interrupt mask is cleared.

Before the context switch is made, the CCR contains the interrupt masks for the interrupted program. The CCR is then pushed onto the stack during the context switch. Normally when the *rti* instruction is executed, the same CCR is pulled from the stack and the interrupt masks are at the same state as before the interrupt occurred. This is usually what we want to happen, but occasionally we may want an interrupt to disable further interrupts. This can be done in the ISR by setting the *I* mask in the CCR that is stored on the stack with

```
bset 0,sp,%00010000
```

This would cause the interrupts controlled by the *I* mask to be disabled when the context is switched back to the interrupted program and the CCR is pulled off the stack. Note that we cannot do this with the *X* mask. Once the *X* mask is cleared, it remains cleared until the system is reset or a context switch to an XIRQ ISR is performed.

Vectors. The next step in the interrupt processing shown in Figure 8.25 is to get to the interrupt service routine. An interrupt vector is two fixed memory locations that contain the address of the interrupt service routine for a given interrupt. The CPU12 interrupt vectors are stored at $FF80–$FFFF. Table 8.1 shows the vector location for each interrupt source on the M68HC912B32. So for example, to load the interrupt vector for the program in Source 8.6, we would use the following code:

```
org $ffee
fdb Ic0Isr
```

This loads the address of the *Ic0Isr* interrupt service routine into $FFEE:$FFEF, which is the vector for *Timer Channel 0*. Now when a rising edge is detected on *PORTT,* bit 0, the CPU saves the context, sets the *I* mask, and loads the program counter with the contents of addresses $FFEE:$FFEF, which is the address of the ISR.

The code snippet above only loads one interrupt vector. For real programs all vectors should be defined, even if they are not used. The vectors for unused interrupts can be loaded with the address of a program that handles unexpected interrupts. In the final product this improves reliability by protecting the system from spurious interrupts. During development it can help detect spurious interrupts.

8.5.3 Using Interrupts with D-Bug12

When we are writing programs that use interrupts under the D-Bug12 monitor, it is impractical to change the actual interrupt vectors. The vectors are stored in ROM with D-Bug12 and some vectors are used by D-Bug12. Even if they are in Flash ROM, it is not practical to reprogram the whole Flash ROM every time we write or change a program. To accommodate interrupts, D-Bug12 includes a utility routine, *SetUserVec(),* that will load the ISR address into a jump table located in monitor RAM. The actual vector contains the address of an interrupt dispatch program in D-Bug12. If there is a user ISR address in the jump table, the dispatch program jumps to that address. If there is not a user ISR address in

TABLE 8.2 *SetUserVec()* V2.0.0 INTERRUPT VECTOR NUMBERS

Interrupt	Number
PORTH Key Wake-up	7
PORTJ Key Wake-up	8
A-to-D	9
SCI #1	10
SCI #0	11
SPI #0	12
Pulse Accum. Edge	13
Pulse Accum. Overflow	14
Timer Overflow	15
Timer Channel 7	16
Timer Channel 6	17
Timer Channel 5	18
Timer Channel 4	19
Timer Channel 3	20
Timer Channel 2	21
Timer Channel 1	22
Timer Channel 0	23
RTI	24
IRQ	25
XIRQ	26
SWI	27
Unimplemented Opcode Trap	28
Jump Table Address	−1

the jump table, the dispatch program displays a message indicating that an unexpected interrupt occurred.

To use *SetUserVec()*, we pass the ISR address and a vector number that corresponds to the interrupt source. The vector numbers for *SetUserVec()* are shown in Table 8.2. For example, if we need to use the *Timer Channel 0* interrupt, the vector number would be 23.

If the pulse counting program in Source 8.6 had been written to run under D-Bug12, we would have had to use the following code to initialize the jump table:

```
*****************************************************************
* Equates
*****************************************************************
USER_TC0    equ 23
SETUSERVEC  equ $f69a
*****************************************************************
ICInt       ldd #Ic0Isr               ;Initialize D-Bug12
            pshd                      ;Jump Table
            ldd #USER_TC0
            jsr [SETUSERVEC,pcr]
            puld

            ...
*****************************************************************
```

To call *SetUserVec( )* in an assembly program, the ISR address is passed on top of the stack and the vector number is passed in ACCD. As you can see the code shown passes the address of *Ic0Isr* and the *Timer Channel 0* vector number to *SetUserVec( )*. This code snippet would be one of the first tasks of the initialization sequence. As we see in the next subsection, using interrupts under D-Bug12 increases the interrupt latency compared to a stand-alone system.

8.5.4 Interrupt Latency

A critical parameter in many real-time designs is the time it takes the CPU to get to an interrupt service routine. This time plays a major role in the response time of the system. The time between when an interrupt event occurs and the time the interrupt service routine is started is called *interrupt latency*. The interrupt latency is

$$T_l = max[T_{i,\ disabled}, T_{instr}] + T_{csw}$$

where

$T_{i,disabled}$ = the longest period in which interrupts are disabled during normal operation

T_{instr} = the longest execution time for any instruction in the program

T_{csw} = the context switch time for an interrupt

Before an interrupt is serviced, the interrupt mask must be cleared and the CPU must finish executing the current instruction. Since we do not know when the interrupt event will occur in the program flow, the worst case is either the longest time that the interrupt is masked or the longest instruction. For the CPU12 the longest instruction that is not interruptible is *emacs*, which requires 13 cycles to execute.

The CPU12 interrupt context switch time is nine cycles. This includes pushing all registers, setting the interrupt masks, and vectoring to the ISR. If interrupts are not disabled during normal operation the worst-case interrupt latency for a CPU12 is

$$T_l = T_{instr} + T_{csw} = (13 + 9)T_E = 2.75\mu s$$

Latency under D-Bug12. As mentioned earlier, if the program is running under D-Bug12, there is an increase in latency. This is because the CPU first vectors to a D-Bug12 program to determine whether a user vector has been entered into the jump table. If there is a user vector, the CPU jumps to the user's ISR. Under D-Bug12 the latency time is

$$T_l = max[T_{i,\ disabled}, T_{instr}] + T_{csw} + T_{DB12}$$

If interrupts are not disabled during normal operation, the latency time under D-Bug12 is

$$T_l = T_{instr} + T_{csw} + T_{DB12} = (13 + 9 + 31)T_E = 6.625\mu s$$

Approximately 3.875μs is added to the latency time by running the program under D-Bug12. For most programs, this would be insignificant. But, for programs that require a very fast interrupt response time, it would be unacceptable.

Interrupt latency is also increased if there are multiple interrupts being used. This is described in the next section.

Interrupt Response Time. One of the primary advantages to using interrupts is the potential for very low response times. This is because we do not have to poll the signal to detect the event. Therefore there is not a sample period component to the response time. In general the response time is

$$T_R = T_l + T_{TD} + T_{CIR}$$

This response time assumes that there are no other interrupts enabled in the system.

8.5.5 Multiple Interrupts and Priority

When a system has more than one interrupt source enabled, it must be able to handle the condition when more than one interrupt is pending at the same time. Since the CPU can only handle a single interrupt at a time, a priority is assigned to each source. The CPU then services the highest priority pending interrupt first.

The first column in Table 8.3 shows the default priorities for each interrupt source on the M68HC912B32 MCU. In this case *RESET* has the highest priority and the BDLC has the lowest priority. The nonmaskable interrupt sources, shown above the double line in Table 8.3, have a fixed priority that cannot be changed. The relative priority for the

TABLE 8.3   68HC912B32 INTERRUPT PRIORITIES

Default Interrupt Priorities	Interrupt Priorities with HPRIO = $EE
RESET	*RESET*
Clock Monitor	Clock Monitor
COP Failure	COP Failure
Unimplemented Instruction Trap	Unimplemented Instruction Trap
SWI	SWI
/XIRQ	/XIRQ
/IRQ	Timer Channel 0
Real-Time Interrupt	/IRQ
Timer Channel 0	Real-Time Interrupt
Timer Channel 1	Timer Channel 1
Timer Channel 2	Timer Channel 2
Timer Channel 3	Timer Channel 3
Timer Channel 4	Timer Channel 4
Timer Channel 5	Timer Channel 5
Timer Channel 6	Timer Channel 6
Timer Channel 7	Timer Channel 7
Timer Overflow	Timer Overflow
Pulse Accumulator Overflow	Pulse Accumulator Overflow
Pulse Accumulator Input Edge	Pulse Accumulator Input Edge
SPI Serial Transfer	SPI Serial Transfer
SCI 0 Serial System	SCI 0 Serial System
ADC	ADC
BDLC	BDLC

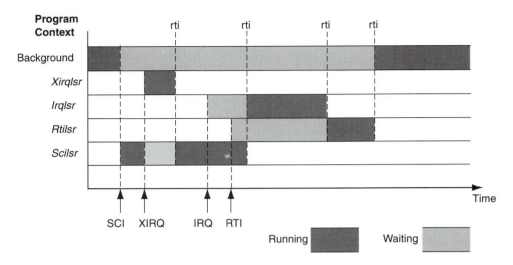

Figure 8.26 Normal CPU12 Interrupt Operation

software interrupt and the unimplemented opcode trap is ambiguous. They are both generated by a program instruction, so they can never occur at the same time.

The order of the maskable interrupt sources below the double line in Table 8.3 is also fixed except that one interrupt source can be promoted to be the highest-priority interrupt by changing the *HPRIO* register. The second column in Table 8.3 shows the priorities with *Timer Channel 0* promoted. With the exception of *Timer Channel 0,* the rest of the interrupt priorities are in the same order. To promote an interrupt source, we write the least-significant byte of the vector address for that source to the *HPRIO* register. Therefore to promote *Timer Channel 0,* $EE was written to *HPRIO.*

Now let's look at how the CPU deals with multiple interrupts. We know the background program will be interrupted, or preempted, by an interrupt. But what happens when we are already in an ISR? Figure 8.26 shows the context diagram for a system that uses four interrupt sources, SCI, XIRQ, IRQ, and RTI. The occurrences of the interrupt events are shown at the bottom of the figure. They are placed close enough to each other so that all but the first SCI interrupt event occurs during the execution of the SCI ISR. This diagram illustrates normal operation in which the interrupt mask remains set throughout each ISR.

First the SCI interrupt immediately preempts the background program. While the SCI ISR is executing, an XIRQ interrupt occurs and then an IRQ and RTI. The XIRQ interrupt does preempt the SCI ISR because the *X* mask is not set by the SCI's interrupt process. While the XIRQ ISR runs, it can only be preempted by a *Fault Tolerance Exception* or a *RESET.* When it is finished, the *rti* instruction returns the CPU to the SCI ISR context. Now when the IRQ occurs, the SCI ISR is not preempted despite the IRQ having a higher priority than the SCI. This is because during the SCI ISR, the *I* mask is set, disabling all other maskable interrupts. The same applies when the RTI interrupt occurs. When the SCI ISR is complete, the *rti* instruction does not return the CPU to the background because there are other interrupts pending. Instead the context is switched to the pending interrupt with the highest priority—in this case IRQ. In fact, even if the IRQ interrupt event occurred after the RTI, it would still be run first because it has a higher priority.

The RTI then has to wait for the IRQ ISR to be completed before the context is switched to the RTI ISR. Finally, after the RTI ISR is completed, the *rti* instruction returns the context to the background because no other interrupts are pending.

Because the maskable interrupts cannot preempt an interrupt service routine, the interrupt latency is greatly increased. For a given interrupt with priority *n,* the latency becomes

$$T_{l,n} = max[T_{i,\,disabled},\ T_{instr}] + T_{csw} \cong max[T_{i,\,n+1},\ldots,\ T_{i,\,l}] + \Sigma(T_{i,\,n-1},\ldots,\ T_{i,\,h}) + T_{csw}$$

where $T_{i,\,n+1},\ldots,T_{i,\,l}$ are all lower priority interrupts and $T_{i,\,n-1},\ldots,\ T_{i,\,h}$ are all higher-priority interrupts that must run first. So the worst-case latency is the longest ISR execution time for all interrupts with lower priority plus the sum of all higher-priority interrupts that become pending before the interrupt can run plus the context switch time.

In the example shown in Figure 8.26, the worst-case latencies for the three interrupts controlled by the *I* mask are

$$T_{l,\,irq} = max[T_{i,\,rti},T_{i,\,sci}] + n_{xirq}T_{i,\,xirq} + T_{csw}$$

$$T_{l,\,rti} = T_{i,\,sci} + n_{xirq}T_{i,\,xirq} + n_{irq}T_{i,\,irq} + T_{csw}$$

$$T_{l,\,sci} = n_{xirq}T_{i,\,xirq} + n_{irq}T_{i,\,irq} + n_{sci}T_{i,\,sci} + T_{csw}$$

The latency for even the highest-priority interrupt, IRQ, is significant, especially if the RTI or SCI ISR have long execution times. It is still shorter than the lower-priority interrupts because it only has to wait for one lower-priority ISR at most to complete and possibly the XIRQ ISR. The lower-priority interrupts must wait until all pending higher-priority interrupts are complete.

We can decrease the latency of the highest-priority interrupt by allowing it to preempt any lower-priority interrupt. To allow the higher-priority interrupt to preempt a lower-priority ISR, we must clear the *I* mask upon entering the lower-priority ISR. Figure 8.27 shows the effects of clearing the *I* mask in the RTI and SCI ISRs.

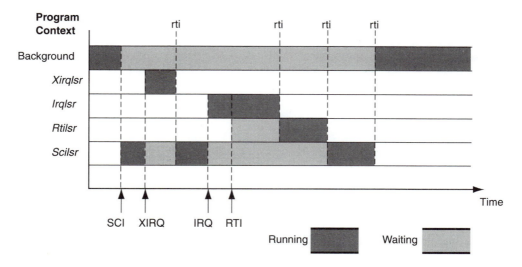

Figure 8.27 CPU12 Interrupt Operation with *I* Cleared in RTI and SCI ISRs

The XIRQ still preempts the SCI ISR as before. This time, however, when the IRQ occurs, the SCI ISR is preempted and the IRQ ISR is executed immediately. Therefore we have reduced the worst-case latency for the IRQ to

$$T_{l,\,irq} = T_{i,\,xirq} + T_{csw}$$

when the RTI interrupt occurs and it does not preempt the IRQ ISR because the I mask remains set. If the I mask was cleared in the IRQ ISR, the RTI would preempt the IRQ even with a lower priority. Therefore this method will work for only one high-priority interrupt. We cannot build a system based on the CPU12 interrupt priorities that fully implement a priority-based preemptive system.

8.5.6 Critical Regions

A *critical region* is a section of code that should not be interrupted. Critical regions normally involve code that changes a shared resource. Changing a shared resource becomes a concern as soon as there is a possibility of the code being preempted. As we have seen in this section, interrupts will preempt a program if the interrupt source is enabled. Later we discover that preemptive multitasking kernels can also preempt a program.

To illustrate the problem let's look at a simple example. Source 8.14 shows parts of a program that collects 16-bit data in the background and displays the data every second in an RTI ISR. The shared resource in this case is the 16-bit global variable *DisplayBuf*. The background program is an endless loop that collects data using the *GetData* subroutine. It then stores the value returned into *DisplayBuf*. The RTI interrupts occur asynchronously to this process and display the contents of *DisplayBuf*.

Source 8.14 A Potential Shared Resource Error

```
*****************************************************************
* while(1){
*     DisplayBuf = GetData();
* }
*****************************************************************
main_lp     jsr GetData
            staa DisplayBuf
            stab DisplayBuf+1
            bra main_lp


*****************************************************************
* RtiIsr(){
*     RTIFLG = RTIF;
*     RtiCnt++;
*     if(RtiCnt == 61){
*         OUTHEXW(DisplayBuf);
*     }
* }
*****************************************************************
RtiIsr      movb #RTIF,RTIFLG    ;clear RTI flag
            ldaa RtiCnt          ;count RTIs for one second
```

```
          inca
          cmpa #61
          bne rti_rtn
          clr RtiCnt            ;if one second display data
          ldd #DisplayBuf
          jsr OUTHEXW
rti_rtn   rti
```

****

The problem occurs if the RTI interrupt preempts the main loop between the *staa* and *stab* instructions. Imagine that the last value in *DisplayBuf* was $0099 and the new value is $0100. If the RTI interrupt preempts the storing process between *staa* and *stab,* the displayed data would be $0199, not $0100 or $0099.

What we have to do is to make sure the data storage process is not preempted. For this simple example there are two solutions. The most obvious solution is to store the data with a single *std* instruction instead of the two *staa* and *stab* instructions. So the main loop becomes

```
main_lp   jsr GetData
          std DisplayBuf
          bra main_lp
```

Since the storage process is completed with a single instruction, it cannot be preempted.

If *DisplayBuf* were longer than 16 bits, however, we could not use a single instruction to complete the storage process. When this is the case, interrupts must be disabled during the storage process. To illustrate this, following is another version of the main loop that protects the critical region by disabling interrupts:

```
main_lp   jsr GetData
          sei                  ;disable interrupts
          staa DisplayBuf      ;critical region
          stab DisplayBuf+1
          cli                  ;enable interrupts
          bra main_lp
```

When interrupts are disabled to protect a critical region, the interrupt latency can be affected. In this case interrupts are only disabled for eight cycles, assuming extended addressing for the store instructions. This is less than the longest instruction, so it would not play a role in the worst-case latency.

8.5.7 External Interrupts

There are two external interrupt pins on the 68HC12 microcontrollers, IRQ and XIRQ. IRQ is an active-low, maskable interrupt request. XIRQ is an active-low, pseudo-nonmaskable interrupt request. These two interrupt pins are typical of those found on many microcontrollers and microprocessors. For microprocessors two pins such as these would be the only sources for interrupt signals.

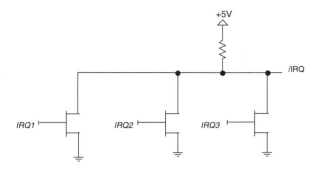

Figure 8.28   Multiple Interrupt Sources Connected As Wired-OR

IRQ is a general-purpose interrupt source. It can be used by any peripheral circuits to interrupt the CPU. Because XIRQ is nonmaskable, it should be reserved for special purposes such as debugging or power-line monitoring. Because many debug monitors and emulators use XIRQ to control the CPU, the use of XIRQ in the program may preclude the use of these tools.

By default the CPU uses active-low level detection for these two interrupt sources. If one of them is low, it will vector to the appropriate ISR. Level detection has its advantages and disadvantages. The advantage of level detection is that we can use a wired-OR circuit to connect several external sources to the same pin. Figure 8.28 shows the logical connection of three interrupt sources, *IRQ1, IRQ2,* and *IRQ3,* to the /IRQ pin on the 68HC12. This is only a logical diagram. The actual circuits more likely would be realized as peripheral ICs that have open-drain outputs for interrupt requests.

Note that the CPU cannot determine the actual source of the interrupt with this circuit alone. There will be one ISR to service all of these interrupt sources. To determine the actual source, the ISR first polls the interrupt sources. This requires either a bus interface or an input port connected to signals that contain the interrupt status for each source.

The disadvantage of using level detection is that it may require additional circuitry to make sure an interrupt event is serviced one time and only one time. Without additional circuitry the level detection results in an *as-long-as* response. So we have to be careful that the interrupt signal is not too long and not too short.

The CPU samples the interrupt signals between each instruction so the signal must be low when the CPU samples it. Therefore, to guarantee it will be detected, the interrupt signal must be low longer than the longest instruction or the longest time that the interrupt is disabled. The interrupt will be serviced as long as the interrupt signal is low. So if the signal is low longer than the ISR, it will be serviced again immediately after the program returns from the ISR. This results in an interrupt signal requirement of

$$max\left[T_{i,\ disabled},\ T_{instr}\right] \le T_{irq} \le T_{ISR} + T_l$$

Without an external circuit to clear the interrupt signal, this can be difficult if not impossible to achieve.

For applications in which level detection is not appropriate, the IRQ interrupt can be set to detect falling edges on the /IRQ pin by setting the *IRQE* bit in the interrupt control register. The advantage to using edge detection is that the event is detected and held pending until the CPU services it, so there is no pulse-width requirement. The disadvantage to using edge detection is that it cannot handle noisy or bouncy signals. These signals may generate extra edges, which would result in multiple interrupts.

8.5.8 Software Interrupts

A software interrupt is an instruction that initiates the interrupt process. In the CPU11 and CPU12, the software interrupt instruction is *swi*. Software interrupts should be used exclusively for debugging tools such as software breakpoints in debug monitors and emulators. Using an *swi* in a program may preclude the use of these tools.

As with other interrupts, when the *swi* instruction is reached, the context is saved, the *I* mask is set, and the CPU vectors to the *swi* ISR. The software interrupt is in effect a single-opcode subroutine call. Because it is only a single opcode, any other instruction opcode can be substituted with an *swi,* which makes it ideal to implement software breakpoints in a debug monitor or emulator.

When a software breakpoint is activated, the monitor program substitutes the instruction at that address with an *swi*. The *swi* ISR then performs the breakpoint program. When the monitor is finished with the breakpoint, the *swi* opcode is replaced with the original opcode. Because the opcodes must be changed, software breakpoints are restricted to programs stored in RAM. Because the *swi* also must be executed in the normal program sequence, it must be placed at an address that contains the first opcode of an instruction.

Another common use for the software breakpoint is to fill all unused ROM locations with an *swi* opcode for fault tolerance. If the CPU program execution gets lost because of a bug, power glitch, or noise, it will land somewhere in the normal program or on an *swi*. The *swi* ISR can then perform the correct action such as a system reset or user warning and trap.

8.5.9 Interrupt Usage Guidelines

Make Sure the Interrupt Source Is Enabled and Configured. Probably the most common reason an interrupt-based system does not work is that all the appropriate enables, masks, and control registers have not been configured. For all on-chip interrupt sources on the 68HC12, there is a local enable that must be set in addition to the clearing *I* mask. Also make sure the interrupt source has been configured correctly. For example, if we are using an interrupt from a timer output compare or input capture, we must first turn on the timer.

Clear the Flag. Another common error is to forget to clear the source's flag bit before exiting the ISR. Remember that the interrupt is pending as long as its flag is set. If you do not clear the flag while in the ISR, the CPU will immediately reenter the ISR after the *rti* instruction. This results in a system lockup that can only be stopped by a *RESET* or non-maskable interrupt.

Also make sure you understand how the flags are cleared. Some are automatically cleared when you read or write to a register related to the source. The timer flags are cleared by writing to the flag register with a one in the bit position of the flag to be cleared. This is covered in more detail in Chapter 9.

Stack Space and Pointer. Interrupts use the stack. Make sure the stack pointer is initialized before any interrupts are enabled. Also make sure there is adequate stack space. For each interrupt nine bytes are stored on the stack. If interrupts are nested, we need nine bytes of stack space for each nesting level along with any additional stack space used by the ISRs.

Be Careful When Nesting. Be careful before you enable interrupts within an ISR so interrupts can be nested. Because of some subtle bugs that can be created, it is best to avoid nesting at all. But, if it is necessary, make sure you have prepared the program for interrupt nesting. For example make sure the flag that caused the interrupt is cleared before enabling further interrupts.

D-Bug12 Latency. When running under D-Bug12, remember that there is almost 4µs additional latency time. Typically this will go unnoticed, but for some applications it can make the difference of the program working or not.

Keep It Short. As a general rule of thumb, always keep your ISRs as short as possible. We saw that the time spent in ISRs directly affects the latency for all other interrupts and the response time for every task in the system. Obviously there should never be a blocking loop of any kind in an ISR. Yes, there are exceptions. Some systems perform all tasks in ISRs and the background loop does nothing. The only time this method should even be considered is when we need to place the CPU in a low-power mode until an event occurs.

Critical Regions. Make sure all critical regions are protected by masking interrupts. Be especially watchful of shared resources. On the other hand realize that the longer the interrupts are disabled, the longer the interrupt latency time. Try to keep critical regions to a minimum and comment them very well.

External Interrupt Signals. The external interrupts use simple level detection. If we require that the event be serviced one time and only one time, make sure the interrupt is active long enough to be detected but short enough not to be detected more than one time. It may require additional external circuitry. If IRQ is set for edge detection, make sure the signal source does not bounce or contain noise that could be seen as event edges.

Reserved Sources for Debugging. The software interrupt *swi* and the pseudo-nonmaskable interrupt XIRQ should be reserved for debugging systems. Using these interrupts in your application may preclude the use of important debugging tools. If you do use these interrupts, it should be for fault tolerance or events that may result in catastrophic failures.

Reinitialize Registers. Do not assume a register contains a certain value. Since interrupts can occur anywhere in the program, assuming the contents of any register requires the register to contain the assumed value throughout the whole program. For example, if you access control registers using an index register and an offset, make sure to initialize the index register in the ISR before using it. Do not assume that it is valid just because you initialized it in the main program. This problem comes up more often when programming with the CPU11 because the control registers are typically accessed by using indexed addressing.

▶ 8.6 BASIC REAL-TIME DEBUGGING

In Chapter 3 we covered the debugging commands for the D-Bug12 monitor. These tools work well as long as the program does not have to respond to real-time events. In this section we look at ways to use a monitor program, along with some new tools and techniques to help debug and analyze real-time programs.

8.6.1 Real-Time Debugging with D-Bug12

The ability to debug a real-time program with D-Bug12 depends on the development system configuration. In this section we cover the single EVB configuration as shown in Chapter 3. This configuration limits the effectiveness of D-Bug12 when working with real-time systems because the monitor program is running on the same processor as the user's program. In Chapter 10 we look at using D-Bug12 in a two-EVB setup that uses the 68HC12's background debug mode. Using the background debug mode and running D-Bug12 on a second EVB allows us to look at and edit memory contents while the program is running. In the single EVB configuration, we cannot do this because the program must be stopped to run D-Bug12.

Using D-Bug12 or any monitor program in a single-EVB configuration is an invasive debugging system. It is invasive because in order to give feedback to the user, the CPU context must be switched from the user's program to the monitor program. Therefore the user's program must be stopped while the monitor program takes over the CPU. This is especially problematic when the user program must respond to real-time events.

For example, imagine we use a D-Bug12 breakpoint in the pulse-counting program in Source 8.6 on page 250. As soon as the breakpoint is reached, the context is switched to the monitor. So interrupts are disabled, the register contents are displayed, and the monitor waits for user input. In the meantime there are more pulses occurring that are supposed to be counted. The first pulse will set the input capture flag, but the rest are missed because the flag is already set, and the interrupt is not serviced because the CPU is in the monitor program with interrupts disabled.

If we try to proceed from the breakpoint with a *G* or *T* command, we always end up in the ISR because of the pulses that occurred while we were staring at the monitor display. If we try to proceed from the breakpoint, it looks like we are stuck in the same place and the program is not working.

For the pulse-counting example, the only consequence is that we will miss pulses. If we were using a real-time interrupt to control a timed multitasking kernel, however, the whole system would probably fail. Imagine the monitor program as just another task. If it is used to debug a time-slice cyclic scheduler and it lasts longer than a time slice, the system will probably fail. Because the monitor's breakpoint routine blocks until the user enters a command, it is a very long task.

About the only thing we can do with the monitor program once interrupts are running is to set a breakpoint to see if we ever reach that point in the program. For example we can set a breakpoint to the start of an ISR to see if it is reached. If the breakpoint is not reached, we know that the vectors are not loaded correctly, the interrupt source is not configured correctly, or the interrupt mask has not been cleared. Once we reach the breakpoint, however, we cannot proceed from that point. We have to restart at the beginning of the program.

8.6.2 Noninvasive Signal Watching

When debugging real-time programs, we want to try to be as noninvasive as possible. That is, we would like to collect as much data as possible while affecting the system timing as little as possible. To do this we either have to collect data by watching external signals while the program is running or collect internal data to look at later. In a real system, there is a trade-off between the amount and type of data we can watch or save and the invasiveness of the debugging tool. This is especially true when working with microcontrollers.

The easiest source of data to watch is the data contained in external signals. Access to these signals can be achieved in a very noninvasive way. We can connect an oscilloscope or logic analyzer to the appropriate signals and watch them without affecting the program at all.

At minimum the external signals on a system will tell us if the system meets its requirements. However, these signals can also help us gain some insight into why the system is not working, especially if we have access to other external signals that contain information about program execution. The most obvious and helpful external signals are the bus signals. If we have access to the bus system, we can use a logic analyzer to see exactly what the program is doing, bus cycle by bus cycle. This is one of the most powerful debugging tools for real-time systems. It is essentially noninvasive, and it provides detailed information about the program's operation. About the only data that cannot be accessed by a logic analyzer connected to the bus system are the contents of the CPU registers. For most debugging problems, this is a minor deficiency.

The problem is that more and more designs today are using single-chip microcontrollers in which the bus is not accessible. For these systems, it is more difficult to gain information about detailed program operation. We can try to gain as much insight as possible from the accessible signals, but if that is not enough, we need to use software and hardware debug helpers. Debug helpers are added program code or hardware resources used exclusively for debugging and testing the system.

The problem of not having access to the bus system has also helped in the emergence of background debugging hardware included in the microcontroller. The CPU12 family of microcontrollers has a background debug system that allows access to internal memory and control of the CPU. We cover the background debug system in Chapter 10.

8.6.3 Hardware and Software Helpers

Software and hardware debug helpers are additional code and hardware resources used solely for the purpose of debugging and analysis. They are not part of the specified function of the system but may remain in the final product for diagnostic purposes. Two basic forms of helpers are those that capture and store data at run time to be accessed later and those that provide additional external signals that can be watched while the program is running.

Because these helpers involve additional software, they are invasive and will affect the program timing. So the goal when writing debug code is to make it as simple and efficient as possible. Many times it can be reduced to a single move or store instruction.

The first type of debug helper is one that captures and stores critical data during run time. At critical points in program operation, debug code is added to save a critical variable that may contain signal data, a number that indicates the current task, or the current state of a state machine. The data are stored in a location reserved for the helper and can be examined later by a monitor program. Several values also can be stored to capture a trace of a critical variable. In this case the data are stored in a ring buffer so the most recent values are available. We also can store additional information with each sample. For example we could store the value of a variable along with a time stamp to indicate when the variable changed. This requires more code though, which will have a bigger effect on the program's timing.

The second type of debug helper is used to add external signals that contain debug information. These signals are normally binary indicators that provide more information about the program's operation. They then can be watched externally in many ways. The signal could be connected to an LED or monitored with an oscilloscope or a logic analyzer. Another way to provide external information is to send data out the serial port for display. This can be very invasive, however, because of the time it takes to send data through the serial port.

EXAMPLE 8.6

Using Helpers to Analyze the Stopwatch Program

In this debugging example we add helper signals to the stopwatch program described in Section 8.4. First a GPIO port is selected for debugging purposes. In this case the BDLC port was selected because it is unlikely that the BDLC will be used. A BDLC bit is reserved for each task in the system, including the periodic timer routine. Following are the equate instructions defining the ports and bits used in this example:

```
DBPORT      equ $fe              ;Debug helper port
DBPDIR      equ $ff              ;Debug port direction
TIMTSK_DB   equ $01              ;Time-slice task debug bit
SWTSK_DB    equ $02              ;Scan switch and mode select task debug bit
UPSTSK_DB   equ $04              ;Update seconds task debug bit
OUTTSK_DB   equ $08              ;Output seconds task debug bit
```

Now for each task code is added so the corresponding output bit goes high when the program is executing that task. To do this we add a *bset* instruction at the start of the task and a *bclr* instruction at the end of the task. Following is the time-slice task with the added debug code:

```
WaitSlice   bset DBPORT,TIMTSK_DB            ;for debug only
            brclr RTIFLG,RTIF,WaitSlice
            movb #RTIF,RTIFLG
            inc RtiCnt
            bclr DBPORT,TIMTSK_DB            ;for debug only
            rts
```

With this code, the *TIME_DB* bit will show how long the program waits for the next slice, in effect showing the available processor load.

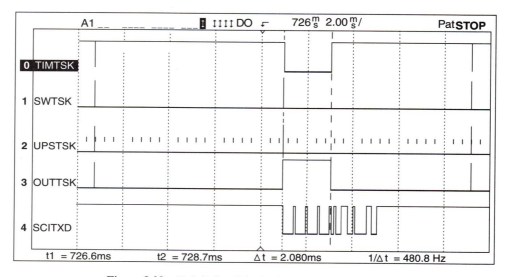

Figure 8.29   Task Debug Display from an HP54645D MSO

Now we can watch the debug signals with an oscilloscope or a logic analyzer. Figure 8.29 shows a trace captured from an HP54645D Mixed-Signal Oscilloscope (MSO). The HP54645D is an ideal tool for debugging microcontrollers because it has two analog channels, 16 digital channels, and deep-capture memory. In this case we are using five digital channels.

The waveforms shown include the four helper bits *TIMTSK, SWTSK, UPSTSK,* and *OUTTSK.* Also, in order to show the data being sent to the display, the output of the SCI is shown as *SCITXD.*

The captured traces show two cycles through the time-slice scheduler. When the *TIMTSK* signal goes low, we know the timer event occurred, and the program exits the time-slice task to execute the task loop. The first time slice shown is a slice when the display does not have to be updated. Each task time is so small that it appears in this display as a glitch. With this MSO we could zoom in and find the actual task times.

The second slice at the center of the trace is one in which the display is updated. Remember that the display task uses the SCI, which requires about 1ms for each word sent to the display. If we send too many characters, the display task may not finish before a new time slice. The display task sends four characters—two backspaces and the value of the *Seconds* variable. In Figure 8.29, we can see that we spend 2.08ms in the display task, *OUTTSK.* This is because the task had to wait for the two backspaces to be sent before sending the next two characters. It then sends the *Seconds* variable, but since we are using the SCI, we do not have to wait for it to be sent before exiting the task. The task returns while the SCI continues to transmit the last character.

From the second slice in the trace, we can find the worst-case total task time. In this case it is less than 2.1ms. This means we have more than 6ms of idle time waiting for the next slice. From these data the peak CPU load for this program is

$$L = \frac{2.1\text{ms}}{8.192\text{ms}} = 0.256$$

We can also determine the response time between the one-second timer event and the display data being sent. The timer event occurred just before the *TIMTSK* signal went low and the action is completed when the SCI is finished transmitting. Using cursor measurements not shown in Figure 8.29, the resulting response time is

$$T_R = 4.12\text{ms}$$

This is just one example in which helper GPIO bits are used to watch the operation of the system. You can imagine there are many other events that can be signaled and watched in this way.

► ## SUMMARY

In this chapter we introduced real-time and multitasking programs. The concepts covered can be difficult to grasp, but they are extremely important for embedded program design. Virtually every embedded program must run more than one task and must do it in a timely manner.

We first covered event detection and response using event loops and interrupts. We saw that most real-time tasks must run periodically, so they are implemented as endless loops. The task period and the time it takes to run a task determine the types of events that can be detected, the CPU load, and the response time.

We then looked at ways to run more than one task concurrently. In this chapter we covered one type of cooperative multitasking kernel—a time-slicing kernel that uses cyclic scheduling. This kernel is simple to design and debug. The kernel requires that all tasks are complete before the next timer event, yet the timer event determines the task periods. So it works well unless we have a task that takes a long time to execute.

We looked at a simple stopwatch example, which illustrated several concepts including the time-slice cyclic scheduler, global variables used to communicate between asynchronous tasks, and state machines implemented in a program.

We then covered interrupts in more detail, specifically the CPU12 interrupt sources. We saw how the CPU12 processed an interrupt and learned some techniques to avoid common interrupt-related errors.

Finally we looked at new techniques for debugging real-time and interrupt-driven programs. We saw that monitor breakpoints and trace commands are no longer very useful for real-time event-driven programs. So we looked at some software and hardware techniques for debugging real-time code. This included adding debug code to capture and store data at run time or to generate additional external signals that can be watched with an LED, an oscilloscope, or a logic analyzer.

EXERCISES

1. A task must be run every 20ms. How long can the task execution time be to keep the CPU load less than 0.7? Assume it is the only task on the system.
2. Write a program that continuously outputs PP0 ⊕ PP1 to PP2. Include data direction initialization code.

Introduction to Real-Time I/O and Multitasking

3. Calculate the CPU load if the program in Exercise 2 is executed in a timed loop with a period of 100µs. Assume an 8MHz E-clock.

4. An engineer attempts to count wheel rotations by using a fixed magnetic reed switch that detects a magnet on the wheel. The minimum activation time is 2ms and the maximum bounce time for activation and release is 0.5ms. Is it possible to use edge detection with noise rejection to detect each wheel rotation one time and only one time? If so, what is the range of sampling rates that can be used?

5. Repeat Example 8.1 on page 243 with a pulse train that is a 20kHz square wave.

6. Repeat Example 8.2 on page 247 with a pulse train that is a 20kHz square wave.

7. Repeat Example 8.3 on page 251 with a pulse train that is a 20kHz square wave. Compare the results with Exercises 5 and 6.

8. Write a code snippet that initializes the SPI interrupt vector to the service routine *SpiIsr*.

9. Repeat Exercise 8 for a program running under D-Bug12.

10. A system uses the *Timer Channel 0* and *Timer Channel 1* interrupts. Find the latency time for both interrupts if they both occur immediately at the start of a critical section of code, during which I is set. The timing parameters are

$$T_{critical}: \qquad 50µs$$

$$T_{isr, TC0}: \qquad 10µs$$

$$T_{isr, TC1}: \qquad 100µs$$

Assume the interrupt masks are not set in the ISRs, the *HPRIO* register is at its *RESET* value, and the system is running under D-Bug12.

9

Microcontroller
I/O Resources

In the last chapter we saw how important on-chip resources can be for reducing the CPU load. In this chapter we describe the basic operation and applications of the I/O resources found on many of the M68HC12 microcontrollers. Included are the general purpose I/O (GPIO) ports, the standard timer module, the real-time interrupt, the pulse-width modulator, the serial peripheral interface (SPI), the serial communications interface (SCI), and the ADC.

One of the most important strengths of a microcontroller is a rich set of on-chip resources. In order to make microcontrollers more application specific, the resources included in a given MCU vary tremendously, even within a single microcontroller family. Therefore this chapter is more specialized to the M68HC912B32 MCU than other chapters. Many of the resources covered are the same as those found on other 68HC12 MCUs, and many are similar to the resources found on the 68HC11 MCUs.

It is not our intention to repeat the material provided in Motorola's *CPU12 Reference Manual* or the *68HC912B32 Technical Summary*. We cover only enough detail to understand the capabilities of the resources and write some basic application code. For example we are not going to cover every bit in every control or status register.

There are also some specifications from Motorola's *Electrical Characteristics* used throughout this chapter. It is important to note that some electrical specifications may change throughout the lifetime of a product—some for better, some for worse. Therefore it is important to check the latest Motorola data before using the numbers found here.

▶ 9.1 GENERAL PURPOSE I/O

In Chapter 1 we were introduced to GPIO ports and we have used GPIO in many of the examples throughout the last few chapters. Now let's dig a little deeper into the electrical characteristics of the M68HC912B32 GPIO ports.

9.1.1 Functional Review

General purpose I/O is the most basic form of CPU I/O. It consists of input pins or output pins that can be accessed by the CPU. Most GPIO pins on the 68HC11 and 68HC12 MCUs can be programmed for use in either direction. Figure 9.1 shows a simplified functional logic diagram of a single bidirectional GPIO bit. There are two registers in the diagram: the data register *PORT* and the data direction register *DDR*. These registers are mapped to the register block of the MCU's memory map. The circuitry associated with the pin's drive level and the internal pull-up logic is not shown.

The data direction register *DDR* determines the direction of the port pin. If the *DDR* output is a one, then the port is an output and the *PORT* register controls the logic level of the pin. If the *DDR* output is a zero, the data register output is disabled and the port pin is placed in a high-impedance state. The logic level of the pin is determined by the external circuit connected to the pin.

The *DDR* output can be changed by writing to the *DDR*, which activates the *WR_DDR* signal. In this case, since this port is connected to bit 0 of the data bus *D0*, it can be changed by writing to the port address with the correct value in bit 0. This means that to change the data direction register with program code, we write the value we want to the register's address. For example, to make bit 0 of *PORTP (PP0)* an output, we would use

```
bset DDRP,$01          ;Set PORTP direction
```

where *DDRP* is the address of the data direction register for *PORTP*.

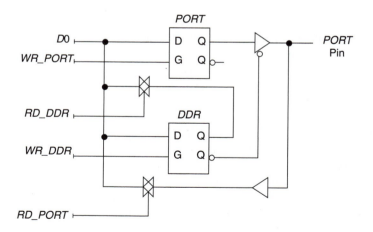

Figure 9.1   Bidirectional GPIO Functional Logic Diagram

A write to the *PORT* register activates the *WR_PORT* signal, which enables the *PORT* register. Note that writing to the *PORT* register has no effect on the port's pin if the port is set to be an input. It does change the *PORT* register output, however. So if the direction is changed, the last value written to the *PORT* register is placed on the pin.

A read of the *PORT* register always returns the state of the pin, regardless of port direction. For example, if *PORTP* bit 0 is configured as an output, we can make the pin go to one by writing a one to bit 0 of the *PORTP* register as follows:

```
bset PORTP,$01          ;Set PORTP bit-0 high.
```

where *PORTP* is set to the address of the data register for *PORTP*.

How to Avoid Port Transients and Glitches.   When the MCU is powered up or reset, all GPIO ports are configured as inputs. This can cause a problem when a port is used as an output in normal operation. The port will remain an input until the software changes the data direction register. In the meantime the circuit will have two inputs connected together, so the node floats to an unknown state. It may go high or it may go low. But if it floats to the active level of the output, the external device connected to the port may do something bad. For example if an output is connected to a motor or solenoid driver, the motor or solenoid may run intermittently during this time and have serious consequences. Two things must be done to solve this problem.

First a pull-up or pull-down resistor must be added to the port so the node will always be pulled to the inactive level instead of floating. Some of the 68HC12 ports have internal pull-up resisters, but unfortunately they are not enabled out of *RESET* so they do not solve this problem.

Second the port must be preset to the inactive level before changing its direction to an output. This is because when the port direction is changed from an input to an output, the output pin goes to the current value on the output of the port's data register. This can cause a temporary glitch on the output, even with the appropriate pull-up or pull-down resistor.

For example, if the active level of the output were a one, we would use a pull-down resistor for the power-up and *RESET* conditions. If the port's data register happens to be a one, however, then the port would go high as soon as the direction is changed.

To avoid this glitch, we can write to the port data register before the direction is changed. For this example we could use the following code:

```
bset PORTP,$00          ;Preset PORTP bit-0 low.
bset DDRP,$01           ;PORTP bit-0 an output
```

The first line presets the port's data register to zero. Then the second line changes the port direction to an output. This may look backward, but it does work and is necessary to avoid a temporary glitch on the output.

Check Your Resources.   It is important to read the MCU documentation when using any GPIO port. Most GPIO ports are shared with other on-chip resources. If the resource is enabled, the GPIO bit associated with that resource is typically not available. This is not always the case, however. For example, when the SPI system is enabled, some of the SPI bits can still be changed as if they were GPIO. Some GPIO ports also have internal pull-ups and a reduced drive option. Again this is only true for some of the ports, so if you need to use one of these functions, you need to check the data carefully. It can be frustrating to get far along in a design before realizing that your MCU does not have enough ports because one you were counting on is not available.

9.1.2 Interfacing

Like any digital device, before we can connect something to an input or output, we need to know the specifications for the interface parameters. In this subsection we examine the interface parameters for the M68HC912B32 I/O pins. These parameters are valid regardless of the function of the pin. For example *PORTA* and *PORTB* can be used as GPIO, or they can be used for the bus system. The parameters given here are valid for both cases.

Voltage Parameters and Noise Margin. The first parameters to consider are the input and output voltage levels and the corresponding noise margins. Figure 9.2 shows a noise margin diagram with the relevant parameters. V_{DD} and V_{SS} are the supply voltages, which are typically +5V and GND. Because these are HCMOS devices, all the voltage parameters are a function of V_{DD}. This means that if V_{DD} changes, so do the voltage parameters.

The output voltage parameters are V_{OL} and V_{OH}. V_{OL} is the output voltage for a zero and V_{OH} is the output level for a one. When we are evaluating the noise margins, the worst case for these parameters are $V_{OL, max}$ and $V_{OH, min}$, which are shown on the figure.

The input voltage parameters are V_{IL} and V_{IH}. $V_{IL, max}$ is the maximum input voltage that is guaranteed to be seen as a zero. $V_{IH, min}$ is the minimum input voltage that is guaranteed to be seen as a one. If an input falls between $V_{IL, max}$ and $V_{IH, min}$, the result is undetermined. It may be a one, or it may be a zero. We obviously need to avoid this region.

For a digital system to work correctly, the output high voltage always must be between $V_{IH, min}$ and V_{DD}, and the output low voltages must always be between zero and $V_{IL, max}$. Since there is always noise on the system, we need to include an extra margin so the noise does not cause a logic error.

The high-level noise margin is defined as

$$m_H = V_{OH, min} - V_{IH, min}$$

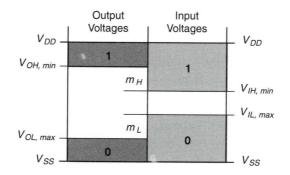

Figure 9.2 General Voltage Margin Diagram

and the low-level noise margin is defined as

$$m_L = V_{IL,\,max} - V_{OH,\,max}$$

Since all signals will contain both ones and zeros, the overall noise margin is defined as

$$m = min[m_H, m_L]$$

For the system to work the noise margin must be greater than zero. The value of the noise margin gives us an indication of the level of noise that can be tolerated by the system.

The output voltage parameters are not only a function of V_{DD}. They also depend on the current being carried by the output. We cover the interface current parameters next. Then we analyze systems with various loads.

Current Parameters and Fanout. The interface current parameters are the output currents, I_{OH} and I_{OL}, and the input leakage current, I_{In}. I_{OH} is the current flowing out of a high output and I_{OL} is the current that is flowing into a low output. I_{In} is the leakage current that flows into or out of an input pin. Some 68HC12 I/O pins can be set to a reduced drive configuration so there are two sets of output current levels, one for normal drive level and one for reduced drive level.

These currents are used to determine the static fanout of a device, that is, the number of inputs that can be connected to one output while preserving the required voltage margins. The static fanout for a low output is

$$n_L = \frac{|I_{OL,\,max}|}{|I_{In}|}$$

and the static fanout for a high output is

$$n_H = \frac{|I_{OH,\,max}|}{|I_{In}|}$$

Again the worst case of these two parameters determines the overall static fanout. So

$$n = min[n_H, n_L]$$

No-Load Interfacing. If we look at the data sheet for a digital device, there are two values listed for the voltage output levels. One is for small loads, so the output current is limited to a very small value. Typically it is 10μA for HCMOS devices. We normally use these voltage specifications when connecting logic devices to other logic devices of the same family. Figure 9.3 shows these voltage parameters for a 68HC12. With these values the noise margin is 0.8V. Note that these output voltages are valid only if the output current is less than 10μA. This represents about 10 HCMOS logic gates or only two 68HC12 inputs.

Interfacing with Loads. When working with GPIO, we often need to connect the I/O pins to loads that require more current than 10μA. For example we may want to connect an LED or a small relay to an output. To do this we need to determine whether we can directly

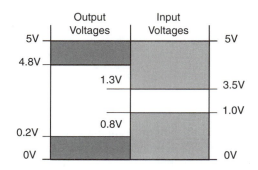

Figure 9.3   Voltage Parameters for 5V No-Load Outputs

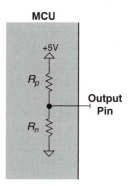

Figure 9.4   Simple Load Model for a General Purpose Output

connect it to an MCU output or we need to add a driver IC to protect the MCU from dangerous current levels or transients.

If we look at the *Maximum Ratings* in the electrical specifications for a 68HC12, there is a maximum current rating per pin of ±25mA. This does not mean we can reliably source or sink 25mA, however. The *Maximum Ratings* give the values that if exceeded may destroy the part. We need to make sure these parameters are not exceeded, but it does not tell us what the output characteristics will be for a given load.

To determine the output characteristics, we need to model the port output as shown in Figure 9.4. R_p is the equivalent resistance of the internal p-channel transistor, and R_n is the equivalent resistance of the internal n-channel transistor.

When the output is high, the p-channel transistor is on and the n-channel transistor is off. When the output is low, the p-channel transistor is off and the n-channel transistor is on. For this model we can assume that the resistance for a transistor that is off is infinite. To find the on resistance, we need to use the parameters given in the electrical characteristics. Electrical specifications for digital devices normally do not specify these values. Therefore we have to derive the values based on the specifications that are given. The on resistance for the p-channel transistor is

$$R_{p(on)max} = \frac{V_{DD} - V_{OH,\,min}}{|I_{OH}|}$$

TABLE 9.1 NORMAL DRIVE STRENGTH OUTPUT CHARACTERISTICS

Parameter	68HC912B32 Value
$V_{OH, min}$ (no-load, $I_{OH} < 10\mu A$)	V_{DD}-0.2V
$V_{OH, min}$ ($I_{OH} = -0.8$mA)	V_{DD}-0.8V
$V_{OL, max}$ (no-load, $I_{OL} < 10\mu A$)	0.2V
$V_{OL, max}$ ($I_{OL} = 1.6$mA)	0.4V
$R_{p(on), max}$	1000Ω
$R_{n(on), max}$	250Ω

and the on resistance for the n-channel transistor is

$$R_{n(on)max} = \frac{V_{OL, max}}{|I_{OH}|}$$

The parameters used for these equations are the loaded output parameters. These parameters for the 68HC912B32 outputs with normal drive strength are shown in Table 9.1. Notice the maximum values of R_p and R_n. They are relatively large values, especially R_p.

There is no way to determine the minimum values of R_p or R_n from the published specifications. This means we do not know the maximum short circuit current for an output tied directly high or low. These high resistance values do provide some protection against output short circuits or logic contention. We cannot be sure if or for how long an output can be shorted without destroying the microcontroller, however.

Another thing to notice is that like most digital logic devices, the outputs can sink more current than they can source. Consequently devices that require significant load current like an LED or opto-isolator should be connected active-low.

EXAMPLE 9.1

Connecting an LED to an Output Port

Determine whether we can drive an active-low LED from an M68HC912B32 output. Assume that a high-efficiency LED with $I_{F, min} = 5$mA and $V_{F, max} = 2$V is used. It has been determined from the LED specifications that if I_L is less than 5mA, the LED will not be bright enough. So first determine a value for R_L so the minimum LED current is greater than 5mA. Then check to see that the maximum current for the output is not exceeded.

Solution

Step 1. Determine R_L. The circuit for the LED is shown in Figure 9.5. The LED is connected active-low so that when the output goes low, the current, I_L, flows through the LED and turns it on. R_L must be set so I_L is at least 5mA.

I_L flows through R_L, the LED, and through $R_{n(on)}$. $R_{p(on)}$ is infinite because the p-channel transistor is off. So

$$R_L = \frac{V_{DD} - I_F \cdot R_{n(on)} - V_F}{I_F} = \frac{5V - 5mA \cdot 250\Omega - 2V}{5mA} = 150\Omega$$

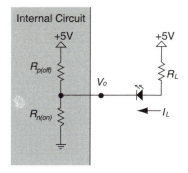

Figure 9.5 Circuit for Example 9.1

The closest 1% standard resistor value that is always less than 150Ω is 147Ω.

Step 2. Maximum Current. Now determine whether a 147Ω resistor is large enough to limit the current to less than 25mA under all conditions. Worst-case conditions occur when I_L is at its maximum value. This occurs when $R_{n(on)}$, V_F, and R_L are all at minimum values. Unfortunately $R_{n(on), min}$ and $V_{F, min}$ are not given in the specifications.

Let's look at two sets of calculations. The first is more conservative; the second is more practical. In both cases assume a supply voltage range of +5V ± 10% and a 1%, 147Ω resistor for R_L.

The most conservative analysis is made by setting $R_{n(on), min} = 0$ and $V_{F, min} = 0$. With these values

$$I_{F, max} = \frac{V_{DD, max}}{R_{L, min}} = \frac{5.5V}{145.5\Omega} = 37.8\text{mA}$$

This exceeds the maximum current specification (25mA) for the M68HC912B32, which means that the MCU could be destroyed. If this was a conservative design, another IC such as a 74AC240 would have to be used to drive the LED.

The cost of adding the extra IC can be prohibitive for some cost-sensitive designs. For these cases a more practical estimation of $R_{n(on), min}$ and $V_{F, min}$ is possible. A reasonable estimation is to use 50% of the maximum values for $R_{n(on), min}$ and $V_{F, min}$. Now

$$I_{F, max} = \frac{V_{DD, max} - V_{F, min}}{R_{L, min} + R_{n(on), min}} = \frac{5.5V - 1V}{145.5\Omega + 125\Omega} = 16.64\text{mA}$$

This is well within the 25mA limit of the output and the maximum current limit of the LED. It appears that we can use a 147Ω, 1% resistor in the circuit and connect it to an M68HC912B32 output.

We are not done yet. This calculation indicated that the maximum current for the output is not exceeded under practical conditions, but it does not address power dissipation in the MCU. This circuit is probably fine for a single LED, but for multiple LEDs we also need to make sure that the power dissipation of the MCU does not cause it to get too hot and be destroyed.

TABLE 9.2 REDUCED DRIVE STRENGTH OUTPUT CHARACTERISTICS

Parameter	68HC912B32 Value
$V_{OH,\ min}$ (no-load, $I_{OH} < 4\mu A$)	V_{DD}-0.2V
$V_{OH,\ min}$ ($I_{OH} = -0.3$mA)	V_{DD}-0.8V
$V_{OL,\ max}$ (no-load, $I_{OL} < 3.6\mu A$)	0.2V
$V_{OL,\ max}$ ($I_{OL} = 0.6$mA)	0.4V
$R_{p(on),\ max}$	2667Ω
$R_{n(on),\ max}$	667Ω

Reduced Drive Strength. Some of the 68HC12 GPIO ports can be set to have reduced drive strength. Reduced drive strength in effect increases the on-resistance associated with the output, which reduces the current-carrying capacity of the output. Reduced drive strength should be used to reduce power consumption and radio frequency interference (RFI). The power dissipation is reduced because the output current is reduced. The RFI is reduced because the higher on-resistance of the output, in combination with the capacitance on the output, will increase the transition times. Slower transitions mean less energy at RFI frequencies. Of course reduced drive strength also reduces the output's fanout along with source and sink currents. Table 9.2 shows the output characteristics for an output that is set for reduced drive strength.

On the 68HC912B32, reduced drive strength is available on Port E, Port A, and Port B. The drive strength for these ports is controlled by the *RDRIV* register. Out of *RESET,* they are all set for full drive strength. These ports are all associated with the bus system. This is because the bus system is normally the primary generator of RFI. It runs at a high rate and many outputs change at the same time. We do not have to be concerned about the bus when the MCU is run in single-chip mode, but there may be other I/O signals that generate RFI. These signals should be connected to one of the ports that can be set for reduced drive strength. For example connecting an LCD display to Port A or B allows it to be interfaced at reduced drive strength.

Internal Pull-up Resistors. Another added feature to the 68HC12 family is internal pull-up resistors on some of the I/O ports. Internal pull-ups reduce the overall system cost by eliminating the need for external pull-up resistors.

Pull-up resistors are normally required when an input is connected to an open-drain device. Pull-ups can also be used to ensure an input powers up at a known state. This requires the pull-up be enabled out of *RESET*, however.

On the 68HC912B32 Port A, Port B, and Port E have optional internal pull-ups. The pull-ups are controlled by the *PUCR* register. The only pull-ups that are active out of *RESET* are on Port E bits 0, 1, 2, 3, and 7.

9.1.3 Power Dissipation Limitations

In the last subsection we looked at voltage and current limitations to the general purpose outputs. We also must check the total power dissipated by the MCU. An increase in power dissipation causes the junction temperature to rise and an increased junction temperature can destroy the MCU or reduce its lifetime. For this text we will use the following rule of thumb:

> For every 10°C that the junction temperature rises above 100°C, the device's failure rate doubles.

If the IC is dissipating too much power, the junction temperature rises. If we assume a maximum ambient temperature of 85°C, the junction temperature for an M68HC912B32 running in single-chip mode in a QFP package is

$$T_J = T_A + P_D \times \Theta_{JA} = 102.1 + P_{I/O} \cdot 76°C$$

where

$$T_J = \text{the junction temperature}$$

$$T_A = \text{the ambient temperature}$$

$$\Theta_{JA} = \text{the junction-to-ambient thermal resistance of the package}$$

$$P_{I/O} = \text{the power dissipated by the output circuits}$$

This means that for every 132mW dissipated in the output circuits, the MCU's expected lifetime would be cut in half.

EXAMPLE 9.2

Power Dissipation Limitation to Example 9.1 LED Circuit

For the LED circuit in Example 9.1, the worst-case power dissipated by the output is

$$P_{I/O} = I_L^2 \cdot R_{n(on)} = (16.64\text{mA})^2 \cdot 125\Omega = 34.6\text{mW}$$

Therefore four LEDs connected as shown in Example 9.1 may cause the lifetime of the MCU to be cut in half. This is another good reason to use separate driver ICs to connect multiple LEDs to the MCU.

9.1.4 GPIO Timing

Sometimes when using GPIO we need to analyze the input timing for a port read operation and the output timing for a port write operation. The input timing analysis checks the setup and hold times. The output timing analysis is used to determine the circuit delays associated with the output port.

Input Timing. If we have an input that changes at a high rate, we need to watch for setup-time and hold-time violations. The result of a setup- or hold-time violation is the potential for a metastable condition. A metastable condition may or may not cause a catastrophic or a noticeable error. It is beyond the scope of this text to go into the details of metastability.

Figure 9.6 shows the timing diagram for a read cycle of an I/O port. It shows that the data on the port signals must remain valid between the setup time before the falling edge of

304

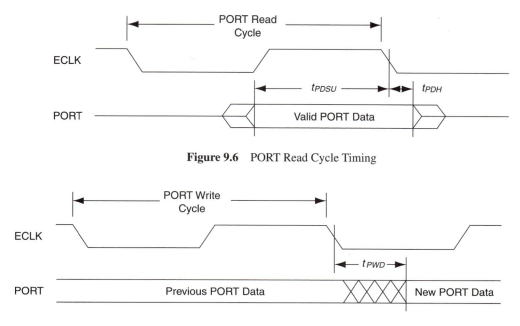

Figure 9.6 PORT Read Cycle Timing

Figure 9.7 PORT Write Cycle Timing

the E-clock, t_{PDSU}, and the hold time after the falling edge of the E-clock, t_{PDH}. If the data change within this region, a metastable condition may result.

  If we look at the *Peripheral Port Timing* section of the *Technical Summary* for the M68HC912B32, we can see that the setup time is quite large. It also depends on the E-clock frequency. A large setup time is not desirable when trying to avoid metastability. To reduce the effective setup time, we can add a D-type register clocked by the E-clock to synchronize the port signals before reaching the port inputs. This reduces the setup time requirement for the signals and significantly reduces the chance of a metastable condition reaching the MCU.

Output Timing. The output timing determines the delay caused by the port circuitry. Figure 9.7 shows the output delay after a write to the port register, t_{PWD}. For a 68HC912B32, t_{PWD} is only 40ns, which is much smaller than most software delays.

  For precise timing relative to the software instruction that causes the write, we need to look at the instruction timing and the port timing. For example if we change *PORTP,* bit 0 with

```
bset PORTP,$01
```

we need to look at the cycle-by-cycle timing for the *bset* instruction. From the *CPU12 Reference Manual,* we have the following bus cycles for a *bset* instruction using direct addressing:

Cycle 1:   Read *PORTP*
Cycle 2:   Program word access
Cycle 3:   Optional program word access
Cycle 4:   Write to *PORTP*

The write cycle is the last cycle for the *bset* instruction. From the start of the *bset* instruction, the delay before the output becomes valid is

$$T_D = T_I + t_{PWD} = 4 \times T_E + t_{PWD} = 500\text{ns} + 40\text{ns} = 540\text{ns}$$

This is assuming a 16-bit data access and no clock stretching.

▶ 9.2 TIMERS

Timers are the most widely used on-chip resource in real-time embedded systems. They are used to generate timed signals, to synchronize the program with timing events, or to capture external time events. In Chapter 8 we saw how using a timer circuit significantly reduced the CPU load for a pulse counting program.

In this section we cover the resources in the 68HC12's *standard timer module* including input captures, output compares, and the pulse accumulator. We also cover the real-time interrupt, which is part of the 68HC12's *clock generation module,* and the pulse-width modulator. The *computer operating properly* (COP) and the *clock monitor* will be covered in Chapter 10.

9.2.1 Timer Flag Model

Before covering the individual timer modules, we need to look at the characteristics of the 68HC12 timer flags. The timer flags do not work in the same way as other event flags. Because of this, they are often the source of program errors, especially for programmers new to the 68HC11 or 68HC12 families.

Figure 9.8 shows a functional model for a timer event flag. When the timer event occurs, the output of the flip-flop goes high. This is the event flag signal. As expected, when we read the flag bit, we will read the flag signal. When we write to the flag bit, however, the write data go to the flip-flop's clear pin, not the flag signal. Therefore if you write to the flag register with a one in the bit position of the flag, the flag will be cleared.

Clearing Timer Event Flags

To clear a timer event flag, you write to the flag register with a one in the bit position of the flag.

A common error is to try to clear the event flag by writing a zero to it. For example the following attempt to clear the timer flag for *Timer Channel 7, C7F,* will not work.

```
bclr TFLG1,$80          ;Flag clearing error
```

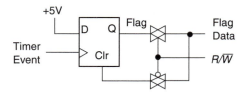

Figure 9.8   CPU12 Timer Flag Functional Model

A zero is written to the *C7F*'s clear pin, which will not change the flag. Worse yet, every other flag in the flag register that was set will be cleared in error. Source 9.1 shows several code snippets that can be used to clear a timer flag.

Source 9.1 Methods Used to Clear a Timer Flag

```
;***********************************************************************
; Equates
;***********************************************************************
TFLG1     equ $008E
C7F       equ $80


;***********************************************************************
; Using the movb instruction
;***********************************************************************
        movb #C7F,TFLG1             ; $80 -> (TFLG1)


;***********************************************************************
; Using 68HC11 compatible instructions
;***********************************************************************
        ldaa #C7F                   ; $80 -> ACCA
        staa TFLG1                  ; ACCA -> (TFLG1)


;***********************************************************************
; Using bclr. Not recommended, trickery
;***********************************************************************
        bclr TFLG1,~C7F             ; (TFLG1) & $7F -> (TFLG1)


;***********************************************************************
; Pseudo-C:
;
;       TFLG1 = C7F;
;
;***********************************************************************
```

Timer Interrupts.   The MCU also uses the timer flag signal to indicate that an interrupt is pending for that resource. If the interrupt for a timer resource is enabled and the *I* bit is cleared, the MCU will sample the flag at the end of each instruction. If the flag is set, the interrupt context switch is initiated. It is important to remember to clear the flag while in the interrupt service routine. If it is not cleared, the interrupt process will be repeated immediately after the end of the service routine because the flag signal is still set. This will cause the MCU to be trapped in an endless loop executing the interrupt service routine. The only way to escape is to reset the CPU or interrupt the CPU with an interrupt that has a higher priority.

Automatic Flags.   Optionally on the 68HC12, we can configure some of the timer flags to be cleared automatically. This eliminates the code required to clear the flag. In general automatic flags are cleared when the associated resource is accessed. For example the output compare flags are cleared automatically on a write to the timer channel register and the input capture flags are cleared on a read of the timer channel register. You have to be careful when using automatic flags. An access to a register could inadvertently clear a flag, making it possible to miss the associated event.

9.2.2 The Real-Time Interrupt

The real-time interrupt (RTI) generates a periodic timer event. A program can be synchronized to the event by either polling the RTI event flag, *RTIF,* or by using the RTI interrupt.

Note
_____

The real-time interrupt function should not be confused with a *real-time clock.* A real-time clock is a hardware circuit that generates data for the time of day and dates. The real-time interrupt only generates a periodic event.

_____

On the 68HC912B32, the registers used to configure the RTI are *RTICTL* and *RTIFLG.* Figure 9.9 shows these two registers, their addresses, and the bit labels. Also shown in the figure is the state of each register out of *RESET.*

RTI Initialization.   The RTI system is relatively easy to use. The only configuration required is to set the RTI rate and if the RTI interrupt is used, enable the interrupt. The RTI rate is configured by setting the RTI rate bits, *RTR2:RTR1:RTR0.* Table 9.3 shows the resulting

RTICTL:

($0014)	RTIE	RSWAI	RSBCK	0	RTBYP	RTR2	RTR1	RTR0
RESET:	0	0	0	0	0	0	0	0

RTIFLG:

($0015)	RTIF	0	0	0	0	0	0	0
RESET:	0	0	0	0	0	0	0	0

Figure 9.9   Real-Time Interrupt Registers

TABLE 9.3   RTI EVENT RATES

RTR2:RTR1:RTR0	E-Clock Divisor	RTI Event Period ($f_E = 8.0$MHz)
000	OFF	OFF
001	2^{13}	1.024ms
010	2^{14}	2.048ms
011	2^{15}	4.096ms
100	2^{16}	8.192ms
101	2^{17}	16.384ms
110	2^{18}	32.768ms
111	2^{19}	65.536ms

RTI periods for the different combinations of the RTI rate bits. Note that the RTI rate is a function of the E-clock frequency. At *RESET* the rate bits are all zero, so the RTI is disabled. To enable the RTI, we only need to set the rate bits for the desired RTI rate. From that point, the RTI flag will be set at the selected rate.

RTI Operation. Once initialized, the RTI generates periodic timer events. To synchronize a program to the RTI events, we can poll the RTI flag, *RTIF,* or we can use the RTI interrupt, *RTII.* If we use the interrupt, it must be enabled by setting the RTI interrupt enable bit, *RTIE,* and clearing the global interrupt mask, *I.*

Polling the RTI Flag. In Chapter 8 we saw three program examples that used the RTI system, Source 8.7, Source 8.8, and Source 8.10. All of these programs used the RTI events to synchronize timed event loops. The RTI code for the stopwatch example is shown again in Source 9.2.

First the RTI initialization code is shown. By writing a $04 to the *RTICTL* register, we disabled the RTI interrupt and set the RTI rate to 8.192ms. Verify this by checking the RTI rate table and the *RTICTL* register description shown previously.

Source 9.2 The RTI Code in the Simple Stopwatch Example

```
;*********************************************************************
; RTI System Initialization.
; Pseudo-C:
; main(){
;     RTICTL = 0x04;                      ;Enable RTI for 8.192ms
;     RTIFLG = RTIF;                      ;Clear RTI flag
;     while(1){
;         WaitSlice();
;           ...
;     }
; }
;*********************************************************************
main:       movb #%00000100,RTICTL       ;Enable RTI for 8.192ms
            movb #RTIF,RTIFLG

sched_lp    jsr WaitSlice                ;Wait for time slice
            ...
            bra sched_lp


;*********************************************************************
;WaitSlice - The time-slice timer routine. It uses the RTI system flag
;            RTIF. The RTI is set for an 8.192ms for the switch debounce.
;
; Pseudo-C:
; WaitSlice(){
;     while(RTICTL & RTIF == 0){}
;     RTIFLG = RTIF;
;     RtiCnt++;
; }
```

```
;*****************************************************************************
WaitSlice    brclr RTIFLG,RTIF,WaitSlice    ;wait for RTI event
             movb #RTIF,RTIFLG              ;clear RTI flag
             inc RtiCnt                     ;increment counter
             rts


;*****************************************************************************
```

The routine *WaitSlice* waits for the next RTI event by polling the RTI flag *RTIF*. Once the RTI event occurs, the program clears the RTI flag, increments a counter, and returns.

Using the RTI Interrupt. We could have implemented the program in Source 9.2 using the RTI interrupt instead of polling for the flag. Source 9.3 shows how the stopwatch program can be changed to use the RTI interrupt.

First in the initialization, D-Bug12's interrupt jump table is initialized so it will jump to *RtiIsr* when an RTI interrupt occurs. Of course if this program were a standalone application, this code would be deleted and the RTI interrupt vector would have to be loaded with the address of *RtiIsr*. To enable the RTI interrupt, an $84 is written to the *RTICTL* register. This sets the *RTIE* bit to one, enabling the interrupt. Once the rest of the initialization tasks are complete, the global interrupt mask is cleared. From this point the code can be interrupted by the RTI interrupt.

Source 9.3 RTI Code for Interrupt-Driven Stopwatch

```
;*****************************************************************************
; A 1-button stop-watch based on the 68HC12 RTI System.
; This version uses the RTI interrupt instead of polling for the flag.
; Controller: 68HC912B32
; Monitor: D-Bug12
; Uses BasicIO module
; Pseudo-C:
; main(){
;     SETUSERVEC(USER_RTI,RtiIsr)
;     RTICTL = 0x84;
;     RTIFLG = RTIF;
;
;       ...
;     ENABLE_INT();
;     while(1){
;         while(!TaskFlag){}
;         TaskFlg = 0;
;
;           ...
;     }
; }
;*****************************************************************************
main:        ldd #RtiIsr                    ;Init. D-Bug12 RTII jump table
             pshd
             ldd #USER_RTI
             jsr [SETUSERVEC,pcr]
             puld
             movb #%10000100,RTICTL         ;Enable RTI for 8.192ms, interrupt
```

```
            movb #RTIF,RTIFLG
            ...
            cli

sched_lp    tst TaskFlag                    ;wait for time slice
            beq sched_lp
            clr TaskFlag
            ...
            bra sched_lp

;*************************************************************************
;RtiIsr - The RTI interrupt service routine. It is used for time slicing.
;         The RTI is set for an 8.192ms for the switch debounce.
; Pseudo-C:
; ISR RtiIsr(){
;     RTIFLG = RTIF;
;     TaskFlag = 1;
;     RtiCnt++;
; }
;
;*************************************************************************
RtiIsr      movb #RTIF,RTIFLG
            movb #1,TaskFlag
            inc RtiCnt
            rti

;*************************************************************************
```

The *WaitSlice* function from Source 9.2 has been replaced by an event flag polling loop and an interrupt service routine. Before the main program makes a pass through the task loop, it waits for *TaskFlag* to be set. *TaskFlag* is set by the *RtiIsr* after each occurrence of an RTI timer event. The interrupt service routine also increments the counter, *RtiCnt*.

In this example there is no benefit in using the RTI interrupt over polling the flag. It is only shown to illustrate the configuration and operations required to use the RTI interrupt.

Using the RTI system is simple and requires no other resources. But it is limited to the rates shown in Table 9.3. This can present a problem if the times required for the design are not integer multiples of the available rates. For example the simple stopwatch program shown is not very accurate because one second is not an integer multiple of any available RTI rate. We used 122 RTI events to approximate one second. Since the RTI rate is 8.192ms, this results in an actual time of 0.99942 seconds, a 0.06% error. This may be adequate for a short-duration timer like a kitchen timer, but it would not be accurate enough for a *time-of-day,* real-time clock. Later we consider how we can use an output compare to generate periodic timer events that are more accurate.

9.2.3 Standard Timer Module Overview

The 68HC12's standard timer module consists of a 16-bit free-running counter, output compares, input captures, and a pulse accumulator. If there is one word to describe the timer module, it is versatility. Because of its versatility, it can also be complicated to configure.

This subsection provides an overview of the timer module. It is followed by subsections that cover the output compares, the input captures, and the pulse accumulator. Figure 9.10 shows the functional diagram for the standard timer module. There are eight channels: seven identical channels that can be configured as an output capture or an input compare and one channel, *Channel 7,* that can be configured to change multiple outputs on an output compare.

As we can see from Figure 9.10, the free-running counter is the heart of the timer system. It is a 16-bit read-only counter and is used for all timer subsystems except for the pulse accumulator. The free-running counter is clocked by the timer clock, which can be set to a variety of sources.

Timer Clock Configuration. Figure 9.11 is a functional diagram that shows the sources that can be used for the timer clock in the M68HC912B32. In most applications it will be set to the prescaled E-clock. But it can also be set to the pulse-accumulator clock, *PACLK,* which may come from the E-clock or an external clock connected to *PT7.* The *PACLK* can also be divided down by the pulse accumulator counter, which allows us to use a very slow clock.

The timer clock selection is made by the pulse-accumulator control register, *PACTL,* which is shown in Figure 9.12. To select the clock source to be used for the timer, we set the *PAEN, CLK1,* and *CLK0* bits as shown in Table 9.4.

When the E-clock is selected as the clock source, the prescaler bits *PR2, PR1,* and *PR0* in the *TMSK2* register can be used to divide the E-clock down to a lower frequency as shown in Table 9.5. Out of *RESET* these bits are all zero, so the free-running counter frequency is equal to the E-clock frequency.

We must be careful about lowering the timer clock frequency. As shown in Table 9.5, lowering the timer frequency increases the timer's range, but it also increases the resolution. Remember this affects all the timer's input captures and output compares. Later if we need to add a timer function that required a lower resolution, we would have to rewrite all the timer code.

By keeping the prescaler bits set to zero, we always have the best timer resolution available for the microcontroller. Range can then be increased by using software counters to keep track of *TCNT* overflows.

Free-Running Counter Configuration and Access. Figure 9.12 shows the registers required to configure and access the free-running counter.

The most important bit in these registers is the timer enable bit, *TEN,* in the *TSCR* register. As we can see from Figures 9.10 and 9.11, *TEN* enables the free running counter clock and the counter itself. In effect *TEN* enables the whole timer system. Forgetting to set this bit is a common source of error, especially when converting timer code written for the 68HC11 to the 68HC12, because the 68HC11's timer module does not have an enable bit.

TCNT contains the current free-running counter value. It is a 16-bit read-only register in normal operating modes. It must be read in a single 16-bit operation to avoid incorrect results. For example to load ACCD with the counter value, we should use

```
ldd TCNT           ;correct way to read TCNT
```

not

```
ldaa TCNT          ;incorrect way to read TCNT
ldab TCNT+1
```

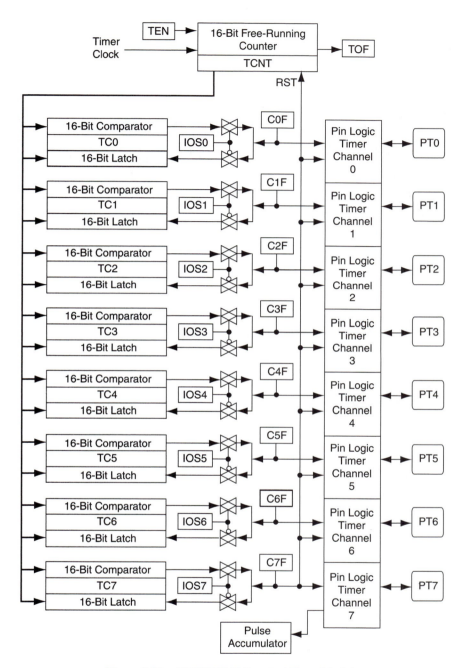

Figure 9.10 68HC912B32 Standard Timer Module

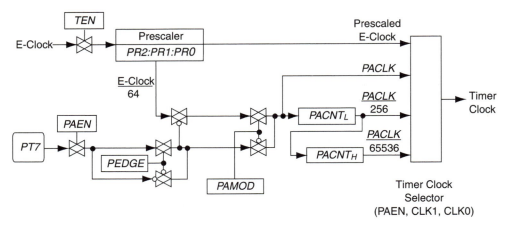

Figure 9.11   Timer Module Clock Selection Circuit and Register

PACTL:

($00A0)	0	PAEN	PAMOD	PEDGE	CLK1	CLK0	PAOVI	PAI
RESET:	0	0	0	0	0	0	0	0

TCNT:

($0084)	BIT15	BIT14	BIT13	BIT12	BIT11	BIT10	BIT9	BIT8
($0085)	BIT7	BIT6	BIT5	BIT4	BIT3	BIT2	BIT1	BIT0

TSCR:

($0086)	TEN	TSWAI	TSBCK	TFFCA	0	0	0	0
RESET:	0	0	0	0	0	0	0	0

TMSK2:

($008D)	TOI	0	PUPT	RDPT	TCRE	PR2	PR1	PR0
RESET:	0	0	0	0	0	0	0	0

TFLG2:

($008F)	TOF	0	0	0	0	0	0	0
RESET:	0	0	0	0	0	0	0	0

Figure 9.12   Control Registers for the Free-Running Counter

TABLE 9.4 TIMER MODULE CLOCK SELECT BITS

PAEN:CLK1:CLK0	Timer Clock
0xx	Prescaled E-Clock
100	Prescaled E-Clock
101	PACLK
110	PACLK/256
111	PACLK/65536

TABLE 9.5 TIMER PRESCALER SETTINGS

PR2:PR1:PR0	Prescale Factor	Counter Frequency	Timer Range $(f_E = 8\text{MHz})$	Timer Resolution $(f_E = 8\text{MHz})$
000	1	8MHz	8.192ms	125ns
001	2	4MHz	16.384ms	250ns
010	4	2MHz	32.768ms	500ns
011	8	1MHz	65.536ms	1.0μs
100	16	500kHz	131.072ms	2.0μs
101	32	250kHz	262.144ms	4.0μs
110	Reserved	—	–	—
111	Reserved	—	–	—

The second example would read *TCNT* at two separate times. Between the two reads, *TCNT* will change to a new count so we would end up with ACCA from one count and ACCB from another count.

The free-running counter is called *free running* because in general it cannot be changed by the user. It starts at *RESET* and continues counting until the next *RESET*. When it reaches $FFFF, the next clock will cause an overflow, and the counter will go to $0000. When the free-running counter overflows, the timer overflow flag *TOF* in *TFLG2* is set. If the timer overflow interrupt enable bit *TOI* is also set, an interrupt will occur.

Timer Channels. The registers that are used to configure and access the timer channels are shown in Figure 9.13. The *TIOS* register contains I/O designator bits for each channel. When a channel's I/O designator bit *IOSn* is zero, the channel is configured as an input capture. When *IOSn* is a one, the channel is configured as an output compare. Notice all timer channels come out of *RESET* as input captures.

The timer channel registers *TC0–TC7* are 16-bit registers that are used for both the input capture and output compare functions for each channel.

TFLG1 contains all the timer event flags and *TMSK1* contains all the interrupt enables for the timer channels. These flags and interrupts are used for input captures and the output compares.

The *CFORC* register allows a program to force an output compare on any channel. By writing a one to one of the *FOCn* bits, the output action for that channel will occur. This register may be used in exceptional cases in which conditions require an immediate action instead of waiting for the free-running counter. It can also be used to cause more than one output compare action to occur simultaneously.

TIOS:

($0080)	IOS7	IOS6	IOS5	IOS4	IOS3	IOS2	IOS1	IOS0
RESET:	0	0	0	0	0	0	0	0

CFORC:

($0081)	FOC7	FOC6	FOC5	FOC4	FOC3	FOC2	FOC1	FOC0
RESET:	0	0	0	0	0	0	0	0

TFLG1:

($008E)	C7F	C6F	C5F	C4F	C3F	C2F	C1F	C0F
RESET:	0	0	0	0	0	0	0	0

TMSK1:

($008C)	C7I	C6I	C5I	C4I	C3I	C2I	C1I	C0I
RESET:	0	0	0	0	0	0	0	0

TC0:

($0090)	BIT15	BIT14	BIT13	BIT12	BIT11	BIT10	BIT9	BIT8
($0091)	BIT7	BIT6	BIT5	BIT4	BIT3	BIT2	BIT1	BIT0

TC1:

($0092)	BIT15	BIT14	BIT13	BIT12	BIT11	BIT10	BIT9	BIT8
($0093)	BIT7	BIT6	BIT5	BIT4	BIT3	BIT2	BIT1	BIT0

TC2:

($0094)	BIT15	BIT14	BIT13	BIT12	BIT11	BIT10	BIT9	BIT8
($0095)	BIT7	BIT6	BIT5	BIT4	BIT3	BIT2	BIT1	BIT0

TC3:

($0096)	BIT15	BIT14	BIT13	BIT12	BIT11	BIT10	BIT9	BIT8
($0097)	BIT7	BIT6	BIT5	BIT4	BIT3	BIT2	BIT1	BIT0

TC4:

($0098)	BIT15	BIT14	BIT13	BIT12	BIT11	BIT10	BIT9	BIT8
($0099)	BIT7	BIT6	BIT5	BIT4	BIT3	BIT2	BIT1	BIT0

Figure 9.13 Timer Channel Registers

TC5:

($009A)	BIT15	BIT14	BIT13	BIT12	BIT11	BIT10	BIT9	BIT8
($009B)	BIT7	BIT6	BIT5	BIT4	BIT3	BIT2	BIT1	BIT0

TC6:

($009C)	BIT15	BIT14	BIT13	BIT12	BIT11	BIT10	BIT9	BIT8
($009D)	BIT7	BIT6	BIT5	BIT4	BIT3	BIT2	BIT1	BIT0

TC7:

($009E)	BIT15	BIT14	BIT13	BIT12	BIT11	BIT10	BIT9	BIT8
($009F)	BIT7	BIT6	BIT5	BIT4	BIT3	BIT2	BIT1	BIT0

Figure 9.13   *(continued)*

9.2.4 Output Compares

An *output compare* generates an output action or timer event that is synchronized to the free-running counter. The output compare function described here applies to the normal output compare channels, *Timer Channel 0* through *Timer Channel 6. Timer Channel 7* has a special output compare function, which we cover later in this section.

Figure 9.14 shows the functional logic diagram for an output compare. It consists of a 16-bit comparator that compares the contents of the timer channel register, *TCn,* and the free-running counter, *TCNT.* If the timer channel is configured as an output compare (*IOSn* = 1) and the two values are equal, the comparator sets the flag *CnF* and initiates an output action on the pin. If the timer channel's interrupt is enabled (*CnI* = 1) an interrupt will be pending as long as the flag is set. We will call this a *compare event.*

Figure 9.15 shows the two registers, *TCTL1* and *TCTL2,* required to configure the output action for an output compare. The output action on the pin depends on the value of the pin action bits, *OMn* and *OLn.* Table 9.6 shows the output action for an output compare pin based on the values of *OMn* and *OLn.*

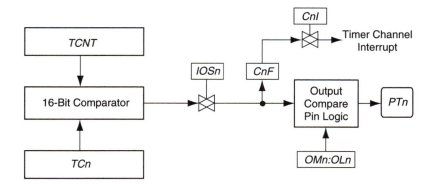

Figure 9.14   Basic Output Compare Functional Diagram

TCTL1:

($0088)	OM7	OL7	OM6	OL6	OM5	OL5	OM4	OL4
RESET:	0	0	0	0	0	0	0	0

TCTL2:

($0089)	OM3	OL3	OM2	OL2	OM1	OL1	OM0	OL0
RESET:	0	0	0	0	0	0	0	0

Figure 9.15   Output Compare Configuration Registers

TABLE 9.6   OUTPUT COMPARE ACTIONS

OMn	OLn	Action
0	0	Timer disconnected from the output pin
0	1	Toggle *OCn* output
1	0	Clear *OCn* output to 0
1	1	Set *OCn* output to 1

Note that when *OMn* and *OLn* are both zero, the output compare is disconnected from the pin. This allows the pin to be used as a general purpose input or output even if the output compare is being used to generate a timer event for software.

Let's look at three programming examples using an output compare. The first example uses an output compare to generate a periodic timer event. The second example shows how the first example can be modified to generate a square wave. The last example uses an output compare for pulse-width modulation.

Generating Periodic Events with an Output Compare.   We have seen how the real-time interrupt RTI can be used to generate a periodic timer event. The problem with the RTI is that there are a limited number of event rates available. The timing for the stopwatch examples shown in Source 9.2 and Source 9.3 was based on the RTI. But since one second is not a multiple of the RTI periods, there is a 0.06% error in the timer. We can make the system more accurate by using an output compare to generate a periodic timer event at a period that is a factor of one second.

Source 9.4 shows the modifications made to the stopwatch program to use a periodic timer event generated by the output compare on *Timer Channel 0*. To meet the switch de-bouncing requirements for the stopwatch, the time-slice period must be between 5 and 10ms. We use 10ms for this example. Therefore, the *UpdateSecs* routine counts 100 time slices for exactly one second. The only timing error remaining in this system is the frequency error due to the MCU crystal.

The first part of the program shown in Source 9.4 is the initialization. It starts by initializing the D-Bug12 interrupt jump table. This is only required if this program is run under

D-Bug12. Then the timer is enabled and *Timer Channel 0* is set to be an output compare by writing a one to bit 0 of the *TIOS* register. Since we are only using the output compare for a periodic interrupt, the pin PT0 can be disconnected and used for GPIO. This is accomplished by writing a $00 to the *TCTL2* register. The *Timer Channel 0* interrupt enable bit, *C0I*, is then set to enable the interrupt.

Source 9.4 Output Compare-Based Timer

```
;****************************************************************************
; A 1-button stopwatch based on an output compare on timer channel 1.
; This version has a debounced switch to stop/reset/start timer.
; Controller: 68HC912B32
; Monitor: D-Bug12
; Uses BasicIO module
;****************************************************************************
main:          ldd #Tc0Isr                   ;Init. D-Bug12 jump table
               pshd
               ldd #USER_TC0
               jsr [SETUSERVEC,pcr]
               puld
               movb #$80,TSCR                ;enable timer clock
               movb #$01,TIOS                ;set TC0 to output compare
               movb #$00,TCTL2               ;disconnect pin from OC0
               movb #C0I,TMSK1               ;enable TC0 interrupt
               movb #C0F,TFLG1               ;clear TC0 flag
                   …

               cli
sched_lp       jsr WaitSlice
                   …
               bra sched_lp


;****************************************************************************
;WaitSlice - The time-slice timer routine. It uses the OC0Cnt to run the
;              task loop every 10ms.
;              It also increments SecCnt for the UpdateSecs routine.
; Pseudo-C:
; WaitSlice(){
;     while(OC0Cnt != 10){}
;     OC0Cnt = 0;
;     SecCnt++;
; }
;****************************************************************************
WaitSlice      ldaa OC0Cnt                   ;Wait for 10 OCs (10ms)
               cmpa #10
               bne WaitSlice
               clr OC0Cnt
```

```
            inc SecCnt                    ;Count slice for seconds update
            rts


;***********************************************************************
; Timer Channel 0 Service Routine
;     - setup for a 1ms periodic interrupt
; Pseudo-C:
; Tc0Isr() {
;     TFLG1 = C0F;
;     OC0Cnt++;
;     TC0 = TC0 + 8000;
; }
;***********************************************************************
;
Tc0Isr      movb #C0F,TFLG1               ;Clear timer flag
            inc OC0Cnt                    ;increment OC counter
            ldd TC0                       ;set up for next compare
            addd #8000
            std TC0
            rti


;***********************************************************************
;
```

After the interrupts are enabled, if a compare event occurs, it generates an interrupt request and the program context is switched to the interrupt service routine, *Tc0Isr. Tc0Isr* increments a counter, *OC0Cnt,* which is used by *WaitSlice* to control execution of the task loop.

To generate a periodic event, the interrupt service routine also sets up *TC0* for the next interrupt. It adds 8000 to *TC0* so the compare event will occur 8000 E-clock cycles after the last compare event. If the E-clock frequency is 8MHz, this corresponds to a 1ms period between compare events.

The cost of using an output compare instead of the RTI is the extra CPU load required to set up *TC0* for the next compare event. The load-add-store instructions used to change *TC0* takes eight clock cycles. For the example in Source 9.4, this corresponds to a CPU load of only 0.001. If the interrupt rate is increased or a slower processor is used, the load required to change *TC0* may become significant.

Generating a Square-Wave. To generate a square-wave with an output compare, we only have to make small modifications to the periodic event program above. The first change is to configure the pin action for the output compare so it will invert the pin every compare event. To do this we need *OM0* = 0 and *OL0* = 1. We can modify the code that writes to *TCTL2* to

```
    movb #$01,TCTL2    ;Toggle PT0 on compare
```

Now every time a compare event occurs, the pin will be inverted. If we write the interrupt service routine for the output compare so it sets up the next compare one-half cycle later, we generate a square-wave with the desired frequency. Following is an interrupt service routine for a square-wave:

```
Tc0Isr      movb #C0F,TFLG1        ~4
            ldd TC0                 3
            addd #HALF_PERIOD       2
            std TC0                 3
            rti                     8
```

where *HALF_PERIOD* is the number of E-clock cycles in one half of the square-wave's period. Since the free-running counter and the timer channel are 16-bit registers, *HALF_PERIOD* must be less than 65,535. This corresponds to a minimum frequency of 61Hz unless the E-clock frequency is reduced or the timer's prescaler is set to a nonzero value.

Note

If we configure the output compare to toggle the output pin in the stopwatch program, we end up with a great debugging helper. We can verify that the periodic interrupt is working by watching the signal on *PT0* with an oscilloscope. If it is a square-wave with a period of two times the interrupt period, we know that the interrupt is working correctly. For the stopwatch example, we would expect to find a 500Hz square-wave on *PT0*.

The maximum square-wave frequency is a bit more difficult to determine. We can count cycles in the ISR and calculate the maximum frequency, assuming that the *rti* instruction must be executed before a new compare event can occur. This would not be a practical value, however, because it does not allow the CPU to do anything else. Having an MCU that only generates a square-wave would not be a cost-effective solution. Instead we can calculate the maximum frequency of the square-wave based on an available CPU load.

The CPU load required to generate the square-wave is

$$L = \frac{T_{csw} + T_{isr}}{\frac{T_{sq}}{2}} = 2(9 + 20)T_E \cdot F_{sq}$$

So if we have 50% of the CPU load available to generate the square-wave, for example, the maximum frequency of the square-wave is about 68.9kHz.

There are two ways to increase the maximum square-wave frequency and reduce the CPU load. One is to use *Timer Channel 7*'s ability to automatically reset the free-running counter. The other way is to use a pulse-width modulator, if one is available. These two methods will be covered later in this section for generating pulse-width-modulated waveforms. A square-wave is just a special case of PWM signal with a fixed 50% duty cycle.

Pulse-Width Modulation Using a Normal Output Compare.   Pulse-width modulation is an important function for controlling DC motors, digital-to-analog conversion, and communications. Figure 9.16 shows the timing parameters for a pulse-width-modulated waveform.

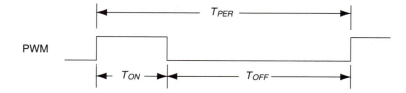

Figure 9.16 Pulse-Width Modulated Waveform

T_{PER} is the period of the pulse train, T_{ON} is the on time, and T_{OFF} is the off time. The duty cycle for a PWM waveform is

$$DutyCycle = \frac{T_{ON}}{T_{PER}} = \frac{T_{ON}}{T_{ON} + T_{OFF}}$$

The average value of the waveform is

$$\langle v(t) \rangle = \frac{T_{ON}}{T_{PER}} V_{pk} = \frac{T_{ON}}{T_{ON} + T_{OFF}} V_{pk}$$

where

V_{pk} is the peak output voltage

So we can change the average value of the PWM waveform by changing the duty cycle. This is especially useful for motor controllers because the motor speed is a function of the average voltage, whereas the torque is a function of the peak voltage. By using a PWM signal to control a motor, we can change the speed of the motor while maintaining a high torque.

In this text we generate PWM signals using three different timer resources. The first example, which is covered here, uses a standard output compare channel. Later we look at examples that use the special output compare on *Timer Channel 7* and the on-chip pulse-width modulator.

Source 9.5 is a demonstration program that uses an output compare to generate a 1kHz pulse-width modulated waveform. It uses the Basic I/O module to receive keyboard commands that are used to increase or decrease the duty cycle in 10% steps. The duty cycle is limited to 10%–90%.

The initialization is similar to the previous output compare examples. The timer is enabled, *Timer Channel 0* is set to be an output compare, and the interrupt is enabled. This time the output action is configured to set the output pin *PT0* to one. The pulse-width modulation on time, *OnTime,* and off time, *OffTime,* are initialized for a 10% duty cycle. Notice that *OffTime* is calculated by subtracting the *OnTime* from the constant period, *PERIOD.* For a 1kHz signal, *PERIOD* is set to 8000 E-clock cycles, which corresponds to 1ms.

Once initialization is complete and interrupts are enabled, the program enters a background loop that executes the task that controls the duty cycle.

Source 9.5 Pulse-Width Modulation Demonstration Using an Output Compare

```
;*********************************************************************
; A PWM demonstration based on an output compare.
; This version uses keyboard input to change the duty cycle.
;
; Controller: 68HC912B32, fE = 8MHz
; Monitor: D-Bug12
; Module: Basic IO
;*********************************************************************
; Equates
;*********************************************************************
TSCR            equ $86             ;Timer constants
TIOS            equ $80
TFLG1           equ $8e
TMSK1           equ $8c
TCTL2           equ $89
OL0             equ %00000001
TC0             equ $90
C0F             equ %00000001
ON_MAX          equ 7200            ;PWM constants
ON_MIN          equ 800
ON_STEP         equ 800
PERIOD          equ 8000
TOUPPER         equ $0fe0           ;Basic IO routines
INPUT           equ $0fb0
SCIOPEN         equ $0fec
SETUSERVEC      equ $f69a           ;D-Bug12 interrupt jump table routine
UserTC0         equ 23


;*********************************************************************
; Program
;*********************************************************************
;Initialize
;     D-Bug12 jump vector init.
;     TC0 as an Output Compare(interrupt enabled,Action -> 1)
;     Initialize SCI
;     Global variables
;*********************************************************************
main            ldd #Tc0Isr                 ;initialize D-Bug12 ISR jump table
                pshd
                ldd #UserTC0
                jsr [SETUSERVEC,pcr]
                puld
                movb #$80,TSCR              ;Enable timer clock
                movb #$01,TIOS              ;Set channel 0 to OC
                movb #$03,TCTL2             ;Set TC0 to set PT0 to one
                movb #C0F,TMSK1             ;Enable TC0 interrupt
```

```
                movb #C0F,TFLG1              ;Clear TC0 flag
                jsr SCIOPEN
                movw #ON_MIN,OnTime          ;Initialize Duty Cycle
                movw #(PERIOD-ON_MIN),OffTime
                cli

BkGndLp         jsr UpdateDC
                bra BkGndLp

;******************************************************************************
; UpdateDC
;     - This routine reads serial input to update duty cycle.
;     - Duty Cycle limited from ON_MIN to ON_MAX
;
; UpdateDutyCycle() {
;     if((UPCASE(INPUT()) == 'D') && (OnTime > ON_MIN)){
;          OnTime = OnTime - ON_STEP;
;          OffTime = PERIOD - OnTime;
;     }else if((UPCASE(INPUT()) == 'U') && (OnTime < ON_MAX)){
;          OnTime = OnTime + ON_STEP;
;          OffTime = PERIOD - OnTime;
;     }
; }
;
;******************************************************************************
UpdateDC        pshd
                jsr INPUT            ;Get keyboard input
                jsr TOUPPER
                beq dc_up_rtn        ;Return if no key is pressed
                cmpb #'D'            ;If 'D' and > ON_MIN, reduce duty cycle
                bne chk_up
step_dwn        ldd OnTime
                cpd #ON_MIN
                bls dc_up_rtn
                subd #ON_STEP
                std OnTime
                bra up_offtime
chk_up          cmpb #'U'            ;If 'U' and < ON_MAX, increase duty cycle
                bne dc_up_rtn
                ldd OnTime
                cpd #ON_MAX
                bhs dc_up_rtn
                addd #ON_STEP
                std OnTime
up_offtime      ldd #PERIOD          ;Set off time here to reduce CPU load
                subd OnTime
                std OffTime
dc_up_rtn       puld                 ; Return to background loop
                rts
```

```
;********************************************************************
; Tc0Isr - Timer Channel 0 Service Routine
;
; Tc0Isr() {
;     if(OL0 = 1){                  /*Pulse is high*/
;         OL0=0;
;         TC0 = TC0 + OnTime;
;     }else{                        /*Pulse is low*/
;         OL0=1;
;         TC0 = TC0 + OffTime;
;     }
;     TFLG1 = C0F;
; }
;
;********************************************************************
Tc0Isr       brclr TCTL2,OL0,tc0_low   ;Pulse is high?
             bclr TCTL2,OL0            ;Set TC0 to clear PT0
             ldd TC0                   ;Update TC0 for next edge
             addd OnTime
             std TC0
             bra tc0_rtn
tc0_low      bset TCTL2,OL0           ;Pulse is low
             ldd TC0                   ;Set TC0 to set PT0
             addd OffTime              ;Update TC0 for next edge
             std TC0
tc0_rtn      movb #C0F,TFLG1           ;Clear TC0 flag
             rti

;********************************************************************
; Variables
;********************************************************************
OnTime       rmb 2    ;Pulse on time
OffTime      rmb 2    ;Pulse off time

;********************************************************************
```

The interrupt service routine for *Timer Channel 0, Tc0Isr,* reconfigures the output compare to generate the next edge. There are two states involved with this routine—the output high state and the output low state. The state variable used to determine the state is the *OL0* bit in *TCTL2*. If this bit is high, we know that the output is currently high because the compare event that just occurred caused the output to go high. If this bit is low, we know the output is low.

If the output is high, we want it to remain high for the required on time. Once the on time has expired, we want the output to go low. To accomplish this we change the output action to clear the output by clearing *OL0*. We then add *OnTime* to the compare register. This sets the output compare to generate a falling edge at the end of the required on time.

If the service routine is entered with the output low, it sets the output action to set the output to one after the required off time has expired.

In this program the minimum pulse-width (high or low) is limited by the interrupt latency. This may be the only alternative, however. For example, if we are using a 68HC11 MCU that does not have an on-chip pulse-width modulator, we need to use this method. For the 68HC12 MCUs, there are two potential alternatives—using the special output compare on *Timer Channel 7* or, if available, using the on-chip pulse-width modulator.

9.2.5 Output Compare 7

On the 68HC12 standard timer module, the output compare function for *Timer Channel 7* is different from the other output compares. It can be configured to generate an output action on any of the timer I/O pins, and it can be configured to reset the free-running counter when a compare event occurs.

Figure 9.17 shows the registers that are used to configure the output actions for *Output Compare 7*. The output compare mask register *OC7M* is used to control which output pins are changed when a compare event occurs. If the bit is set, the action defined by the data register *OC7D* occurs on that pin when a compare event occurs on *Output Compare 7*.

The output compare data register *OC7D* determines the output action for each pin when a compare event occurs. If the bit is one, the pin is set when the compare event occurs. If the bit is zero, the pin is cleared when the compare event occurs. Only the bits with a one in the mask register *OC7M* are changed. For example

```
movb #$0F,OC7M
movb #$0A,OC7D
```

will cause the timer to clear *PT0* and *PT2* and set *PT1* and *PT3* when a compare event occurs on *Timer Channel 7*.

The *TCRE* bit in the *TMSK2* register controls the free-running counter reset. If *TCRE* is one, a compare event on *Timer Channel 7* will cause the free-running counter to be reset to zero. This can be a useful function, but we need to be careful when resetting the free-running counter. Many timer programs assume that the free-running counter is *free running*. For example, the periodic event program in Source 9.4 will not work as expected if the free-running counter is reset.

OC7M:

($0082)	OC7M7	OC7M6	OC7M5	OC7M4	OC7M3	OC7M2	OC7M1	OC7M0
RESET:	0	0	0	0	0	0	0	0

OC7D:

($0083)	OC7D7	OC7D6	OC7D5	OC7D4	OC7D3	OC7D2	OC7D1	OC7D0
RESET:	0	0	0	0	0	0	0	0

Figure 9.17 Output Compare 7 Configuration Registers

PWM Program Using Output Compare 7. The program in Source 9.6 uses the special output compare in *Timer Channel 7, OC7,* with the output compare in *Timer Channel 0, OC0,* to generate a PWM signal on *PT0.* It generates the same 1kHz PWM signal as the program shown in Source 9.5, but by using *OC7* and *OC0,* it places no load on the CPU except to change duty cycle.

The initialization starts by configuring *Timer Channel 7* and *Timer Channel 0* as output compares. It then sets the *TCRE* bit in the *TMSK2* register so that the free-running counter will be reset when a compare event occurs on *OC7.* It also sets *OC7M* and *OC7D* to $01, which causes *PT0* to be set to one when a compare event occurs on *OC7.* The *OC7* compare register, *TC7,* is set to *PERIOD.* For a 1kHz PWM signal, *PERIOD* is equal to 8,000.

With this configuration, the free-running counter will count from zero to 8,000. At that point *OC7* sets *PT0* to one and resets the free-running counter to zero.

The initialization code then configures *Timer Channel 0* to clear *PT0* when its compare event occurs. The minimum pulse-width, *ON_MIN,* is then stored in *TC0.* The minimum duty cycle is 10%, so *ON_MIN* is equal to 800. Now whenever the free-running counter reaches 800, a compare event on *OC0* occurs and it clears *PT0.*

So the free-running counter starts at zero with *PT0* high. When it reaches 800, a compare event occurs on *OC0.* This in turn clears *PT0.* The free-running counter continues until it reaches 8,000. At that point a compare event occurs on *OC7,* which then clears the free-running counter and sets *PT0* back to one. This cycle repeats itself indefinitely, without intervention from the CPU.

To change the duty cycle, the *UpdateDC* routine changes the value stored in *TC0,* which in effect changes the on time of the signal.

Source 9.6 Pulse-width Modulation Demo Using Output Compare 7

```
;*********************************************************************
;
; A PWM demonstration based on OC7 and OC0.
; This version uses keyboard input to change the duty cycle.
;
;
; Controller: 68HC912B32
; Monitor: D-Bug12
; Module: Basic IO
;*********************************************************************
;
; Equates
;*********************************************************************
TSCR        equ $86             ;Timer constants
TIOS        equ $80
OC7M        equ $82
OC7D        equ $83
TMSK2       equ $8d
TCRE        equ $08
TCTL2       equ $89
TC0         equ $90
TC7         equ $9e
ON_MAX      equ 7200            ;PWM constants
```

```
ON_MIN          equ 800
ON_STEP         equ 800
PERIOD          equ 8000
TOUPPER         equ $0fe0           ;Basic IO routines
INPUT           equ $0fb0
SCIOPEN         equ $0fec
;********************************************************************
; Program
;********************************************************************
;Initialize
;    Initialize SCI
;    Reset TCNT on OC7 event
;    TC0 and TC7 output compares
;    OC7 sets PT0, OC0 clears PT0
;    PERIOD -> TC7, on-time -> TC0
;********************************************************************
main            movb #$81,TIOS              ;Set channels 7 and 0 to OC
                movb #TCRE,TMSK2            ;Enable TCNT reset from OC7
                movb #$01,OC7M             ;OC7 action on PT0
                movb #$01,OC7D             ;PT0 -> 1 when OC7 occurs
                movw #PERIOD,TC7           ;Setup PWM period for OC7
                movb #$02,TCTL2            ;Set OC0 to clear PT0
                movw #ON_MIN,TC0           ;Initialize on time for OC0
                movb #$80,TSCR             ;Enable timer clock
                jsr SCIOPEN                ;Initialize SCI

BkGndLp         jsr UpdateDC
                bra BkGndLp


;********************************************************************
; UpdateDC
;    - This routine reads serial input to update duty cycle.
;    - Duty Cycle limited from ON_MIN to ON_MAX
;
; UpdateDutyCycle() {
;     if((UPCASE(INPUT()) == 'D') && (TC0 > ON_MIN)){
;         TC0 -= ON_STEP;
;     }else if((UPCASE(INPUT()) == 'U') && (TC0 < ON_MAX)){
;         TC0 += ON_STEP;
;     }
; }
;********************************************************************
UpdateDC        pshd
                jsr INPUT           ;Get keyboard input
                jsr TOUPPER
                beq dc_up_rtn       ;Return if no key is pressed
                cmpb #'D'           ;If 'D' and > ON_MIN, reduce duty cycle
                bne chk_up
step_dwn        ldd TC0
                cpd #ON_MIN
```

```
            bls dc_up_rtn
            subd #ON_STEP
            std TC0
            bra dc_up_rtn
chk_up      cmpb #'U'               ;If 'U' and < ON_MAX, increase duty cycle
            bne dc_up_rtn
            ldd TC0
            cpd #ON_MAX
            bhs dc_up_rtn
            addd #ON_STEP
            std TC0
dc_up_rtn   puld                    ; Return to background loop
            rts
;**********************************************************************
```

Once configured, the timer runs on its own so the CPU load is zero. In addition the pulse's widths can be as small as a single E-clock cycle. For example if *TC0* is set to zero and *TC7* is set to one, this program generates a 4MHz square-wave.

Note _____

A Word of Caution

When *OC7* is configured to reset the free-running counter, other programs that rely on the counter may no longer function correctly. If an on-chip pulse-width modulator is available, it is preferable to use it instead of this method.

9.2.6 Input Captures

The input-capture function allows us to capture the free-running counter value when an input event occurs. Applications of input captures include event detection, event counting, pulse-width measurement, and frequency measurement. As shown in Figure 9.10 on page 313, all eight standard timer channels can be configured as either an output compare or an input capture. Figure 9.18 shows the functional block diagram for a single input

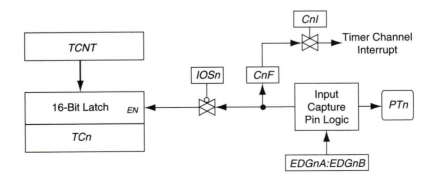

Figure 9.18 Input Capture Functional Block Diagram

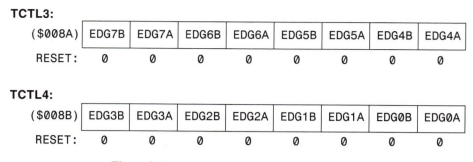

TCTL3:

($008A)	EDG7B	EDG7A	EDG6B	EDG6A	EDG5B	EDG5A	EDG4B	EDG4A
RESET:	0	0	0	0	0	0	0	0

TCTL4:

($008B)	EDG3B	EDG3A	EDG2B	EDG2A	EDG1B	EDG1A	EDG0B	EDG0A
RESET:	0	0	0	0	0	0	0	0

Figure 9.19 Input Capture Configuration Registers

TABLE 9.7 INPUT CAPTURE EVENT SETTINGS

EDGnB	EDGnA	Event
0	0	Input Capture Disabled
0	1	Capture Rising Edges
1	0	Capture Falling Edges
1	1	Capture Any Edge

capture. An event on the input pin causes the timer channel's flag to be set and the current *TCNT* to be latched in the timer channel register. If the timer channel's interrupt is enabled, an input event will also cause the channel's interrupt to occur.

Figure 9.19 shows *TCTL3* and *TCTL4*. These registers are used to configure the input event to initiate a capture. They contain eight pairs of event detection bits—one for each timer channel.

The input events and the corresponding event detection bit values are shown in Table 9.7. The *RESET* value, $00, means the input capture function is disabled. The other three bit combinations allow us to capture on rising edges, falling edges, or all edges. Since the input captures depend on edge detection, we need to make sure to debounce the input signals.

In Section 8.3 we saw two program examples that used an input capture to count pulses by detecting rising edges. One example polled the timer channel flag and the other used the interrupt. We will not repeat these examples here, although it may be a good idea to review the source code and make sure you understand how those examples work. Here we cover two other applications: pulse-width measurement and frequency measurement.

Pulse-Width Measurement Using an Input Capture.   To measure the pulse-width of a signal using an input capture, we need to capture *TCNT* at the start of the pulse and the end of the pulse. The pulse-width is calculated by finding the difference between the two captured values.

Source 9.7 shows a program that measures the pulse-width of a signal tied to *PT0*. The initialization configures *Timer Channel 0* as an input capture that detects rising edges. The program then enters an endless background loop.

Source 9.7 Pulse-Width Measurement Using an Input Capture

```
;*********************************************************************
; Using an input capture to measure active-high pulse-widths.
;*********************************************************************
; Equates
;*********************************************************************
TSCR        equ $86
TFLG1       equ $8e
TMSK1       equ $8c
TCTL4       equ $8b
TC0         equ $90
C0F         equ %00000001
EDG0B       equ %00000010
SETUSERVEC  equ $f69a
UserTC0     equ 23
;*********************************************************************
; Program
;    Initialize TC0 D-Bug12 vector, and TC0 for input capture on
;    rising edges, enable TC0 interrupts.
;*********************************************************************
main         ldd #Tc0Isr        ;D-Bug12 Jump Table
             pshd
             ldd #UserTC0
             jsr [SETUSERVEC,pcr]
             puld
             movb #$80,TSCR       ;turn on timer
             movb #$01,TCTL4      ;start by capturing first rising edge
             movb #C0F,TMSK1      ;enable TC0 interrupts
             movb #C0F,TFLG1
             cli


poll_lp      bra poll_lp


;*********************************************************************
; IC0 Interrupt service routine.
;
;     - On rising edge current TCNT is saved and IC is configured for
;       any edge. (This is faster than setting for falling edge.)
;     - On falling edge PulseWidth is calculated and IC is set for
;       rising edge.
;
; Pseudo-C
; Tc0Isr(){
;                                     /* Clear channel 0 flag */
;     if((TCTL4 & EDG0B) == 0x00){    /* Rising Edge ? */
;         REdge = TC0;                /* Save rising edge time */
;         TCTL4 |= EDG0B;             /* Set EDG0B bit */
;     }else{
```

```
;            PulseWidth = TC0 - REdge;      /* Save pulse-width */
;            TCTL4 &= ~EDG0B;               /* Clear EDG0B bit */
;      }
;   TFLG1 = C0F;
; }
;**********************************************************************
TcOIsr        ldaa  TCTL4                 ; Rising Edge ?
              anda  #EDG0B
              bne   tc0_fedge
              movw  TC0,REdge             ; Yes: save time
              bset  TCTL4,EDG0B           ; set IC0 for any edge
              bra   tc0_rtn
tc0_fedge     ldd   TC0                   ; No: Calculate PulseWidth
              subd  REdge
              std   PulseWidth
              bclr  TCTL4,EDG0B           ; Set IC0 for rising edge
tc0_rtn       movb  #C0F,TFLG1            ; Clear channel 0 flag
              rti


;**********************************************************************
; Variables
;**********************************************************************
PulseWidth    rmb 2
REdge         rmb 2
;**********************************************************************
```

The pulse-width measurement is made in the interrupt service routine for *Timer Channel 0, TcOIsr.* Since the input capture is initialized for rising edges, the first time the ISR is entered it is because a rising edge was detected. The value of *TC0* is saved in *REdge* and the input capture is reconfigured to detect any edge. At this point, the program is supposed to detect the falling edge. It is configured to detect any edge because we only have to set the *EDG0B* bit. If we configure it for falling edges, we would have to set *EDG0B* and clear *EDG0A*.

The next time the ISR is entered, it is because the falling edge has been detected. This time the ISR calculates the pulse-width by subtracting the rising-edge time, *REdge,* from the falling-edge time captured in *TC0*. The result is stored in *PulseWidth*. If the pulses do not occur too fast, *PulseWidth* will contain the number of E-clock cycles in the last pulse.

The program in Source 9.7 only measured the pulse-width in E-clock cycles. It did not do anything with the result, so by itself it is not a useful program. Before we can do anything with the result, we need to calculate the CPU load placed on the processor by this program. We also need to determine the minimum and maximum pulse-width that this routine can measure.

Minimum Pulse-Width. This routine will only work if the falling edge of the pulse occurs after the rising edge has been saved, the input capture is reconfigured, and the timer flag is cleared. Therefore the minimum pulse-width is

$$T_{pulse,\,min} = T_l + T_{srv,\,r} = 2.75\mu s + 2.75\mu s = 5.5\mu s$$

The service time, $T_{srv, r}$, is determined by counting instruction clock cycles. A narrower pulse would cause an erroneous value to be stored in *PulseWidth*.

Pulse Measuring CPU Load. What we can do with the resulting value in *PulseWidth* depends on the CPU load required by the pulse measurement. The CPU load depends on the time required to measure the pulses and the pulse sample rate. Some applications require every pulse to be measured, while others only require an occasional pulse measurement.

The pulse measurement time is the sum of the time it takes to service a rising edge and the time it takes to service the falling edge. For the previous example, the pulse measurement time is

$$T_{T, pulse} = 2(T_{csw} + T_{rti}) + T_{srv, r} + T_{srv, f} = 4.25\mu s + 2.75\mu s + 3.0\mu s = 10\mu s$$

and the CPU load is

$$L = \frac{T_{T, pulse}}{T_S} = \frac{10\mu s}{T_S}$$

where

$$T_S = \text{the pulse sample period}$$

EXAMPLE 9.3

Pulse-Width Measurement CPU Load

Calculate the CPU load that is required to measure pulse-widths with the program in Source 9.7. The pulse signal to be measured is the 1kHz PWM waveform generated by the program in Source 9.6.

$$L = \frac{10\mu s}{T_S} = \frac{13.875\mu s}{1ms} = 0.01$$

In this case the CPU load required for the pulse measurement is very small. Also because the minimum pulse-width for the signal is 100µs, the minimum pulse-width requirement is easily met.

___

Frequency Measurement with an Input Capture. There are two ways to measure the frequency of a signal using an input capture. In the first method we count cycles for a predetermined time. To count cycles, an input capture is used to detect the rising edges or falling edges of the signal. An output compare or real-time interrupt is then used to start and stop the counting. The frequency range and resolution are then functions of the count time and the size of the counter variable. This is the most flexible method for determining frequency because we can easily control the frequency range and resolution. We look at an example that uses this method in the next subsection on the pulse accumulator.

The second method used to measure frequency is to measure the signal period and calculate the frequency by dividing the E-clock frequency by the pulse-width in E-clock cycles:

$$f = \frac{f_E}{N_T}$$

where N_T is the number of E-clock cycles per signal period.

With the 68HC12 the easiest way to do this is to use the extended division instruction, *EDIV.* Source 9.8 shows an interrupt service routine for *Timer Channel 0* that calculates the frequency of the signal connected to *PT0.* The rest of the program can be identical to Source 9.7.

Timer Channel 0 is configured as an input capture that detects rising edges. When an edge is detected, the service routine subtracts the last edge time, *LastEdge,* from the current edge time, *TC0,* to find the period. The frequency is then calculated by dividing the E-clock frequency, 8MHz, by the measured period.

Source 9.8 Frequency Measurement Routine

```
;*************************************************************************
;  IC0 Interrupt service routine.
;
;  Pseudo-C
;  Tc0Isr(){
;      Period = TC0 - LastEdge;          /*Calculate Period */
;      if(Period >= 123){               /*Calculate Frequency if < 65535Hz*/
;          Freq = 8000000/Period;
;      }
;      LastEdge = TC0;                   /* Save Last Edge */
;      TFLG1 = C0F;                      /* Clear channel 0 flag */
;  }
;
;*************************************************************************
Tc0Isr      ldd TC0                      ;Calculate period
            subd LastEdge
            std Period                   ;Save period (optional)
            cpd #123                     ;< 65535Hz?
            blo tc0_rtn                  ;No: skip frequency calculation
            tfr d,x                      ;Calculate frequency
            ldy #$7a                     ;8,000,000 -> Y:ACCD
            ldd #$1200
            ediv                         ;8MHZ/pulsewidth
            sty Freq
tc0_rtn     movw TC0,LastEdge            ;Save edge for next cycle
            movb #C0F,TFLG1              ;Clear channel 0 flag
            rti


;*************************************************************************
```

```
; Variables
;**********************************************************************
Period          rmb 2
LastEdge        rmb 2
Freq            rmb 2
;**********************************************************************
```

The frequency range that this program can measure is 123Hz to 65,535Hz. The minimum frequency is the point at which the measured signal period is greater than 65,535 E-clock cycles, which is the maximum value for the 16-bit variable and 16-bit counter. The maximum frequency is determined by the size of the frequency variable, which is also limited to 16 bits by the *EDIV* instruction.

9.2.7 Pulse Accumulator

The pulse accumulator consists of a 16-bit counter that can be used to count events on *PT7* or measure the time between events on *PT7*. The functional logic diagram and the control registers for the pulse accumulator are shown in Figure 9.20.

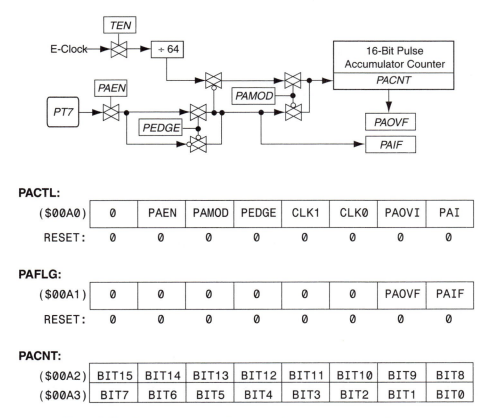

PACTL:

($00A0)	0	PAEN	PAMOD	PEDGE	CLK1	CLK0	PAOVI	PAI
RESET:	0	0	0	0	0	0	0	0

PAFLG:

($00A1)	0	0	0	0	0	0	PAOVF	PAIF
RESET:	0	0	0	0	0	0	0	0

PACNT:

($00A2)	BIT15	BIT14	BIT13	BIT12	BIT11	BIT10	BIT9	BIT8
($00A3)	BIT7	BIT6	BIT5	BIT4	BIT3	BIT2	BIT1	BIT0

Figure 9.20 Pulse Accumulator Functional Diagram and Control Registers

The mode of the pulse accumulator is controlled by the *PAMOD* bit in the *PACTL* register. For event counting mode, the *PAMOD* bit is set to zero. This sets the counter clock source to the *PT7* input pin. For gated-time mode, the *PAMOD* bit is set to one. This changes the counter clock source to the E-clock divided by 64. This clock is gated by the signal on *PT7*.

Event Counter Mode.   When in the event counting mode, the pulse accumulator counter counts events on the *PT7* pin as long as the pulse accumulator is enabled, *PAEN* = 1. *PEDGE* controls the event that will be counted. If *PEDGE* is zero, falling edges are counted, and if *PEDGE* is one, rising edges are counted. Each time an event edge is detected, the pulse accumulator input flag, *PAOVF* is set. If the *PAI* bit is also set, the *PAI* interrupt will occur.

Gated-Time Accumulation Mode.   In the gated-time mode, the pulse accumulator clock is set to the E-clock divided by 64. The signal on *PT7* is used to enable the clock, so in effect the counter measures the length of the enable signal on *PT7*. If *PEDGE* is zero, a one on the input enables counting. If *PEDGE* is one, a zero on the input enables counting. The pulse accumulator input flag, *PAIF,* is set on the trailing edge of the enable signal.

The *PAOVF* flag is set when the pulse accumulator counter overflows. This flag can be used to extend the counter to more bits using software. The *CLK1* and *CLK0* bits in *PACTL* are used to control the timer module clock. This was covered previously in the subsection on the standard timer module.

The pulse accumulator is most useful as an event counter for high-speed signals. We have seen how to count input events with the input capture. The problem with using the input capture is the large CPU load required for high event rates. When using the pulse accumulator, there is no CPU load while counting events. The only limitation on the event rate is the *Pulse accumulator pulse width* specification, which is typically close to the E-clock period.

Source 9.9 shows a program that measures frequency using the pulse accumulator and an output compare. The count time is set to 100ms so the program's resolution is ±10Hz. The measured value is stored in a 16-bit variable so the frequency range is 0 to 655,350Hz. We could also add code to detect a pulse accumulator overflow to add an additional bit to the frequency variable.

Source 9.9 Pulse Accumulator Frequency Measurement Program

```
;**************************************************************************
; A frequency measurement program using the pulse accumulator. An output
; compare is used to set a sample period of 100ms.
; Controller: 68HC912B32
; Monitor: D-Bug12
; Uses BasicIO module
;**************************************************************************
; Equates
;**************************************************************************
TSCR        equ $86         ;Timer definitions
TIOS        equ $80
TCNT        equ $84
```

```
TFLG1        equ $8e
TMSK1        equ $8c
TCTL2        equ $89
TC0          equ $90
C0F          equ %00000001
PACTL        equ $A0            ;pulse accumulator definitions
PACNT        equ $A2
PAEN         equ %01000000
USER_TC0     equ 23             ;D-Bug12 interrupt definitions
SETUSERVEC   equ $f69a


;**********************************************************************
; main - Timer init and Frequency measurement loop.
; Pseudo-C:
; main(){
;     SETUSERVEC(USER_TC1,OC1Isr)
;     TSCR = 0x80;            /* Initialize Timer and OC0 */
;     TIOS = 0x01;
;     TCTL2 = 0x01;           /* Toggle OC0 pin for debug only */
;     TMSK1 = C0F;
;     TFLG1 = C0F;
;     PACTL = 0x40;           /* Enable pulse accumulator */
;     ENABLE_INT();
;
;     while(1){
;         OC0Cnt = 0;              /* Set up for sample time */
;         TC0 = TCNT+8000;
;         PACNT = 0;               /* Clear pulse accum. counter */
;         while(OC0Cnt != 100){ /* Wait for sample complete */
;         }
;         TenthFreq = TempFreq; /* Save frequency */
;     }
; }
;**********************************************************************
main:        ldd #Tc0Isr                  ;Init. D-Bug12 RTII jump table
             pshd
             ldd #USER_TC0
             jsr [SETUSERVEC,pcr]
             puld
             movb #$80,TSCR               ;enable timer clock
             movb #$01,TIOS               ;set TC0 to output compare
             movb #$01,TCTL2              ;invert pin on OC0 for debug
             movb #C0F,TMSK1              ;enable TC0 interrupt
             movb #C0F,TFLG1              ;clear TC0 flag
             movb #$40,PACTL              ;enable pulse acummulator
             cli                          ;enable interrupts

SampLoop     clr OC0Cnt                   ;set up for sample time
```

```
                ldd TCNT
                addd #8000
                std TC0
                movw #0,PACNT              ;reset pulse accum. counter
wait_ts         ldaa OC0Cnt               ;wait for 100ms sample done
                cmpa #100
                bne wait_ts
                movw TempFreq,TenthFreq    ;save frequency
                bra SampLoop


;**********************************************************************
; Timer channel 0 Service Routine
;    - capture pulse accumulator count
;    - setup for a 1ms periodic interrupt
; Pseudo-C:
; Tc0Isr() {
;     TempFreq = PACNT;
;     OC0Cnt++;
;     TC0 = TC0 + 8000;
;     TFLG1 = C0F;
; }
;**********************************************************************
Tc0Isr          ldd PACNT                 ;capture pulse accum. counter
                std TempFreq
                inc OC0Cnt                ;increment 1ms counter
                ldd TC0                   ;set up for next OC
                addd #8000
                std TC0
                movb #C0F,TFLG1           ;clear TC0 flag
                rti


;**********************************************************************
; Variables
;**********************************************************************
OC0Cnt          rmb 1          ;Output compare counts
TenthFreq       rmb 2          ;Frequency/10
TempFreq        rmb 2          ;Temporary frequency capture
;**********************************************************************
```

The *Timer Channel 0* output compare is set up to generate a 1ms periodic interrupt. In addition it immediately captures the pulse accumulator's counter value. In the main background loop, the sample time is set up by resetting *OC0Cnt* and synchronizing the output compare to the current value of the free-running counter. In this way, when the program counts 100 interrupts, we know that 100ms has occurred. Without this synchronization the sample time could be as low as 99ms. Once the sample time is set, the *PACNT* is cleared to start a new count. The main loop then waits for 100 interrupts to occur before saving the *PACNT* value captured by the output compare ISR, *TempFreq*.

9.2.8 Pulse-Width Modulator

The M68HC912B32 microcontroller has an on-chip, pulse-width modulation (PWM) module that can generate up to four PWM signals with no CPU loading. Once the PWM module is configured, the PWM signals are generated without software intervention except to change the duty cycle. The PWM module can be configured as four 8-bit PWM channels, two 16-bit PWM channels, or one 16-bit and two 8-bit PWM channels. Figure 9.21 shows the PWM module configured as four 8-bit PWM channels. On the M68HC912B32 MCU, the PWM channels use PORTP, bits 0–3.

The four channels are grouped into two pairs, *PWM0–PWM1* and *PWM2–PWM3*. Each pair shares common clock sources, *Clock A* or *S0* for *PWM0* and *PWM1*, and *Clock B* or *S1* for *PWM2* and *PWM3*.

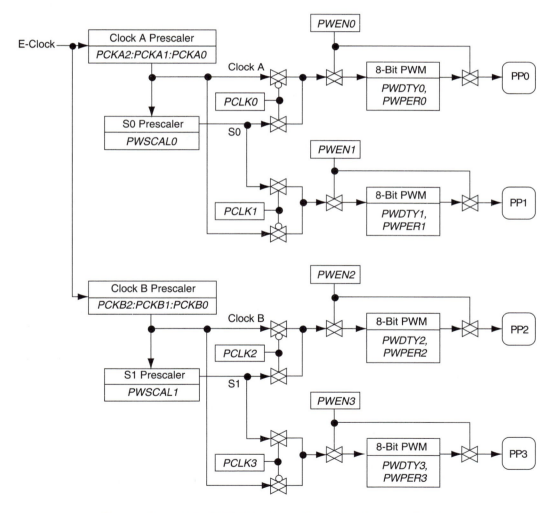

Figure 9.21 68HC912B32 Pulse-Width Modulation Module, 8-Bit Mode

TABLE 9.8 PULSE-WIDTH MODULATOR CONFIGURATION

Function	Register	Control Bits	Scope
Configure the PWM to generate left-aligned or center-aligned PWM waveforms.	PWCTL	CENTR	module
Configure channel pairs into two 8-bit or one 16-bit modulator.	PWCLK	CON23,CON01	pair
Set the clock frequency of the PWM clocks, Clock A and Clock B.	PWCLK	PCKAx,PCKBx	pair
Set the clock frequency of the prescaled PWM clocks, S0 and S1.	PWSCAL0, PWSCAL1	All	pair
Select the PWM clock for each channel.	PWPOL	PCLKn	channel
Set the polarity of each channel.	PWPOL	PPOLn	channel
Enable PWM channels.	PWEN	PWENn	channel
Set PWM period.	PWPERn	All	channel
Set PWM duty cycle.	PWDTYn	All	channel

PWCTL:

($0054)	0	0	0	PSWAI	CENTR	RDPP	PUPP	PSBCK

PWCLK:

($0040)	CON23	CON01	PCKA2	PCKA1	PCKA0	PCKB2	PCKB1	PCKB0

PWSCAL0:

($0044)	Bit 7	Bit 6	Bit 5	Bit 4	Bit 3	Bit 2	Bit 1	Bit 0

PWSCAL1:

($0046)	Bit 7	Bit 6	Bit 5	Bit 4	Bit 3	Bit 2	Bit 1	Bit 0
RESET:	0	0	0	0	0	0	0	0

Figure 9.22 PWM Module Configuration Registers

The configuration steps required to use a pulse-width modulator are shown in Table 9.8. It also shows which registers are used for each step and the scope of the configuration.

We first cover the configuration steps that affect the whole PWM module or a channel pair. Then we go over the configuration steps for the individual PWM channels. We consider only the registers required for a typical PWM configuration. For example the PWM counters are accessible, but they are not required for most PWM applications.

The registers used to configure the PWM module are shown in Figure 9.22. They include the PWM control register, *PWCTL,* the PWM clock scale and concatenate register, *PWCLK,* and the two PWM clock prescale registers, *PWSCAL0* and *PWSCAL1.*

PWM Alignment. The only bit in the PWM control register we look at is the *CENTR* bit. If this bit is clear, the PWM waveforms will be left-aligned. If it is set to one, they will be center-aligned. Figure 9.23 shows the difference between left-aligned and center-aligned PWM waveforms.

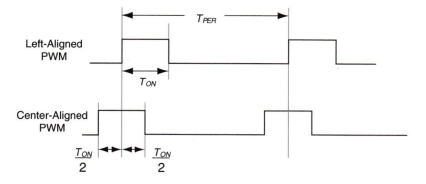

Figure 9.23 PWM Waveform Alignment

Channel Concatenation. The *CON23* and *CON01* bits in the *PWCLK* register are used to concatenate a PWM pair into a single 16-bit PWM. If *CON23* is set, channels 2 and 3 are configured as a single 16-bit PWM. If *CON01* is set, channels 0 and 1 are configured as a single 16-bit PWM.

PWM Clock Configuration. The duty cycle and period for the PWM channels are derived from the PWM clocks. By setting the clock frequency, we can control the PWM frequency range and the pulse-width resolution.

As shown in Figure 9.21, each modulator pair has two potential clock sources—a channel clock, *Clock A* or *Clock B,* and a scaled clock, *S0* or *S1. PWM0* and *PWM1* can use either *Clock A* or *S0,* and *PWM2* and *PWM3* can use either *Clock B* or *S1.*

The *PCKAx* and *PCKBx* bits in the *PWCLK* register are used to set the frequency of *Clock A* and *Clock B.* Table 9.9 shows the clock frequencies for values of the prescaler bits.

The frequency of the prescaled clocks *S0* and *S1* is set by the values stored in the scale registers *PWSCAL0* and *PWSCAL1.* The scaled frequencies are

$$f_{S0} = \frac{f_{ClockA}}{2(PWSCAL0 + 1)}, \qquad f_{S1} = \frac{f_{ClockB}}{2(PWSCAL1 + 1)}$$

TABLE 9.9 PWM CLOCK A AND CLOCK B PRESCALER SETTINGS

PCKA2:PCKA1:PCKA0 (PCHB2:PCKB1:PCKB0)	Clock A Frequency (Clock B Frequency)
000	f_E
001	$f_E \div 2$
010	$f_E \div 4$
011	$f_E \div 8$
100	$f_E \div 16$
101	$f_E \div 32$
110	$f_E \div 64$
111	$f_E \div 128$

PWEN:

| ($0042) | 0 | 0 | 0 | 0 | PWEN3 | PWEN2 | PWEN1 | PWEN0 |

PWPOL:

| ($0041) | PCLK3 | PCLK2 | PCLK1 | PCLK0 | PPOL3 | PPOL2 | PPOL1 | PPOL0 |
| RESET: | 0 | 0 | 0 | 0 | 0 | 0 | 0 | 0 |

PWPER0:

| ($004C) | bit 7 | bit 6 | bit 5 | bit 4 | bit 3 | bit 2 | bit 1 | bit 0 |

PWPER1:

| ($004D) | bit 7 | bit 6 | bit 5 | bit 4 | bit 3 | bit 2 | bit 1 | bit 0 |

PWPER2:

| ($004E) | bit 7 | bit 6 | bit 5 | bit 4 | bit 3 | bit 2 | bit 1 | bit 0 |

PWPER3:

| ($004F) | bit 7 | bit 6 | bit 5 | bit 4 | bit 3 | bit 2 | bit 1 | bit 0 |
| RESET: | 1 | 1 | 1 | 1 | 1 | 1 | 1 | 1 |

PWDTY0:

| ($0050) | bit 7 | bit 6 | bit 5 | bit 4 | bit 3 | bit 2 | bit 1 | bit 0 |

PWDTY1:

| ($0051) | bit 7 | bit 6 | bit 5 | bit 4 | bit 3 | bit 2 | bit 1 | bit 0 |

PWDTY2:

| ($0052) | bit 7 | bit 6 | bit 5 | bit 4 | bit 3 | bit 2 | bit 1 | bit 0 |

PWDTY3:

| ($0053) | bit 7 | bit 6 | bit 5 | bit 4 | bit 3 | bit 2 | bit 1 | bit 0 |
| RESET: | 1 | 1 | 1 | 1 | 1 | 1 | 1 | 1 |

Figure 9.24 PWM Channel Configuration Registers

Channel Configuration.   Once the module configuration is complete, we can configure and enable each PWM channel. The PWM parameters that need to be configured are the clock source, the polarity, the period, and the duty cycle. The registers required for channel configuration are shown in Figure 9.24.

To select the clock for each PWM channel, we need to set the *PCLKn* bit in the *PWPOL* register. If the *PCLKn* bit is cleared, the channel clock for that channel pair is used. If the *PCLKn* bit is set, the scaled clock is used. For example if we execute the following instruction:

```
movb #%00011111,PWPOL
```

PWM2 and *PWM3* will use *Clock B*, *PWM1* will use *Clock A,* and *PWM0* will use *S0.*

The *PWPOL* register also sets the polarity for each channel. If the *PPOLn* bit is high, the PWM generates active-high pulses. If it is low, it generates active-low pulses. The code example given previously sets all four channels to generate active-high pulses.

To configure the PWM period and duty cycle, we load the appropriate values into the period and duty cycle registers. Recall that the duty-cycle for an active-high pulse train is

$$DutyCycle = \frac{T_{ON}}{T_{PER}}$$

where T_{ON} is the pulse width and T_{PER} is the pulse period. Normally to vary the duty cycle, the period is kept constant and the pulse width is changed. If an active-low pulse is used, the duty cycle becomes

$$DutyCycle = 1 - \frac{T_{ON}}{T_{PER}}$$

In the PWM module, the *PWPERn* registers determine the pulse periods for each channel and the *PWDTYn* registers determine the duty cycle by changing the pulse-width for each channel. For left-aligned waveforms, the period for modulator n is

$$T_{PERn} = T_{CLK}(PWPERn + 1)$$

and the pulse width is

$$T_{ONn} = T_{CLK}(PWDTYn + 1)$$

This results in a duty cycle for an active-high pulse of

$$DutyCycle = \frac{PWDTYn + 1}{PWPERn + 1}$$

Center-aligned waveforms require a clock frequency that is two times greater than the equivalent left-aligned signal. For a center-aligned signal, the period is

$$T_{PERn} = 2 \cdot T_{CLK} \cdot PWPERn$$

and the pulse-width is

$$T_{ONn} = 2 \cdot T_{CLK} \cdot PWDTYn$$

This results in a duty cycle of

$$DutyCycle = \frac{PWDTYn}{PWPERn}$$

assuming an active-high pulse.

Once the module configuration and channel configuration is complete, the PWM channel can be enabled. The *PWEN* register is straightforward. A one written to one of the *PWENn* bits will enable that PWM channel. Once enabled, the PWM channel's clock is enabled and the output pin is connected to the PWM. When two 8-bit channels are concatenated, both enable bits must be set. For example if we concatenate *PWM2* and *PWM3,* we need to set both *PWEN2* and *PWEN3* to one.

PWM Example Using the On-Chip PWM Module. In this example we use the pulse-width modulator to implement a demonstration program that works the same way as the PWM demonstrations shown in Source 9.5 and Source 9.6. This program generates a PWM signal with a frequency of 1kHz and an adjustable duty cycle from 10% to 90% in 10% steps. The signal pulse is active-high and left-aligned. The complete source for this example is shown in Source 9.10.

There are many different ways that we can configure the PWM module to meet the requirements for this example. Let's go over some of the thought processes that should take place when selecting a PWM module configuration.

The required dynamic range for this signal is only 10. An 8-bit PWM has more than enough dynamic range, so we use the 8-bit PWM on channel 0, *PWM0*.

Next we need to determine an acceptable PWM clock frequency. The minimum clock frequency is determined by the required pulse-width resolution, that is, the change in pulse-width for a single step. In this case each pulse-width step is 10% of the period, which is 100µs. So the clock frequency must be a multiple of 10kHz. The maximum clock frequency is determined by the size of the PWM registers—in this case eight bits.

$$10 \cdot f_{PWM} = 10\text{kHz} \leq f_{CLK} \leq 250 \cdot f_{PWM} = 250\text{kHz}$$

Which clock frequency is selected depends on other system requirements. We need to consider what the requirements are for *PWM1* because it shares clock sources with *PWM0*. If we select 250kHz, we preserve the best possible resolution for *PWM1*. If we select 10kHz, lower PWM frequencies can be used for *PWM1*. In this example we use 100kHz so each pulse-width step is equal to a 1% step in duty cycle. This frequency was only selected for convenience. It provides far more dynamic range (100) than is actually required.

Now we have to determine how to reduce the E-clock frequency down to 100kHz. For an 8MHz E-clock frequency, we need a total divisor of 80. The following combinations of *Clock A* and *S0* frequencies can be used:

PCK2:PCK1:PCK0	Clock A	PWSCAL0 (Divisor)	S0
000	8MHz	39 (80)	100kHz
001	4MHz	19 (40)	100kHz
010	2MHz	9 (20)	100kHz
011	1MHz	4 (10)	100kHz

Again selecting a higher *Clock A* frequency allows for better pulse-width resolution on *PWM1*, whereas a lower frequency allows for lower PWM frequencies. In this example *Clock A* is configured for 8MHz to preserve resolution. The module and channel pair configuration registers are set as follows:

```
$00 -> PWCTL        ; Left-Aligned pulse.
$00 -> PWCLK        ; Keep Clock A at 8MHz
 39 -> PWSCAL0      ; fS0 = FA/2(39+1) = 100kHz
```

Notice that the *PWCTL* and *PWCLK* registers are set to their *RESET* state.

Now we need to configure *PWM0*. This involves selecting the clock source, polarity, period, and duty cycle.

```
$11 -> PWPOL                ;PWM Ch0: S0 clock,active high
 99 -> PWPER0               ;Period = (99 + 1)TSO = 1ms
  9 -> PWDTY0               ;Duty = 9+1/Period = 10%
```

A duty cycle of 10% was selected for the initial duty cycle. This is the duty cycle that will be generated when the system is started. Be careful to use the correct initial value. You probably would not want a motor to start running immediately out of *RESET*.

Finally we can enable the *PWM0:*

```
$01 -> PWEN                 ;Enable PWM Channel 0
```

Source 9.10 PWM Demonstration Using the On-Chip PWM

```
;**************************************************************************
; A PWM demonstration based on 68HC912B32 PWM.
;    This is a 1kHz active-high pulse. The duty cycle can
;    range from 10% to 90%. It uses a single 8-bit PWM, PW0.
;
; Controller: 68HC912B32, fE = 8MHz
; Monitor: D-Bug12
; Module: Basic IO
;
;**************************************************************************
; Equates
;**************************************************************************
PWCLK           equ $40         ;PWM module constants
PWPOL           equ $41
PWEN            equ $42
PWSCAL0         equ $44
PWCNT0          equ $48
PWPER0          equ $4c
PWDTY0          equ $50
PWCTL           equ $54
TOUPPER         equ $0fe0       ;Basic IO routines
INPUT           equ $0fb0
SCIOPEN         equ $0fec


;**************************************************************************
; main - Initialize PWM. Forever update pulse-width based on SCI input.
;
; Pseudo-C:
; main(){
;     SCIOPEN();               /* Initialize SCI for demo.*/
;     PWCTL = 0x00;            /* Left-Aligned pulse. */
;     PWCLK = 0x00;            /* Keep Clock A at 8MHz */
;     PWSCAL0 = 39;            /* fSO = FA/2(39+1) = 100kHz */
```

```
;       PWPOL = 0x11;               /* PWM Ch0: S0 clock,active high */
;       PWPER0 = 99;               /* Period = (99 + 1)TSO = 1ms */
;       PWDTY0 = 9;                /* Duty = 9+1/Period = 10% */
;       PWEN = 0x01;               /* Enable PWM Channel 0 */
;       while(1) {
;           UpdateDC();
;       }
; }
;*************************************************************************
main        jsr SCIOPEN                 ;Initialize SCI for demo.
            movb #$00,PWCTL             ;Left-Aligned pulse
            movb #$00,PWCLK             ;Set Clock A to 8MHz
            movb #39,PWSCAL0            ;fSO = FA/2(39+1) = 100kHz
            movb #$11,PWPOL             ;PWM Ch0: S0 clock,active high
            movb #99,PWPER0             ;Period = (99 + 1)TSO = 1ms
            movb #9,PWDTY0              ;Duty = 9+1/Period = 10%
            movb #$01,PWEN              ;Enable PWM Channel 0

BkGndLp     jsr UpdateDC
            bra BkGndLp


;*************************************************************************
; UpdateDC
;    - This routine reads serial input to update duty cycle.
;    - Duty Cycle limited from 10% to 80%
;
; Pseudo-C:
; UpdateDutyCycle() {
;     if((UPCASE(INPUT()) == 'D') && (PWDTY0 > 9)){
;         PWDTY0 -= 10;
;     }else if((UPCASE(INPUT()) == 'U') && (PWDTY0 < 80)){
;         PWDTY0 += 10;
;     }
; }
;
;*************************************************************************
UpdateDC    pshd
            jsr INPUT              ;Get keyboard input
            jsr TOUPPER
            beq dc_up_rtn         ;Return if no key is pressed
            cmpb #'D'             ;If 'D' and > 9, reduce duty cycle
            bne chk_up
step_dwn    ldab PWDTY0
            cmpb #9
            bls dc_up_rtn
            subb #10
            stab PWDTY0
            bra dc_up_rtn
chk_up      cmpb #'U'            ;If 'U' and < 80, increase duty cycle
            bne dc_up_rtn
```

```
        ldab PWDTY0
        cmpb #80
        bhs dc_up_rtn
        addb #10
        stab PWDTY0
dc_up_rtn  puld                  ; Return to background loop
        rts


;**********************************************************************
```

PWM Summary.   We have looked at three ways to generate PWM waveforms: a normal output compare, output compare 7, and the PWM module. Which option you use depends on the on-chip resources available. On most 68HC11 microcontrollers, you must use a normal output compare. On the M68HC812A4, you can either use a normal output compare or output compare 7. On the M68HC912B32, all three methods are available. Normally if a PWM module is available, it is the preferred method because it does not affect the free-running counter and requires no CPU load.

▶ 9.3 SERIAL I/O

Serial I/O modules are very important resources for long-distance communications and on-board system expansion. The Serial Communications Interface (SCI) is normally used for point-to-point communications over standard asynchronous serial ports such as the COM ports on a personal computer or a modem. The Serial Peripheral Interface (SPI) is normally used for in-system communications for distributed processing or for on-board system expansion. The SPI is becoming more important for system expansion because more designs are using single-chip MCUs where the CPU bus is not available. There are many serial peripheral devices available now including DACs, temperature sensors, programmable logic devices, and displays. Many specialized processors such as DSPs also use a serial interface.

In this chapter we first go over some background and serial communications techniques. Then we look at the SCI and SPI modules for the 68HC12 family.

9.3.1 Serial I/O Background

First let's go over different ways to communicate data over a serial link. Figure 9.25 shows a simplified model of a serial communications system. In this system the transmitter encodes a data signal to be sent to the receiver. The timing of the data signal is based on the transmitter clock, f_T.

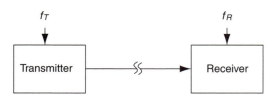

Figure 9.25   Serial Communications Link

The receiver samples the serial signal to detect and decode the data. The timing of the receiver sampling is based on a receiver clock signal, f_R. In order to catch every data bit one time and only one time, the receiver must know when and how often to sample the signal. This means that the receiver sampling times must be synchronized to the transmitted signal in some way. It is impossible to reliably detect and recover the data without some form of synchronization.

Here we look at two different types of serial communication systems—synchronous and asynchronous. *Synchronous* communications systems are those systems that transmit a clock signal to synchronize the receiver to each bit time. *Asynchronous* communications systems are those systems that use the *Start-Stop* protocol to synchronize the receiver at the start of each character frame.

Synchronous Serial Communications.   Synchronous communications systems always transmit a clock signal with the data to synchronize the receiver to each bit time. The clock is provided as a separate clock signal, or it can be embedded in the data signal itself.

Two common ways to embed a clock into the data signal are to use Manchester, or biphase, signaling or to use variable pulse-width signaling. Most computer networks, such as Ethernet, use a Manchester protocol. Figure 9.26 shows an example of a Manchester encoded bit stream. Notice that there is a signal change in the middle of every bit time. This signal change is used by the receiver to synchronize the sampling process.

Systems that use variable pulse-widths include some network systems such as the J1850 network implemented in the 68HC912B32 BDLC module. Many other systems use a variable pulse-width to minimize the number of signal lines. Examples include the 68HC12's Background Debug Module (BDM) and Dallas Semiconductor's One-Wire Bus. Figure 9.26 shows a variable pulse-width signal similar to the data signal for the BDM module. Again there is a signal change for every bit time. The BDM signal always has a falling edge at the start of each bit time. This edge is used to synchronize the receiver so it will sample the signal in the middle of the bit time. The sampled value represents the data value for that bit.

Manchester encoding requires a more complex transmitter and receiver than other synchronous techniques. The variable pulse-width systems use a simpler circuit but not as

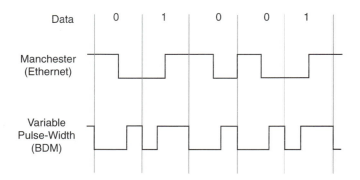

Figure 9.26   Examples of Manchester and Variable Pulse-Width Encoded Signals

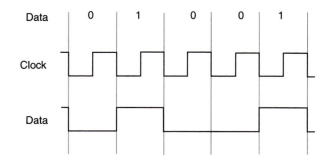

Figure 9.27   Synchronous Data and Separate Clock Signals

simple as a synchronous system with a separate clock. Both of these systems require that the transmitter and receiver be set to a fixed data rate because the receiver must know the bit time. The primary advantage to these systems is that they only require a single signal channel.

Other synchronous serial communication systems send the synchronization clock as a separate clock signal. Figure 9.27 shows data and clock signals for a typical synchronous system that uses a separate clock signal. In this case the clock's rising edge always falls in the center of the data bit time.

The advantage of using a synchronous system with a separate clock is circuit simplicity. The receiver circuit can be as simple as a rising-edge triggered shift register. Also the data rate does not have to be fixed. In fact the data rate can change bit-to-bit as long as the clock's rising edge is always during a valid bit time. The only speed limitation is the rate at which the shift register can be clocked, so these systems can transmit data at a higher rate than asynchronous systems using the same technology.

The disadvantage of using these synchronous systems is that a separate clock signal is required. For long distances this can be prohibitively expensive. In addition, over long distances care must be taken to make sure the clock signal remains in phase with the data. Because these systems are good for short-distance, high-speed applications, they are ideal for an inexpensive alternative to the MCU bus system for system expansion. Some common protocols that use this method include Motorola's SPI interface, I²C, and National's MicroWire.

Asynchronous Serial Communications.   Of course, as we have already discussed, no communication system can be completely asynchronous. So the term *asynchronous* is a little misleading. Asynchronous communication systems use the standard *Start-Stop* protocol to synchronize the receiver to the start of each character frame. Figure 9.28 shows an asynchronous data signal and the detail for one character. The signal is high (idle line) until data are transmitted in character-sized frames. Each frame has one *Start* bit; a number of data bits, least-significant bit first; an optional parity bit, *P;* and one or more *Stop* bits.

The receiver detects the falling edge of the start bit and then attempts to sample in the center of each bit time. To do this the receiver must know the data rate so it knows where to sample the next bit. It also must know how many data bits are in each character, the type of parity used if any, and the number of *Stop* bits.

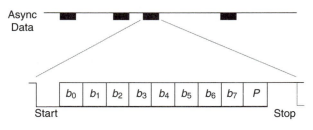

Figure 9.28 Asynchronous Data Signal

This system works well as long as the character frame contains a limited number of bits. This is because the receiver samples are based on a receiver clock, which is at a slightly different frequency than the transmitter clock. Remember that there is no way to make these two clocks have the exact same frequency. Because of this each successive sample is offset more from the center of the data bit time. If the receiver clock is faster than the transmitting clock, it may sample the same data two times. If the receiver clock is slower, it may miss a data bit. This places an accuracy requirement on the two clocks. The accuracy will depend on the size of the data frame. The larger the frame, the more accurate the clocks must be. For a typical character frame that has 9 or 10 bits, the receiver clock must be within 4% of the transmitter clock.

Asynchronous systems are the most commonly available serial interface. They are inexpensive and add a lot of value to a design, even if it is not part of the design's primary function. An asynchronous serial port can be used for debugging, servicing, or analyzing the system.

Now let's look at the 68HC12 SCI and SPI modules in detail. Remember that the SCI is an asynchronous interface and the SPI is a synchronous interface with a separate clock signal.

9.3.2 Serial Communications Interface

The Serial Communications Interface (SCI) is a general-purpose, asynchronous transceiver. Another common name for this type of transceiver is a UART. The SCI is normally used to communicate to computers, modems, and other systems over a standard RS-232 link. Unlike most UART peripherals, the Motorola SCI does not include RS-232 hardware handshaking signals.

In addition to normal point-to-point communications, the SCI can be configured in 9-bit mode for a multipoint system, or in loop mode for loopback testing. We will only cover the basic use of the SCI for normal, point-to-point computer communications.

Baud-Rate Generation.   The first step in configuring the SCI is to set the baud rate. When selecting a baud rate, we must consider existing asynchronous system standards and the MCU clock frequency. There is a standard set of baud rates for asynchronous communication systems, most notably 2,400, 4,800, 9,600, and 19,200bps. You are probably familiar with these rates from configuring your PC's serial port or modem. It would be very difficult to justify configuring the SCI to a nonstandard baud rate because it means it would no longer be able to communicate with most serial systems.

SC0BDH:

($00C0)	BTST	BSPL	BRLD	SBR12	SBR11	SBR10	SBR9	SBR8

SC0BDL:

($00C1)	SBR7	SBR6	SBR5	SBR4	SBR3	SBR2	SBR1	SBR0
RESET:	0	0	0	0	0	0	0	0

SC0CR1:

($00C2)	LOOPS	WOMS	RSRC	M	WAKE	ILT	PE	PT

SC0CR2:

($00C3)	TIE	TCIE	RIE	ILIE	TE	RE	RWU	SBK
RESET:	0	0	0	0	0	0	0	0

SC0SR1:

($00C4)	TDRE	TC	RDRF	IDLE	OR	NF	FE	PF
RESET:	1	1	0	0	0	0	0	0

SC0SR2:

($00C5)	0	0	0	0	0	0	0	RAF
RESET:	0	0	0	0	0	0	0	0

SC0DRH:

($00C6)	R8	T8	0	0	0	0	0	0
RESET:	U	U	0	0	0	0	0	0

SC0DRL:

($00C7)	R7T7	R6T6	R5T5	R4T4	R3T3	R2T2	R1T1	R0T0
RESET:	0	0	0	0	0	0	0	0

Figure 9.29 SCI Control and Status Registers

Figure 9.29 shows the two baud-rate registers, *SC0BDH* and *SC0BDL.* The baud rate is controlled by the bits, *SBR12–SBR0,* and the equation

$$SCI\ Baud\ Rate = \frac{f_P}{16 \cdot BR}$$

where *BR* is the value stored in *SBR12–SBR0* and f_p is the frequency of the MCU P-clock. The P-clock is an internal MCU clock that has the same frequency as the E-clock, $f_{XTAL}/2$, but is 90° out of phase. Table 9.10 shows the required baud-rate register settings for standard baud rates and a P-clock frequency of 8MHz.

TABLE 9.10 BAUD-RATE REGISTER
SETTINGS FOR $F_p = 8$MHZ

Bit Rate (bps)	SBR12–SBR0
2,400	208
4,800	104
9,600	52
14,400	35
19,200	26
38,400	13

Mode Configuration. The next step in configuring the SCI is to set up the operating mode by loading a value into the first SCI control register, *SC0CR1*. The most important *SC0CR1* bits have the following functions:

LOOPS: Loop mode bit
      0 = Normal Mode
      1 = Loop Mode
M: Mode bit
      0 = Normal 8-bit mode
      1 = 9-bit mode
WAKE: Wake-up Mode
      0 = Wake up by idle line
      1 = Wake up by ninth data bit
PE: Parity enable bit
      0 = No parity
      1 = A ninth bit is used for parity
PT: Parity type bit
      0 = Even parity
      1 = Odd parity

For most systems this register can remain in its *RESET* state, $00. This configures the SCI to run in normal operating mode with one start bit, eight data bits, one stop bit, and no parity.

The final step required to configure the SCI is to enable appropriate interrupts and enable the transmitter and receiver. This is done in the second SCI control register, *SC0CR2*. The interrupts available for the SCI include the *Transmit Data Register Empty* interrupt, *TIE;* the *Transmit Complete* interrupt, *TCIE;* the *Receiver Data Register Full* interrupt, *RIE;* and the *Idle Line* interrupt, *ILIE*. These interrupts are pending once the corresponding flag in the status register *SC0SR1* is set.

Source 9.11 shows a set of basic SCI routines, *sci_open, sci_read,* and *sci_write*. These routines are the lowest-level routines that are required for the Basic I/O module described in Chapter 7. *sci_open* is the routine for initializing the SCI. It must be called before the other SCI routines can be used.

Transmitting SCI Data.   The simplified SCI transmitter is shown in the upper half of Figure 9.30. It consists of a transmit data register, *SC0DRL,* which acts as a buffer, and a 10- or 11-bit shift register, depending on the SCI mode. When a byte is written to *SC0DRL,* that byte, along with the *Start* bit, the *Stop* bit, and optionally a parity bit, is transferred to the transmit shift register. After the transfer, the shift register shifts the data out the *TxD* pin.

The progress of the transmitter can be determined by the two transmitter flags in *SC0SR1.* These two flags function as follows:

TDRE: Transmit Data Register Empty Flag

Set after the data have been transferred to the shift register.

Cleared by a read of *SC0SR1* followed by a write to *SC0DRL.*

TC: Transmit Complete Flag

Set when the shift register completes shifting out the data.

Cleared by a read of *SC0SR1* followed by a write to *SC0DRL.*

The *TDRE* flag is normally used by software to transmit a character. The *TC* flag is used to detect when the transmitter is finished transmitting data.

The *sci_write* routine in Source 9.11 is an example of a routine to transmit one character. Before a byte can be written to the transmit data register, *SC0DRL,* the program must first make sure the register is empty. This is done by polling the *TDRE* flag. As *sci_write* is written, it may have to wait as long as 10 or 11 bit times before loading the transmit data register.

Receiving SCI Data.   The receiver section of the SCI is shown on the lower half of Figure 9.30. The receiver is more complicated because it not only has to shift the data in, but it also has to detect various receiver errors. Shown in the figure are the receiver shift register and the receiver data register, *SC0DRL.* Notice that this is the same register name

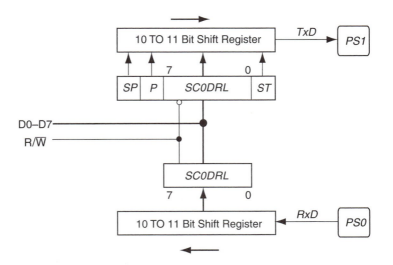

Figure 9.30   Simplified Functional Block Diagram of the SCI

as the transmit register. *SCODRL* is actually made up of two registers controlled by the $R/\overline{W}$ signal. If data are written to *SCODRL,* the data go to the transmit register. If data are read from *SCODRL,* the data come from the receive data register.

To shift the data into the receiver shift register, the receiver must first detect an idle line and then detect the falling edge of the *RxD* signal. The receiver clock is set to 16 times the bit rate. The falling edge is detected by the receiver detecting three ones and one zero. Once the falling edge is detected, it takes three samples seven, eight, and nine clock cycles later. These three samples land in what should be the center of the *Start* bit. If two or more samples are zero, the *Start* bit is verified. If not, the noise flag is set and the receiver waits for the next *Start* bit. After the *Start* bit is verified, the receiver counts 14 more clocks cycles and then takes three more samples. The majority of the three samples determine the accepted data. In addition the SCI detects any falling edge in the data and resynchronizes the sampler.

There are seven status flags controlled by the receiver. These flags are in the two SCI status registers, *SC0SR1* and *SC0SR2*. The receiver flags function as follows:

RDRF: Receive Register Full Flag

> Set when a character is received and transferred into *SC0DRL.*

> Cleared by a read of *SC0SR1* followed by a read of *SC0DRL.*

IDLE: Idle Line Flag

> Set when an idle line is detected.

> Cleared by a read of *SC0SR1* followed by a read of *SC0DRL.*

OR: Overrun Flag

> Set when a character is shifted into the receiver but cannot be transferred because the receive register is already full.

> Cleared by a read of *SC0SR1* followed by a read of *SC0DRL.*

NF: Noise Flag

> Set when noise is detected by the receiver.

> Cleared by a read of *SC0SR1* followed by a read of *SC0DRL.*

FE: Framing Error Flag

> Set when a zero is detected during the *Stop* bit time.

> Cleared by a read of *SC0SR1* followed by a read of *SC0DRL.*

PF: Parity Error Flag

> Set when the incorrect parity is detected by the receiver.

> Cleared by a read of *SC0SR1* followed by a read of *SC0DRL.*

RAF: Receiver Active Flag

> Set when a character is being received.

> Cleared when an idle state is detected.

The *sci_read* routine in Source 9.11 is an example of a routine to read a received character. *sci_read* samples the *RDRF* flag, and if it is set, it reads the receive data register, *SC0DRL.* The act of reading the *RDRF* flag and reading *SC0DRL* automatically clears the *RDRF* flag. Notice that *sci_read* does not wait for a character to be received. If it did, it might block the system for a very long time, which may be unacceptable. The *getchar* function in the Basic I/O module described in Chapter 7 calls *sci_read* until a character is received.

Source 9.11 SCI Driver Routines

```
;*********************************************************************
; sci_open - Initializes SCI.
;    Normal 8-bit mode, 9600bps,no interrupts
;        -Assumes 8MHz Eclk
; Pseudo-C:
; sci_open() {
;     SC0BD = 52;              /* 9600bps */
;     SC0CR1 = 0;              /* Normal 8-bit mode, no parity */
;     SC0CR2 = 0x0c;           /* No interrupts */
; }
;*********************************************************************
sci_open      movw #52,SC0BD            ;9600bps @ 8MHz Eclk
              movb #$00,SC0CR1          ;Normal 8-bit mode, no parity
              movb #$0c,SC0CR2          ;No ints
              rts


;*********************************************************************
; sci_read() - Read sci.
;    Returns ACCB=char or 0 if no character has been received.
; Pseudo-C:
; sci_read() {
;     if((SC0SR1 & RDRF) == RDRF){
;         return SC0DRL;
;     }
; }
;*********************************************************************
sci_read          clrb
                  brclr SC0SR1,RDRF,sci_read_rtn  ; get status
                  ldab SC0DRL                     ; get data
sci_read_rtn      rts


;*********************************************************************
; sci_write() - Write to sci.
;    Outputs the character passed in ACCB.
;    Blocks until the transmit register is empty ( <=1 character time).
; Pseudo-C:
; sci_write(c){
;     while((SC0SR1 & TDRE) == TDRE){}
;     SC0SRL = c;
; }
;*********************************************************************
sci_write         brclr SC0SR1,TDRE,sci_write  ; wait for TDRE
                  stab SC0DRL                  ; send data
                  rts


;*********************************************************************
;
```

9.3.3 Serial Peripheral Interface

The 68HC12's Serial Peripheral Interface (SPI) is a synchronous serial communications system with a clock signal, two data signals, and a slave select signal. The SPI can be used for point-to-point communications between two devices or for multipoint communications over an SPI bus. For bus access, the SPI uses a master-slave protocol, which means one device, the master, controls all bus access by all other devices.

The 68HC12's SPI system has some small but important improvements over the SPI system in the 68HC11 microcontrollers. These include a more functional slave select signal, a single-wire mode, and selectable data bit order. All of these functions can be implemented with the 68HC11 SPI but it requires more and sometimes cumbersome code and a higher CPU load.

The SPI system has four signals: Master-In-Slave-Out, *MISO;* Master-Out-Slave-In, *MOSI;* the SPI clock, *SCK;* and an active-low slave select pin, */SS.* The SPI can also be configured to use a single, bidirectional data signal. In this mode the *MOSI* pin is used for the data line on the master and the *MISO* pin is used for the data line on the slave.

To understand how the SPI works, let's first look at connecting two 68HC12 MCU's in a point-to-point configuration. The functional logic diagram for this system is shown in Figure 9.31. Each SPI module has an 8-bit shift register connected to the SPI data register *SP0DR.* To connect the master to the slave, the four SPI signals are connected straight across to the slave device—*MISO* to *MISO, MOSI* to *MOSI,* and so on. The synchronization clock is always generated by the master device. In effect this creates a distributed 16-bit shift register connected in a data loop.

When data are written to the master's *SP0DR* register, the SPI shifts the eight bits out of the *MOSI* pin while shifting in eight bits from the MISO pin. At the same time the slave shift register shifts the master data in while shifting the slave data out. This results in the master data and the slave data being exchanged. (Figure 9.33 shows timing diagrams for a single 8-bit transfer.)

Notice that since the slave device does not generate the clock, a write to the slave *SP0DR* cannot initiate the data exchange. To get the slave data to the master, the master

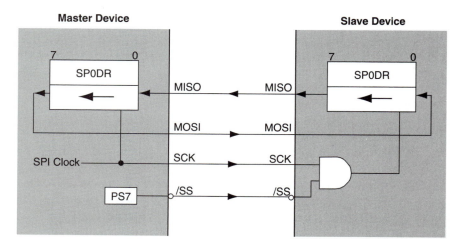

Figure 9.31 SPI Communication System

SP0BR:

($00D2)	0	0	0	0	0	SPR2	SPR1	SPR0
RESET:	0	0	0	0	0	0	0	0

SP0CR1:

($00D0)	SPIE	SPE	SWOM	MSTR	CPOL	CPHA	SSOE	LSBF
RESET:	0	0	0	0	0	1	0	0

SP0CR2:

($00D1)	0	0	0	0	PUPS	RDS	0	SPC0
RESET:	0	0	0	0	1	0	0	0

Figure 9.32 SPI Configuration Registers

must write to its *SP0DR*. This in turn generates eight clock cycles, which shift the slave data into the master. This is true even for output-only slave devices. It may seem odd to write data to the SPI to transfer data from the slave, but it is required to generate an SPI clock.

SPI Configuration. To configure the SPI, we need to set the baud rate, configure the SPI clock format, set the data mode, and, unlike most on-chip resources, set the pin direction in the data direction register for PORTS. The SPI configuration registers are shown in Figure 9.32. They include a baud-rate register, *SP0BR,* and two control registers, *SP0CR1* and *SP0CR2.*

SPI Baud Rate. The SPI clock frequency and the baud rate are controlled by the *SPR2, SPR1,* and *SPR0* bits in the baud-rate register, *SP0BR.* The SPI clock frequency and data rate are

$$R_{SPI} = f_{SCK} = \frac{f_P}{2^{(SPR+1)}}$$

where *SPR* is the value stored in the *SPR2, SPR1,* and *SPR0* bits and f_p is the P-clock frequency, which is the same as the E-clock frequency.

For example if the P-clock frequency is 8MHz and *SPR2, SPR1,* and *SPR0* are zero, the SPI data rate is 4Mbps. If the baud rate bits are all one (*SPR* = 7), the data rate is 31.3Kbps. Notice how much higher the SPI data rates are compared to the SCI rates. This is because the SPI shifts data out at the SPI clock rate while the SCI must shift data out at the SCI clock rate divided by 16. Standard baud rates are also not required with the SPI because a separate clock signal is used for bit synchronization.

SPI Clock Format. A large number of devices in the market today use synchronous serial communications. The problem is that many of these devices have different timing requirements. To make the SPI compatible with as many devices as possible, it is designed to be versatile by providing programmable clock formats, bit order, and select/enable logic. Because of this, the SPI is compatible with many non-SPI devices.

One of the most important parameters for compatibility with these systems is the clock format. To make the SPI compatible with a synchronous device, we need to configure the clock format so that the relationship between the clock and data meets the requirements of the device. The clock format is controlled by two bits in the *SP0CR1* register, the clock polarity bit *CPOL*, and the clock phase bit *CPHA*. Configuring the clock format is probably the most difficult step in configuring the SPI. A careful timing analysis first must be made to determine the requirements for the device. Those requirements must then be met by correctly setting the *CPOL* and *CPHA* bits. Figure 9.33 shows the clock formats for the four combinations of *CPOL* and *CPHA*.

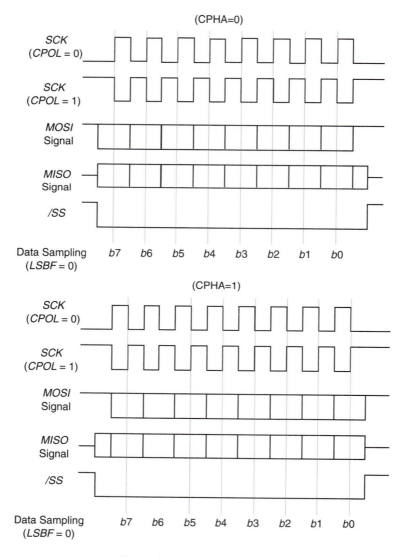

Figure 9.33   SPI Clock Formats

The polarity bit, *CPOL,* determines the clock edge used to sample the data. Determining the correct clock phase is more subtle, however.

The top timing diagram in Figure 9.33 shows the relationship between the clock and data when the clock phase bit *CPHA* is zero. In this mode the data out of an SPI master or SPI slave are valid before the first clock edge. The SPI starts sampling the incoming data on the first clock edge so it assumes the data are already valid. This mode cannot be used for reading data from most non-SPI output devices because they require a clock to output the data in the first place.

The second timing diagram in Figure 9.33 shows the relationship between the clock and the data when *CPHA* is one. In this mode the input data are sampled on the second clock edge, so the first edge can be used by the slave device to output data before sampling. This mode can be used for a simple slave device that shifts data out at the first clock edge. We look at some code examples that illustrate this at the end of this section.

Miscellaneous SPI Configuration. In addition to the clock format, other parameters can be programmed to configure the SPI to meet the requirements of the connected device. These are all programmed by bit values in the two control registers, *SP0CR1* and *SP0CR2.* Here is a list of all configuration bits for the SPI:

SPIE:	SPI Interrupt Enable
	0 = SPI interrupt disabled
	1 = SPI interrupt occurs when SPIF or MODF flags are set
SPE:	SPI System Enable
	0 = SPI system disabled
	1 = SPI system enabled
SWOM:	Port S Wired-OR mode
	0 = Normal Port S outputs
	1 = *PS4–PS7* are open drain outputs
MSTR:	SPI Master/Slave Mode
	0 = Slave mode
	1 = Master mode
CPOL, CPHA:	Clock Format Bits
	See previous description
SSOE:	Slave Select Output Enable Bit (master mode only)
	0 = */SS* pin is used for the MODF function or for GPIO
	1 = */SS* function is enabled
LSBF:	LSB First Enable Bit
	0 = Data are sent MSB first
	1 = Data are sent LSB first

PUPS:	Port S Pull-up Enable
	0 = No internal pull-ups on Port S
	1 = All Port S input pins have active pull-up
RDS:	Reduce Drive for Port S
	0 = Normal output drive levels
	1 = Reduced drive levels on Port S outputs
SPC0:	Serial Pin Control
	0 = Normal data lines *(MISO, MOSI)*
	1 = Bidirectional data line

Port S Configuration.   The last and often missed step in configuring the SPI system is to set the data direction for the SPI pins on *PORTS.* Unlike most on-chip peripherals, the SPI does not automatically take over the pin function when it is enabled. This means that the pins corresponding to SPI outputs must be configured as outputs in the Port S data direction register, *DDRS.* For a master device, *PS6/SCK, PS5/MOSI,* and probably *PS7/SS* need to be configured as outputs. For a slave device, only *PS4/MISO* needs to be configured as an output.

Note _____

A common problem encountered by novices using the 68HC11 and 68HC12 SPI is not to set the */SS* pin as an output on the master. When */SS* is configured as an input on the master device, it functions as a mode fault indicator. When the input is floating, false mode faults are detected, disabling the SPI function. This can be detected by reading the status register and checking the *MODF* flag.

SPI Data Transfers.   To transfer data to or from a slave device, a program must make use of the SPI status register, *SP0SR,* and the SPI data register, *SP0DR.* Figure 9.34 shows these two registers.

To send data, the program must test the SPI transfer complete flag, *SPIF,* and write the data to the SPI data register, *SP0DR.* To receive data from a slave device, the program uses the same procedure except that the data written to the *SP0DR* may not be used by the slave device. This process is often called a dummy write because the written data are not used.

SP0SR:

($00D3)	SPIF	WCOL	0	MODF	0	0	0	0
RESET:	0	0	0	0	0	0	0	0

SP0DR:

($00D5)	Bit 7	Bit 6	Bit 5	Bit 4	Bit 3	Bit 2	Bit 1	Bit 0

Figure 9.34   SPI Status Register and Data Register

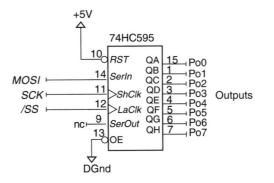

Figure 9.35 General Purpose Output Expansion Using the SPI and a 74HC595

Using the SPI to Add an 8-Bit Output Port. A good way to add additional outputs to a microcontroller in single-chip mode is to use the SPI to shift data into a serial-in, parallel-out shift register. Let's look at an example that uses the SPI and a 74HC595 to add eight general purpose outputs.

The 74HC595. The 74HC595 is an 8-bit serial-input, serial- or parallel-output shift register with latched tristate outputs. Because it has latched outputs, the outputs are not affected during the shifting process. The schematic for this design is shown in Figure 9.35. The *MOSI* signal from the MCU is connected to the serial input, *SerIn*. The *SCK* is connected to the shift register clock, *ShClk*, and the slave select pin, */SS*, is connected to the latch clock, *LaClk*. If we refer to Figure 9.33, we can see that the normal process for transferring data is as follows:

1. */SS* goes low at the start of the transfer cycle. This has no effect on the 74HC595 because */SS* is connected to a rising-edge-triggered latch clock.
2. *MOSI* and *SCK* transfer the byte.
3. */SS* goes high at the end of the transfer cycle. The rising edge of */SS* causes the 74HC595 shift-register contents to be transferred to the output latch. At this point the new outputs become valid.

Clock Configuration. The 74HC595 has a maximum clock frequency of 24MHz. Therefore we can set the SPI for the highest baud rate if we assume the 74HC595 is located near the MCU. If we keep the baud-rate register *SP0BR* at its *RESET* value of $00, the data rate will be 4Mbps and the complete transfer will take 2µs.

To find the correct clock format, we need to look at the timing requirements for the 74HC595. Figure 9.36 shows the critical timing parameters for this design. The shift register clock is rising-edge triggered, so the *MOSI* data must be valid during that edge. If we compare this to the clock formats shown in Figure 9.33, we find that we can set *CPOL* = *CPHA* = 0 or *CPOL* = *CPHA* = 1. The difference between these two combinations is the signal level of *SCK* while it is idle. If 00 is used, the idle *SCK* is zero. If 11 is used, the idle *SCK* is one. For this example either setting will work.

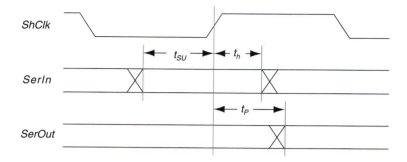

Figure 9.36 Timing Requirements for the 74HC595

While we are looking at the clock formats, let's look at the timing requirements to transfer data from the shift register to the MCU. To shift the data back into the SPI from the shift register, we need to compare the relationship between *SerOut* and the clock with the formats shown in Figure 9.33. Since *SerOut* changes on the rising edge of the shift register clock, we need the SPI to sample the data on the falling edge. The only format that works for this is *CPOL* = 0 and *CPHA* = 1. This precludes us from reading data from the shift register on the same transfer cycle as a write to the shift register. In fact if we try to read the shift register data with *CPOL* = 0 and *CPHA* = 1, we would end up transferring garbage to the outputs.

Because of the somewhat confusing nature of the clock format configuration, a lot of testing should be done to verify the design. This can be done with a standard oscilloscope, but a logic analyzer or a multichannel digital scope like the Hewlett-Packard 54645D is better suited for testing this type of system.

Output Program Code.   The program for this example is shown in Source 9.12. This program starts by initializing the SPI with the routine *SpiInit*. It then goes into an endless loop that outputs a counting variable using *SpiOut*.

Source 9.12 SPI Output Expansion Demonstration Program

```
;*****************************************************************************
; SpiOutPort - Demonstrates the use of the 68HC12 SPI for output expansion
;              using a 74HC595.
; MCU: 68HC912B32, fE = 8MHz
;*****************************************************************************
; Equates
;*****************************************************************************
PORTS        equ $d6
DDRS         equ $d7
SS           equ %10000000
MOSI         equ %00100000
SCK          equ %01000000
SP0CR1       equ $d0
SP0BR        equ $d2
```

```
SP0SR        equ $d3
SP0DR        equ $d5
SPIF         equ $80


;*********************************************************************
;main - Simple test loop
;Pseudo-C:
; main(){
;     SpiInit();
;     b = 0;
;     while(1){
;         SpiOut(b);
;         b++;
;     }
; }
;*********************************************************************
main         jsr SpiInit
             ldab #0
out_nxt      jsr SpiOut
             incb
             bra out_nxt


;*********************************************************************
;SpiInit - Initialize SPI
;*********************************************************************
SpiInit      bset PORTS,SS           ;Set /SS pin
             bset DDRS,SS|MOSI|SCK    ;Initialize PortS outputs
             movb #$00,SP0BR          ;Set bit rate to 4Mbps
             movb #$5E,SP0CR1         ;Enable,Mstr,CPOL:CPHA=11,MSB first
             movb #0,SP0DR            ;Initialize output and SPI flag
             rts


;*********************************************************************
;SpiOut - Output routine for a 74HC595. It uses SS as the
;            latch clock.
;    The data to be sent out are passed in ACCB
;Pseudo-C:
; SpiOut(b){
;     while((SP0SR & SPIF) == 0){}
;     SP0DR = b;
; }
;*********************************************************************
SpiOut
spio_w       brclr SP0SR,SPIF,spio_w  ;Wait for shift complete
             stab SP0DR               ;Output data
             rts


;*********************************************************************
;
```

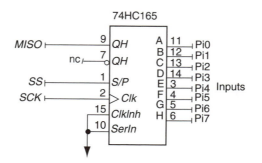

Figure 9.37 Circuit for 8-Bit Input Port Using the SPI and a 74HC165

As you can see from this example, once you get the correct configuration down, the program is simple. This is one of those applications in which you want to do a lot of careful analysis before you jump in and start writing code.

Using the SPI to Add an 8-Bit Input Port. We can add inputs by using the SPI to shift data from a parallel-in, serial-out shift register. Let's look at an example to add an 8-bit general purpose input port using the SPI and a 74HC165. The 74HC165 is a serial-in or parallel-in, serial-out shift register. Figure 9.37 shows the circuit that will be used for this example. The *MISO* signal from the MCU is connected to the serial output, *QH;* the SPI clock is connected to the shift register clock, and the */SS* pin is connected to the serial shift/parallel load input, *S/P.* Because of the way the *S/P* pin works, we have to configure the */SS* pin as a general purpose output for this example. To transfer data from this shift register, the transfer process is as follows:

1. */SS* goes low to load the shift register with the inputs.
2. */SS* goes high to start the shifting.
3. *MOSI* and *SCK* transfer the byte.

To do this we need to make sure the *SSOE* bit in *SP0CR1* is cleared.

The timing relationship between the output data and the clock is the same as the serial output shown in Figure 9.36 for the 74HC595. This means we need to set *CPOL* = 1 and *CPHA* = 0. The program for this example is shown in Source 9.13.

Source 9.13 A Program to Add an 8-bit Input Port Using the SPI

```
;*********************************************************************
; SpiInPort - Demonstrates the use of the 68HC12 SPI and a 74HC165 to
;             add an 8-bit input port.
;
;
; MCU: 68HC912B32
;*********************************************************************
; Equates
```

```
;***********************************************************************
PORTS       equ $d6
DDRS        equ $d7
SS          equ %10000000
MOSI        equ %00100000
SCK         equ %01000000
SP0CR1      equ $d0
SP0CR2      equ $d1
SP0BR       equ $d2
SP0SR       equ $d3
SP0DR       equ $d5
SPIF        equ $80

;***********************************************************************
; Main interface loop
;   A test loop that reads from the 74HC165 and stores it in InData
;***********************************************************************
main        jsr SpiInit
next_in     jsr SpiIn
            stab InData
            bra next_in

;***********************************************************************
; SpiInit - Initialize SPI
;***********************************************************************
SpiInit     bset PORTS,SS            ;Initialize the SR load signal
            bset DDRS,SS|MOSI|SCK    ;Set SPI outputs direction
            movb #$00,SP0BR          ;Set for 4 Mbps
            movb #$58,SP0CR1         ;Enable,Mstr,CPOL:CPHA=10,MS first
            movb #0,SP0DR            ;Dummy write to initialize SPI flag
            rts

;***********************************************************************
; SpiIn - Input routine for a 74HC165. It uses SS as the
;       latch clock.
;   The dummy data to be sent out are passed in ACCB. Input
;   value is returned in ACCB.
;Pseudo-C:
; SpiInt(d){
;     PORTS &= ~SS;                     /* Transfer Latch -> SR */
;     PORTS |= SS;
;     SP0DR = d;                        /*write dummy data to start xfer */
;     while((SP0SR & SPIF) == SPIF){}   /*wait for transfer complete */
;     return SP0DR;                     /*return data from SR */
; }
;***********************************************************************
SpiIn       bclr PORTS,SS               ;Transfer Latch -> SR
            bset PORTS,SS
            stab SP0DR                  ;Dummy write to start transfer
```

```
spiin_w2      brclr SP0SR,SPIF,spiin_w2      ;Wait for shift complete
              ldab SP0DR                     ;return data from SR
              rts

;************************************************************************
; Variables
;************************************************************************
InData        rmb 1
;************************************************************************
```

▶ 9.4 A-TO-D CONVERSION

In this section we look at the analog-to-digital (A-to-D) conversion process and specifically the ADC found in 68HC12 MCUs. So far all the signals we have covered have been digital signals, and of course the CPU can only deal with digital signals. Since we live in an analog world, however, most embedded systems require an ADC to sample analog signals and a DAC to generate analog signals.

9.4.1 A-to-D Conversion Background

The A-to-D conversion process described here is called pulse-code modulation, PCM. PCM involves a two step process. First the analog signal is sampled at a fixed rate called the *sample rate*. The minimum sample rate is determined by the bandwidth or frequencies contained in the analog signal. Then the sample amplitudes are rounded to the closest digital value by a quantizer. This step determines the dynamic range and the signal-to-noise ratio of the system.

Sampling Analog Signals.   Figure 9.38 shows the two-step sampling process in the time domain. In the figure, a sample is taken every sample time, T_S. The sampling frequency is then

$$f_S = \frac{1}{T_S}$$

Every sample is then approximated to the closest digital word. In this case the word size is three bits, so there are only eight possible sample values. The shaded region for each sample indicates the error introduced because of the finite word size. The digital values that represent the samples shown in this figure are 000, 010, 100, 101, 111, 111, 111, 101, 011, 001, 001, and 000.

Sampling theory states that if samples are taken of an analog signal, the original analog signal can be fully recovered from the samples if the sampling rate is greater than two times the bandwidth of the analog signal.

$$f_S > 2B$$

This is an important statement and is the theoretical bases for all sampled systems. To help understand this better, let's look at the frequency spectrum of a sampled signal. Figure 9.39

366   Chapter 9

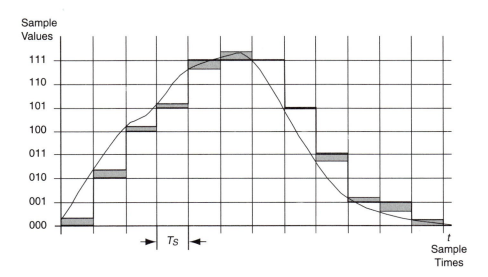

Figure 9.38 A-to-D Conversion Samples in the Time Domain

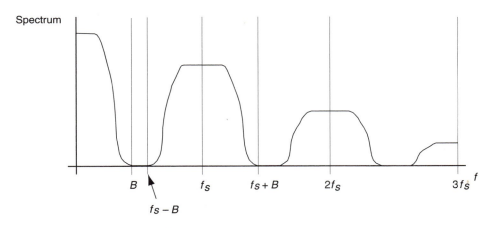

Figure 9.39 Frequency Spectrum of a Sampled Signal

shows the spectrum of a sampled signal with a sample rate f_S and an analog signal bandwidth B. The spectrum of the original analog signal is the component from DC to B. When sampled this component is repeated every integer multiple of the sample rate as shown. This means that if we design a low-pass recovery filter with a pass-band for frequencies less than B and a stop-band for frequencies greater than $f_S–B$, we will end up with the original spectrum and therefore the original analog signal. The difference between B and $f_S–B$ determines the complexity and cost of the recovery filter. So we want f_S to be as large as possible.

However, the sample rate of a real system is limited by the conversion time of the ADC, by the memory required to save samples, and by the amount of processing that must be performed on each sample.

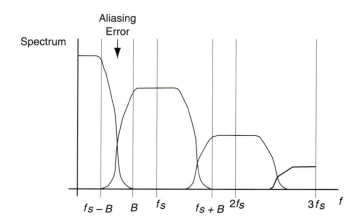

Figure 9.40 A Sampled Signal with Aliasing Error

If the sample rate is less than two times the bandwidth, aliasing error will occur. Figure 9.40 shows the spectrum of a sampled signal with f_s less than two times the bandwidth. In this case the spectrum components overlap and there is no way to recover the original signal.

Also notice that the analog signal must be band-limited to B. If there are components above B, they may introduce aliasing error. To limit an analog input signal, a band-limiting filter is often required. Again the cost and complexity of this filter depends on the sample rate.

A sample rate that is equal to two times the bandwidth only works in theory. We must set a sample rate based on the acceptable cost of the band-limiting and recovery filters.

Signal-to-Noise Ratio and Dynamic Range. Because each sample must be rounded to the nearest digital word, the A-to-D conversion process introduces noise and limits the dynamic range.

The average signal-to-noise ratio of a sampled signal is

$$\left(\frac{S}{N}\right)_{ave} = \frac{Signal\ Power}{Noise\ Power} = 2^{2n}$$

where n is the number of bits in each digital word. This is assuming that the signal is evenly distributed over the full range of the ADC. Audio systems normally require a minimum signal-to-noise ratio based on the desired audio quality. For example audio CDs require high-fidelity audio. The sample size for an audio CD is 16 bits. This results in

$$\left(\frac{S}{N}\right)_{ave} = 2^{2n} = 4.29 \times 10^9 = 96.3 \text{dB}$$

Of course this assumes that the noise in the original analog signal is smaller than the noise introduced by the sampler. For systems greater than eight bits, this is not a simple task.

The dynamic range of a sampled signal is

$$DR = \frac{Range}{Resolution} = 2^n$$

where n is the number of bits in each digital word. The dynamic range determines the required word size of an ADC. For example a digital thermometer with a range of –40° to 120°F and a resolution of 0.5°F requires a dynamic range of 320. This in turn would require at least a 9-bit ADC.

It is important to note that in a real system, it may not be possible to take advantage of the full range of the ADC. Most external circuits such as operational amplifiers cannot reach their supply rails. So depending on the op-amp type and supply voltage, some voltages close to zero and close to the full-scale voltage cannot be reached. This situation has improved tremendously with the advent of CMOS operational amplifiers designed for 5V or even 3.3V supplies.

9.4.2 The 68HC12 On-Chip ADC

The M68HC912B32 MCU has an eight-channel, 10-bit, ADC connected to *PAD0–PAD7*. It uses a charge distribution process to convert the samples, so an external sample and hold circuit is not required. Significant improvements have been made compared to the 68HC11 ADC including a 10-bit sample mode, an A-to-D interrupt, and more result registers to allow conversions of all eight channels in one conversion sequence.

Figure 9.41 shows the logic diagram for the ADC. The required analog signals are *VRH, VRL,* and *ANx,* where *ANx* is one of the eight input channels. The two reference voltages, *VRH* and *VRL,* determine the input voltage range and step size. *ADRx* is the n-bit conversion result, which is stored in one of the result registers.

When a conversion is completed, the resulting value of the digital word is

$$ADRx = \frac{ANx - VRL}{VRH - VRL} \cdot 2^n$$

where the word *ADRx* is limited to

$$0 \le ADRx \le 2^n - 1$$

If *ANx* is equal to *VRL,* the resulting digital word is zero. If *ANx* is equal to *VRH,* the resulting digital word is $2^n - 1$.

On most systems the low reference voltage *VRL* is connected to ground. When this is the case, this equation becomes

$$ADRx = \frac{ANx}{VRH} \cdot 2^n$$

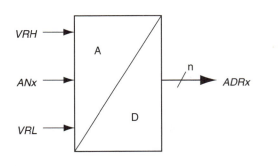

Figure 9.41 A-to-D System Signals

From these equations we can see that the reference voltages have a direct effect on the resulting word, so we have to make sure to minimize noise on these signals. An RC low-pass filter on *VRH* is normally required.

The voltage step size, or resolution, *q,* is equal to

$$q = \frac{VRH - VRL}{2^n}$$

This value determines the minimum noise that is allowable on the analog input signal. If the noise level exceeds this value, the last bit in the digital word is meaningless because it is simply measuring noise. For example in a 5V system, the step size for an 8-bit ADC is 19.5mV. It is relatively simple to keep noise levels below this. The step size for a 10-bit ADC is only 4.88mV and for a 16-bit ADC is only 76.3μV, however. These small noise levels require much more attention to noise reduction.

A-to-D Configuration. To configure the 68HC12 ADC, we must power up the converter module and set the conversion time. We also need to configure the A-to-D module's mode, including 8-bit or 10-bit conversion mode; one-channel, four-channel, or eight-channel mode; and continuous *scan* or single conversion mode. Figure 9.42 shows the registers used to configure the ADC on the M68HC912B32. There are more configuration registers but these are the ones required for normal A-to-D operation.

The *ATDCTL2* register contains some module level control bits. The A-to-D power-up bit *ADPU* is the most important. To save power consumption on the MCU, the A-to-D module is not powered up out of *RESET*. This means we have to set this bit before we can use the ADC. In addition it takes up to 10μs for the A-to-D system to stabilize after the *ADPU* bit is set.

The *AFFC* bit is used for fast flag clearing mode. In this mode a channel's conversion complete flag is cleared when that channel's result register is read. Otherwise the status flag register and a result register must be read to clear the flag.

ATDCTL2:

($0062)	ADPU	AFFC	AWAI	0	0	0	ASCIE	ASCIF
RESET:	0	0	0	0	0	0	0	0

ATDCTL4:

($0064)	S10BM	SMP1	SMP0	PRS4	PRS3	PRS2	PRS1	PRS0
RESET:	0	0	0	0	0	0	0	1

ATDCTL5:

($0065)		S8CM	SCAN	MULT	CD	CC	CB	CA
RESET:	0	0	0	0	0	0	0	0

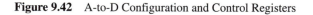

Figure 9.42   A-to-D Configuration and Control Registers

The *ASCIE* and *ASCIF* are the conversion sequence complete interrupt enable and flag. The A-to-D conversion complete flag *ASCIF* is set when an A-to-D conversion sequence is complete. If *ASCIE* is also set, an interrupt occurs.

The *ATDCTL4* register is primarily used for setting the A-to-D conversion time, which in turn helps determine the minimum sample rate. The 10-bit mode bit *S10BM* also affects the conversion time. When this bit is set, 10-bit conversions will be made. Otherwise 8-bit conversions are made.

The time it takes to convert each analog channel depends on the sample time bits, *SMP1–SMP0,* and the clock prescaler bits, *PRS4–PRS0.* For the fastest conversion times, *SMP1* and *SMP0* are cleared to zero. With this setting, an 8-bit conversion takes 18 clock cycles, and a 10-bit conversion takes 20 clock cycles. Note that these times are for a single channel, not a complete conversion sequence, which consists of either four- or eight-channel conversions.

The A-to-D clock frequency is determined by the clock prescaler bits, *PRS4–PRS0,* and the P-clock frequency. The P-clock frequency is the same as the E-clock frequency, which is the crystal frequency divided by two. The A-to-D clock frequency is

$$f_{ATD,\,CLK} = \frac{f_P}{2(PRS + 1)}$$

where *PRS* is the value stored in the five prescaler bits. However, the frequency is also limited to

$$500\text{kHz} \le f_{ATD,\,CLK} \le 2\text{MHz}$$

so not all prescaler bit combinations can be used. If the P-clock frequency is 8MHz, the prescale bits can only range from 1 to 7.

The fastest conversion time possible on the 68HC12 is 9μs for an 8-bit conversion and 10μs for a 10-bit conversion. This results in maximum sample rates of 111Ksps for 8-bit samples and 100Ksps for 10-bit samples. Of course, these sample rates are normally not practical because there would be little time to do anything with the samples.

The rest of the A-to-D configuration is made using the *ATDCTL5* register. The first thing to understand is that an A-to-D conversion sequence is always in sets of four or eight conversions, depending on the value of the *S8CM* bit. When *S8CM* is zero, four conversions are made for each sequence. If *S8CM* is one, eight conversions are made in each sequence.

The channels on which the conversions are made depend on the *MULT* bit and the channel select bits *CD, CC, CB,* and *CA.* If *MULT* is zero, all conversions are made on the channel selected by *CC, CB,* and *CA.* If *MULT* is one, conversions are made on the four channels selected by *CC* or all eight channels. Table 9.11 shows the different conversion modes determined by *MULT, S8CM, CD, CC, CB,* and *CA.* Note that in all cases, *CD* is cleared. The additional modes when *CD* is one are either reserved or for testing.

The last bit in the *ATDCTL5* register that we cover is the *SCAN* bit. When *SCAN* is one, the ADC samples continuously. That is, once it finishes with one conversion sequence, it immediately starts another. When using the continuous mode, we have to make sure to grab the samples before the next conversion sequence writes over them. When the *SCAN* bit is zero, a single conversion sequence is executed.

TABLE 9.11 A-TO-D CONVERSION MODES

MULT	S8CM	CD	CC	CB	CA	Description
0	0	0	Channel			Four conversions of one channel selected by *CC:CB:CA*
0	1	0	Channel			Eight conversions of one channel selected by *CC:CB:CA*
1	0	0	*	d	d	One conversion made on four channels *AN0–AN3 if CC = 0 *AN4–AN7 if CC = 1 d = do not care
1	1	0	d	d	d	One conversion made on all eight channels d = do not care

The Conversion Process. Any write to the *ATDCTL5* register will start a new conversion sequence. If a conversion is in progress, it is aborted and restarted. A conversion sequence is a sequence of four or eight conversions depending on the setting of the *S8CM* bit.

During a conversion sequence, the ADC stores the resulting samples in the result registers *ADR0–ADR7*. These registers are shown in Figure 9.43. As each channel conversion is completed, the converter sets the corresponding conversion complete flag, *CCF7–CCF0*, in the A-to-D status register *ATDSTAT*. After a complete conversion sequence is completed, the *SCF* and *ASCIF* flags are set.

The result register used to store a conversion depends on the mode. In all cases the result registers are used in the order that the conversions are made. For example if *MULT* is zero, *S8CM* is zero, and *AN0* is selected, the result registers, *ADR0, ADR1, ADR2,* and *ADR3* will contain a sample from *AN0*. If *S8CM* is set, all eight result registers will contain a sample from *AN0*.

As shown in Figure 9.44, the samples in the result registers are left-justified, which is a bit strange. This means that in 8-bit mode, the samples are stored in the high byte of the result register. In 10-bit mode, the eight most-significant bits are stored in the high byte and bits 0 and 1 are stored in the bit-6 and bit-7 position of the low byte.

A Simple A-to-D Example. Source 9.14 is a simple A-to-D conversion program to demonstrate or test the M68HC912B32 ADC. It samples *AN0* every second and displays the hex value on a terminal.

For a simple A-to-D test we can connect a potentiometer to *AN0* as shown in Figure 9.45. It is best to use a 10-turn potentiometer for this because it provides enough resolution to easily watch single-bit changes.

In the main loop the program writes $00 to *ATDCTL5*. This starts a conversion sequence that takes four samples of *AN0* and saves them in *ADR0, ADR1, ADR2,* and *ADR3*. Since the A-to-D clock is set to 2MHz, each conversion takes 9µs, and the complete conversion sequence takes 36µs.

ATDSTAT:

($0066)	SCF	0	0	0	0	CC2	CC1	CC0
RESET:	0	0	0	0	0	0	0	0

ATDSTAT:

($0067)	CCF7	CCF6	CCF5	CCF4	CCF3	CCF2	CCF1	CCF0
RESET:	0	0	0	0	0	0	0	0

ADR0:

($0070)	BIT15	BIT14	BIT13	BIT12	BIT11	BIT10	BIT9	BIT8
($0071)	BIT7	BIT6	BIT5	BIT4	BIT3	BIT2	BIT1	BIT0

ADR1:

($0072)	BIT15	BIT14	BIT13	BIT12	BIT11	BIT10	BIT9	BIT8
($0073)	BIT7	BIT6	BIT5	BIT4	BIT3	BIT2	BIT1	BIT0

ADR2:

($0074)	BIT15	BIT14	BIT13	BIT12	BIT11	BIT10	BIT9	BIT8
($0075)	BIT7	BIT6	BIT5	BIT4	BIT3	BIT2	BIT1	BIT0

ADR3:

($0076)	BIT15	BIT14	BIT13	BIT12	BIT11	BIT10	BIT9	BIT8
($0077)	BIT7	BIT6	BIT5	BIT4	BIT3	BIT2	BIT1	BIT0

ADR4:

($0078)	BIT15	BIT14	BIT13	BIT12	BIT11	BIT10	BIT9	BIT8
($0079)	BIT7	BIT6	BIT5	BIT4	BIT3	BIT2	BIT1	BIT0

ADR5:

($007A)	BIT15	BIT14	BIT13	BIT12	BIT11	BIT10	BIT9	BIT8
($007B)	BIT7	BIT6	BIT5	BIT4	BIT3	BIT2	BIT1	BIT0

ADR6:

($007C)	BIT15	BIT14	BIT13	BIT12	BIT11	BIT10	BIT9	BIT8
($007D)	BIT7	BIT6	BIT5	BIT4	BIT3	BIT2	BIT1	BIT0

ADR7:

($007E)	BIT15	BIT14	BIT13	BIT12	BIT11	BIT10	BIT9	BIT8
($007F)	BIT7	BIT6	BIT5	BIT4	BIT3	BIT2	BIT1	BIT0

Figure 9.43   A-to-D Status and Result Registers

ADR with a 8-bit sample:

	BIT7	BIT6	BIT5	BIT4	BIT3	BIT2	BIT1	BIT0
($0070)	BIT7	BIT6	BIT5	BIT4	BIT3	BIT2	BIT1	BIT0
($0071)	–	–	–	–	–	–	–	–

ADR with a 10-bit sample:

($0070)	BIT9	BIT8	BIT7	BIT6	BIT5	BIT4	BIT3	BIT2
($0071)	BIT1	BIT0	–	–	–	–	–	–

Figure 9.44   Result Register Alignments

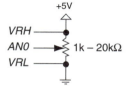

Figure 9.45   Potentiometer Circuit for A-to-D Demonstration

The program then waits for approximately one second by waiting for the RTI counter to reach 61. After one second the program displays the sample stored in *ADR0*. Because one second is much greater than 36µs, there is no need to check the conversion complete flag.

Source 9.14 An A-to-D Demonstration Program

```
;************************************************************************
; AtoDDemo - A simple 68HC912B32 A-to-D demonstration that samples AN0
;            every second.
;          - The RTI is used for timing (0.9994 Sec)
;
; Controller: 68HC912B32
; Monitor: D-Bug12
; Modules: BasicIO
;************************************************************************
; Equates
;************************************************************************
PUTCHAR     equ $0fb6          ;BasicIO constants
SCIOPEN     equ $0fec
OUTCRLF     equ $0fc8
OUTHEXB     equ $0fbc
CR          equ $0d
ATDCTL2     equ $62            ;A-to-D Definitions
ATDCTL4     equ $64
ATDCTL5     equ $65
ADR0H       equ $70
RTIFLG      equ $15            ;RTI Definitions
```

```
RTICTL        equ $14
RTIF          equ %10000000
UserRTI       equ 24
SETUSERVEC    equ $f69a
ONESEC        equ 61


;**********************************************************************
; main - Initialization and background loop.
;Pseudo-C:
; main() {
;     /* Initialization */
;     while(1){
;         ATCDCTL5 = 0x00;
;         while(RtiCnt != ONESEC){}
;         RtiCnt = 0;
;         DispSample();
;     }
; }
;**********************************************************************
main
              movb #$80,ATDCTL2       ;Power-up ATD early
              jsr InitRti             ;Initialize RTI
              jsr SCIOPEN             ;Initialize SCI
              movb #$01,ATDCTL4       ;2MHz A-to-D clock
              jsr OUTCRLF
              cli                     ;Enable interrupts for RTI

main_lp       movb #$00,ATDCTL5       ;Start single 4 sample sequence on AN0
main_w        ldaa RtiCnt             ;Wait 1 second
              cmpa #ONESEC
              bne main_w
              clr RtiCnt
              jsr DispSample          ;Display A-to-D result
              bra main_lp


;**********************************************************************
; DispSample() - Displays the sample contained in ADR0H in hex.
;     - CR only so results are displayed on the same line.
;Pseudo-C:
; DispSample(){
;     PUTCHAR(CR);
;     OUTHEXB(ADR0H);
; }
;**********************************************************************
DispSample    pshd                    ;Return to column 1
              ldab #CR
              jsr PUTCHAR
              ldd #ADR0H              ;Display sample
              jsr OUTHEXB
```

```
                puld
                rts

;*************************************************************************
; InitRti
;    -Initializes the D-Bug12 interrupt jump table
;    -Initializes RTI to interrupt every 16.34ms
;*************************************************************************
InitRti         ldd #rti_isr                     ;D-Bug12 Jump Table
                pshd
                ldd #UserRTI
                jsr [SETUSERVEC,pcr]
                puld
                movb #%10000101,RTICTL           ;Init. RTI
                movb #RTIF,RTIFLG
                clr RtiCnt
                rts

;*************************************************************************
; RTI Service Routine
;*************************************************************************
rti_isr         inc RtiCnt
                movb #RTIF,RTIFLG
                rti

;*************************************************************************
; Variables
;*************************************************************************
RtiCnt          rmb 1
;*************************************************************************
```

► SUMMARY

In this chapter we briefly covered the on-chip I/O resources for the M68HC912B32 and we looked at some simple applications. As you can imagine, there are many more applications for these resources than shown here. With this material as a basis you should be able to come up with the required configuration to adapt a resource to other applications.

This material was very specific to the M68HC912B32. Many of the resources are the same in other 68HC12 MCUs, however. Make sure to always refer to the latest Motorola technical data for the microcontroller you are using.

EXERCISES

1. A relay coil with a coil resistance of 200Ω is connected active-low to a 68HC912B32 general purpose output. Find the maximum load current when the coil is on. Assume there is adequate protection to disregard transients.

2. Repeat Exercise 1 if the output is configured for reduced drive.

3. What is the power dissipated by the MCU I/O circuitry for Exercise 1?

4. Show the instruction(s) required to clear *RTIFLG*.

5. Write a code snippet to configure the RTI for an event period of 1.024ms. Assume an 8MHz E-clock.

6. Write a code snippet to configure the timer module clock source so it comes from a clock signal applied to *PT7*.

7. Write a code snippet to configure the timer prescaler so the timer has as long a range as possible while maintaining a resolution better than or equal to 1μs.

8. Write an interrupt service routine for *Timer Channel 2* configured as an output compare that generates an interrupt every 350μs.

9. Use *Timer Channel 7* and *Timer Channel 0* to generate a 1kHz pulse train with a 20% duty cycle. Write the configuration code so there is no load on the CPU.

10. Repeat Exercise 9 using the pulse-width modulator.

11. Write the configuration code to set the SCI baud rate to 19,200bps.

12. What is the minimum required sample rate for an voice audio sound system with a bandwidth of 3kHz?

13. What is the number of bits required for the voice audio system to have a signal-to-noise ratio better than 30dB?

14. What is the dynamic range for the system in Exercise 13?

15. Write the code to configure the ADC to take 8-bit samples on channel 0 continuously at a rate that is as close to 5kHz as the A-to-D clock will allow.

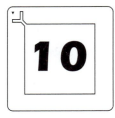

The Final Product

So far we have developed programs that were designed to run under the D-Bug12 monitor program on an evaluation board. In this chapter we cover the material required to design the hardware and software typically found in a stand-alone final product.

Every final product eventually reaches the point at which it must stand alone and run without the aid of development tools. This is the final integration and prototyping stage shown in Figure 1.6. The transition from a system running on development tools to a final product may be significant, depending on which development tools are being used and the requirements placed on the final product. The goal for any development system is to make this transition as simple as possible while providing the tools needed to give insight into the system's operation.

We have constructed our programs using the Motorola 68EVB912B32 development board with the D-Bug12 monitor resident on the board. This environment was introduced first because it is simple to learn, and we could concentrate on learning assembly code, the MCU programming model, and the MCU on-chip resources. The programs constructed with this process are far from a final product, however. We need to consider several items before we begin to create programs that are stand-alone final products.

First, all of our programs so far were downloaded and executed out of RAM, which is limited to 512 bytes and is volatile. The program code for a stand-alone product requires nonvolatile memory, so it is there when the power is turned on. Most programs are also much larger than 512 bytes.

Second, our programs did not include MCU and system initialization code. Because the programs are executed by the monitor program, we assumed that the monitor performed all system initialization tasks. A stand-alone program must include all the code required from the moment the system is turned on.

Third, we used a resident monitor program, D-Bug12, to debug the program. If we want to have the same capabilities on the final product, we would need to include the monitor program with our final code. On many final products the memory is not large enough for both the product code and the monitor code. This would be true especially for the D-Bug12 monitor, which requires a relatively large amount of memory. If we want to have the equivalent debugging capabilities on our final product, we would have to use an emulator, or if available, an on-chip debugging system.

Finally, we have not looked at the hardware required for the stand-alone system. When using the evaluation board (EVB), we only had to add external circuitry required for I/O. If the system uses a single-chip MCU, this may be as simple as designing the power supply, *RESET,* and clock circuitry. If an expanded mode MCU is required, however, it may involve bus interfacing and designing logic to decode and demultiplex the bus system. In this chapter we cover the basic requirements for a single-chip design. Bus expansion and decoding will be covered in Chapter 11.

Section 10.1 starts us off with the basic hardware design issues for the power source and the clock. In Section 10.2 we cover the hardware and MCU operations for *RESET* and the fault tolerance exceptions. Then in Section 10.3 we look at the MCU operating modes. Section 10.4 introduces start-up code, and finally in Section 10.5 we cover some of the issues we are faced with and tools that we can use when developing a final product.

▶ 10.1 MCU HARDWARE DESIGN

One of the benefits of using a microcontroller is that so much of the hardware is on chip. This means the hardware design for the MCU core can be relatively simple—consisting of the power supply, a clock source, and a reset circuit. In this section we look at the power supply and clock design. Section 10.2 covers the hardware required for the reset exceptions. All the hardware descriptions here are for the M68HC912B32 microcontroller in single-chip mode.

At first it seems straightforward to design the circuitry for the supply and clock. There are some issues that can greatly affect the reliability of the system, however; in addition, if the system is battery operated, a lot of care must be taken to extend battery life.

10.1.1 Power Supplies

First it is important to know that the M68HC912B32, like most microcontrollers at this point, is a 5-volt part. Although 5V MCUs are the most common parts on the market today, there are many more low-voltage parts being introduced. The most common low-voltage parts have 3.3V supplies, but some parts can operate with supply voltages down to 2V. Lower-voltage parts are becoming more common for two reasons: (1) to meet the requirements for low-power portable devices and (2) because of smaller geometry used in the IC layout.

The M68HC912B32 has 10 power supply pins for normal operation plus one pin for programming the on-chip Flash EEPROM. This is very different from the two supply

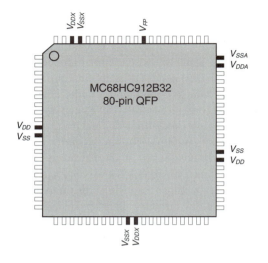

Figure 10.1 M68HC912B32 Power Supply Pin Layout

pins traditionally found on the corners of the ICs. The multiple supply pins allow for better power distribution within the part and allow for separate bypassing and routing for each supply. This helps reduce noise induced on the supply due to switching currents. Although there are several power supply pins, they should all be connected to the same voltage source.

As shown in Figure 10.1, the supply pins are evenly distributed on the IC. There are separate supply pins for three different parts of the MCU internal circuitry. V_{DD} and V_{SS} are the +5V and ground sources for the internal MCU circuitry. There are two sets of these pins to improve the on-chip power distribution. V_{DDX} and V_{SSX} are the sources for the on-chip I/O driver circuitry. There are also two sets of these pins. They are separated from the MCU internal supplies because the I/O drivers may drive high current loads. By separating the supply source, it allows us to provide separate bypass circuitry and layout so the noise generated by the I/O switching is not coupled into the supply for the internal circuitry. V_{DDA} and V_{SSA} are the sources for the ADC. It is especially important for these sources to be isolated from the other two relatively noisy digital supply sources.

By carefully designing the PCB layout and using separate bypassing for these supply pins, we can reduce the amount of digital noise induced on the analog signals. The last supply pin on the M68HC912B32 is the Flash EEPROM programming supply voltage, V_{FP}. In normal operation, when the Flash EEPROM is a read-only memory, V_{FP} must be connected to V_{DD}. During erase and programming operations, V_{FP} must be connected to a separate +12V supply.

At minimum there should be a separate 0.1–1.0μF capacitor across the two sets of V_{DD}-V_{SS} supply pins and the two sets of V_{DDX}-V_{SSX} supply pins. These capacitors should be physically located close to the IC pins and have good high-frequency characteristics. Typically monolithic ceramic capacitors are used for this application. In addition, if the ADC is used, additional bypassing capacitors are required for the V_{DDA}-V_{SSA} supply pins.

Most systems will have all four V_{SS} and V_{SSX} pins connected to a ground plane or to wide, low-impedance ground traces. If the ADC is not used, or the analog noise requirements are not significant, the V_{SSA} pin also will be connected to the same ground. On a system that requires low-noise analog signals or 10-bit analog samples, however, a separate ground plane or ground trace should be used for V_{SSA}. The two grounds, analog and digital, should only be connected at the ground source.

The V_{DD}, V_{DDX}, and V_{DDA} supply pins also should use separate traces to the source. Alternatively a single power plane for V_{DD} and V_{DDX} can be used, but this would require a four-layer PCB, which significantly increases the cost of the board.

A good example of PCB layout for bypassing the M68HC912B32 supply pins is shown in Figure 10.2. The figure shows the back side, or solder side, of the Motorola M68EVB912B32 EVB board. Capacitors $C18$ through $C21$ in the center of the image are surface-mount bypass capacitors for the two V_{DD} supply pins and the two V_{DDX} supply pins. They are located directly under the MCU so they are close to the supply pins.

Also notice the ground plane created by filling in the PCB area that does not require traces. This provides a very low-impedance path from all ground connections to the ground

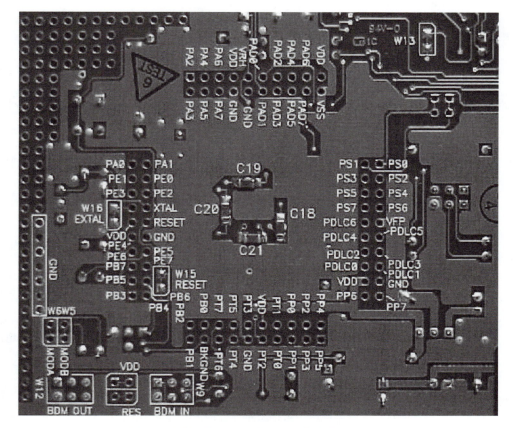

Figure 10.2   Motorola 68EVB912B32 PCB Layout

source. A ground plane like this one is more effective if as many signal traces as possible are placed on the other side of the board. It requires more work on the PCB layout, but it can make the difference between using an inexpensive two-layer board or having to go to an expensive four-layer board.

Supply Voltage Requirements.   There are two supply voltage requirements that must be met, the *maximum rating* and the *specified operation range.* All supply voltage ratings are relative to ground.

Note _____

From this point on, the parameter V_{DD} will be used to refer to all the +5V supply pins. The term *ground* or *GND* will refer to all the V_{SS} pins. This is consistent with the fact that all the V_{DD} pins are connected to the same 5V source and all the ground pins are connected to the same ground. The only difference is in the layout of the PCB and the placement of bypass capacitors.

The Maximum Rating.   The maximum rating specifies the safe voltage range for V_{DD} with respect to ground. If the maximum rating is exceeded, the MCU may be destroyed. For the M68HC912B32, the *maximum supply voltage* is

$$-0.3V < V_{DD} < +6.5V$$

One important thing to note about this range is that the maximum value of +6.5V is greater than four 1.5V battery cells. This allows us to supply the MCU directly from four battery cells if necessary.

The Specified Operating Range.   The specified operating range is the supply voltage range in which the rest of the MCU electrical specifications are guaranteed. For the M68HC912B32 the specified operating range is

$$V_{DD} = +5.0V \pm 10\%$$

The most important part of this specification is the lower limit, +4.5V. If the supply drops lower than +4.5V, the correct operation of the CPU is not guaranteed. This means that the CPU may not execute the code correctly, which in turn could result in a catastrophic failure. For this reason an MCU design should include a low-voltage *RESET* circuit. These circuits will be covered in the next section.

If the MCU is operated between +5.5V and +6.5V, it will execute the code correctly, but the power dissipation will be higher and the timing characteristics may be different than those listed in the specifications.

10.1.2 Power Dissipation

Power is dissipated by the MCU internal circuitry and the I/O circuitry when the MCU is driving loads. In Chapter 9 we covered the power dissipated by the I/O circuitry. Here we look at the power dissipated by the MCU internal circuitry.

TABLE 10.1   SUPPLY CURRENT REQUIREMENTS FOR THE M68HC912B32

MCU Mode	Supply Current (f_E = 8MHz)
Single-chip mode, Run	45mA
Single-chip mode, *WAIT*	5mA
Single-chip mode, *STOP*	10μA
Expanded mode, Run	70mA
Expanded mode, *WAIT*	10mA
Expanded mode, *STOP*	10μA

The M68HC912B32 internal power dissipation is specified as the *maximum total supply current, I_{DD},* and it is measured with no loads on the I/O pins. The internal power dissipation then is found by using

$$P_{INT} = V_{DD} \cdot I_{DD}$$

and the total power dissipation is given by

$$P_D = P_{INT} + P_{I/O}$$

The supply current required by the MCU depends on several parameters including the supply voltage, the operating mode, and the clock frequency. If we assume that the supply voltage is within the specified operating range and the E-clock frequency is 8MHz, maximum supply currents for the various M68HC912B32 modes are shown in Table 10.1.

As we see in Section 10.3, the MCU can be placed in several different operating modes, and as shown in the table, the supply current depends on the selected mode. There are two types of operating modes, bus-expansion modes and low-power modes. We cover these modes in more detail in Section 10.3.

As far as power dissipation is concerned, the *expanded mode* requires more supply current in order to drive the external address and data bus. There are two low-power modes, *STOP* and *WAIT.* In the *WAIT* mode the clock signals to some of the MCU modules are turned off to reduce the required supply current. In the *STOP* mode all the MCU clocks are turned off to reduce the supply current to the MCU's quiescent current, which in this case is only 10μA.

Because the M68HC912B32 is a CMOS IC, the supply current also depends on the clock frequency. If we run the MCU at a frequency other than the one that is specified, we can approximate the actual supply current by assuming a linear relationship between E-clock frequency and supply current. In the *single-chip mode* we end up with the following equation for normal operation:

$$I_{DD}(f_E) = 5 \times f_E + 5 \text{(mA)}$$

where

$$f_E = \text{the E-clock frequency in MHz}$$

$$I_{DD} = \text{the supply current in mA}$$

This means we can reduce the supply current by reducing the clock frequency. Of course reducing the clock frequency also slows the CPU execution speed, but many times this is an acceptable trade-off because an application may not require the CPU to run at full speed.

Considerations for Battery Sources. Designing a system for battery-powered operation is significantly more difficult than you might think. If you are using small batteries, every small current can result in a drained battery over time. We look at some of the general problems involved and some simple solutions. Covering all of the options and issues involved in battery-powered systems is beyond the scope of this book.

First many of the parts that we take for granted cannot be used in a battery-powered device because of their quiescent supply currents. A classic example is the Motorola low-voltage *RESET* circuit, the MC34064. It is shown in practically every example schematic using a M68HC11 or M68HC12 MCU. If it is used in a small battery application like a television remote, however, its 500μA quiescent current will drain a set of AAA batteries in less than 30 days. This drain occurs just from the detector circuit itself, without even considering any other circuitry.

On-off switches are rarely a good solution because the user can forget to turn the unit off. You have probably experienced this yourself—maybe you found that an ohmmeter or camera was dead when you tried to use it, and you discovered you had forgotten to turn it off.

A better solution is to place the system in a low-power mode after a period of inactivity. There are several ways to do this, but many of them can be tricky. The reason is that you cannot turn off the clock or the power supply with a circuit that relies on that clock or supply itself. Because of this, it is probably best to use an off-the-shelf power management IC that is designed for battery operation instead of designing your own system.

Since the advent of CMOS, many CPUs and MCUs include low-power modes, in which the clock to parts of the MCU can be turned off to conserve power. The 68HC11 and 68HC12 families have two low-power modes, *STOP* and *WAIT*. The *STOP* mode is much more useful for battery operation because the supply current is very low. Once the MCU is in the *STOP* mode, however, the only way to wake it up is to assert the RESET, XIRQ, or IRQ pin.

Some microcontrollers such as the M68HC812A4 also have *key wake-up* input ports. Key wake-ups can be used as general purpose inputs and/or IRQ sources, so they can be used to bring the CPU out of the *STOP* mode. They are ideal for applications that must remain in the low-power mode until one of many buttons is pressed. Most cellular telephones are examples of this type of system.

Other applications such as remote monitors require that the MCU operate periodically without human intervention. In these cases we need to add a periodic interrupt signal from an external circuit to wake up the MCU. *Watchdog timer* ICs and many *real-time clock* ICs have an interrupt output that can be used to periodically wake up the MCU every second, once a day, or at specific times each week. Since many of these ICs are available as nonvolatile modules, they require very little supply current or can even run with the main supply turned off.

10.1.3 Clocks

Like any synchronous sequential digital circuit, a microcontroller requires a system clock. The clock affects the CPU execution speed, power dissipation, and the configuration for some of the on-chip resources. Most modern MCUs contain an internal oscillator circuit that only requires an external crystal circuit, ceramic resonator, or clock signal.

The M68HC12 MCUs have two clock oscillator pins, XTAL and EXTAL. The frequency at these pins must be two times the desired E-clock frequency. There are three common circuits used for these two pins—a ceramic resonator, a crystal circuit, and a clock signal from a separate oscillator circuit. Let's quickly go over these options from the least expensive to the most accurate and reliable.

Ceramic Resonators.   Using a ceramic resonator as the resonating device for the internal oscillator is the least-expensive solution, and with internal capacitors, it is a single-device solution.

Ceramic resonators typically have an initial frequency accuracy of ±0.5% with frequency drift due to temperature or age of about ±0.3%. This accuracy is not adequate for many applications. Imagine trying to use this as a timing source for a real-time clock. The resulting clock could have an error of more than an hour in one six-month period! They are, however, adequate for many cost-sensitive designs that can tolerate a clock with this accuracy.

Crystals.   Using a crystal as the resonating device for the internal oscillator is the most common solution. Figure 10.3 shows the common crystal circuit used for most MCUs. This circuit is very sensitive to PCB layout and noise. A lot of care must be taken to place this circuit close to the EXTAL and XTAL pins. The capacitor values depend on the crystal type and the layout of the circuit.

Crystals come with initial frequency tolerances of ±20ppm to ±100ppm, where *ppm* is parts per million. In percentages this corresponds to ±0.002% to ±0.01%, which is much more accurate than a ceramic resonator. Again, if we use the real-time clock example, the 20ppm crystal would result in an error of 5 minutes in six months. The 100ppm crystals would cause an error of 25 minutes in six months, which is not accurate enough for most real-time clock applications, but it is adequate for most other applications.

Crystals are also relatively inexpensive. We must also take into account the cost of the resistor and two capacitors, however.

The problems with these external crystal circuits is that they are sensitive to the layout and cannot reliably drive other circuits. Some MCUs provide a buffered clock signal output that is at the same frequency as the crystal frequency. More often, however, a signal like the E-clock is provided, which is one-half the crystal frequency.

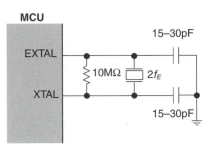

Figure 10.3   Crystal Circuit

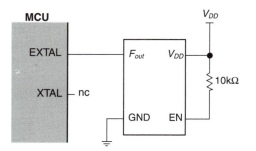

Figure 10.4 Clock Oscillator Circuit

External Oscillator ICs.   The last solution we cover is an external clock oscillator. External clock oscillators are hybrid circuits that include the oscillator circuit, the crystal, and a buffered output in a single package. Because these devices are in a single package and the output is buffered, we do not have to be concerned about the layout and we can drive multiple CMOS loads.

Some of these devices also have an output enable, which can be used in systems that require testing at a lower frequency than normal operation. Figure 10.4 shows a circuit for a clock oscillator with an enable. The pull-up resistor on the enable pin allows a test fixture to pull the pin low and insert another clock signal on EXTAL.

Using an external clock oscillator is the most reliable design, but it also costs more. In addition clock oscillators dissipate more power because they do not use the internal oscillator circuit already in the MCU. A typical 16MHz oscillator may require a supply current of 25mA. That is almost one-half the supply current required by the whole MCU.

Overclocking

Overclocking refers to the practice of running a CPU or MCU at a clock frequency higher than the maximum frequency specified. Occasionally you may hear about someone who runs a microcontroller at a clock speed much higher than specified by increasing the power supply voltage beyond the specified operating range. This is an interesting experiment, but it should never be done on a real product. The reliability is seriously compromised, it may not work on every part, and the manufacturer will not help you when the systems fail. To design a reliable product, you must always design within worst-case specifications.

▶ 10.2 RESET EXCEPTIONS

Once the power supply and the clock circuits are completed, the next step in the final product design is to design a reset circuit. A *reset* is used to initialize the CPU and the MCU on-chip resources to a known state called the *reset state*. A reset is always required to establish the reset state when the MCU is powered up. Resets can also be used to establish the reset state when an MCU fault is detected or when a user needs to manually reset the system.

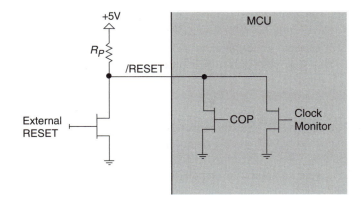

Figure 10.5 Simplified Functional Diagram for /RESET

Resets are considered a form of exception, because when a reset is detected, the next instruction is not executed by the CPU. Instead the CPU processes the exception. Resets are quite different from the interrupt exceptions that we covered in Chapter 8, however.

One difference is that a reset exception is immediately recognized asynchronously to the clock. Once a reset occurs, the MCU immediately transitions to the reset state and stays there until the reset condition is removed. An interrupt on the other hand is only sampled by the CPU at the end of each instruction sequence.

In this section we cover the function of the M68HC12 reset exceptions and look at some external reset circuits that can be used. First we look at the two external resets, the *power-on reset* and the *external reset*. Then we cover the two fault tolerance resets, the *clock monitor* and the *computer operating properly* (COP). These two fault tolerance resets are also called *internal resets* because they are generated by circuitry inside the MCU.

In the 68HC11 and 68HC12 microcontrollers, all the reset conditions involve the /RESET pin. A simplified functional diagram for the /RESET pin is shown in Figure 10.5. When /RESET is held low either by an external reset or an internal reset, the MCU is held in the reset state. When /RESET is released by the reset source and the MCU, it goes high and the CPU starts a reset exit sequence. Once the exit sequence is completed, the MCU starts executing program code.

10.2.1 Reset Source Determination

Since all the reset sources, external and internal, indicate a reset condition by pulling the /RESET pin low, how does the MCU know which source caused the reset? The 68HC11 and the 68HC12 microcontrollers use a similar method of determining the reset source by checking the time that /RESET is held low.

First an active circuit internal to the MCU detects a falling edge on the /RESET signal and actively pulls it low. For the 68HC12 microcontrollers, it holds /RESET low for 16 E-clock cycles. After the /RESET signal is released by the MCU, it waits eight E-clock cycles and then samples the /RESET signal again. If the signal is high, the CPU assumes that the reset was caused by an internal source and internal circuitry is used to determine whether it was a COP reset or a clock monitor reset. If the signal is still low, the MCU assumes that the reset was caused by an external reset source. For this reason, on a 68HC12 microcontroller an external reset pulse must be greater than 32 E-clock cycles. Otherwise it may be mistaken as an internal reset.

10.2.2 External Resets

The two types of external resets are the *power-on reset* and the *external reset*. A power-on reset occurs when a circuit inside the MCU detects the rising edge on V_{DD}. An external reset occurs when the /RESET pin is pulled low by an external circuit.

Figure 10.6 shows the steps taken by the MCU when an external reset condition is removed. For a power-on reset, a voltage detector detects a rising edge on V_{DD}. To initialize the hardware and provide time for the MCU clocks to stabilize, the MCU then waits 4,098 E-clock cycles before starting the reset exit sequence. For an external reset, the rising edge of /RESET is detected and the processor proceeds to the reset exit sequence. Note that the hardware is initialized while the CPU is in the reset state.

The Reset Exit Sequence.   Once the reset condition is removed, the CPU executes a short exit sequence to start executing the program code. The exit sequence starts by setting the X and I interrupt masks. This is required to disable all interrupts so an interrupt does not occur before the system is properly initialized. It then starts loading the instruction queue with the program code pointed to by the reset vector. The reset vector is located at addresses $FFFE:$FFFF, which must contain the address of the program to be started out

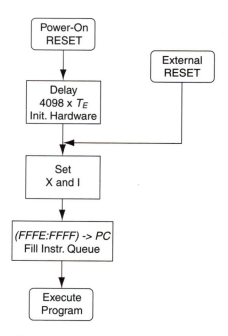

Figure 10.6   External and Power-On RESET Function

of reset. Once the instruction queue is filled, the CPU executes program code until another exception occurs.

This is how the CPU knows where to start executing code when a system is turned on. Recall that in the previous chapters, we started all of our programs by using D-Bug12's *GO* command. In a final product the system does not even have a monitor, so there must be another way to designate what program must be executed.

For example if our program code starts at the label *Start,* we need to add the following code to the program in order to load the reset vector with the starting address:

```
org $fffe
fdb Start                       ;Program start address
```

Note that the *org* directive as shown is normally not used. It is there to indicate that the stored constant is to be placed in $FFFE:$FFFF. We will cover this code in more detail in Section 10.4.

10.2.3 Fault Tolerance Exceptions

The term *fault tolerance* refers to a system that includes extra hardware and/or software that enables it to continue operating correctly in spite of certain system failures. You may remember the reports from NASA on the activities of the Mars Sojourner while it was crawling around the surface of Mars looking for interesting rocks to test. Occasionally the Sojourner reported that there was a system failure and it had to reset itself. This is a classic example of an application that requires significant fault-tolerant design because it is operating on the surface of Mars. It would be impossible to manually reset the system, or because of the one-way light time to Mars, even remotely reset the system from Earth. It had to detect the failure itself and recover from it automatically.

Of course a system does not have to be a mobile robot on a distant planet to benefit from a fault tolerant design. There are many reasons to include some form of fault tolerance in an embedded system. Avoiding catastrophic failures is an obvious reason, but improving user satisfaction and reducing service calls are also important for consumer products.

The response to a failure depends on the application. If a life-supporting medical instrument has a failure, the response should be to first continue the life-support operation as well as possible, and second to notify the operator that the unit needs to be serviced immediately. A cell phone on the other hand may want to recover from the error as best it can without the user noticing that anything went wrong at all. In the medical instrument, the fault-tolerant design can save a life. In the cell phone, the fault-tolerant design is used to improve the user's confidence in the product and to reduce service calls.

The 68HC11 and 68HC12 microcontrollers include two fault-tolerance exceptions: the *clock monitor* and the *computer operating properly timer.*

The Clock Monitor. The clock monitor includes an internal RC circuit that periodically checks for a clock edge. If a clock edge does not occur within the clock monitor RC time delay, a clock monitor reset occurs. This is an internal reset that pulls the /RESET pin low. When /RESET is released, a clock monitor reset vector is fetched from $FFFC:$FFFD.

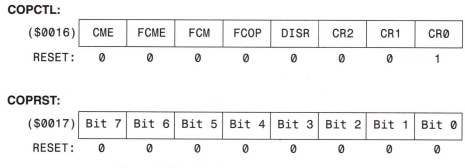

COPCTL:

($0016)	CME	FCME	FCM	FCOP	DISR	CR2	CR1	CR0
RESET:	0	0	0	0	0	0	0	1

COPRST:

($0017)	Bit 7	Bit 6	Bit 5	Bit 4	Bit 3	Bit 2	Bit 1	Bit 0
RESET:	0	0	0	0	0	0	0	0

Figure 10.7 COP Control and Reset Registers

Because the clock monitor pulls the /RESET pin low, it may be possible to reset the external system producing the clock. As we saw in Section 10.1, however, most systems do not have a clock source that can be reset. In these systems, a clock monitor reset pulls /RESET low and stops. It cannot do anything else because the reset exit sequence is a clocked process. If there is no clock, the CPU cannot proceed.

Note that when the system is placed in the *STOP* mode, the clocks are stopped. Therefore if the clock monitor is enabled, the *STOP* mode will cause a clock monitor reset.

Out of an external reset, the clock monitor is disabled. It can be enabled by using the *CME* and *FCME* bits in the *COPCTL* register, which is shown in Figure 10.7. The clock monitor enable bit *CME* is set to one to enable the clock monitor. It can be written to at any time. The force clock monitor enable bit *FCME* is used to permanently turn on the clock monitor. If it is zero, the clock monitor is controlled by the *CME* bit and can be turned on or off at any time. If *FCME* is set to one, the clock monitor is enabled regardless of the value of *CME*. Since the *FCME* bit can only be written one time, the *FCME* bit protects the clock monitor from being accidentally disabled.

The Computer Operating Properly Timer, COP. The M68HC12 COP timer is the most popular fault-tolerance system in embedded systems. It is also called a *watchdog timer.* A watchdog timer normally consists of a retriggerable one-shot that will generate a reset pulse unless it is retriggered by the CPU. If the program fails to retrigger the one-shot, the reset pulse will occur and the CPU will be reset.

The system's program is designed to reset the watchdog during its normal operation. If a software bug or a blocking routine prevents the program from reaching the COP reset code in time, the COP reset will occur. The COP reset program is then designed to recover from the failure.

The M68HC12 COP timer is an internal circuit that can be set to one of several reset times. To keep the COP from resetting the processor, the program must write a $55 and $AA to the *COPRST* register before the COP reset time expires. If a COP reset does occur, the /RESET pin will be pulled low, causing a reset. When /RESET is released, the CPU will fetch the address stored in the COP reset vector, $FFFA:$FFFB.

The COP watchdog times are set by the COP timer rate bits *CR2:CR1:CR0* in the *COPCTL* register. Table 10.2 shows the COP reset times for the different combinations of the COP rate bits. Notice that the actual reset time is a function of the E-clock frequency.

TABLE 10.2 COP RESET TIMES

CR2:CR1:CR0	E-Clock Divisor	COP Reset Time (f_E = 8.0MHz)
000	OFF	OFF
001	2^{13}	1.024 ms
010	2^{15}	4.096 ms
011	2^{17}	16.384 ms
100	2^{19}	65.536 ms
101	2^{21}	262.144 ms
110	2^{22}	524.288 ms
111	2^{23}	1.048576 s

Unlike the clock monitor, the COP is enabled out of reset in normal modes. As shown in Figure 10.7, the reset state of the *COPCTL* register is $01. Therefore the COP is on and will reset the MCU if it is not disabled or reset within 1.024ms (assuming an 8MHz E-clock). We address this more in Section 10.4.

In Chapter 8 we saw that most embedded systems are implemented as endless loops. When using the COP, a reset time is selected that is greater than the scheduling loop time. The COP reset code is then executed each time through the task loop.

Again the response to a COP reset depends on the application. In many systems the response may be as simple as jumping to the normal external reset code. Other systems may go through a complex testing and recording process whenever a COP reset occurs.

10.2.4 Reset Circuitry

As we saw in Figure 10.5, all the 68HC11 and 68HC12 resets affect the /RESET pin. Therefore the /RESET pin is a bidirectional pin, and because the pin is bidirectional, all devices connected to /RESET must have open-collector outputs. In addition there must also be a pull-up resistor to pull the pin high when there is no reset.

Figure 10.8 shows the minimal external reset circuit. This circuit can be used for some cost-sensitive designs, but it is inadequate for most systems. Most systems require low-voltage detection and possibly a user reset switch or a delayed reset.

Sometimes a system may require a delayed reset when the system is powered up. For example some BBRAM devices cannot be read immediately after power is turned on. If the BBRAM contains the program code, the CPU must wait before reading the reset vector from the device. One traditional way to cause a delayed reset is to add a capacitor to the reset circuit. Figure 10.9 shows the traditional circuit used for delayed resets.

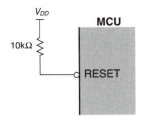

Figure 10.8 Minimal Reset Circuit

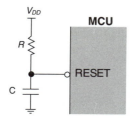

Figure 10.9 Minimal Delayed Reset Circuit

This circuit is not adequate for most designs for several reasons. First it makes it impossible for the MCU to distinguish between an internal reset and an external reset. This RC circuit will not only delay the external reset, it will also delay an internal reset, making the MCU think that the reset was caused externally. If an RC circuit is used, we can no longer use the internal fault tolerance exceptions.

Second the /RESET pin is a standard CMOS input, so the actual delay cannot be precisely controlled. Standard CMOS inputs have a large undetermined region in their input voltage specification, so we do not know exactly when /RESET will be seen as a one.

The third problem with this circuit is the slow rise time of the /RESET signal. Because the rise time is so slow, there may be many actual resets for a single reset activation. What happens is that when the MCU comes out of reset, there is an additional load placed on the supply. This in turn drops the /RESET signal slightly to cause another reset. This cycle may occur many times before the MCU finally comes out of reset for good. Having multiple resets is probably not a problem during normal operation. It can be very irritating when trying to debug the system with a logic analyzer, however, because we cannot trigger on the vector fetch or the first few instructions out of reset. Because of the multiple resets, there may be several trigger events before the program is finally executed.

The circuit shown in Figure 10.9 is also not adequate for many designs because it does not include a low-voltage detection circuit to protect against a drop in supply voltage.

Low-Voltage Detection Reset Circuitry. In the last section we saw that if the supply voltage falls outside its specified operating range, the MCU may not operate as expected. If V_{DD} falls below the minimum value, in this case +4.5V, the CPU may run unpredictably and in turn cause a catastrophic error. For this reason all but the most cost-sensitive designs must include a low-voltage reset circuit. Figure 10.10 shows a low-voltage detector build with discrete parts. $U1$ is a comparator that has an open-collector output. Notice that since an open-collector output is required, most op-amps cannot be used to implement this comparator.

The negative comparator input $V-$ is connected to a voltage reference IC so it is always equal to V_{ref}. For 5V systems, this reference voltage is typically set around 2.5V. The positive input $V+$ is connected to a voltage divider across V_{DD} and ground. $R1$ and $R2$ are set so the positive input voltage is equal to V_{ref} when V_{DD} is at the minimum specified value. So if V_{DD} is within the specified operating range, $V+$ is greater than $V-$ and the comparator output is off. When the comparator output is off, Rp pulls the /RESET pin high and the MCU runs normally. If V_{DD} drops below its minimum value, $V+$ becomes less than $V-$ and the comparator output is turned on. This pulls the /RESET pin low and resets the MCU before something bad can happen.

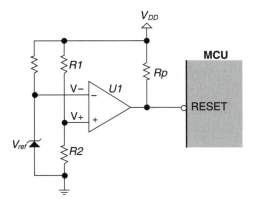

Figure 10.10 Discrete Low-Voltage Detector

We normally do not use a discrete circuit for the low-voltage detector because there are several ICs available that are specifically designed for this purpose. These ICs are normally called *microprocessor supervisory* ICs.

Figure 10.11 shows the typical reset circuit using Motorola's MC34064 low-voltage detector. There are several versions of the MC34064, each with a different input trigger level. This circuit is good for most stand-alone products that do not require a delayed reset. We could add a capacitor like the one in Figure 10.9, but it would also preclude the use of the internal resets. We can also add a push-button switch from the /RESET pin to ground for a user reset.

If we need a delayed reset, low power consumption, or a small footprint, a good solution is to use Maxim's MAX6314. The MAX6314 family of microprocessor supervisors were designed specifically for the 68HC11 microcontrollers. Figure 10.12 shows the complete circuit, which includes an optional manual reset button. There are many versions of the MAX6314 with different input trigger levels and different reset delays. The delays are set internally so no external capacitors are required.

Notice that we do not even need a pull-up resistor on the /RESET signal. The MAX6314 has an internal active pull-up that not only pulls the /RESET signal high but also pulls it high quickly. In this way the MAX6314 makes sure that an internal reset can always be distinguished from an external reset.

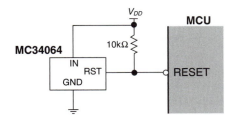

Figure 10.11 Low-Voltage RESET Circuit Using an MC34064

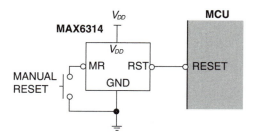

Figure 10.12 Low-Voltage RESET Circuit Using a MAX6314

The MR pin is for other reset sources, such as a manual reset switch. It also includes an internal pull-up. Because there are no external components required and the MAX6314 comes in a small surface-mount package, it is a very small solution. It is also a better solution for battery-powered designs because the quiescent supply current is only 12μA.

Note

Manual Reset Switches in Final Products

In general it is not a good idea to have a manual reset switch on an embedded system final product. Think about it. An embedded system runs a single program out of ROM. The only time the user would need to reset the system would be when there is a software or hardware failure. Since the software is firmware that comes with the product, there should not be any software errors. This is in contrast to a computer system in which the system is designed to run many different software programs, good and bad. Imagine buying a refrigerator that had a reset button. By the time that you realize that you need to use it, it would probably be too late.

A manual reset button on an embedded system is usually a sign of a poor design. A design strategy for failures should include using fault-tolerance exceptions, and as a last resort the system's power can be cycled. Manual reset buttons, however, are appropriate and useful for development systems because it is meant to execute many different programs that may not be fully debugged, and a reset switch is more convenient than cycling power.

▶ 10.3 M68HC912B32 OPERATING MODES

The configuration of the microcontroller when it is reset depends on the operating mode in which it is reset. The M68HC912B32 microcontroller has eight different operating modes. When the MCU is powered on or reset, the mode is determined by the state of the BKGD, MODA, and MODB pins. Table 10.3 shows the operating modes for the different values of these three pins. In this chapter we focus on the two single-chip modes, *special single chip* and *normal single chip.*

The MODA and MODB are bidirectional pins that are used to set the operating mode at reset, and then after reset they can be used as GPIO or as the instruction queue tracking signals IPIPE1 and IPIPE0. These tracking signals are used to recreate the actual instruction execution when debugging the system with a logic analyzer. MODA and MODB also have internal active pull-downs. If there is no external circuitry connected, they will be seen as zeros, which corresponds to one of the single-chip modes.

TABLE 10.3   M68HC912B32 OPERATING MODES

BKGD	MODB	MODA	Operating Mode
0	0	0	Special single chip
0	0	1	Special expanded narrow
0	1	0	Special peripheral
0	1	1	Special expanded wide
1	0	0	Normal single chip
1	0	1	Normal expanded narrow
1	1	0	Reserved
1	1	1	Normal expanded wide

MODE:

($000B)	SMODN	MODB	MODA	ESTR	IVIS	EBSWAI	0	EME
RESET State:								
Nrm Expd Nrw	1	0	1	1	0	0	0	0
Nrm Expd Wide	1	1	1	1	0	0	0	0
Sp Expd Nrw	0	0	1	1	1	0	0	1
Sp Expd Wide	0	1	1	1	1	0	0	1
Peripheral	0	1	0	1	1	0	0	1
Nrm Single Chip	1	0	0	1	0	0	0	0
Sp Single Chip	0	0	0	1	1	0	0	1

Figure 10.13   *MODE* Register

The BKGD pin is also bidirectional. At reset it sets the operating mode to a normal mode or a special mode. After reset it is used as a communications channel to send commands to the on-chip background debug circuitry. It has an internal pull-up, so if there is no external circuitry connected to the BKGD pin, the MCU resets in a normal mode.

With no external circuitry on BKGD, MODA, or MODB, the M68HC912B32 starts in the normal single-chip mode. Because of its internal Flash ROM, this is the most common mode used for this microcontroller. Also since all three of these pins are bidirectional, they should never be tied directly to power or ground.

Once the MCU is running, we can determine and sometimes change the mode by accessing the *MODE* register. Figure 10.13 shows the contents of the *MODE* register for each mode.

10.3.1 Normal Single-Chip Mode

The normal single-chip mode is the most common operating mode for the M68HC912B32. When in this mode, the bus system is not available externally and *PORTA* and *PORTB* can be used as GPIO. The memory map out of reset is shown in Figure 10.14.

$0000	
	Internal Registers
$01FF	
$0200	
	Unused
$07FF	
$0800	
	On-Chip RAM
$0BFF	
$0C00	
	Unused
$0CFF	
$0D00	
	Byte-Erasable EEPROM
$0FFF	
$1000	
	Unused
$7FFF	
$8000	
	Flash EEPROM (Enabled)
$FFFF	

Figure 10.14   Memory Map for Normal Single-Chip Mode

The interrupt and reset vectors are located from $FF00 to $FFFF; therefore, they must be loaded into the Flash EEPROM.

10.3.2 Special Single-Chip Mode

Special single-chip mode is the same as the normal single-chip mode except the background debug circuit is enabled and active. In this mode there is a BDM ROM block located from $FF00 to $FFFF. When the MCU is reset in this mode, the vectors are not fetched. Instead the MCU waits for BDM commands, which are received on the BKGD pin.

As we see in Section 10.5, the special single-chip mode is used to develop programs designed for the normal single-chip mode. We will be debugging our code in special single-chip mode and then running the final product in normal single-chip mode. Using this method, the transition from development code to the final product code is simple.

10.3.3 Expanded Modes

In all expanded modes, *PORTA* and *PORTB* are used for the multiplexed address and data bus. Since they can no longer be used as GPIO, the trade-off between single-chip and expanded modes includes a reduced number of GPIO bits.

In expanded wide mode, the external data bus is 16 bits wide. The upper byte of the address bus is multiplexed with the upper byte of the data bus on *PORTA*. The lower byte of the address bus is multiplexed with the lower byte of the data bus on *PORTB*.

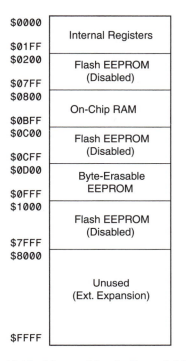

$0000	Internal Registers
$01FF	
$0200	Flash EEPROM (Disabled)
$07FF	
$0800	On-Chip RAM
$0BFF	
$0C00	Flash EEPROM (Disabled)
$0CFF	
$0D00	Byte-Erasable EEPROM
$0FFF	
$1000	Flash EEPROM (Disabled)
$7FFF	
$8000	Unused (Ext. Expansion)
$FFFF	

Figure 10.15   Memory Map for Expanded Modes

In expanded narrow mode, the external data bus is only 8 bits wide. This mode results in slower instruction execution, but it allows for less-expensive external devices. The 8-bit data bus is multiplexed with the upper byte of the address bus on *PORTA*. Narrow mode only affects external accesses. Internal accesses remain 16 bits wide.

Expanded modes also have a different memory map. Figure 10.15 shows the memory map for all the expanded modes on the M68HC912B32. The Flash EEPROM is moved to $0000–$7FFF and is disabled out of reset in all the expanded modes. It can then be moved and enabled by setting the *MAPROM* and *ROMON* bits in the *MISC* register.

The reset and interrupt vectors are still located from $FF00 to $FFFF, so they must be loaded into an external ROM device mapped to that space.

10.3.4 Changing Memory Maps

In all modes, the internal resources—the RAM, the registers, the byte-erasable EEPROM, and the Flash EEPROM—can be moved to different memory locations after reset. The registers used to move these resources are shown in Figure 10.16. The registers and the RAM can be moved to any 2K-byte boundary ($xx00) in the memory map and the byte-erasable EEPROM can be moved to any 4K-byte boundary ($xD00) in the memory map. The Flash EEPROM is moved by changing the *MAPROM* bit in the *MISC* register.

All three of these registers are write-once registers in normal modes. This means that after you write to these registers, they are protected from any further changes. A write to

INITRG:

($0011)	REG15	REG14	REG13	REG12	REG11	0	0	MMSWAI
RESET:	0	0	0	0	0	0	0	0

INITRM:

($0010)	RAM15	RAM14	RAM13	RAM12	RAM11	0	0	0
RESET:	0	0	0	0	1	0	0	0

INITEE:

($0012)	EE15	EE14	EE13	EE12	0	0	0	EEON
RESET:	0	0	0	0	0	0	0	1

MISC:

($0013)	0	NDRF	RFSTR1	RFSTR0	EXSTR1	EXSTR0	MAPROM	ROMON
RESET:								
Exp. Modes	0	0	0	0	1	1	0	0
Single-Chip	0	0	0	0	1	1	1	1

Figure 10.16 Resource Location Registers

these registers should also be followed by a *nop* instruction to make sure that the resource has finished being moved before it is accessed.

EXAMPLE 10.1

Changing the Internal Memory Map

Just to be different, a team of designers decides to move the on-chip resources for their design. They decide to move the RAM so it starts at $0000, move the registers so they start at $0800, and move the byte-erasable EEPROM so it starts at $7D00.

Solution

To do this, the following changes are required:

```
$08 -> INITRG          ;Move registers to $0800
$00 -> INITRM          ;Move RAM to $0000
$71 -> INITEE          ;Move EEPROM to $7D00 and turn on
```

The initialization code required to make these changes is as follows:

```
movb #$80,INITRG     ;Move registers to $0800
nop                  ;Wait for registers to be moved
movb #$00,INITRM     ;Move RAM to $0000
nop                  ;Wait for RAM (not required here)
movb #$71,INITEE     ;Move EEPROM to $7D00
nop
```

Notice that *nop* instructions are used after each memory map change. Also notice that after the register block is moved but before the RAM is moved, they are at the same location ($0800). This is okay because the internal chip-select logic gives priority to the registers. Therefore we can still write to the registers but we cannot use the RAM until it is moved.

After these changes the resulting memory map is

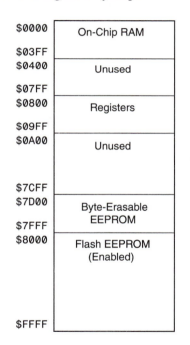

Of course a change like this is not justified by just wanting to be different. These changes may benefit a design by reducing the code space required for RAM access because direct addressing can be used. And these changes make the byte-erasable EEPROM and Flash EEPROM contiguous. But because the programs designed for this system would be incompatible with most other systems, it would be rare that changing the memory map can be justified at all.

The M68HC912B32 is normally run in normal single-chip mode, with special single-chip mode used for development. Other 68HC12 MCUs such as the M68HC812A4 are designed for expanded mode systems. These will be covered in more detail in Chapter 11.

▶ 10.4 CONFIGURATION AND START-UP CODE

All the programs that we wrote in previous chapters were designed to be executed by a resident monitor program. Specifically we used an M68EVB912B32 board with the D-Bug12 monitor resident in Flash EEPROM. The programs were all loaded into on-chip RAM and executed by using D-Bug12's *GO* instruction.

A final product, however, must stand alone without a monitor program, so the application code must include everything that is required from the time the MCU is reset. In this section we introduce the additional program code required for a final product.

10.4.1 Program Organization and Memory Map

In Chapter 3 a type of program organization was introduced for programs that are executed by the D-Bug12 monitor. That organization only required one *org* statement, because all the code was loaded in the on-chip RAM. A program designed for a final product is very different. It must be separated so the different sections of code go into the proper memory types and locations.

To illustrate the additional code required for a final product, let's look at a new, stand-alone version of the *demo1* program that was introduced in Chapter 3. It will be designed for a M68HC912B32 that is running in normal single-chip mode. The source code for this version of *demo1* is shown in Source 10.1.

Refer to the memory map shown in Figure 10.14 on page 396 to find that we need to locate the program, the stored constants, and the strings at the start of Flash EEPROM ($8000). The variables need to be in on-chip RAM ($0800), and the exception vectors need to be at the end of the Flash EEPROM ($FFD0).

In order to place these three sections of code in the correct locations, we need three *org* directives. Notice that there are three different *org* directives in the new *demo1* program. The first one places the program code in Flash EEPROM, starting at $8000. The second *org* directive places the variables in the internal RAM at $0800, and the last *org* directive places the start of the vectors at $FFD0.

Source 10.1 demo1 Program Designed as a Final Product

```
;**********************************************************************
; A simple demonstration program to see the HC12 run.
; It generates 60 1-second, active-low pulses on PP0.
; This version is designed as a final product with the default single-chip
; memory map.
;
; MCU: 68HC912B32EVB, E=8MHz
;**********************************************************************
; Equates
;**********************************************************************
DDRP            equ $57
PORTP           equ $56
```

```
BIT0                    equ %00000001
TC1MS                   equ 1996            ;Delay count for 1ms
INTCR                   equ $001E           ;Interrupt Control Register
MODE                    equ $000B           ;Mode Register
INITRG                  equ $0011           ;Register Initialization
INITRM                  equ $0010           ;RAM Initialization Register
INITEE                  equ $0012           ;EEPROM Initialization Register
MISC                    equ $0013           ;Miscellaneous Mapping Control
PEAR                    equ $000A           ;Port E Assignment Register
RDRIV                   equ $000D           ;Reduced Drive Register
EEMCR                   equ $00F0           ;EEPROM Configuration Register
RTICTL                  equ $0014           ;RTI Control Register
COPCTL                  equ $0016           ;COP Control Register
;********************************************************************
; Program
;********************************************************************
                        org $8000
start                   lds #$0C00                  ;Initialize stack pointer
                        jsr MCUInit                 ;Configure and protect MCU
main                    bset PORTP,BIT0             ;Initialize PORTP, BIT0
                        bset DDRP,BIT0
                        movb InitCnt,CurCnt         ;Initialize pulse counter


;********************************************************************
; Main loop for output pulse generation
;********************************************************************
pulse                   bclr PORTP,BIT0             ;Turn pulse on.
                        ldd #250                    ;Wait 250ms
                        jsr WaitDmS
                        bset PORTP,BIT0             ;Turn pulse off.
                        ldd #750                    ;Wait 750ms
                        jsr WaitDmS
                        dec CurCnt                  ;Count pulses?
                        bne pulse                   ;  No: Another pulse
                        bra *                       ;  Yes: Trap here
;********************************************************************
; Subroutine WaitDmS - A programmable delay in ms.
;       Arguments: The number of ms is passed in ACCD.
;       Registers: Preserves all registers except CCR.
;       Stack Reqs: 6 bytes stack space
;       Req. Subs: Dly1ms
;********************************************************************
WaitDms                 pshd                        ;preserve ACCD
msdlp                   jsr Dly1ms                  ;execute 1ms ACCD times
                        subd #1
                        bne msdlp
                        puld                        ;recover ACCD
                        rts
```

```
;********************************************************************************
; Subroutine Dly1ms - 1ms delay loop.
;     MCU: 68HC12, E=8MHz, no clock stretching, 16-bit bus
;     Registers: Preserves all registers except CCR.
;     Stack Reqs: 2 bytes stack space
;********************************************************************************
Dly1ms          pshx                    ;preserve IX
                ldx #TC1MS              ;execute loop TC1MS times
d1mslp          dex
                bne d1mslp
                pulx                    ;recover IX
                rts


;********************************************************************************
; MCU Write-once Register Initialization
;********************************************************************************
; Required initialization for all M68HC912B32 stand-alone systems
;    - COP must either be turned off or reset.

MCUInit         movb #$00,COPCTL        ;Turn COP off, clock mon off


;****************************************************************
; Recommended initialization for all stand-alone systems

                movb #$90,MODE          ;Protect from mode change
                movb #$0F,MISC          ;Prot. flash from turn off
                movb #$00,INITRG        ;Protect regs from move
                movb #$08,INITRM        ;Protect RAM from move
                movb #$01,INITEE        ;Protect EEPROM from move
                movb #$FE,EEMCR         ;Lock EEPROM Protection bits


;****************************************************************
; Optional initialization for some stand-alone systems

                movb #$90,PEAR          ;PortE always GPIO
                movb #$00,RDRIV         ;Ports E, B, A full drive
                movb #$60,INTCR         ;Prot. IRQ mode, stop delay
                movb #$00,RTICTL        ;COP, RTI run in wait, bkgd
                rts


;********************************************************************************
; Minimal Unexpected Interrupt Trap
;********************************************************************************
UnexpInt        bra *           ;Endless trap


;********************************************************************************
; Minimal Unexpected Internal RESET Trap
;********************************************************************************
UnexpRst        movb #$00,COPCTL        ;Turn COP off and clock monitor off
```

```
                bra start

;*************************************************************************
; Constants
;*************************************************************************
InitCnt         fcb 60          ;Initial pulse count

;*************************************************************************
; Variables
;*************************************************************************
                org $0800       ;Locate in on-chip RAM

CurCnt          rmb 1           ;LED flash counter

;*************************************************************************
; Vectors
;*************************************************************************
                org $ffd0       ;Locate at start of 68HC912B32 vectors

                fdb UnexpInt    ;BDLC interrupt vector
                fdb UnexpInt    ;ATD interrupt vector
                fdb UnexpInt    ;reserved vector
                fdb UnexpInt    ;SCI interrupt vector
                fdb UnexpInt    ;SPI interrupt vector
                fdb UnexpInt    ;Pulse Accumulator interrupt vector
                fdb UnexpInt    ;Pulse Accumulator overflow vector
                fdb UnexpInt    ;Timer overflow vector
                fdb UnexpInt    ;Timer Channel 7 interrupt vector
                fdb UnexpInt    ;Timer Channel 6 interrupt vector
                fdb UnexpInt    ;Timer Channel 5 interrupt vector
                fdb UnexpInt    ;Timer Channel 4 interrupt vector
                fdb UnexpInt    ;Timer Channel 3 interrupt vector
                fdb UnexpInt    ;Timer Channel 2 interrupt vector
                fdb UnexpInt    ;Timer Channel 1 interrupt vector
                fdb UnexpInt    ;Timer Channel 0 interrupt vector
                fdb UnexpInt    ;Real-time interrupt vector
                fdb UnexpInt    ;IRQ interrupt vector
                fdb UnexpInt    ;XIRQ interrupt vector
                fdb UnexpInt    ;SWI interrupt vector
                fdb UnexpInt    ;Unimplemented opcode trap vector
                fdb UnexpRst    ;COP failure reset vector
                fdb UnexpRst    ;Clock Monitor reset vector
                fdb start       ;System reset vector

;*************************************************************************
```

As we see in Chapter 12, we could also use the Introl *section* directive instead of an *org* directive. The *section* directive allows us to name the code sections, which is required when writing modular programs in assembly or C.

10.4.2 Exception Vectors

In Section 10.2 we saw that the reset exception is used to start the MCU at a known state. When the system is powered on or reset, the CPU jumps to the address stored in the reset vector and starts executing code from that point. In addition, in Chapter 8 we saw that when an interrupt occurs, the CPU jumps to the address stored in the corresponding interrupt vector and starts executing code from that point.

When we ran a program using D-Bug12, the exception vectors were part of the D-Bug12 code. When a reset occurred, the CPU jumped to D-Bug12, which in turn executed the required start-up tasks, displayed a prompt, and waited for user input. To change an interrupt vector, we had to use the D-Bug12 utility routine *SetUserVec()* so the monitor could redirect program control to the user's interrupt service routine.

When we write a program for a final product, we need to include the assembly code that defines the exception vectors. In Source 10.1, a block of code was added at the end of the program to define the exception vectors.

For each *fdb* directive, the assembler stores the value of the symbol in memory. For example the reset vector is defined by storing the value of the symbol *start* at $FFFE:$FFFF. *start* is a label that was defined at the start of the initialization code. In this case *start* is equal to $8000. So when the MCU is reset, it jumps to the code at *start* because $8000 is stored in addresses $FFFE:$FFFF.

Be careful when defining the vectors. It is easy to add an extra line or miss a line, resulting in some vectors being placed in the incorrect addresses. It is always a good idea to look at the listing file to verify that all the vectors are correctly placed.

Unused Exceptions. Notice that there is a vector defined for every exception, whether that exception is used or not. All stand-alone programs should have every vector defined so the CPU will not jump to some unknown location, execute the program or data there, and possibly cause a more serious failure. The *demo1* program in Source 10.1 does not use any interrupts. Therefore all interrupt vectors are loaded with the address of *UnexpInt()*, a routine added to handle unexpected interrupts. Likewise *UnexpRst()* is used to handle an unexpected internal reset.

These routines are traps that should only be reached if an unexpected interrupt occurs or an internal reset occurs. They are minimal solutions that can only be used in systems that do not use the clock monitor or the COP.

When one of these routines is reached, we do not know which interrupt or reset occurred, but at least we know why the program failed. To be able to identify the specific exception that occurred, we would need to write a separate routine for each exception source. These routines are typically short routines that identify the source and then jump to a general unexpected exception routine. This routine can then recover from the unexpected event appropriately, notify the user, and/or make a record of the failure in EEPROM.

A well-written routine for unexpected events or failures can reduce troubleshooting time dramatically. Instead of the system mysteriously crashing, a well-written routine may display or record the source of the error. In that way we immediately know that a failure occurred and what the source of the failure was. The routines in *demo1* provide little help for debugging.

10.4.3 Configuration and Initialization

Again, because a monitor program is not used on the final product, we must include all the start-up tasks required before executing the main program. These start-up tasks include stack pointer initialization, MCU configuration, and program initialization.

Stack Pointer Initialization. In Source 10.1 the first code that is executed out of *RESET* is the stack pointer initialization. If the stack pointer is not initialized, we cannot use subroutines or interrupt service routines.

In this case the stack pointer is initialized to $0C00, which is the byte after the end of a RAM block. Since stacks expand toward lower addresses, by initializing the stack pointer at the end of the RAM block, we keep it as far away from the variables as possible.

Note

Beware when initializing the stack pointer using a 68HC11 MCU. Since the 68HC11's stack pointer points to the next available space on the stack, it must be initialized so it points to the last byte in RAM, not the byte following the last byte. If you are porting code between the 68HC12 and 68HC11, you should initialize the stack pointer for the 68HC11. That way it will work with both MCUs.

Write-Once Register Initialization. After the stack pointer has been initialized, the program calls *MCUInit( )*, which is a subroutine that performs all the MCU configuration that involves write-once registers. Each of these registers controls the configuration of a critical piece of on-chip hardware. To protect these resources from accidental changes, Motorola has made its control registers write-once registers. Write-once registers can only be written to one time. After that they cannot be changed without resetting the MCU. Therefore to protect these resources, we need to write to the register during the start-up sequence even if the reset state of the register is correct for our application.

Notice that in Source 10.1, *MCUInit( )* is divided into three snippets of code. The first snippet disables the COP timer. Since the COP timer is enabled out of reset, it must be either disabled or reset during the initialization process. The second snippet includes code that is recommended for all designs. A system will work without this code but it is not as fault tolerant. The last code snippet includes initialization that is only needed for some designs. It does not hurt to add this section, but it is not always necessary.

Resource and Program Initialization. The rest of the MCU configuration code and program initialization code should be placed into separate routines. For example if we are using *Timer Channel 0,* it should be configured in a separate routine that performs that initialization. Then to call the routine, the start-up code may look like this:

```
start        lds #$0C00          ;Initialize stack pointer
             jsr MCUInit         ;Configure and protect MCU
             jsr TC0Init         ;Configure the Timer Channel 0
main         ...
```

By keeping these routines separate, it is easier to reuse the code in a different project. Notice that in Source 10.1, *PORTP* was initialized in the main program, not in *MCUInit()*. We see more examples of this in Chapters 12 and 14 when we cover modular programming.

Program initialization is normally completed last. It should include initialization for all global variables. In *demo1* there is only one variable, *CurCnt,* which is initialized to the stored constant *InitCnt.*

▶ 10.5 FINAL PRODUCT DEVELOPMENT

The ultimate goal of any development system is to provide an environment that enables us to efficiently construct reliable programs. In this section we look at the tools required to execute and test a program in a final product. The system must be able provide insight into the detailed operation of the program while being as noninvasive to the final product as possible.

In the last several chapters, we tested and debugged our programs using the D-Bug12 monitor program, which was contained in the MCU Flash EEPROM. D-Bug12 provided the tools necessary to load machine code into target memory, execute the code, and view the results.

This system provided a simple and effective way to develop small programs, routines, and code snippets. We cannot use it for a final product, however, because the final product must run the application code immediately out of reset. In addition the application code must be in nonvolatile memory.

At some point we have to cut the ties to the monitor program and run the final product completely from our own software. The next four subsections cover some of the tools and processes involved in making the transition from a development environment to the final product. We start with a traditional process that uses an EPROM for the program code without the aid of a monitor. We then cover different techniques and technologies that can make the transition from development to final product easier. Finally we cover a development system based on two M68EVB912B32 development boards using the 68HC12's Background Debug Mode, BDM, and Flash EEPROM.

10.5.1 The Traditional Process

Let's start with the traditional process of programming an EPROM with the final product code, placing the EPROM in the target board, and running the code out of reset. By itself this development system does not include any way to control the CPU execution besides *RESET.* This represents the simplest system, but it is also the least efficient.

Figure 10.17 shows the construction process for a final product that uses an EPROM. To get the program into target memory, the machine code is loaded into a device programmer, which is then used to program an EPROM device. The EPROM is then inserted into a socket on the target board, and the program is executed by turning on the power or initiating a reset.

There are a couple of problems that make this process very inefficient. First the EPROM programming process is manual and time consuming. The EPROM must be erased using an ultraviolet light source and then physically placed into the programmer to be programmed.

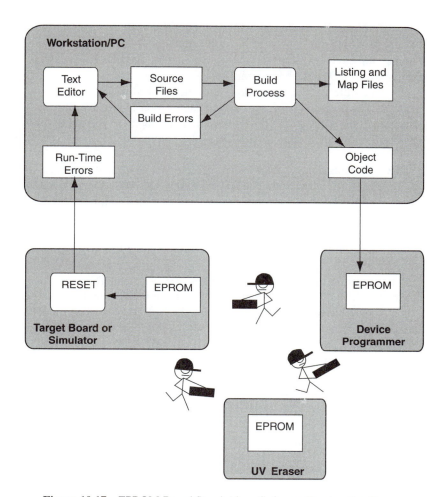

Figure 10.17 EPROM-Based Stand-Alone Software Construction Process

Once it is programmed, it must be moved to the target board. If the code does not work, the cycle is repeated. Because of the time required to erase an EPROM, it is necessary to use several EPROM devices so there is at least one erased EPROM ready for the next iteration.

The second reason that this process is so inefficient is the lack of debugging tools. The target system no longer has a monitor to control execution or to examine memory and register contents. We can develop as much of the code as possible using the development process. Then once the code is as close to a working final product as possible, we make the transition to the stand-alone process. How close we can get to the final stand-alone program depends on the capabilities of the development system. In some development systems we can get very close to the final product. In others it can be a big jump.

In addition to this process being inefficient, a device programmer is required, and it is limited to microcontrollers that run in expanded mode. We can use this process for single-chip microcontrollers that have an on-chip EPROM, but the same inefficiencies apply, and these microcontrollers tend to be expensive and hard to get.

Traditionally, to design a final product that uses a single-chip microcontroller, we would have to use an expensive in-circuit emulator or an evaluation board that emulates the single-chip mode by using a port replacement unit. The final product would then use an on-chip masked ROM for the program code.

Since the manufacturer programs the masked ROM, we cannot use an iterative process. The program that is sent to the manufacturer to produce parts had better be correct. Because of the expense and time involved in development, the use of masked ROM is only practical for mature products with large sales volumes or expensive specialized products.

EPROM Emulation. One way to automate this process of programming the EPROM is to replace the EPROM with a battery-backed RAM, BBRAM, or a RAM during development. The BBRAM is nonvolatile, so it better emulates an EPROM because it retains its contents when power is removed. The RAM of course loses its contents when power is removed, but it can be quickly reloaded.

Since most memories follow a standard pinout, a RAM or BBRAM can normally be found with the same memory size and practically the same pinout as the EPROM. The RAM can be placed in the EPROM socket during development. Then once the program is completed, an EPROM can be programmed with the same program that was used in the RAM.

To be able to use a RAM to emulate an EPROM, we need to make some hardware considerations. Although a standard pinout is used for both devices, there are some differences between an EPROM pinout and a RAM pinout. For example let's compare a 27C64 8K-byte EPROM with a 6264 8K-byte RAM. The following three pins are different between the two ICs:

Pin	EPROM	RAM	Connect
1	V_{pp}	NC	+5V
26	NC	CS2	+5V
27	/PGM	/WE	+5V or $R/\overline{W}$

Pins 1 and 26 can be directly connected to +5V. This is the correct value for V_{PP} on the EPROM and for CS2 on the RAM.

The most significant difference is on pin 27. An EPROM does not have a $R/\overline{W}$ signal because it is a read-only device. Pin 27 on the EPROM is /PGM, which must be held high during normal operation. If we want to be able to write to the RAM to load the program, however, we need to connect the $R/\overline{W}$ signal to /WE.

The best solution is to use a jumper that selects between $R/\overline{W}$ and +5V. The jumper is placed in the $R/\overline{W}$ position to load the program into the RAM. Once the program is loaded, the jumper is changed to the +5V position. This is the correct signal for the EPROM, and if we are executing the program out of RAM, it keeps a program bug or other failure from causing the CPU to write on top of the program.

Dual-Boot System. Using RAM helps make the development process for a final product more efficient. Some complications remain, however.

First the monitor cannot reside in the same memory that we are trying to program. Unfortunately the monitor program normally needs to reside in program ROM. This conflicts with the final product code, because it must also reside in program ROM.

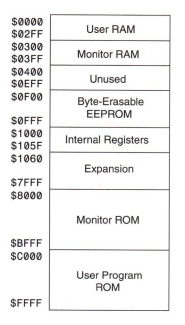

$0000	
	User RAM
$02FF	
$0300	
	Monitor RAM
$03FF	
$0400	
	Unused
$0EFF	
$0F00	
	Byte-Erasable EEPROM
$0FFF	
$1000	
	Internal Registers
$105F	
$1060	
	Expansion
$7FFF	
$8000	
	Monitor ROM
$BFFF	
$C000	
	User Program ROM
$FFFF	

Figure 10.18   68HC11 Development Memory Map

Another problem is that we need to control the reset vector for the MCU. While developing or loading code, we need the MCU to jump to the monitor program when it is reset. On the other hand the final product must jump directly to the final product code when it is reset. On most microcontrollers this is handled by using different operating modes that use different exception vectors or different memory maps.

For example it is easy to set up a dual-boot development system for the 68HC11 microcontrollers using two memory devices—one for the monitor and one for the final product code. The memory map for the system can be designed as shown in Figure 10.18.

The monitor ROM includes exception vectors for the monitor program at $BFC0–$BFFF and the user program includes vectors for the final product at $FFC0–$FFFF. Now to reset into the user program, the MCU is reset with the mode select pins set for the normal expanded mode. In this mode, a 68HC11 MCU uses the vectors located at $FFC0–$FFFF. To use the monitor, the MCU is reset with the mode select pins set for the special test mode. In this mode a 68HC11 MCU uses the vectors located at $BFC0–$BFFF.

If we use a RAM to emulate the program ROM, we can reset in special test mode to get to the monitor and load the user program into the program RAM. Then we can run the user program either by using the monitor's *GO* instruction or by resetting the MCU in normal expanded mode.

The limitation to this system is that the program for the final product is limited to 16K-bytes ($C000–$FFFF) and some RAM must be reserved for the monitor. It can be an efficient process, however, because the final product code can be loaded and debugged with the monitor using essentially the same process we have used previously.

Another solution is to use a special bootstrap program. A *bootstrap program* consists of a simple loader that reads the machine code being downloaded via the SCI and loads the program into the correct memory. A bootstrap program enables us to load a program into memory, but it does not include the debug capabilities that a monitor program has. Most microcontrollers with on-chip EEPROM also provide a bootstrap mode that allows the MCU to be reset directly into the bootstrap program, or the bootstrap program can be temporarily loaded into RAM and executed to download and program an on-chip EEPROM.

10.5.2 Using On-Chip EEPROM

Again the techniques described in the previous subsections are normally restricted to microcontrollers that are run in an expanded mode. One of the most important technologies that allow users to program their own single-chip designs is electrically erasable read-only memory (EEPROM). These memories can be programmed on-chip by using a monitor program, a special bootstrap program, a device programmer, or on-chip debugging circuitry.

There are two types of EEPROM devices, byte-erasable EEPROM and Flash EEPROM. Flash EEPROM is the most important for program storage. It uses less chip area, so manufacturers can put a larger Flash EEPROM into an MCU. In order to reduce the chip size, Flash EEPROM devices must be erased in blocks. Byte-erasable EEPROM requires much more chip area, but they can be erased by individual bytes. Byte-erasable EEPROM is normally limited to a few kilobytes.

10.5.3 Background Debug Systems

Development of a single-chip design can be dramatically improved by using a microcontroller that combines the use of EEPROM and internal debugging circuitry. The 68HC12 microcontrollers all have Motorola's Background Debug Mode (BDM). The BDM involves a different operating mode, on-chip circuitry, a small ROM program, and a serial communications port for debugging. When we use a system that combines the BDM and EEPROM, the transition from the debugging process to the final product is trivial. Yet at the same time the BDM provides debugging tools that are better for debugging real-time systems than a resident monitor is.

Let's quickly go over how the 68HC12 BDM works before covering the specific development environments that will be used throughout the rest of this text. Again, for the detailed operation of the BDM system, refer to the *Technical Summary* for the 68HC12 microcontroller. The description here assumes that we will be using a development system designed for the BDM, which makes the detailed BDM operations transparent to the user.

The BDM circuitry enables the system to execute a set of hardware commands, firmware commands, communicate to a host over the BKGD pin, and on some MCUs, implement hardware breakpoints. Because extra hardware circuitry is used, this system is relatively noninvasive to the CPU. To execute the hardware commands, the BDM circuit uses bus cycles that are not used by the CPU in normal operations. Occasionally it may have to steal a cycle or two, but in general it allows the CPU to run at real time. This is a significant improvement over a resident monitor program, which requires the CPU to stop executing the user program for any monitor command.

The BDM hardware commands are shown in Table 10.4. They are used for basic BDM initialization and memory access. These commands are sent by a host connected to the BKGD pin. The hosts we cover here are designed for the BDM, so we will never see

TABLE 10.4 M68HC912B32 BDM HARDWARE COMMANDS

Command	Description
BACKGROUND	Enter background mode
READ_BD_BYTE	Read 8-bit word from memory with BDM ROM in map
STATUS	Read BDM status register
READ_BD_WORD	Read 16-bit word from memory with BDM ROM in map
READ_BYTE	Read 8-bit word without BDM ROM in map
READ_WORD	Read 16-bit word without BDM ROM in map
WRITE_BD_BYTE	Write 8-bit word to memory with BDM ROM in map
ENABLE_FIRMWARE	Set ENBDM bit in BDM status register
WRITE_BD_WORD	Write 16-bit word to memory with BDM ROM in map
WRITE_BYTE	Write 8-bit word to memory without BDM ROM in map
WRITE_WORD	Write 16-bit word to memory without BDM ROM in map

TABLE 10.5 M68HC912B32 BDM FIRMWARE COMMANDS

Command	Description
READ_NEXT	Read next 16-bit word pointed to by IX
READ_PC	Read program counter, PC
READ_D	Read accumulator D, ACCD
READ_X	Read index register, IX
READ_Y	Read index register, IY
READ_SP	Read stack pointer, SP
WRITE_NEXT	Read next 16-bit word pointed to by IX
WRITE_PC	Write to program counter, PC
WRITE_D	Write to accumulator D, ACCD
WRITE_X	Write to index register, IX
WRITE_Y	Write to index register, IY
WRITE_SP	Write to stack pointer, SP
GO	Go to user program
TRACE1	Execute one instruction and then return to BDM
TAGGO	Enable tagging and go to user program

any of these commands. Instead we enter commands like the D-Bug12 memory modify command, *MM*. The host system then converts that request into the required BDM commands and sends them to the MCU through the BKGD pin.

The BDM also includes a small ROM that contains BDM firmware. The firmware includes commands that cannot be executed by the BDM circuitry so they have to be executed by the CPU. This means that we cannot use these commands while the CPU is executing a program in real time. Again these commands are sent by a host through the BKGD pin. The BDM firmware commands are listed in Table 10.5. Notice that the commands used to access the registers require the CPU. This means that we have to stop the CPU to read the register contents.

The BDM can be used while the MCU is in any mode. We will only cover a BDM development system for a single-chip design using the M68HC912B32, however. This is the normal process used in the system described in the next subsection.

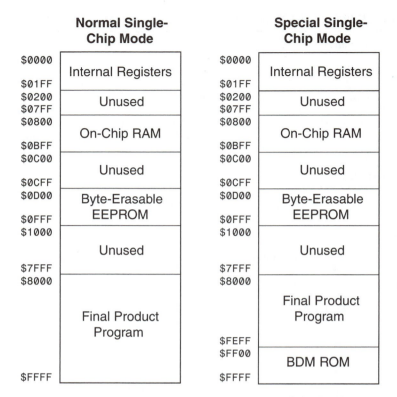

Normal Single-Chip Mode

Address	Region
$0000 – $01FF	Internal Registers
$0200 – $07FF	Unused
$0800 – $0BFF	On-Chip RAM
$0C00 – $0CFF	Unused
$0D00 – $0FFF	Byte-Erasable EEPROM
$1000 – $7FFF	Unused
$8000 – $FFFF	Final Product Program

Special Single-Chip Mode

Address	Region
$0000 – $01FF	Internal Registers
$0200 – $07FF	Unused
$0800 – $0BFF	On-Chip RAM
$0C00 – $0CFF	Unused
$0D00 – $0FFF	Byte-Erasable EEPROM
$1000 – $7FFF	Unused
$8000 – $FEFF	Final Product Program
$FF00 – $FFFF	BDM ROM

Figure 10.19   Single-Chip Memory Maps for the M68HC912B32

As we saw in Section 10.3, there are two single-chip modes, normal single-chip and special single-chip. They are identical except that in the special single-chip mode, the reset vector is not fetched and the BDM is enabled and active. Figure 10.19 shows the two memory maps for these two modes. Notice that the BDM ROM is mapped at addresses $FF00–$FFFF. This allows the firmware commands to be executed.

During the debugging process, the MCU is reset in the special single-chip mode. Once reset, the CPU is not running and the BDM is waiting for a command to be sent by a host on the BKGD pin. The host can send command sequences that load a program into RAM, program an EEPROM, or execute a program using the *GO* command.

We can use this mode to develop the program until it appears to be operating to specification. At that point we can reset the MCU in normal single-chip mode. This will run the same code except that it uses the user's exception vectors.

10.5.4 A BDM-Based Debugging System

There are several BDM debugging systems available for the 68HC12 family. In this subsection we cover a system that uses the same Motorola EVB boards that we used in Chapter 3. Another BDM debugging system that uses the Noral Flex BDM debugging pod will be covered in Chapter 12. It is a more expensive system, but it is much more powerful, especially when developing programs in C.

The debugging system described here uses two Motorola M68EVB912B32 boards—one as the final product (target) and the other as a BDM pod. It is relatively inexpensive and uses the same D-Bug12 command interface as described in Chapter 3. The difference between this system and the one in Chapter 3 is that D-Bug12 is no longer resident on the target board. It runs on the POD board and uses the BDM interface to communicate to the target. In this way we get all the advantages of using the BDM system along with the familiar D-Bug12 command interface.

Hardware Configuration. Figure 10.20 shows a typical system that uses two M68EVB912B32 evaluation boards. One is configured as a BDM POD, and the other is configured as a target. The POD is connected to a PC COM port and the target's BDM connector. The POD runs the D-Bug12 command interface for the user on the PC. It then translates the D-Bug12 commands into the appropriate BDM commands and sends them to the target.

The POD EVB is configured in POD mode while the target EVB is configured in EVB mode. To configure the EVB modes, we set the mode jumpers as follows:

Mode	W3	W4
EVB	0	0
POD	0	1

A BDM cable is then connected from the POD board's *BDM OUT* connector to the target board's *BDM IN* connector.

With this configuration, when the reset button is pressed on the target board, it will reset in normal single-chip mode, and it uses the user exception vectors located from $FF00–$FFFF. When the target is reset by the POD, it is reset in the special single-chip mode and waits for BDM commands from the POD to operate.

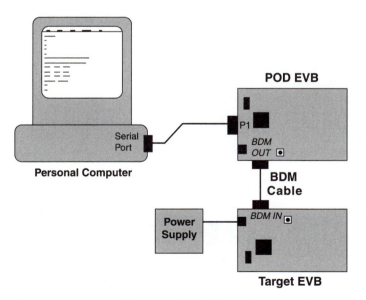

Figure 10.20 M68EVB912B32 BDM-Based Debugging System

Basic Operation. To send commands to this system, we use a terminal emulator, in this case HyperTerminal, to communicate to the POD board over a serial COM port. We used this same PC configuration in Chapter 3.

When the POD is reset, we get the following D-Bug12 prompt on the terminal window:

```
D-Bug12 v2.0.2
Copyright 1996 - 1997 Motorola Semiconductor
For Commands type "Help"

R>
```

Notice that it is the same D-Bug12 prompt as shown in Chapter 3 except there is an *R>* instead of the normal prompt, >. This indicates the status of the target board. There are two possibilities, as follows:

R>   The target is executing a program.

S>   The target is stopped.

If we type *RESET* at the prompt, the target will be reset in special single-chip mode so it will respond with a *S>* prompt. We can also use the *STOP* command to stop the CPU and show the current state of the CPU.

We can determine the characteristics of the target board MCU by typing *DEVICE* at the *S>* prompt. The *DEVICE* command will return the MCU type and the memory map of the target. For example

```
S>device

Device: 912B32
EEPROM: $0D00 - $0FFF
Flash: $8000 - $FFFF
RAM: $0800 - $0BFF
I/O Regs: $0000
```

Once the target is stopped, we can execute a program on the target by using the D-Bug12 *G* command. For example if we have a program at $8000, we can run it with the following sequence.

```
R>reset
Target Processor Has Been Reset
S>g 8000
R>
```

Notice the new prompt is *R>* because the target board is now executing a program.

When we use this system, the D-Bug12 commands that we send to the POD are actually executed as BDM commands on the target. So for example, if we run the following command:

```
S> md 0000
```

the resulting memory dump would be from the target memory, not the POD memory.

Programming EEPROM.   Now that the D-Bug12 monitor program is not resident in the target board, all memory is free to be used for the application code. We can load programs into RAM as we did in Chapter 3, or we can program one of the on-chip EEPROM memories. In general a final product's application code will be loaded into the Flash EEPROM.

To program the Flash EEPROM, we use the *FBULK* and *FLOAD* commands. We also must apply +12V to the V_{FP} pin of the MCU. To do this, +12V are connected to the target's V_{PP} node. Then jumper *W7* must be set to V_{PP}.

Note
___

Some early mask sets of the M68HC912B32 required a lower programming voltage than +12V. Make sure to check the documentation for the MCU mask set being used; including the errata sheets published for each mask set.

___

Once V_{PP} is applied to V_{FP}, we can use *FBULK* to erase the Flash EEPROM and *FLOAD* to load the machine code into the Flash EEPROM. The following command sequence erases the Flash EEPROM in the target MCU:

```
S>fbulk

S>
```

To program the flash, some additional configuration settings must be made on the terminal emulator. It takes up to 3.5ms to program each Flash EEPROM byte. If we send the S-Records to the POD too quickly, an error will occur because the programming cannot keep up. The POD does buffer each S-Record line, so we can send the characters at normal speed as long as we add a delay before sending each line. If supported by the terminal emulator, we could also wait for the asterisk (*) to be returned before sending the next line. HyperTerminal can be set up with a line delay by accessing the *Properties → Settings → ASCII Settings* dialog box. To account for the worst-case programming delay, the *Line delay* should be set to 300ms.

Once the line delay has been configured, we can load our program code using the *FLOAD* command. The *FLOAD* command works in the same way as the *LOAD* command described in Chapter 3. We type *FLOAD* at the D-Bug12 prompt, select *Transfer → Send Text File...*, and select the S-Record file to be loaded. While the Flash EEPROM is being programmed, an asterisk (*) is returned for each S-Record line. Once the S9 record at the end of the file is reached, the *S>* prompt is returned.

Following is the sequence to program the Flash EEPROM with the *demo1* program in Source 10.1:

```
S>fload
*******
S>
S>
```

Once the Flash EEPROM is programmed, the program can be executed by typing *G 8000* as shown earlier. Or to test the final product code including the exception vectors, we can press the Reset button on the target board.

Using Hardware Breakpoints. The breakpoints normally used by D-Bug12 are called *software breakpoints*. To stop the CPU at the desired instruction, a software breakpoint replaces that instruction with an *swi* instruction. Now that our program is in Flash EEPROM, however, this is not possible. To resolve this problem, some versions of the 68HC12 include BDM circuitry for hardware breakpoints. This circuitry watches the address bus, and when it is equal to the breakpoint address, it tags the instruction so the CPU will stop before executing it.

To tell D-Bug12 to use hardware breakpoints, we use the *USEHBR* command as follows:

```
S>usehbr
Using Hardware Breakpoints
S>
```

From this point on hardware breakpoints will be used when a *BR* command is executed. Note that on the M68HC912B32, we are limited to two hardware breakpoints.

Final Product Debugging Summary. Other than the differences we have described here, all the other D-Bug12 operations work the same as described in Chapter 3. Remember that the *RESET* command from the POD resets the target in special single-chip mode, while a reset of the target itself resets it into normal single-chip mode. Also remember that we can access memory while the target is running, but we have to stop the target to access registers. Accessing memory while the target is running allows us to watch variables in real time.

► SUMMARY

In this chapter we covered the basic requirements for developing a final product. The design for microcontrollers running in single-chip mode was emphasized here.

We looked at the MCU core hardware design, which included the power supply, the clock circuitry, and the reset circuitry. The core design can be very simple. It can be complex, however, if we are designing a battery-operated system or a system with significant fault tolerance.

We also looked at how resets and the mode select pins are used to start the MCU at a known state. Resets are used for power up, user resets, or for fault tolerance such as low-voltage detection, the clock monitor, and the COP timer.

Then we covered the additional code required for a program that must stand alone in a final product. This included stack pointer initialization, write-once registers initialization, MCU configuration and code initialization, and the exception vectors. This additional code is not large or complex; however, we need to be very careful and not overlook the details. This code is the source of most errors that occur when the transition is made from development code to a final product.

Finally we looked at the development process for programming and debugging the code on a final product. The system is complicated by the limited amount of memory available and the fact that the program for a final product must be programmed into a ROM device.

In addition a monitor program becomes very invasive in a final product because it requires a significant amount of memory and it must be executed out of reset. We looked at one system that used a dual-boot method for developing a final product while also using a monitor program.

We then looked at a system that combines the modern technologies of Flash EEPROM and background debugging circuitry. The system described uses two M68EVB92B32 evaluation boards—one as a POD device and the other as the final product. With this system, the transition from development to the final product was trivial.

EXERCISES

1. A new +3V version of a +5V microcontroller is introduced. If the supply current for both parts is the same, find the power savings due to the change in supply voltages.

2. A M68HC912B32 running with an E-clock frequency of 8MHz consumes an average of 225mW. An engineer designing a portable application has determined that the power dissipation of the MCU must be reduced to 100mW in order to have an acceptable battery life.
   (a) If the power dissipation is reduced by only lowering the E-clock frequency, find the required E-clock frequency.
   (b) If the power dissipation is reduced by periodically running in the *STOP* mode, find the percentage of time that the MCU must be in the *STOP* mode.

3. An embedded product includes a kitchen timer that must have an accuracy of ±5minutes every 6 hours. Can the designer get away with using the ceramic resonator that is described in Section 10.1? Assume that the frequency error of the ceramic resonator is the only error in the system.

4. Determine the minimum reset pulse-width for a M68HC912B32 that is running with an E-clock frequency of 2MHz. The MCU must be able to distinguish between an internal and external reset.

5. What is the time delay between the power supply turning on and the first memory read (reset vector fetch)? Assume a M68HC912B32 MCU running with an E-clock frequency of 8MHz and the reset circuit shown in Figure 10.8 on page 391.

6. Write a two-instruction sequence that resets the COP timer.

7. Write a code snippet that configures the COP timer with a reset time of more than one second. Assume an 8MHz E-clock frequency.

8. Calculate R_1 and R_2 in the low-voltage reset circuit shown in Figure 10.10 (on page 393) for $V_{DD, min}$ = 4.75V and V_{REF} = 2.5V.

9. Repeat Exercise 8 with $V_{DD, min}$ = 2.85V and V_{REF} = 1.2V.

10. Show the three *org* directives that are required for final product code designed for the memory map in Example 10.1 on page 398.

11. Show the stack pointer initialization required for the memory map in Example 10.1.

System Expansion

In this chapter we look at the basic design process for adding external resources to the microprocessor bus system. Adding resources to the microprocessor bus is one of the reasons for the flexibility of a microcomputer system. A CPU system can be customized by adding only the required resources for the application on the bus. Before microcontrollers were produced with a large number of variants, microprocessor-based systems were more flexible than using a microcontroller. Now, however, technology allows more resources on a single-chip microcontroller, as well as a variety of configuration choices, and important new resources like Flash EEPROM and background debug circuitry.

Despite these new technologies, there are still times when we can provide the required resources only by adding them to the bus system. To do this we need to use a CPU, or more likely a microcontroller running in expanded mode.

Some microcontrollers are designed for bus expansion, whereas others such as the 68HC912B32 are specifically designed to run in single-chip mode. The 68HC912B32 can run in expanded mode, but it is limited to a multiplexed bus system with only a 16-bit address, and it contains no internal chip-select logic. In addition, because of the cost involved in the on-chip Flash EEPROM, it is more expensive than other MCUs. Contrast this to the 68HC812A4 microcontroller, which has a nonmultiplexed bus with a 22-bit address that can be used to address up to 4M-bytes of address space. It also includes seven on-chip programmable chip-selects, which makes it possible for most designs to require no external parts beyond the peripherals themselves.

We cover only the basic design tasks involved in using a 64K-byte bus system. Most of the material is general enough to apply to any system, but some is specific to the

418

68HC812A4. These design tasks include chip-select logic, read/write logic, and bus timing analysis. There are other complications if we are using a multiplexed bus or if we are using an 8-bit peripheral with a 16-bit data bus, but we do not cover these problems here.

▶ 11.1 THE BUS CYCLE

In Chapter 1 we saw that the bus system is made up of an address bus, a data bus, and bus control signals.

The address bus is used to select the device and the specific location within the device to be accessed. The 68HC812A4 has a 22-bit address bus. In this text, however, we only look at the design of a 64K-byte, nonpaged system, so are concerned only with 16 address bits.

The data bus contains the data transferred between the CPU to the addressed location. The 68HC12 family has an internal 16-bit data bus. Externally the data bus is 8 bits wide if the MCU is in expanded narrow mode and 16 bits if the MCU is in the expanded wide mode. Expanded narrow mode is used for low-cost systems that only use 8-bit peripherals. When a 16-bit read or write is required, however, the MCU needs to use two bus cycles instead of one. Expanded wide mode allows for faster execution time because a 16-bit access requires only one bus cycle unless the word is not aligned.

The bus control signals are used for bus direction, bus timing, byte selection, demultiplexing, and bus access. We focus on the two primary bus control signals in the 68HC11 and 68HC12 families—$R/\overline{W}$ and E. $R/\overline{W}$ is used to control the data bus direction and the E-clock or *(E)* is used for bus timing.

11.1.1 The 68HC12 Read Cycle

Let's first look at a functional description of the 68HC12 read cycle, which is functionally the same as the 68HC11 read cycle and is shown in Figure 11.1. Recall that during a read cycle, the data are coming from the external device to the MCU.

The first step, **A,** is the start of a bus cycle. The falling edge of the E-clock always starts the bus cycle. As we will soon see, the start of one bus cycle is the end of the last cycle.

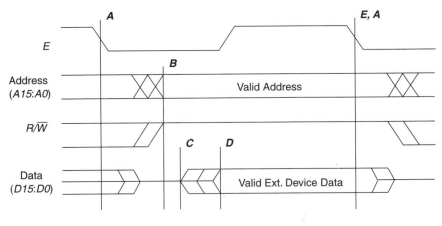

Figure 11.1 68HC12 Read Cycle

After some delay, at step **B,** the MCU places the address on the address bus and sets $R/\overline{W}$ high. This tells the external circuitry what location to access and that the cycle is a read cycle.

Once the address and $R/\overline{W}$ are valid, the external circuitry starts to react. First at **C** the chip-select logic enables the selected external device. At this point, the device may put data on the bus. The data are not guaranteed at this point, but we need to make sure data from another device have been removed, or there may be a bus collision.

After a delay called the device's *access time,* at **D** the external device is guaranteed to have valid data on the bus.

Finally it is up to the MCU to latch the data while they are valid. On the 68HC11 and 68HC12 processors, this happens on the falling edge of the E-clock, **E.** Notice that this is also the start of the next bus cycle. We see later that the data must meet the setup and hold-time requirements of the MCU to be reliably read.

11.1.2 The 68HC12 Write Cycle

Now let's look at a functional description of the 68HC12 write cycle, shown in Figure 11.2. It is also the same as the 68HC11. Recall that during a write cycle, the data are coming from the MCU to the selected external device.

The first step, **A,** is the start of a bus cycle, which again is the falling edge of the E-clock.

After some delay, at step **B** the MCU places the address on the address bus and resets $R/\overline{W}$ low. This tells the external circuitry what location to access and that the cycle is a write cycle.

After **B,** the external chip-select logic selects the addressed external device. After the rising edge of the E-clock, the 68HC12 places the write data on the bus, **C.**

Now that data are on the bus, the external device can latch them. The time at which the data are latched depends on the external circuit design. Typically the latch occurs when

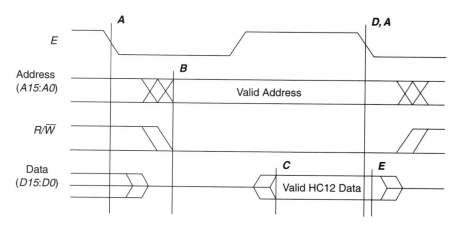

Figure 11.2 68HC12 Write Cycle

the external device is deselected, which is usually based on the falling edge of the E-clock, *D*. Figure 11.2 shows the data being latched at *E*, after a small delay from *D*.

It is important to understand the functional sequence for these two bus cycles before embarking on the more detailed design tasks.

▶ 11.2 CHIP-SELECT LOGIC

Chip-select logic is used to select an external device connected to the bus. The chip-select logic selects the device by decoding the address. If the address is within the block defined for the device, the select line for that device is activated. At that point the device decodes the rest of the address to select the specific location within the device to be accessed.

The most important rule when designing chip-select logic is that only one device can be selected at a time. If more than one device is selected, there will be two outputs connected to each other, which can result in a short circuit. This is often called *bus contention* or a *bus collision*.

Chip-select logic essentially divides the memory map into blocks. Since there can be only one device selected at a time, all devices must have a limited block size. The chip-select logic determines a system's memory map.

Chip-select logic can be implemented as external logic circuits or as on-chip programmable chip selects. First let's go through the steps used to design external chip-select logic by designing an example system. Table 11.1 shows the requirements for the system that we will design. Most modern systems would use much larger devices and consequently would be much easier to design.

11.2.1 Determine Device Block Locations

The first step is to determine the desired locations for each device block. Two main considerations must be made when selecting block locations—MCU mapping requirements and defining device block boundaries by the most-significant bits of the address. In addition to these considerations, we should try to keep large contiguous blocks together so that if we need to expand a memory device, we can use two adjacent blocks.

MCU Mapping Requirements. The microcontroller makes some assumptions about the memory map that must be considered when we place our devices. The most important assumption made by the 68HC812A4 is that the reset and interrupt vectors are located in the block $FF00–$FFFF. This means that our program ROM must include this block.

TABLE 11.1 EXAMPLE SYSTEM REQUIREMENTS

Device	Device Size
Program EPROM:	8 K-bytes
RAM:	2 K-bytes
Modem:	8 bytes

The locations of the on-chip resources must also be considered. As we saw in the last chapter, the on-chip resources for the 68HC12 can be moved, but it would be at a cost of compatibility. So we want to keep the on-chip resources where they are at reset. The on-chip resources for the 68HC812A4 are as follows:

Registers: $0000–$01FF
On-chip RAM: $0800–$0BFF
EEPROM: $1000–$1FFF

The rest of the memory map is free to use for our external devices.

Device Block Boundaries.   When placing our device blocks, we cannot just put them anywhere that they fit. Haphazard placement could mean complex address decoding logic. Device blocks must be placed at a location that is determined by the most-significant bits of the address.

To see how this works, refer to Figure 11.3. The figure shows three memory maps divided up into equal-sized blocks. The first map is divided into two 32K-byte blocks. The first block is defined by $A15 = 0$ and the second block is defined by $A15 = 1$. That is, the most-significant bit of the address defines the block boundaries. This means a 32K-byte device can only be placed in $0000–$7FFF or $8000–$FFFF. We should never try to place a 32K-byte device in $2000–$A000, for example.

The second memory map shows the four legal 16K-byte blocks. These four blocks are defined by the two most-significant addresses. Finally, the third map shows the eight

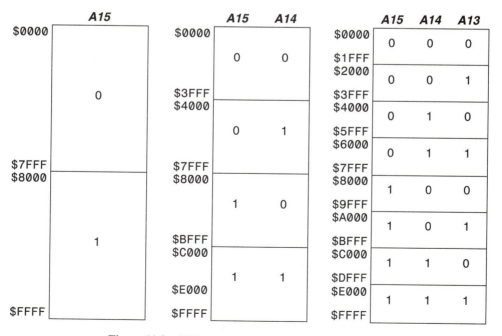

Figure 11.3   32K-, 16K-, and 8K-byte Block Boundaries

422

TABLE 11.2 EXAMPLE SYSTEM DEVICE STARTING ADDRESSES

Device	Device Size	Starting Address
Program EPROM:	8K-bytes	$E000
RAM:	2K-bytes	$2000
Modem:	8 bytes	$0400

legal 8K-byte blocks defined by the three most-significant addresses. It is not hard to imagine how we can continue to divide the memory map into smaller blocks.

Taking into account the MCU requirements and the block boundaries, we will use the starting addresses shown in Table 11.2.

The program EPROM had to be located in the block $E000–$FFFF because it had to include the reset and interrupt vectors. The RAM was placed at $2000 so it would be as far from the ROM as possible. In this way we can expand to a larger RAM or ROM, and they would still be contiguous blocks.

11.2.2 Chip-Select Logic Equations, Full Decoding

The next step is to determine the address decoding logic equations for each device. This is done by determining the bit combinations for the most-significant address bits that define the device block.

EPROM Address Decoding Logic. The EPROM in our example system requires an 8K-byte block. We have already determined that we would use the $E000–$FFFF block. An 8K-byte EPROM has 13 address lines, which leaves three address lines to be decoded. As discussed in the last subsection, the decoding logic will use the three most-significant addresses, *A15–A13*. So the address decoding is broken down as follows:

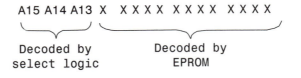

If we want the EPROM to include $FFFF, then the bit combinations of the addresses decoded by the select logic is

$$A15 = A14 = A13 = 1$$

If we assume an active-low chip select, then the address decoding equation is

$$/CS_{EPROM} = (A15 \cdot A14 \cdot A13)'$$

This could be implemented with a single three-input NAND gate.

To determine the starting and ending addresses of the block, we look at the address range in which the most-significant bits select the device. For the first address in the block, all bits decoded by the EPROM are zero. In the last address in the block, all bits decoded by the EPROM are one. As shown in the following bit patterns, this corresponds to the block \$E000–\$FFFF:

```
1 1 1│0  0 0 0 0  0 0 0 0  0 0 0 0  -> $E000
1 1 1│1  1 1 1 1  1 1 1 1  1 1 1 1  -> $FFFF
```

RAM Address Decoding Logic.   The RAM in our example system requires a 2K-byte block. We have already determined that we would start the RAM at \$2000. A 2K-byte RAM has 11 address lines, which leaves five address lines to be decoded. Again the decoding logic will use the five most-significant addresses, *A15–A11*. So the address decoding is broken down as follows:

If we want the RAM to start at \$2000, then the bit combinations of the addresses decoded by the select logic are

$$A15 = 0, A14 = 0, A13 = 1, A12 = 0, A11 = 0$$

If we assume an active-low chip select, then the address decoding equation is

$$/CS_{RAM} = (A15' \cdot A14' \cdot A13 \cdot A12' \cdot A11')'$$

This is starting to be a little messy to implement with discrete logic. PLDs are popular for this application because they provide a single-chip solution and have fast propagation delays, which are important for decoding logic.

To determine the starting and ending addresses of the block, we look at the address range in which the most-significant bits select the device. As shown in the following bit patterns, this corresponds to the block \$2000–\$27FF:

```
0 0 1 0  0│0 0 0  0 0 0 0  0 0 0 0  -> $2000
0 0 1 0  0│1 1 1  1 1 1 1  1 1 1 1  -> $27FF
```

Modem Address Decoding Logic.   The modem in our example system requires an 8-byte block. We have already determined that we would start the modem at \$0400. An 8-byte modem has three address lines, which leaves 13 address lines to be decoded. The decoding logic will use the 13 most-significant addresses, *A15–A3*. So the address decoding is broken down as follows:

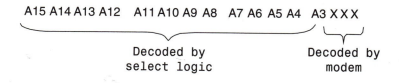

$0000 $01FF	Internal Registers
$0200 $03FF	Unused
$0400 $0407	Modem
$0408 $07FF	Unused
$0800 $0BFF	On-Chip RAM
$0C00 $0CFF	Unused
$0D00 $0FFF	Byte-Erasable EEPROM
$1000 $1FFF	Unused
$2000 $27FF	RAM
$2800 $DFFF	Unused
$E000 $FFFF	EPROM

Figure 11.4 Memory Map for Example System Using Full Decoding

If we want the modem to start at $0400, then the address decoding equation is

$$/CS_{modem} = (A15' \cdot A14' \cdot A13' \cdot A12' \cdot A11' \cdot A10 \cdot A9' \cdot A8' \cdot A7' \cdot A6' \cdot A5' \cdot A4' \cdot A3')'$$

This ridiculous equation is not even done cheaply with a PLD. You would need a PLD with 13 inputs. The address range for this device $0400–$0407:

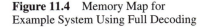

```
0 0 0 0   0 1 0 0   0 0 0 0   0|0 0 0  -> $0400
0 0 0 0   0 1 0 0   0 0 0 0   0|1 1 1  -> $0407
```

This is an example in which a small device ends up requiring a large decoding logic equation. The process we used in this section is called *full address decoding.* Full address decoding requires the memory block size to be the same as the device block size. Figure 11.4 shows the resulting memory map.

11.2.3 Chip-Select Logic Equations, Partial Decoding

Notice that the logic equations in the last subsection got more complex as the device got smaller. This is because there were more address lines to decode. We can trade logic complexity, which represents a per-unit hardware cost, for address space. In this case we have plenty of address space remaining, so this would make sense. To make this trade, we make the memory block larger than the device block. This is called *partial decoding.*

There is plenty of spare memory space in our example system, so let's simplify the logic by using partial decoding. The largest device in this case is the 8K-byte EPROM, so we will partial decode the RAM and modem into 8K-byte blocks.

First the modem will not fit at address $0400 with an 8K-byte block, so we have to move it. In this case we will move it to the $6000–$7FFF block. The RAM and EPROM will remain at their current starting addresses. This results in the following address decoding equations:

$$/CS_{EPROM} = (A15 \cdot A14 \cdot A13)'$$

$$/CS_{RAM} = (A15' \cdot A14' \cdot A13)'$$

$$/CS_{modem} = (A15' \cdot A14 \cdot A13)'$$

This is a perfect application for the ever-popular 74HC138, 3-to-8 decoder. If we connect the decoder as shown in Figure 11.5, each output represents an 8K-byte block. And, since the outputs are active low, no other decoding logic is required. This circuit also provides room for growth. We could add additional 8K-byte memory devices without adding any more decoding logic.

The new memory map is shown in Figure 11.6. This map, however, is a bit deceptive. By looking at the map, it appears that we have an 8K-byte RAM device and modem. It is true that these devices require that much space in the memory map, but we did not just increase the size of our RAM by changing the decoding. It is still a 2K-byte RAM device.

Let's look at the decoding for the RAM block and RAM device in more detail to see what happens. Here is how the bits are decoded:

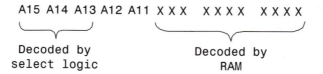

$A15$, $A14$, and $A13$ are decoded by the chip-select logic and the RAM decodes address $A10$–$A0$. So, that leaves addresses $A12$ and $A11$. These two bits are not decoded by anything so they are ignored. Now let's run two write instructions to this space:

```
staa $2001
staa $3801
```

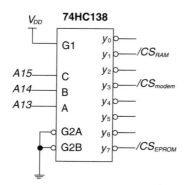

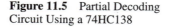

Figure 11.5 Partial Decoding Circuit Using a 74HC138

Address	Region
$0000 – $01FF	Internal Registers
$0200 – $07FF	Unused
$0800 – $0BFF	On-Chip RAM
$0C00 – $0CFF	Unused
$0D00 – $0FFF	Byte-Erasable EEPROM
$1000 – $1FFF	Unused
$2000 – $3FFF	RAM
$4000 – $5FFF	Unused
$6000 – $7FFF	Modem
$8000 – $DFFF	Unused
$E000 – $FFFF	EPROM

Figure 11.6 Memory Map for Example System Using Partial Decoding

The first instruction writes to address $2001, which results in the following decoding:

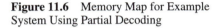

```
0 0 1  0  0 0 0 0  0 0 0 0   0 0 0 1
 └─┬─┘           └──────┬──────┘
  RAM                  RAM
selected          decodes to
                     $001
```

Because *A15, A14,* and *A13* have the correct bit pattern, the RAM is selected. Then the RAM decodes address *A10–A0* and stores the data in RAM address $001.

Now the second instruction writes to address $3801, which is still located in the 8K-byte RAM block. The decoding is

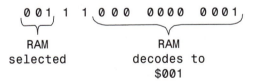

```
0 0 1  1  1 0 0 0  0 0 0 0  0 0 0 1
 └─┬─┘           └──────┬──────┘
  RAM                  RAM
selected          decodes to
                     $001
```

The RAM was selected again, and when the RAM decoded *A10–A0,* it stored the data into RAM address $001 again. This means that we just wrote on top of the previous data that were stored in address $2001.

Because *A12* and *A11* are not decoded, there are four 2K-byte blocks, or images, in the 8K-byte block. Each 2K-byte block is identical to the other and represents the contents of the RAM. So we need to document this by showing one 2K-byte RAM block and three images. The memory map should look like this:

$2000 $27FF	RAM
$2800 $2FFF	RAM Image
$3000 $37FF	RAM Image
$3800 $3FFF	RAM Image

Trading memory space for simplified decoding logic is a common practice. One famous PC manufacturer had two serial ports that consisted of four bytes, mapped to a 2M-byte space. Fortunately more microcontrollers and CPUs are being designed today with on-chip programmable chip-select logic, which can eliminate the need for any external circuitry used for address decoding.

11.2.4 68HC812A4 Programmable Chip Selects

In this subsection we implement the example system described in the last subsection with the 68HC812A4 programmable chip selects. We also go over the start-up code required for the normal expanded narrow mode. The narrow mode is used because the external devices in this case are all 8-bit devices.

The Chip Selects. There are seven programmable chip selects, which use *PORTF* on the 68HC812A4. The pins on *PORTF* that are not used as chip selects can be used for GPIO. The registers that are used to configure the chip selects are shown in Figure 11.7. There are three types of chips selects: peripheral chip selects, data chip selects, and program chip selects.

The peripheral chip selects include *CS0, CS1, CS2,* and *CS3.* They are used for devices with small device blocks such as peripheral I/O devices and are mapped as shown in Figure 11.8. *CS0, CS1,* and *CS2* always follow the register block, whereas *CS3* can be configured to follow the register block or not. This means that if we move the registers, these chip-select blocks will also move.

There is one data chip select, *CSD.* It can also be referred to as a general purpose chip select because it can be used for either data or additional program space. Because it is not enabled out of reset, however, it cannot be used as the only program chip select.

CSD can be mapped for the 32K-byte block $0000–$7FFF or to the 4K-byte block $7000–$7FFF, depending on the *CSDHF* bit in *CSCTL1.* If *CSDHF* is zero, *CSD* is mapped to $7000–$7FFF.

Notice that there appears to be a conflict between the *CSD* chip-select space and several other devices. When we were designing our own chip-select logic, this meant that more than one device would be selected at one time. When using the internal chip selects or internal resources, the MCU makes sure only one device is selected by assigning each device a priority. If the space includes two devices, the device with the highest priority will

PEAR:

($000A)	ARSIE	CDLTE	PIPOE	NECLK	LSTRE	RDWE	0	0
RESET:	0	0	0	0	0	0	0	0

MISC:

($0013)	EWDIR	NDRC	0	0	0	0	0	0
RESET:	0	0	0	0	0	0	0	0

PORTF:

($0030)	0	CSP1	CSP0	CSD	CS3	CS2	CS1	CS0
RESET:	0	0	–	0	0	0	0	0

CSCTL0:

($003C)	0	CSPIE	CSP0E	CSDE	CS3E	CS2E	CS1E	CS0E
RESET:	0	0	1	0	0	0	0	0

CSCTL1:

($003D)	0	CSP1FL	CSPA21	CSDHF	CS3EP	0	0	0
RESET:	0	0	0	0	0	0	0	0

CSSTR0:

($003E)	0	0	SRP1A	SRP1B	SRP0A	SRP0B	STRDA	STRDB
RESET:	0	0	1	1	1	1	1	1

CSSTR1:

($003F)	STR3A	STR3B	STR2A	STR2B	STR1A	STR1B	STR0A	STR0B
RESET:	1	1	1	1	1	1	1	1

Figure 11.7   Chip-Select-Related Control Registers

$0000 $01FF	Registers
$0200 $02FF	CS0 (256 bytes)
$0300 $037F	CS1 (128 bytes)
$0380 $03FF	CS2 (128 bytes)
$0400 $07FF	CS3 (1K-bytes)
$0800 $0BFF	On-Chip RAM

Figure 11.8   Peripheral Chip-Select Memory Blocks

be selected. In effect what this does to the memory map is put the higher-priority device on top of the conflicting device. The priorities for the 68HC812A4 are as follows:

	Device/Chip Select
Highest Priority	Registers
	BDM
	On-Chip RAM
	On-Chip EEPROM
	CS3
	CS2
	CS1
	CS0
	CSP0
	CSD
	CSP1
Lowest Priority	Other External Devices

So if we had *CSD* configured for $0000–$7FFF, it would result in the memory map shown in Figure 11.9. Notice that the *CSD* chip select fills in any holes between the registers, the peripheral chip selects, the on-chip RAM, and the on-chip EEPROM. If any of the peripheral chip selects were not enabled, the *CSD* chip select would also fill that space.

There are two program chip selects, *CSP1* and *CSP0*. *CSP0* maps to the 32K-byte block $8000–$FFFF. It is the only chip select that is enabled out of reset, which is required so it can be used to access the reset vectors and run the start-up code. Therefore *CSP0* is always used for the program device that contains the reset and interrupt vectors. *CSP1* is an extra program chip select that can be used to fill the remainder of the memory map. The configuration for *CSP1* is controlled by the *CSCTL1* register.

Address	Block
$0000 – $01FF	Registers
$0200 – $02FF	CS0 (256 bytes)
$0300 – $037F	CS1 (128 bytes)
$0380 – $03FF	CS2 (128 bytes)
$0400 – $07FF	CS3 (1K-bytes)
$0800 – $0BFF	On-Chip RAM
$0C00 – $0FFF	CSD
$1000 – $1FFF	On-Chip EEPROM
$2000 – $7FFF	CSD

Figure 11.9 Memory Blocks with CSD Mapped to $0000–$7FFF

The chip selects are enabled by setting the appropriate bit in the *CSCTL0* register. When a chip select is connected to a slow external device, it can be configured to cause the bus cycles that access that device to be stretched. The *CSSTR0* and *CSSTR1* registers set the number of additional E-clock cycles that are added for each chip select. Out of reset, all the stretch settings are set to the maximum value of 3. So if these registers are not changed, every external access will take four E-clock cycles instead of one.

Example System.   To demonstrate the use of the programmable chip selects, we will re-design our example system. The following descriptions are given for each device. The clock stretch values are determined by analyzing the bus timing as described in the next section.

Device	Description
8K-byte EPROM	Program ROM, active-low enable, clock stretch: 2
2K-byte RAM	Active-low enable, clock stretch:1
8-byte Modem	Active-low enable, clock stretch: 1

Given these characteristics we can assign the devices to the chip selects as follows:

Device	Chip Select, Address Range
8K-byte EPROM	*CSP0*, $8000–$FFFF
2K-byte RAM	*CSD,* $7000–$7FFF
8-byte Modem	*CS1,* $0300–$037F

CSP0 was selected for the EPROM because it is the program memory. *CSD* was selected for the 2K-byte RAM because it is the only chip select that is not a program chip select and maps to a large enough block size. We also could have used *CSP1*, but that should be reserved for additional program memory that may be added in the future. *CS1* is used for the modem because it has one of the smallest block sizes. We could have also used *CS2*. The resulting memory map for this system is shown in Figure 11.10.

Now we need to determine the register settings that are required to configure these chip selects correctly. Following are the required register settings:

Register	Required Setting	Description
PEAR	%00000100	Enable $R/\overline{W}$
MISC	%01000000	*CS1* controls an 8-bit device
CSCTL0	%00110010	Enable *CSP0, CSD,* and *CS1*
CSCTL1	%00000000	*CSD* set to $7000–$7FFF
CSSTR0	%00111001	Stretch *CSP0* two cycles, *CSD* one cycle
CSSTR1	%11110111	Stretch *CS1* one cycle

All other registers that relate to the chip selects and memory expansion must remain at their reset state. This means all memory expansion windows are disabled, so we only have a 64K-byte memory map. It also means that the chip select pins on *PORTF* have active pull-ups. This is required because all chip selects except *CSP0* are disable out of reset, which means that they are general purpose inputs and can float to an unknown level. This can inadvertently enable a device, which in turn may place data on the bus.

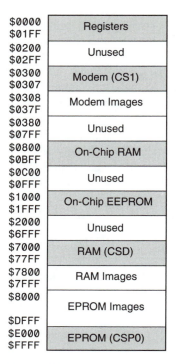

$0000 $01FF	Registers
$0200 $02FF	Unused
$0300 $0307	Modem (CS1)
$0308 $037F	Modem Images
$0380 $07FF	Unused
$0800 $0BFF	On-Chip RAM
$0C00 $0FFF	Unused
$1000 $1FFF	On-Chip EEPROM
$2000 $6FFF	Unused
$7000 $77FF	RAM (CSD)
$7800 $7FFF	RAM Images
$8000 $DFFF	EPROM Images
$E000 $FFFF	EPROM (CSP0)

Figure 11.10   Memory Map Using Programmable Chip Selects

Start-Up Code.   Source 11.1 shows a routine that performs all the start-up code for the MCU configuration, including the chip select configuration. As shown in the last chapter, the routine should be called immediately after reset and stack pointer initialization. Notice that the chip-select configuration is completed before the chip selects are enabled.

Source 11.1 Start-Up Code for Chip-Select Example

```
;************************************************************************
; MCU Write-once Register and Chip Select Initialization
;************************************************************************
; Required initialization for all M68HC812A4 stand-alone systems
;    - COP must either be turned off or reset.

MCUInit           movb #$00,COPCTL     ;Turn COP off, clock monitor off


;***************************************************************
; Recommended initialization for all stand-alone systems

                  movb #$B0,MODE       ;Protect from mode change
                  movb #$40,MISC       ;CS1 connected to 8-bit device
                  movb #$00,INITRG      ;Protect regs from move
                  movb #$08,INITRM      ;Protect RAM from move
                  movb #$11,INITEE      ;Protect EEPROM from move
```

```
                movb #$FE,EEMCR         ;Lock EEPROM Protection bits

;*************************************************************
; Optional initialization for some stand-alone systems

                movb #$04,PEAR          ;Enable R/W pin
                movb #$60,INTCR         ;Prot. IRQ mode, stop delay
                movb #$00,RTICTL        ;COP, RTI run in wait, bkgd

;*************************************************************
; Chip-Select Initialization

                movb #$39,CSSTR0        ;Stretch CSP0 2, CSD one cycle
                movb #$F7,CSSTR1        ;Stretch CS1 one cycle
                movb #$32,CSCTR0        ;Enable CSP0, CS1, CSD
                rts

;*****************************************************************
```

The functional logic diagram for this system is shown in Figure 11.11. This logic diagram only includes the chip-select circuitry and the bus system.

The memories used in this example are small by today's standards. In fact they would probably be difficult to find. They were used to illustrate the process for designing chip-select logic. If we look at the memory map in Figure 11.10, we can see that there is plenty of space used by images of the memory devices. We could easily replace these devices with larger parts. The most common solution would include a 32K-byte RAM mapped by *CSD* from $0000–$7FFF. Because $0000–$1FFF is used by other devices, only 24K-bytes of the RAM are in a contiguous block. The EPROM would also be replaced with a 32K-byte part. This would result in a 32K-byte block from $8000–$FFFF.

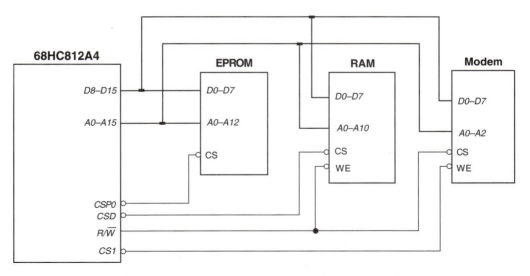

Figure 11.11 Chip-Select Circuitry for Example System

Read-Write Logic. The last signal to consider when selecting a device is the $R/\overline{W}$ signal. If the memory or peripheral is writable, there is normally a write-enable pin called /WE, which should be connected directly to $R/\overline{W}$. If the pin is high, the part knows that it is a read cycle and places data onto the bus. If the pin is low, the part knows it is a write cycle and will latch the data on the bus.

▶ 11.3 BUS TIMING ANALYSIS

If there is one design task that is viewed upon with fear, it is bus timing analysis. It is essential if we are to ensure that our systems will run reliably, yet it is often skipped or glossed over by many "engineers."

The timing analysis for a design cannot be accomplished by taking measurements with an oscilloscope or a logic analyzer. Timing characteristics for all parts depend on temperature, IC fabrication lot, load, and supply voltage. Measurements taken in a lab only give the timing for those parts, in that environment. The system may act completely different in the field.

In fact timing errors are one of the most nefarious bugs that a system can have. Timing errors normally produce soft failures that are difficult to reproduce. Typically they show up in the field and then disappear in the shop. Or they occur consistently until you place logic probes on the bus to see what is causing the problem. The load caused by the logic probes can make the problem go away.

To keep these bugs from occurring, we need to analyze the worst-case timing parameters for the bus interface, especially when the design includes parts that were not designed for the same type of bus system.

To illustrate the process to verify the primary timing parameters, let's look at a simple example of a 6264 RAM chip connected to the *CSD* chip select. The RAM is connected to the 68HC812A4 as shown in Figure 11.11.

The values shown in this section are from Motorola's *Electrical Characteristics* for the M68HC812A4 and from Hitachi's data sheet for the HM6264LP-10 static RAM. It would be a good idea to have this data available as a reference while reading this section.

11.3.1 Read-Cycle Timing

When analyzing the read-cycle timing, we need to make sure that the microcontroller setup and hold-time requirements are met. Figure 11.12 shows the applicable timing parameters for the read cycle.

Read Data Setup Time. First, we must verify that the data setup time is met. The data setup time is the time before the falling edge of the E-clock that the device data must be valid. This is labeled ⑪ in the timing diagram. The requirement for the 68HC812A4 with an 8MHz E-clock is that the setup time be greater than 30ns.

To verify this we need to understand what causes the data to become valid in the first place. For this example the RAM will place the correct data on the bus after it has the correct address, after /WE is high, and after the /CS chip select goes low. This represents three timing paths, each of which can cause the data to be delayed. The setup time requirement must be met for all three of these paths.

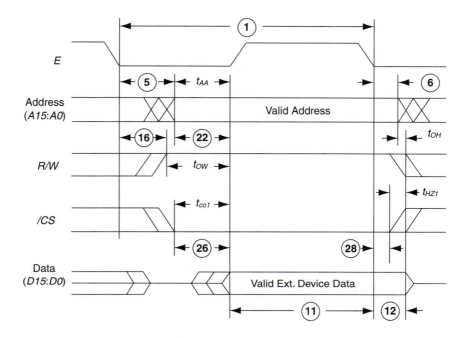

Figure 11.12 68HC12 Read Cycle with Timing Parameters

Address to data delay path: $t_{AA, max}$ must be $< \textcircled{22}_{max}$

$$t_{AA, max} = 100ns \not< 35ns$$

This does not meet the requirement. So we need to add clock stretching to the chip select for this part. If we add one clock cycle, $\textcircled{22}_{max}$ becomes 160ns. So the setup time for this path can only be met if we add one clock stretch cycle to the chip select.

/WE to data path: $\textcircled{1}_{min} - \textcircled{16}_{max} - t_{OW, max}$ must be greater than $\textcircled{11}_{min}$.

$$125ns - 49ns - 5ns = 71ns > 30ns$$

This requirement is met, even without clock stretching.

Chip select to data path: $t_{col, max}$ must be $< \textcircled{26}_{max}$.

$$100ns \not> 60ns$$

Again this path violates the data setup time if we do not have clock stretching. If we add the one cycle of clock stretching, $\textcircled{26}_{max}$ is 185ns. So the setup time requirement for this path will only be met if we stretch the chip select by one clock cycle.

Read Data Hold Time. The next read cycle parameter that we need to verify is the data hold time. The data hold time is the length of time after E-clock falling edge that device data must be held. The RAM will remove the data from the bus if either the address changes or the device is deselected. This represents two paths that we have to verify.

Address hold time path: $\textcircled{6}_{min} + t_{OH, min}$ must be $> \textcircled{12}_{min}$

$$20\text{ns} + 10\text{ns} > 0\text{ns}$$

This requirement is met because the address is held by both the MCU and the RAM.
Chip deselect to high-Z: $\textcircled{28}_{min} + t_{HZ1, min}$ must be $> \textcircled{12}_{min}$.

$$0\text{ns} + 0\text{ns} > 0\text{ns}$$

In this case there is no hold time for the chip select or the RAM. Is this okay? The best way to decide is to consider how this process works. The chip select goes high at the same time as the E-clock falling edge. Then because the chip select went high, the RAM outputs were disabled. Because the chip-select signal caused the outputs to be disabled, there is no way that it can happen in zero time. There must be some delay before the outputs go to the high-Z state. Therefore we can consider that the data are held longer than 0ns.

Overall the data hold time and the data setup time can be met as long as we stretch the chip select one cycle to meet the data setup time requirement.

11.3.2 Write-Cycle Timing

The parameters we will verify for the write cycle are the data setup time, the data hold time, and the address hold time. Figure 11.13 shows the relevant timing parameters for the write cycle.

Write-Cycle Latch Time.   The first step in determining the write-cycle timing is to find the time when the RAM will latch the data. All the parameters are based on this time.

The 6264 RAM write state is defined as $/CS1' /WE' CS2$. So the write state ends and the data are latched at the first occurrence of $/CS1 = 1$, $/WE = 1$, or $CS2 = 0$. Because $\textcircled{28}$ is always less than $\textcircled{18}$, $/CS1 = 1$ will always occur sooner than $/WE = 1$. Therefore the data are latched at the rising edge of $/CS1$, which is labeled /CS in Figure 11.13.

Write Data Setup Time.   The data setup time for the write cycle is the time that the data must be valid before the external device latches it. In Figure 11.13 this is shown as t_{DW}. The requirement is that t_{DW} must be greater than 40ns. This means $\textcircled{15}_{min} + \textcircled{28}_{min}$ must be greater than 40ns.

$$30\text{ns} + 0\text{ns} \not> 40\text{ns}$$

This shows that the setup time is violated. But if we add the clock stretch that was required for the read cycle, $\textcircled{15}_{min}$ becomes 155ns and the requirement is easily met.

Write Data Hold Time.   This is the time after RAM's latch time that the MCU data must be held. It is shown as t_{DH} in the timing diagram. The requirement is that t_{DH} is greater than 0ns. This means that $\textcircled{14}_{min} - \textcircled{28}_{max}$ must be greater than 0ns.

$$30\text{ns} - 10\text{ns} > 0\text{ns}$$

This requirement has been met because of the MCU chip-select timing relative to the write data timing.

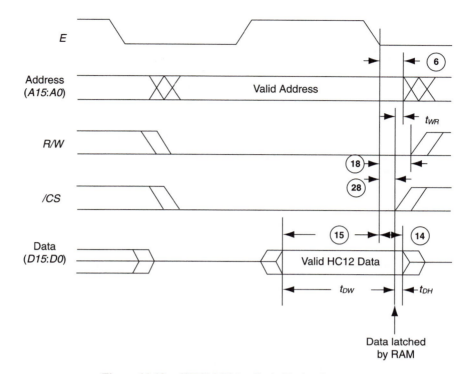

Figure 11.13 68HC12 Write-Cycle Timing Parameters

Address Hold Time.   The address hold time is the time that address must be held after the RAM latches the data. If the address changes before this time, it is possible for data to be written to another location in the RAM. The address hold time is labeled t_{WR} in the timing diagram.

The requirement is that t_{WR} is greater than 5ns. This means that $\textcircled{6}_{min} - \textcircled{28}_{max}$ must be greater than 5ns. The actual values are

$$20\text{ns} - 10\text{ns} > 5\text{ns}$$

This requirement is also met, which means the write-cycle timing meets all requirements as long as we stretch the chip select by one clock cycle.

This was only one simple example of bus timing analysis. Most systems are more complex than this, especially when external decoding logic is used. The timing specifications for microprocessors and microcontrollers have improved over the years. Some older chips can be difficult to analyze because of the way the parameters are specified.

► ## SUMMARY

In this chapter we were introduced to the design techniques and processes involved in expanding a system by adding parts to the bus. We only covered the basic issues for a 64K-byte memory map. Memory paging and mixed bus sizes were not covered.

The two primary design tasks that we covered are chip-select logic and timing analysis. Chip-select logic is especially important to understand how resources are mapped in the memory space. We looked at external chip-select logic for both full decoding and partial decoding. We then looked at the 68HC812A4 programmable chip selects. By using programmable chip selects, we were able to build a system with no extra circuitry for address decoding.

We then went through a bus timing example. Bus timing analysis is not as difficult as it is tedious. It is important to be methodical and detailed. Take your time and recognize that the time you spend doing timing analysis now can be much less painful than troubleshooting a system with a timing problem.

EXERCISES

1. A stand-alone embedded system has the following requirements:

Device	Device Size	Decode Block Size	Start Address	Device Direction
ROM1	8K-bytes	8K-bytes	$c000	Read Only
ROM2	8K-bytes	8K-bytes	$e000	Read Only
RAM	16K-bytes	16K-bytes	$4000	Read/Write
Modem	8 bytes	4K-bytes	$2000	Read/Write

(a) Give the Boolean equations for the partial decoding logic required for each device. Use the Decode Block Size indicated. All device chip selects are active low.

(b) Show the memory map of the system. Only include the devices given in the table. Do not include the microcontroller resources.

2. Show the new memory map if the example system described in Section 11.2 used a 32K-byte RAM and a 32K-byte EPROM. Assume *CSD* is used for the RAM and is configured for $0000–$7FFF and *CSP0* is used for the EPROM.

3. Repeat the timing analysis example in Section 11.3 with a 68HC812A4 running with a 4MHz E-clock.

Modular and
C Code
Construction

In this chapter we look at the contents of a C source program, introduce the modular build process, and introduce techniques for debugging C code. This material is the background needed for modular C code construction.

First we summarize some of the more important parts of a C source file. This information is not intended to be a complete description of the C language. For a complete description, one of the many good books covering the C programming language should always be handy as a reference.

We then look at a process for building modular programs. This build process is required for programs written in C or assembly programs with multiple source files. The material covers the Introl-CODE development system, but the concepts are the same and the techniques are similar to other tools. Both the command-line interface and GUI (graphical user interface) for the Introl-CODE development system are covered. We do not cover the Introl-CODE system entirely, however. For complete documentation, refer to the on-line Introl-CODE manual.

The last section covers the debugging process for C programs. We look at two different systems—a manual debugging system that uses the same tools we covered in Chapters 3 and 10, and a full-featured source-level debugger and BDM pod from Noral Micrologics. The focus is on methods to relate the CPU operation back to the C source code.

Once you finish with the material in this chapter, you should be able to construct a modular program written in assembly, C, or both.

▶ 12.1 C SOURCE CODE

In this section we look at using C source code to program embedded systems. We start with some of the advantages and disadvantages of using a high-level language and then look at the contents of a C source file.

12.1.1 Comparing C and Assembly

In Chapter 2 we introduced programming languages and covered the fundamental differences between programming in assembly and programming in a high-level language. In this text we use C as the high-level language. Let's look at some of the design trade-offs in more detail. Table 12.1 shows some important characteristics of C and assembly.

Assembly. Assembly code precisely controls the operation of the CPU. There is a one-to-one correspondence between an assembly instruction and the CPU operation. This is the fundamental advantage to programming in assembly. It allows the programmer to be precise when writing code, and the debugging process is straightforward.

Assembly language is also unsurpassed as an educational tool. To understand how a CPU works or to understand the strengths and weaknesses of a CPU, we must understand its instruction set, register set, and detailed operation. Learning to use a CPU in assembly language helps us understand how the CPU works.

In general, because of its precise nature, we can write smaller and faster code in assembly. This is not as important an advantage as many would have us believe, however. Yes, we can usually improve both code size and speed by writing assembly code. Many of the methods used to do this sacrifice program structure, portability, and readability, however. Because most programming involves revising existing code, it may be cheaper to use a faster CPU with more memory to accommodate a C program than to rewrite assembly code.

That brings us to the negative aspects of writing programs in assembly. We saw in Chapter 2 that our assembly code includes all the implementation details of a program. These details not only made the code CPU dependent, but they also cluttered the program and required that we always use the CPU software model in detail.

In addition programming in assembly allows the programmer free rein over structure and style. You may have noticed that the only rules in assembly language programming are syntax rules. This gives the programmer total freedom to write a program in any way—good

TABLE 12.1   CHARACTERISTICS OF C AND ASSEMBLY LANGUAGES

C	Assembly
Efficient development due to abstraction	Precise control of the CPU
	Typically requires less code space and is faster
Can be made CPU independent	Always CPU dependent
Cannot access all CPU instructions	Access to all CPU instructions
Some enforced program structure and type-checking	No program-content rules
More difficult to debug without source-level debugger	Simple development process for small systems

or bad. As soon as even a small amount of complexity is introduced, conventions and techniques must be learned and followed to keep things manageable. Writing a quality assembly language program is much more than simply learning an instruction set. It involves learning techniques and styles that are actually enforced or built into many high-level languages.

Regardless of whether we are writing in assembly or C, when we are writing code for a small microcontroller, we have to constantly keep an eye out on the resulting assembly code. Just because we use a C compiler to generate our code does not mean we do not have to know assembly code. We need to constantly look at the code generated by the compiler to make sure we are not introducing unintended or unexpected inefficiencies.

Why C? Using a high-level language has two important advantages—built-in structure and type-checking and abstraction. Because of these characteristics, it is much easier for a programmer to write quality programs that are easier to read, revise, and port to a different system.

There are both practical and technical reasons C is covered instead of other high-level languages. First of all, on the practical side, C is used in an overwhelming majority of embedded designs, which means more prewritten libraries and code modules are available, along with a better selection of development tools.

Technically C also has its strengths. It is often referred to as a *low-level* high-level language because it includes data types and operations that apply directly to bits, bytes, and addresses. These items must be directly controlled in small, embedded systems. Therefore C code can be more efficient than other high-level languages.

C also has most of the important components of a good high-level language such as structured program constructs and data typing. Because data typing is not strictly enforced and there are ways for a programmer to write unstructured programs, however, there have been complaints about C compared with other high-level languages. Actually C lands right between assembly code and other high-level languages. It is not as strict as other high-level languages, but is much more strict than assembly. It is not as efficient as assembly language, but is more efficient than most other high-level languages. It is a compromise—good or bad.

12.1.2 C Program Contents and Organization

We cover the contents of a C file by looking at some examples. The first example is shown in Source 12.1, which is a C program that implements the same function as the *demo1* program in Chapter 3. It is a small program, but it does serve as a good illustration of the components being covered. A version of this program will be part of the *demo1* project that we use throughout this chapter.

Source 12.1 pulse1.c: A C Source File

```
/*****************************************************************
* A simple demonstration program.
* It generates 60 active-low pulses at a rate of 1Hz
* out of PORTP, bit 0.
*****************************************************************
* Include MCU definitions
```

```
*****************************************************************/
#include "hc912b32.h"

/****************************************************************
* Defined constants
*****************************************************************/
#define ON_TIME 250
#define OFF_TIME 750

/****************************************************************
* Function Prototypes
*****************************************************************/
extern void msDelay(unsigned int);

/****************************************************************
* Global variable definitions
*****************************************************************/
unsigned char PulseCnt;

/****************************************************************/
void main(void) {

    _H12PORTP |= 0x01;                  /* Turn off */
    _H12DDRP |= 0x01;                   /*Init. PP0 as output */

    for(PulseCnt = 60; PulseCnt > 0; PulseCnt--) {
        _H12PORTP |= 0x01;              /* Turn PP0 off */
        mSDelay(OFF_TIME);
        _H12PORTP &= ~0x01;             /* Turn PP0 on */
        mSDelay(ON_TIME);
    }
    _H12PORTP |= 0x01;
    while(1){}
}
/****************************************************************/
```

The first thing to notice about *pulse1.c* is that it is not a complete program by itself. It is dependent on at least three other files—the header file *hc912b32.h,* another module that contains the external function *mSDelay(),* and initialization code and information about where this program should go in memory.

This is a characteristic of all C programs. When we write a program in C, we must use a modular program. A modular program is one that is derived from several source files. Each source file is called a *module.* In fact a large part of learning to program embedded systems in C involves learning a new, relatively complex, modular build process. The modular build process is covered next in Section 12.2.

The collection of files and processes required for code construction is called a *project.* The C program in Source 12.1 is only one of several source files for the project. It cannot be built or executed without being combined with the other files that are part of the project.

The *main()* Function.   If a project is written in C it must have one *main()* function in one of its C source files. The *main()* function is the default name of the function that is called after the program is started and system initialization is complete. *pulse1.c* in Source 12.1 contains the *main()* function.

Note that this is not to say that in order to use C functions, you must have a *main()*. For example you may have a program written in assembly that calls C functions, in which case you do not need a *main()* function.

Organization.   If you look at the overall organization of the code in Source 12.1, you will see that different parts of the program are clearly identified and separated with comments and blank lines. Unlike assembly source code, C does not have strict requirements about white space. This means we can write some very obscure and cryptic C programs. As an extreme example of how bad a C program can get, look at Source 12.2. This program actually works (hint: Merry Christmas). In fact it is really cool, if your goal is to write a totally unreadable, tricky program.

Source 12.2 A C Program with Really Bad Organization

```
#include <stdio.h>
main(t,_,a)
char *a;
{
return!0<t?t<3?main(-79,-13,a+main(-87,1-_,main(-86,0,a+1)+a)):
1,t<_?main(t+1,_,a):3,main(-94,-27+t,a)&&t==2?_<13?
main(2,_+1,"%s %d %d\n"):9:16:t<0?t<-72?main(_,t,
"@n'+,#'/*{}w+/w#cdnr/+,{}r/*de}+,/*{*+,/w{%+,/w#q#n+,/#{l,+,/n{n+,/+#n+,/#\
;#q#n+,/+k#;*+,/'r :'d*'3,}{w+K w'K:'+}e#';dq#'l \
q#'+d'K#!/+k#;q#'r}eKK#}w'r}eKK{nl]'/#;#q#n'){)#}w')){)#}w')){)nl]'/+#n';d}rw'
i;# \
){nl]!/n{n#'; r{#w'r nc{nl]'/#{l,+'K {rw' iK{;[{nl]'/w#q#n'wk nw' \
iwk{KK{nl]!/w{%'l##w#' i; :{nl]'/*{q#'ld;r'}{nlwb!/*de}'c \
;;{nl'-{}rw]'/+,}##'*}#nc,',#nw]'/+kd'+e}+;#'rdq#w! nr'/ ') }+}{rl#'{n'
')# \
}'+}##(!!/")
  :t<-50?_==*a?putchar(31[a]):main(-65,_,a+1):main((*a=='/')+t,_,a+1)
    :0<t?main(2,2,"%s"):*a=='/'||main(0,main(-61,*a,
"!ek;dc i@bK'(q)-[w]*%n+r3#l,{}:\nuwloca-O;m .vpbks,fxntdCeghiry"),a+1);

}
```

This program, however cool, would be totally unacceptable for an actual product. Can you imagine the poor engineer assigned to debug or revise this program? It would be far better to start over and rewrite it in a readable form.

The program in Source 12.2 is an extreme example, but some programmers take pride in writing cryptic programs. Maybe they are thinking about job security. They are certainly not thinking about the success of the project. The goal to program organization and style is productivity. We want the code to be readable, portable, and reliable.

One of the best ways to avoid poorly written programs is to follow a set of programming conventions. The employer should establish a set of conventions that must be followed by all programmers. If the employer does not have a set of code conventions, you should create your own so at least your code is consistent.

Do not get too caught up in arguments over code conventions. The conventions are about producing a readable, portable, and reliable result. They are not about defining the *best* way to write a program. People can argue forever about what is the best way. Conventions are more about consistency from programmer to programmer and from project to project. So you should write your code conventions with productivity in mind. A set of coding conventions will also vary with the tools used for software development. For example the conventions for a development system based on the VI editor may be very different than one based on a full-featured programming editor. Appendix A shows one set of programming conventions that can be used as a starting point to create your own.

12.1.3 Syntax and Tokens

In this subsection we look at how the compiler interprets a C source file. This includes syntax rules and the interpretation of words or *tokens*.

There are two kinds of program lines for a C program—preprocessor lines and C program lines. A preprocessor line always starts with a pound symbol (#) in column 1. If this symbol does not appear in column 1, the line is interpreted as a C program line. We will discuss preprocessor commands later.

White space in a C program is treated differently than in an assembly program. First there are no specific white-space requirements at the beginning of a line. This allows indentation to be used to visually indicate the program's structure.

Another difference is that an *end-of-line* character is interpreted as white space in a C program. This allows you to have a single command line on multiple lines of the source file. The program in Source 12.2 abused this capability. But it can also be used to help make a program more readable when a long expression or a long parameter list is required.

The compiler parses the program into a string of tokens—words separated with white space. There are several types of tokens including comment delimiters, identifiers, keywords, constants, and operators.

Since the end-of-line character is seen as white space, there needs to be another way to identify the end of a C statement. In C a semicolon indicates the end of a statement. Since it is good practice to have one statement per line, most lines end with a semicolon. Look at the following example:

```
x = y * z;
```

There also needs to be a way to group statements for a control structure or function. In C braces are used to group statements:

```
while(1) {
    _H12PORTP ^= _H12PORTP ^ 0x01;
    count++;
}
```

There are many ways to place the braces, but there is no best way. Be consistent by following a code convention that results in clear, readable code. In this book the first brace is always placed at the end of the line that contains the control statement or function. The closing brace is aligned with the control statement or function name, so it is easy to find the start and end of the block.

The braces are optional if there is zero or one statement in the block. It is a good practice to always use the braces, however. For example, for an endless trap, we can use

```
while(1) {}
```

instead of

```
while(1);
```

Comments. The first tokens we cover are the comment delimiters. They identify the beginning and ending of a comment. The commenting delimiters in C are annoying. To create a comment, it must be enclosed by /* and */. For example,

```
/* This is a one-line C comment */
```

and

```
/* This is a two-line
 C comment */
```

The compiler looks for /*. Once it finds it, everything following is a comment until it reaches the */. Unfortunately these delimiters cannot be nested, so we cannot comment out a block of code that already contains comments. We can use conditional preprocessor commands to exclude a block like this, or we can use C++ type comments.

Using C++ comments is one alternative that is becoming more practical. A C++ comment is everything on the line following a //. So, using the C++ method, the comment lines become

```
// This is a one-line C++ comment.
```

and

```
// This is a two-line
// C++ comment.
```

The reason it is becoming more practical to use C++ comments is that so many new compilers are designed for both C and C++ that they except either commenting method. Be careful, however; you might have to change compilers to one that only accepts C comments. It is best to reserve C++ comments for excluding blocks of code during the debugging process. The Introl-CODE system does accept both commenting styles.

Commenting in a C program is just as important as in an assembly program. There may be fewer comments because the C code is at a higher level than assembly, but there still needs to be an explanation of the intent of a program. As you can see from Source 12.1, comments are used to document the code and to help separate the code into blocks to make it more readable.

Identifiers and Keywords.   *Identifiers* in C are equivalent to symbols in assembly. They are the names we use to identify constants, macros, data, and programs. They must start with an alphabetic character or underscore, "_." *Keywords* are identifiers that are reserved for the C language. They include the following:

```
auto        double      int         struct
break       else        long        switch
case        enum        register    typedef
char        extern      return      union
const       float       short       unsigned
continue    for         signed      void
default     goto        sizeof      volatile
do          if          static      while
```

There also may be additional keywords specific to a compiler. The Introl-CODE compiler reserves the following keywords:

```
__mod1__        __mod2__        __interrupt
```

Of course we cannot use a keyword as an identifier for something we create. For example we cannot create a function called *sizeof* or a variable called *register*.

Constants.   Here we are talking about defined constants, not stored constants. Defined constants are not stored in target memory. We discuss stored constants in Chapter 13. There are several types of defined constants—integer constants, floating constants, character constants, enumerated constants, and constant strings. Enumerated constants and constant strings are normally stored constants, so they are covered in Chapter 13.

Integer and Floating Constants.   In C an integer and floating-point constant is an identifier that starts with a number. If the number contains a decimal point, an *e* or *E,* or it has a floating-point suffix, it is a floating-point constant. Otherwise it is an integer.

An integer is decimal unless it starts with a *0* or *0x*. A number with a preceding *0* is interpreted as an octal number and a number preceded by a *0x* is a hex number. Integer and floating constants have optional suffixes to specify the data type to be used. Valid suffixes include *u, U, l, L, f,* or *F* for unsigned, long, and float data types, respectively. For example

    37   is a decimal integer with a value of 37
    037   is an octal integer with a decimal value of 31
    0x37   is a hex integer with a decimal value of 55
    37.0   is a floating-point constant with a decimal value of 37

Character Constants.   A character constant in C is an identifier that contains one ASCII character enclosed in single quotation marks. For example *'0'* is equal to the ASCII code for a zero, which is *0x30*. Because of this format, we need to use a special escape sequence to represent a single quotation mark or nonprintable ASCII characters. An escape sequence in C always starts with a backslash (\). Following is a list of some standard escape sequences:

\a	Alert (bell) character	\\	Backslash
\b	Backspace	\?	Question mark
\f	Formfeed	\'	Single quotation
\n	Newline	\"	Double quotation
\r	Carriage return	\ooo	where *ooo* is a three-digit octal number
\t	Horizontal tab	\xhh	where *hh* is a two-digit hex number
\v	Vertical tab		

These escape sequences are only valid when they are enclosed in single quotation marks or are contained in a string. For example '\x30' is the same as 0x30 or '0'.

Operators. Table 12.2 shows the operators in the C language. The table lists the operators in order from those with the highest precedence to those with the lowest precedence.

Because it is not always obvious what the precedence is for a given expression, it is always wise to use parentheses. Parentheses not only ensure that the order of execution is correct, they also help make the code more readable.

Let's quickly go over some of these operators, focusing on aspects important to programming embedded microcontrollers.

Arithmetic Operators. The arithmetic operators include addition (+), subtraction (−), multiplication (*), division (/), and modulus (%). There are also the unary sign operators, + and −. The arithmetic operators work as you would expect and have the normal arithmetic precedence.

In programming small microcontrollers, we need to be careful when using the multiplication, division, and modulus operators. They are convenient in a program, but you have

TABLE 12.2 C OPERATORS

Grouping Operators:	( )
Array Operators:	[ ]
Structure References:	-> .
Unary Operators:	+ - * ! ~ ++ — & (type) sizeof
Multiplication and Division:	* / %
Addition and Subtraction:	+ -
Shifting:	<< >>
Relational Operators:	< <= > >= == !=
Bitwise AND Operator:	&
Bitwise EX-OR Operator:	^
Bitwise OR Operator:		
Logical AND Operator:	&&
Logical OR Operator:	||
Conditional Operator:	? :
Assignment Expressions:	= += -= *= /= %= &= ^= |= <<= >>=
Comma Operator:	,

to think about the code that will be generated. As we saw in Chapter 7, a division routine may require a large program and take a long time to execute, especially if the microcontroller does not have a division instruction. This can be true when using the modulus operator also, because it requires a division operation. Modulus operators are convenient for executing something every *N* counts. For example the code

```
if(count % 4 == 0){
    /* Do something */
}
```

looks like a good way to do something every fourth count. Depending on the data type of *count,* the modulus operator is going to be relatively slow. A better way to do this is

```
if((count & 0x03) == 0){
    /* Do something */
}
```

This program only requires masking with a bitwise AND operation, which is a very fast operation compared to a division.

Bitwise Boolean Operators versus Boolean Logic Operators. Make sure to use the correct Boolean operator. It is important to understand the difference between the bitwise operators and the logical operators. This problem crops up more when programming embedded systems because they require many bit operations. Following is a list of the AND, OR, and NOT operators:

Bitwise AND	&
Logical AND	&&
Bitwise OR	|
Logical OR	||
Bitwise NOT	~
Logical NOT	!

A bitwise operator performs the operation on each pair of aligned bits in the two arguments. A logical operator on the other hand performs the operation on the complete word and returns a value of $01 (TRUE) or $00 (FALSE). To illustrate the difference, let's compare the results of two AND operations. First let's do a bitwise AND:

```
0x81 & 0x80 -> 0x80
```

The bitwise AND returns 0x80 after the following operation:

$$\begin{array}{r} 1000\ 0001 \\ \cdot\ \underline{1000\ 0000} \\ 1000\ 0000 \end{array}$$

The logical AND

```
0x81 && 0x80 -> 0x01
(TRUE) && (TRUE) -> (TRUE)
```

returns a 0x01 (TRUE) because both 0x81 and 0x80 are nonzero (TRUE).

The difference between NOT operators is similar. A bitwise NOT of 0x80 is

```
~0x80 -> 0x7F
```

because it inverts each bit in the argument. The logic NOT of 0x80 is

```
!0x80 -> 0
```

because 0x80 is TRUE (nonzero), so the logical NOT returns FALSE (zero).

Relational Equals versus Assignment Equals.   Another two operators that can be confused is the relational "equal to," ==, and the assignment operator, =. The relational operator, ==, returns a one if the two arguments are equal. The assignment operator, =, sets the left-hand argument equal to the right-hand expression.

It is common to accidentally use the assignment operator in a conditional expression. The problem with this is that C allows assignment operators in conditional expressions. An assignment operator in a conditional expression is not considered a good programming practice because it may have been placed there by mistake and because it produces a side effect. A common code convention is not to allow an assignment operator in a conditional expression unless it is used in conjunction with the comma operator.

Assignment Expressions.   Assignment expressions are a shorthand notation for a read-modify-write operation, that is, an argument that is set equal to some function of itself. For example in the *pulse1.c* program, there are several assignment operators used. Two of them are

```
_H12PORTP |= 0x01;
_H12PORTP &= ~0x01;
```

These are equivalent to

```
_H12PORTP = _H12PORTP | 0x01;
_H12PORTP = _H12PORTP & ~0x01;
```

Although not everyone understands assignment operators, they can help make a program more readable by reducing the size of an expression.

Note

For those readers who have been following the pseudo-C examples in the previous chapters, you probably noticed the *PORTP* symbol name in this example is *_H12PORTP* instead of *PORTP*. This is a convention specific to the Introl-CODE compiler. It adds the

_H12_ prefix to all microcontroller registers to identify them as MCU specific. These symbols come from the include file _hc912b32.h._

12.1.4 Preprocessor Commands

Preprocessor commands are commands for the compiler's preprocessor, not for the CPU. They are the C equivalent to assembler directives. Preprocessor commands are used primarily to define constants, define macros, conditionally compile blocks of code, and insert code from another file. A preprocessor line is distinguished from a C line by starting with a pound sign, #, at column 1. They also differ because they do not end with a semicolon.

The scope of a preprocessor command includes the file in which it is contained or included. So if a program has two C source files that use the same preprocessor definitions, they both must contain the preprocessor commands.

Defined Constants.   The preprocessor _#define_ instruction can be used to create defined constants. For example in _pulse1.c_ there are two defined constants:

```
#define ON_TIME 250
#define OFF_TIME 750
```

The _#define_ causes the compiler to replace all instances of _ON_TIME_ and _OFF_TIME_ with the numbers 250 and 750 before compiling the program. Defined constants are equivalent to equated constants in assembly code. Like assembly equates, defined constants are not stored in target memory. They help make the code more readable by replacing magic numbers with symbols.

The convention used in this book is that defined constants use all uppercase characters with underscores separating words.

Defined Macros.   A _macro_ is a snippet of code that has a name. When the name is used in a program, the compiler replaces it with the code snippet. Actually they are identical to defined constants except that they contain program statements. For example some common macros are

```
#define FOREVER()      while(1)
#define TURN_LED_ON()  LED_PORT &= ~LED_BIT
#define TRAP()         while(1){}
```

The convention used in this book is to add parentheses to macro names even if you do not use them for parameters. This is to distinguish a macro, which _does_ something, from a constant, which _is_ something.

Macros also make code more readable. They are different from functions because the code is placed in line every place that it is used. The advantage of using macros is that they do not require the overhead of a function call. They can be used without concern about slowing the code down. The disadvantage of using macros is that the overall code size is larger because it places a copy of the code every place that it is used. Functions require only one copy of the function code. Macros are also not practical for code that has more than a couple of lines or that requires more than one parameter.

450

Conditional Compilation. Preprocessor commands can be used to exclude blocks of code from being compiled. The three most common applications of these commands are to exclude debugging code, avoid multiple definitions, or select different code to port a program to a different CPU. The commands are *#if, #elif, #else, #endif, #ifdef,* and *#ifndef.*

Some examples of using conditional assembly are shown in Source 12.4 later in this section. First two *#ifdef...#endif* commands are used to port the program to different microcontrollers. To use an M68HC912B32 we define *HC912B32*. To use the M68HC11F1 we define *HC11F1.*

Another good use of an *#if* command is to exclude blocks of code for debugging. If there is a block of code we want to exclude, we can do the following:

```
#if 0
  …
block of code to exclude
  …
#endif
```

This is one way to exclude a block of code that contains comments. Remember that the C comments cannot be nested, so you cannot just add a comment start, */*, at the beginning of the block and a comment end, **/, at the end of the block.

File Inclusion. The *#include* preprocessor command can be used to include the contents of another file. The syntax is

```
#include "filename"
```

or

```
#include <filename>
```

The compiler replaces the *#include* line with the contents of *filename.* The file included is called a *header file* or an *include file* and normally has a file extension of *.h.* The first form is generally used for project header files and the second form for system header files.

12.1.5 Header Files

Header files are C source files that are added in line to a C source file with the *#include* preprocessor command. They are also commonly called *include files.* A large project may have a master header file, which is included in all C source modules in the project. In addition there may be module header files for other modules used in the program and system header files that are common across different projects.

Header files help make it easier to maintain a program or port a program to a different target system. They also help decrease the clutter that is required at the start of a C source file.

Although header files are more important for large projects, they can also benefit small projects. As an example, Source 12.3 shows *pulse2.c,* a revision of the *pulse1.c* program from Source 12.1. It is smaller, processor independent, and much easier to read.

Source 12.3 Revised demo1 C Source That Uses a Header File

```
/****************************************************************
 * pulse2.c - A simple demonstration program.
 * It generates 60 active-low pulses at a rate of 1Hz out of
 * LED. This version has been made MCU independent and more
 * readable by using header files.
 ****************************************************************
 /* Include Master Header File
 ****************************************************************/
#include "includes.h"

 /****************************************************************
 * Global variable definitions
 ****************************************************************/
UBYTE PulseCnt;                         /* The number of pulses remaining */

 /****************************************************************/
void main(void) {

    TURN_LED_OFF();                     /* Initialize LED */
    INIT_LED_DIR();

    for(PulseCnt = 60; PulseCnt > 0; PulseCnt-) {
        TURN_LED_OFF();
        msDelay(OFF_TIME);
        TURN_LED_ON();
        msDelay(ON_TIME);
            }

    TURN_LED_OFF();                     /* Turn LED off and trap */
    TRAP();
}
 /****************************************************************/
```

The program uses a header file, *includes.h,* to define constants, macros, and declarations while reducing clutter in the source file. All the definitions are contained in the project's master header file, *includes.h,* shown in Source 12.4. The compiler replaces the line *#include "includes.h"* with the code contained in the header file.

Source 12.4 The demo1 Project Header File, includes.h

```
/****************************************************************
 * includes.h - Master header file for demo1 project.
 *
 ****************************************************************
 * General type definitions
 ****************************************************************/
typedef unsigned char    INT8U;
typedef signed char      INT8S;
typedef unsigned short   INT16U;
```

```
typedef signed short    INT16S;
typedef unsigned long   INT32U;
typedef signed long     INT32S;

#define UBYTE INT8U
#define SBYTE INT8S
#define UWYDE INT16U
#define SWYDE INT16S

#define ISR __interrupt void

/*************************************************************************
* General Defined Constants
*************************************************************************/
#define FALSE     0
#define TRUE      1

/*************************************************************************
* General Defined Macros
*************************************************************************/
#define FOREVER()       while(1)
#define TRAP()          while(1){}

/*************************************************************************
* MCU Specific Definitions
*************************************************************************/
#include "hc912b32.h"            /* CODE 68HC912b32 register defines    */

#define   SWI()         asm("\tswi\n")
#define   ENABLE_INT()  asm("\tcli\n")
#define   DISABLE_INT() asm("\tsei\n")
#define   LED_PORT_DIR  _H12DDRP                /* LED Port Definitions  */
#define   LED_PORT      _H12PORTP
#define   LED_BIT       0x01

/*************************************************************************
* Project Constant and Macro Definitions
*************************************************************************/
#define ON_TIME   250
#define OFF_TIME 750
#define TURN_LED_ON()  LED_PORT &= ~LED_BIT
#define TURN_LED_OFF() LED_PORT |= LED_BIT
#define INIT_LED_DIR() LED_PORT_DIR |= LED_BIT

/*************************************************************************
* System Module Header Files
*************************************************************************/

/*************************************************************************
* Project Module Header Files or Declarations
```

Modular and C Code Construction   453

```
*******************************************************************/
extern void msDelay(UWYDE);       /* Delay routine declaration.
                                     Defined in msdelay.s12            */

/*******************************************************************/
```

The name *includes.h* is the naming convention used in Chapter 14 for a master header file. Because this program is so simple and only includes one C module, we also could have called this header file *demo1.h*.

The first part of the header file defines system-level names. They are the definitions required for all programs written by an organization. They must come first because they are required for all the code that follows. This part of the header file goes a long way toward implementing a set of company-wide program conventions.

The next section of the header file contains processor-dependent definitions. By removing all references to MCU-specific resources from the source, we have made *pulse2.c* processor independent. To use a different processor, we would only have to change this part of the header file. For larger programs this section of *includes.h* may be contained in another header file that is included here. In that way we would have one header file that makes all MCU-dependent definitions. These definitions also make the program easier to read. For example, instead of referring to *PORTP*, bit 0, we refer directly to the LED. When reading the code, we do not have to look at the schematic to see what is connected to a port. We already know by reading the code that it controls the LED.

The next section of *includes.h* contains more project-wide definitions that are used by multiple project modules. The idea is that the project-wide definitions can be defined in one place. For this simple program, there is not really a need for this section because there is only one C source module, *pulse2.c*. Because these definitions are only used in *pulse2.c*, this part of the header file could optionally be placed in *pulse2.c*.

The last part of *includes.h* includes the header files for other modules used in the program. These may include system modules such as *math.h* or other project modules. Module header files normally contain the *extern* declarations required to use that module. In this program there is only one C module, *pulse2.c*, so no header files are included. There is another module, *msdelay.s12*, used by this program, which contains the *msDelay()* function. Since it is so small, we will place the function declaration directly in *includes.h* instead of creating another header file.

Header files also can have their disadvantages. If we add too many header files, the program would be too spread out, making it more difficult to maintain. Do not create a header file just to create a header file. Every header file should be created for a specific purpose. We look at the use of header files in more detail in Chapter 14 when we cover modular program design.

▶ 12.2 THE MODULAR BUILD PROCESS

In all the previous chapters, we used the simple build process described in Chapter 3 for our programs. This process allowed us to write small assembly programs contained in a single file. In this section we introduce a more complex build process that is designed for modular programs. Modular programs are programs in which the source code is contained in multiple files. Each source file is called a *module*. For example in the *demo1* project described in the last section, there are two source modules, *pulse2.c* and *msdelay.s12*.

In order to program an embedded system in C, we must use a modular build process, because there must be at least one assembly source file for system initialization in addition to the C source file. Using a modular build process also enables us to use precompiled modules called libraries.

The material covered in this section is just as applicable for programs written in assembly as it is for C programs. Because a modular build process is required for C programs, this is how many designers are first exposed to this process.

There are two ways to run the build process in the Introl-CODE system—we can use a command-line interface or we can use the graphical interface. Typically the command-line interface is used on UNIX systems, and the graphical interface is used on a Windows system. We can use either interface on both operating systems, however. In this text the graphical interface is emphasized. The command-line interface is covered briefly at the end of this section.

Modular Code Benefits.   Many of the benefits attributed to writing programs in a high-level language are actually due to writing modular programs. Modular programming allows the design to be partitioned into modules of related routines. For example the Basic I/O module described in Chapter 7 is an assembly module with a set of related I/O routines. It could be used for the terminal I/O portion of any project. The project would then have other modules for the other project subsystems. This is much more manageable and efficient than dumping every routine into a single long file. Designed well, modules have many of the advantages of objects in an object-oriented language, including information hiding and portability.

The modular build process also enables us to use software libraries. Libraries are precompiled modules, stored in object form. Again many of the perceived advantages of using a high-level language are actually due to the use of libraries. Since libraries also can be used with modular assembly code, this is actually a benefit provided by using a modular build process. For example the *math.h* library is generally associated with C programs. As long as the parameter passing details are understood, however, there is nothing to prevent us from using this library with an assembly program.

12.2.1 The Project Directory

A project directory contains all the files associated with a single design or project. It is primarily an organizational tool used to keep project-related files in a single location. Table 12.3 shows the typical files found in a project directory.

The files are separated into three types—source files, build and configuration files, and generated files. The source files are the programs we write to run on the target system. The build and configuration files define the build process and the characteristics of the target system. The generated files are the resulting files from the build process.

Source Files.   The source files are the programs we write for the project. Source files may consist of any number of assembly modules or C modules. Remember that the modular build process is just as useful when all the source files are written in assembly.

TABLE 12.3 TYPICAL PROJECT FILES

Source Files	Build/Configuration Files	Generated Files

C Source:
`module.c`
 	
C Project Headers:
`includes.h,`
`module.h`
 	
Assembler Source:
`module.s12`
 	
Start-Up File:
`projectstart.s12`
	Command Line Interface
	Linker Command File:
	`project.ld`
	
	Build File:
	`Makefile`
	
	Graphical Interface
	Combined Build and
	Configuration File:
	`project.code`
		Relocatable Object Files:
		`module.o12`
		
		Relocatable Listing Files:
		`module.lst`
		
		Executable Object File:
		`project.e12`
		
		Absolute Listing File:
		`project.adr`
		
		Project Map:
		`project.map`
		
		Project S-Record:
		`project.s19`

The start-up file is a required assembly program that performs all the required initialization before the main program is executed. It can be written by the programmer or generated by the development system. In the examples shown, we use Introl-CODE to generate the start-up file.

Creating a New *demo1* Project Directory. To demonstrate the process of creating a new project, let's use the source files from the last section, *pulse2.c, msdelay.s12,* and *includes.h,* to create a project called *demo1.*

To start let's create a new project directory, and using the Introl-CODE graphical interface, a new project called *demo1.* Figure 12.1 shows the project directory for *demo1* after creating the source files and configuring the project with the *Edit Project* dialog box shown in Figure 12.4 on page 464. The source files in the *demo1* project include one C source file, *pulse2.c;* one assembly code file, *msdelay.s12;* and one C header file, *includes.h.*

We have not seen two of these files before: *demo1.code* and *msdelay.s12.* The *demo1.code* file is the build and configuration file, which we cover in the next subsection on the build process. Another file we have not seen yet is *msdelay.s12.* The main program, *pulse2.c,* uses a software delay to time the LED flashes. Software delays are designed by counting instruction execution time so they should be written in assembly—it makes no sense to write them in C. So the project uses an assembly source module, *msdelay.s12,* for the software delay routine.

The module *msdelay.s12* is shown in Source 12.5. It has two subroutines, *msDelay* and *Dly1ms.* The only routine accessible to other modules is *msDelay,* because it is exported. *msDelay* is called a *public function. Dly1ms* is called a *private* function because it is not exported and can only be accessed within the module.

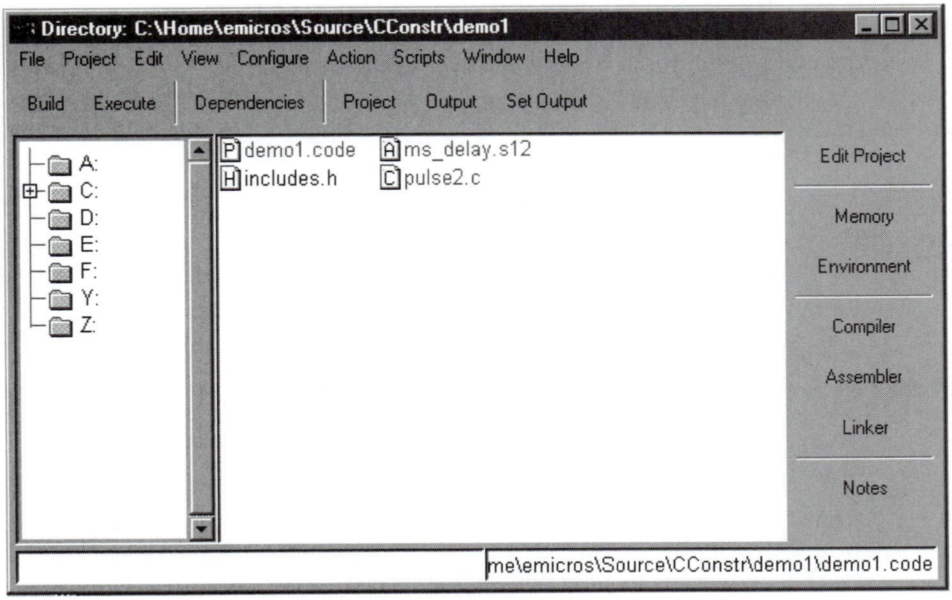

Figure 12.1 Directory Listing for *demo1* Project before Build

Source 12.5 A Software Delay Assembly Module Used by demo1

```
;**************************************************************************
; Module - msdelay.s12
; Description: Software delay routines.
; MCU: 68HC12, E=8MHz
;**************************************************************************
; Export public functions
;**************************************************************************
          export msDelay
;**************************************************************************
; Subroutine msDelay - A software delay routine that delays
;    n milliseconds.
;    - void msDelay(UWYDE nms)
;    - nms passed in ACCD
;    - preserves all registers except CCR
;    - requires 4 bytes stack space
;    - public
;**************************************************************************
          section .text
msDelay:
msdlp     bsr Dly1ms        ;run Dly1ms N times
          subd #1
          bne msdlp
          rts
```

```
;********************************************************************
; Subroutine d1ms - 1ms delay loop.
;    - preserves all registers except CCR
;    - requires 2 bytes stack space
;    - private
;********************************************************************
Dly1ms      pshx
            ldx #1997       ;loop for 1ms
d1mslp      dex
            bne d1mslp
            pulx
            rts


;********************************************************************
```

As you can see from the code in *pulse2.c, msDelay* is called as a function in the C code. This demonstrates how both C and assembly code can be used in a project. Remember that assembly code does not allow the program to be portable to different CPUs. In this case it does not matter because software delays are inherently dependent on the CPU. If we wanted to build this project for another MCU, we would have to change the header file, change the build process, and use a different delay module.

The Start-Up Module.   A start-up module is required by all C programs. It is not shown in Figure 12.1 because it is generated automatically by the Introl-CODE system. When we build the project, Introl-CODE will create a new source module called *demo1start.s12* that will contain all the required start-up code. The code contained in this file depends on the settings we make in the *Edit Project* dialog box.

Because *demo1start.s12* is a source file, we cover it here. Source 12.6 shows the start-up file. Notice that it does not have all the nice formatting and organization that our source files have. This is because it is created by a program. Typically we would never have to edit this file, although we can edit it in cases when a special start-up is required that cannot be defined in the graphical interface.

In this case there is not a lot of start-up code because the *demo1* project was designed to run under D-Bug12. Most of the required initialization has already been done by the monitor. Start-up code is discussed in more detail in Chapter 14.

Source 12.6 Introl-CODE Generated Start-Up File

```
; demo1start.s12 - startup file for the project demo1.
; This file was automatically generated by Introl-CODE. DO NOT EDIT.
;
                section                 .start0
; The program entry point (at RESET)
__start:
;
;                stackinit - initialize the stack pointer
;
```

```
                        import      __stackend

                        section     .start4
__stackinit:            lds         #__stackend         ; point to the system stack
;
; bssinit - initialize bss area
;
; Clear the uninitialized RAM area.  This is required
; so that uninitialized variables have a value of 0.
;
                        section     .start3
__bssinit:
                        import      __ramstart,__ramend

                        ldx         #__ramstart         ; point to start of data

                        bra         1l
2                       clr         1,x+
1                       cpx         #__ramend           ; compare to end address
                        blo         2s                  ; continue if not finished
;
; datainit - copy non-constant initialized data to ram
;
                        section     .start3
__datainit:

                        import      __datastart,__initstart,__initend

                        ldy         #__datastart        ; target of the copy in RAM
                        ldx         #__initstart        ; source of the copy in RAM
                        bra         1l
2                       movb        1,x+,1,y+
1                       cpx         #__initend          ; at the end?
                        blo         2s                  ; jump if not

                        section     .startX
                        import      main
                        jsr         main                ; Call the program
;
; exithalt.s - program termination point
;
; __exit          - the program termination point

                        section     .startZ
__exithalt:

__exit::                sei                             ; mask interrupts
loop                    bra         loop                ; and loop forever
```

Dependencies. Header files and libraries are not considered sources to the project build process. They are files that the project depends on. Sometimes these files are added automatically by the build process, and other times we need to add them manually to a project. What this means is that it is possible for the build process to add code that our sources did not contain or include.

For example in the *demo1* project, there is one dependency not shown, *libgen.a12*. Introl-CODE added this library for register definitions and start-up code definitions. Other libraries that are not known by the Introl-CODE system must be added manually.

12.2.2 The Build Process

Once we have created the source files, we need to build the project to generate the object files, listings, and S-Record file. Figure 12.2 shows the data flow diagram for a modular build process. This process is the underlying build process used by virtually all software development systems for high-level languages, including those with a graphical user interface.

Compiler. If there are C source files, the first process to run is the cross-compiler to compile the C source. The cross-compiler translates the C code into assembly code, which in turn is assembled to generate an object file. Normally the assembly code created by the compiler is automatically deleted after being assembled. You can configure the build process to save these files, but it is rarely needed.

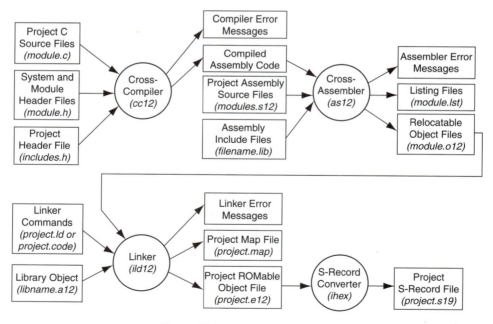

Figure 12.2 Modular Build Process

When we want to look at the assembly code generated by the compiler, we look at the listing files generated by the assembler. The listing files contain both the assembly code and the assembled machine code.

The Assembler.   The assembler translates assembly source to relocatable object files. The relocatable object files produced by the assembler contain the binary machine code but not the address locations for the code or the values of external references.

If desired, listing files also are generated by the assembler. There is a listing file for each C and assembly source module. These listing files are also relocatable and do not contain the absolute location information. They are valuable because they allow you to see what the compiler and assembler actually created.

In an ideal world, you would never have to look at the listing files because they are generated by the compiler and the assembler from your source. If your source is correct, they also should be correct. Right? No, not necessarily. By looking at the listing file, you can see what the compiler generated. This may bring up bugs in the compiler, highlight inefficiencies created by the compiler, or bring up a misunderstanding you have about the C programming language. The compiler may be following ANSI-C rules, yet it still may generate something different from what you expected.

The Linker.   The linker combines the relocatable object files and dependent library objects into one *executable* object file. It also reconciles external references. After the linking process, the code is associated with absolute memory locations. The executable object code is also referred to as *ROMable* object code because it includes absolute addresses so it can be loaded into the target memory.

The linker converts relocatable object code into object code that is built for a specific target configuration. Therefore the linker must know the target requirements and parameters, including start-up code requirements, register descriptions and locations, and memory maps. The linker gets this information from the project's configuration file. For the Introl-CODE graphical interface, this information is contained in the *project.code* file. The *project.code* file is generated by selecting parameters in the *Edit Project* dialog box. We can also use a linker command file. Linker command files are covered in the subsection on using the command-line interface.

12.2.3 Section Mapping

The Introl-CODE system uses the concept of sections to place different parts of the code into the appropriate memory type and location. The assembler translates the assembly code from source files into relocatable object files. The object code in the relocatable objects files is divided into named input sections. The linker then sorts the input sections into output sections corresponding to the actual target memory. Figure 12.3 shows how the sections are mapped by the linker. Note that the input sections in the relocatable files do not have actual addresses, whereas the output sections in the executable file have absolute addresses determined by the definition for the section groups.

For example in a stand-alone system using the M68HC912B32, the ROM group would start at $8000 and end at $FFFF to correspond to the Flash EEPROM. The RAM group would start at $0800 and end at $0BFF to correspond to the on-chip RAM.

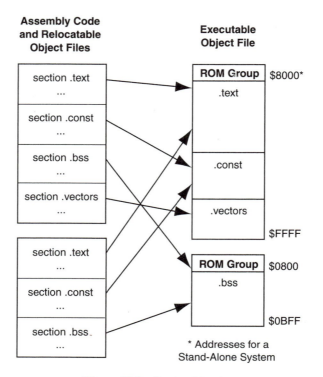

Figure 12.3 Section Mapping

The order in which the code is mapped from a relocatable file into the executable file cannot be determined before linking. For example in the figure above, we do not know if the *.text* code from the first file will end up before or after the *.text* code from the second file.

Defining Sections in Modular Assembly Source. When writing an assembly module for the modular build process, we must use the *section* assembler directive to place code into named sections. This is different from the simple build process described in Chapter 3, in which we used the *org* directive to place code into specific locations.

The *section* directive has the following syntax:

```
section section_name
```

where *section_name* is one of the section names shown in Table 12.4. For example the section directive that should precede program code is

```
section .text
```

We can see examples of using the section directive in Source 12.5 on page 457 and Source 12.6 on page 458.

Section Mapping for C Source. We have seen how assembly source code is mapped into sections. What about C source code? When using C, the source file does not contain

TABLE 12.4 INTROL-CODE SECTION MAPPING

Description	Section Name	Memory Type	Comment
Executable code	*.text*	ROM	Program code
Static variables (uninitialized)	*.bss*	RAM	Includes static variables defined within a block or global variables defined outside all blocks
Static variables (initialized)	*.init* (ROM) *.data* (RAM)	ROM and RAM	*.init* contents are copied to *.data* at run time during the start-up process
Stored constants	*.const*	ROM	Data with *const* modifier
Strings	*.strings*	ROM	Constant strings defined with "..."
Large return data	*.retm*	RAM	When return data are too large to fit in ACCD, this space is used
__mod1__ data	*.mod1*	Special (EEPROM, BBRAM)	Used to designate special memory types like EEPROM or BBRAM (data only)
__mod2__ data	*.mod2*	Special (EEPROM, BBRAM)	Used to designate special memory types like EEPROM or BBRAM (data only)
__mod1__ __mod2__	*.base*	Base page RAM	Used for Motorola base page, $00–$FF

any information about the memory locations or memory types to be used for the code. This is appropriate because C is a high-level language that should be independent of the target hardware.

C source code is automatically mapped into sections by the compiler. Table 12.4 shows the memory type and sections allocated for different parts of a C program. There are additional sections defined when more than 64K-bytes of memory are used. These additional sections are called *far* sections, and they correspond to code that is placed beyond $FFFF.

We can also create our own named sections. By convention all section names in the Introl-CODE system start with a period. To avoid confusion, symbols in the program code should never start with a period.

12.2.4 Libraries

Libraries are precompiled modules that are stored in a binary archive format. When a library function is called by a C source program, the linker inserts the object for that function into the executable object by extracting it from the library archive file.

A good development system will only insert the object for the function that is used. Beware, some compilers will insert the complete library module, which can make the memory requirements explode. You can imagine the wasted space if the complete *stdio* C library was added to a program that only used *fopen(), putc(),* and *getc()*. Instead of tens of bytes of code, you would get thousands of bytes of code. It is important to check for this when purchasing a compiler.

Associated with each library should be a header file. This header file contains the *extern* declarations for the public functions and data in the library module. To use a library function, you only have to include the library's header file in the program. For example to use the standard C math library, we would add

```
#include <math.h>
```

The decision to convert a C module into a library is based on the stability of the code and the general usefulness of the code as is. Once a module is converted to a library, you no longer have to compile the code, but since it is already compiled, you can no longer change the code. To go back, change the source of a library, and rebuild it must be done carefully because there may be many projects dependent on that library. If it is changed, it may break a project that has been working for years.

To create a library archive file using the Introl-CODE system, select *Library* as the *Project Result* in the *Edit Project* dialog box shown in Figure 12.4.

12.2.5 Executing the Build Process

Because the modular build process is so complex, especially for large projects, the execution of the build process must be automated. To automate the process we must first define the process to be automated. Then there must be a simple way to execute the process. The method for defining the process and executing the process depends on which interface we use—the command-line interface or the graphical interface.

When using the Introl-CODE graphical interface, the build process is specified by setting parameters in the *Edit Project* dialog box. Figure 12.4 shows the *Edit Project* dialog

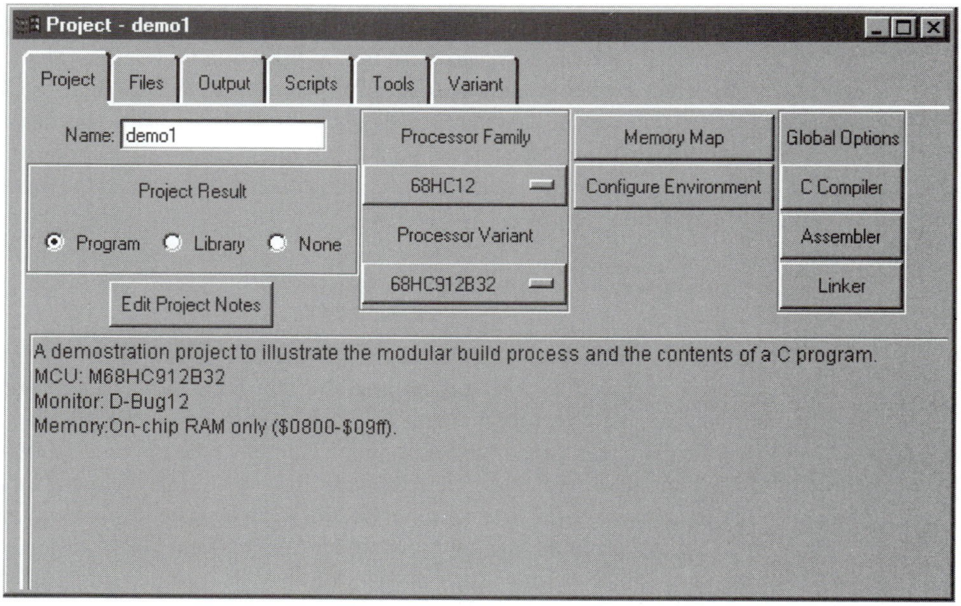

Figure 12.4 Introl-CODE Project Configuration Dialog Box for *demo1*

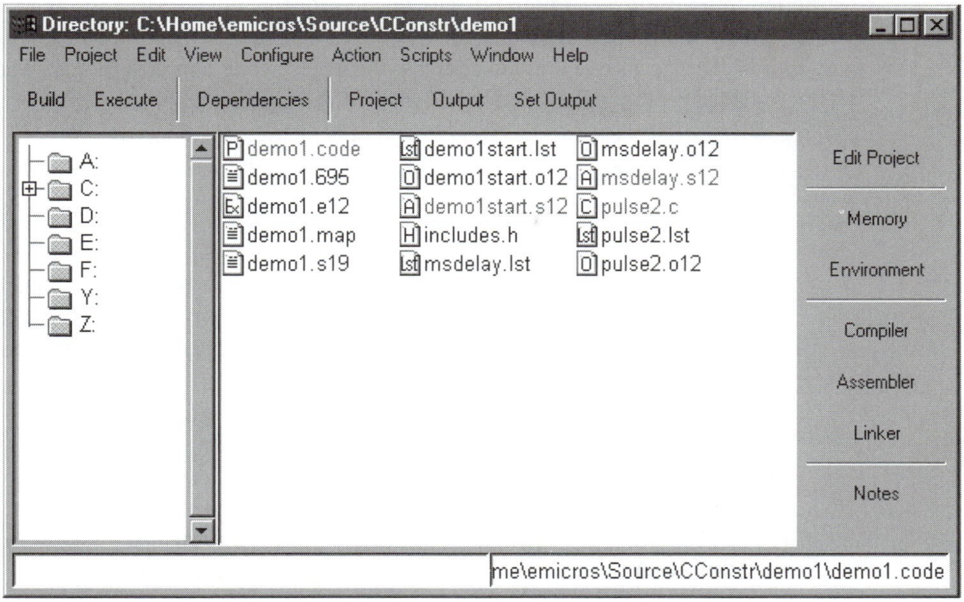

Figure 12.5 Project Directory Contents for *demo1* after *build*

box for the *demo1* project. This dialog box determines both the configuration of the target and specifies the build process. The part shown in the figure is only one page of the dialog box. The other pages are accessed by the tabs along the top.

The parameters set in the *Edit Project* dialog box are saved in the *project.code* file. The *project.code* file is a text file, but in general there is no need to view or edit it with a text editor. To execute the build process, we simply press the *Build* button. Figure 12.5 shows a window that contains the *Build* button.

12.2.6 The Generated Files

Depending on your build parameters, there are several files generated by a build command. Figure 12.5 shows the project directory for *demo1* after the project is built. The most obvious files to be generated are the object files and the S-Record file. These are the files required to execute the program. There are also several other files generated for documentation and debugging. These include the relocatable listing files with the *.lst* extension, the absolute listing file with the *.adr* extension, the map file with the *.map* extension, and an IEEE695 file with the *.695* extension.

Relocatable Listing Files. Listing 12.1 shows part of *pulse2.lst,* which is the listing file for the *pulse2.c* module. It contains the assembly code that the compiler generated along with the object created by the assembler. It also has a lot of other debugging information embedded in the code.

Notice that some lines in the object code are preceded by a ">." This means that the code in that line has unresolved references. For example, in line 21, there is a *movb* instruction to load *PulseCnt* with 60. The opcode for the *movb* instruction is $180B. This is followed by the immediate value for 60 ($3C). The actual location of *PulseCnt* is not known, however, so in place of its address, the object code contains $0000. These zeros will be replaced by the linker once the address of *PulseCnt* is resolved.

Listing 12.1 pulse2.lst, the Relocatable Listing File from pulse2.c

```
pulse2.s12                  Intrrol C compilation of    'pulse2.c'
    2                                         file       pulse2.c
    3                                         import.s   _H12PORTP
    4                                         import.s   _H12DDRP
    5                                         section    .bss,bss
    6                                         line       16
    7 00000000                  PulseCnt: ds.b       1
    8                                         type       ubyte_t
   10                                         section    .text
   11                                         line       19
   12 00000000                  main:       fbegin
   13 00000000                              fentry
   14                                         func       void
   15                                         endl
   16                                         line       21
   17 00000000 >4c0001                       bset       <_H12DDRP,#$1
   18                                         line       22
   19 00000003 >4c0001                       bset       <_H12PORTP,#$1
   20                                         line       24
   21 00000006 >180b3c0000                   movb       #60,PulseCnt
   22 0000000b   2015                         bra        ?0.12
   23 0000000d                  ?0.7
   24                                         line       25
   25 0000000d >4c0001                       bset       <_H12PORTP,#$1
   26                                         line       26
   27 00000010   cc02ee                       ldd        #750
   28 00000013 >160000                        jsr        msDelay
   29                                         line       27
   30 00000016 >4d0001                        bclr       <_H12PORTP,#$1
   31                                         line       28
   32 00000019   cc00fa                       ldd        #250
   33 0000001c >160000                        jsr        msDelay
   34                                         line       24
   35 0000001f >730000                        dec        PulseCnt
   36 00000022 >f60000          ?0.12        ldab       PulseCnt
   37 00000025   26e6                         bne        ?0.7
   38                                         line       31
   39 00000027 >4c0001                        bset       <_H12PORTP,#$1
   40 0000002a                  ?0.14
```

```
41                                            line    32
42 0000002a   20fe                            bra     ?0.14
43                                            line    33
44 0000002c                                   fexit
45 0000002c   3d                              rts
46 0000002d                                   fend
```

The addresses in Listing 12.1 are all relative to the beginning of the current section in the current module. So if we wanted to find the actual address for code in Listing 12.1, we must first find the absolute address for the start of the module. In this case the address of *main* is at the start of this module, so we can add the address of *main* to the offset listed in this file. We can find the address of *main* by looking in the project map file, *demo1.map*.

For example the address of *main* is 0x0838. Therefore the start of the pulse loop is

$$0x0838 + 0x0d = 0x0845$$

We then could use this address for a breakpoint if we want to break each time through the loop.

Absolute Listing File.   Listing 12.2 shows part of the absolute listing file for *demo1*. The complete file shows all the sources files. Listing 12.2 only shows the listing for the *pulse2.c* source file. Notice it only shows the addresses, not the object code. This file is very helpful for finding addresses to set breakpoints. For the previous example we calculated the start of the LED loop to be at 0x0845. If we look at Listing 12.2, we can confirm this calculation.

Listing 12.2 The Absolute Address Listing for demo1

```
pulse2.c           Introl Iadr Listing for 'demo1.e12'

     1                     /*************************************************************

     2                      * pulse2.c - A simple demonstration program.
     3                      * It generates 60 active-low pulses at a rate of 1Hz out of
     4                      * LED. This version has been made MCU independent and more
     5                      * readable by using header files.
     6                      *
     7                      * Todd Morton, 6/14/99
     8                      *************************************************************
     9                     /* Include Project Header File
    10                     *************************************************************
    11                     #include "demo1.h"

    12

    13                     /*************************************************************
    14                      * Global variable definitions
    15                     *************************************************************
    16 00000865           ubyte_t PulseCnt;   /* The number of pulses remaining   */
    17
    18                     /*************************************************************
```

```
19 00000838        void main(void) {
20
21 00000838            INIT_LED_DIR(); /* Initialize LED as output and turn off*/
22 0000083b            TURN_LED_OFF();
23
24 0000083e            for(PulseCnt = 60; PulseCnt > 0; PulseCnt--) {
25 00000845                TURN_LED_OFF();
26 00000848                mSDelay(OFF_TIME);
27 0000084e                TURN_LED_ON();
28 00000851                mSDelay(ON_TIME);
29                        }
30
31 0000085f            TURN_LED_OFF();      /* Turn LED off and trap  */
32 00000862            TRAP();
33 00000864        }
34            /***********************************************************
35
```

The Map File.   The map file, *demo1.map,* lists all the section information and symbol values. It is extremely valuable for debugging the program. Figure 12.6 shows most of the map file for *demo1*.

```
*** Input File(s) Processed ***
Command File: <stdin>
Object File: msdelay.o12
Object File: pulse2.o12
Object File: demo1start.o12
Archive File: Y:/CODE400/lib/libgen.a12
Member(s) Used:
 hc912b32.o12

*** Section Map ***
Group IO
    Output Section: '.chip', Size: 512 (0x200)
                Origin: 0 (0x0)
                BSS (non-initialized) data area
                Input Section(s):
                '.chip', Size: 512 (0x200)
End of Group IO
Group ROM
    Output Section: '.start0', Size: 0 (0x0)
                Origin: 2048 (0x800)
                Input Section(s):
                '.start0', Size: 0 (0x0)
    Output Section: '.start1', Size: 0 (0x0)
                Origin: 2048 (0x800)
                Has no input sections
    Output Section: '.start2', Size: 0 (0x0)
```

Figure 12.6   The *demo1* Map File

```
                    Origin: 2048 (0x800)
                    Has no input sections
Output Section: '.start3', Size: 29 (0x1d)
                    Origin: 2048 (0x800)
                    Input Section(s):
                    '.start3', Size: 29 (0x1d)
Output Section: '.start4', Size: 3 (0x3)
                    Origin: 2077 (0x81d)
                    Input Section(s):
                    '.start4', Size: 3 (0x3)
Output Section: '.start5', Size: 0 (0x0)
                    Origin: 2080 (0x820)
                    Has no input sections
Output Section: '.start6', Size: 0 (0x0)
                    Origin: 2080 (0x820)
                    Has no input sections
Output Section: '.start7', Size: 0 (0x0)
                    Origin: 2080 (0x820)
                    Has no input sections
Output Section: '.start8', Size: 0 (0x0)
                    Origin: 2080 (0x820)
                    Has no input sections
Output Section: '.start9', Size: 0 (0x0)
                    Origin: 2080 (0x820)
                    Has no input sections
Output Section: '.startX', Size: 3 (0x3)
                    Origin: 2080 (0x820)
                    Input Section(s):
                    '.startX', Size: 3 (0x3)
Output Section: '.startZ', Size: 4 (0x4)
                    Origin: 2083 (0x823)
                    Input Section(s):
                    '.startZ', Size: 4 (0x4)
Output Section: 'ROM.text', Size: 62 (0x3e)
                    Origin: 2087 (0x827)
                    Input Section(s):
                    '.text', Size: 62 (0x3e)
Output Section: 'ROM.const', Size: 0 (0x0)
                    Origin: 2149 (0x865)
                    Input Section(s):
                    '.const', Size: 0 (0x0)
                    '.strings', Size: 0 (0x0)
Output Section: '.init', Size: 0 (0x0)
                    Origin: 2149 (0x865)
                    Has no input sections
Output Section: '.bss', Size: 1 (0x1)
                    Origin: 2149 (0x865)
                    BSS (non-initialized) data area
                    Input Section(s):
                    '.bss', Size: 1 (0x1)
Output Section: '.data', Size: 0 (0x0)
```

Figure 12.6 *(continued)*

```
                    Copied from section .init
                    Origin: 2150 (0x866)
                    Has no input sections
        Output Section: '.retm', Size: 0 (0x0) (Overlay)
                    Origin: 2150 (0x866)
                    Has no input sections
        Output Section: '.heap', Size: 0 (0x0)
                    Origin: 2150 (0x866)
                    Has no input sections
        Output Section: '.stack', Size: 410 (0x19a)
                    Origin: 2150 (0x866)
                    Has no input sections
        Output Section: '.base', Size: 0 (0x0)
                    Origin: 2560 (0xa00)
                    Has no input sections
End of Group ROM

*** Symbol Table ***
In File: 'ms_delay.o12'
                .text           2087 (0x827), section: 14
                mSDelay         2087 (0x827), section: 14
                msdlp           2087 (0x827), section: 14
                d1ms            2095 (0x82f), section: 14
                d1mslp          2099 (0x833), section: 14
In File: 'pulse2.o12'
                .bss            2149 (0x865), section: 17
                PulseCnt        2149 (0x865), section: 17
                .text           2104 (0x838), section: 14
                main            2104 (0x838), section: 14
                                2104 (0x838), section: 14
                                2148 (0x864), section: 14
                                2149 (0x865), section: 14
In File: 'demo1start.o12'
                .start0         2048 (0x800), section: 2
                __start         2048 (0x800), section: 2
                .start4         2077 (0x81d), section: 6
                __stackinit     2077 (0x81d), section: 6
                .start3         2048 (0x800), section: 5
                __bssinit       2048 (0x800), section: 5
                2               2053 (0x805), section: 5
                1               2055 (0x807), section: 5
                __datainit      2060 (0x80c), section: 5
                2               2068 (0x814), section: 5
                1               2072 (0x818), section: 5
                .startX         2080 (0x820), section: 12
                .startZ         2083 (0x823), section: 13
                __exithalt      2083 (0x823), section: 13
                __exit          2083 (0x823), section: 13
                loop            2085 (0x825), section: 13
In File: 'Y:/CODE400/lib/libgen.a12:hc912b32.o12'
...
```

Figure 12.6   *(continued)*

It first lists all the object files that are part of the project. In this case it shows *pulse2.o12, msdelay.o12, demo1start.o12,* and *libgen.a12.* After a library file like *libgen.a12,* it shows the functions that were used out of that library. In this case *hc912b32.o12* was used. This library code defines the MCU registers in the M68HC912B32.

After the input files, the map file shows the *Section Map,* which includes the size and location for every code section. It is important to keep an eye on the Section Map as you develop your code. It will tell you how much memory space the program is using.

For the *demo1* project, the sum of all the sections is 102 bytes. This is the total amount of memory used. If the program were a stand-alone design, we would need to distinguish between ROM and RAM requirements. For the *demo1* program, only one byte of RAM was used; the one byte was in the *.bss* section and resulted from the variable *PulseCnt.*

After the Section Map is the *Symbol Table.* The Symbol Table shows the values assigned to all the symbols in the program. It is organized by input file, so you can distinguish between two symbols with the same name from different modules. The Symbol Table is helpful for debugging. If you need to know where a variable is or what the location of a function is, you can find it in the Symbol Table. For example from Figure 12.6 we can see that the *main* function is located at 0x0838.

12.2.7 The Command-Line Interface

Although the world is flocking to graphical interfaces, the command-line interface is a viable and efficient way to develop modular software. The basic ways that the command-line build process differs from the same process using a graphical interface are the way the target configuration is passed to the linker and the way the overall build process is executed. For the target configuration, we create a linker command file. This file contains the information required to import library functions, generate start-up code, define MCU registers, and define all the sections. To automate the build process, we use another program, *make.* The *make* program is widely available and is a powerful tool.

The Linker Command File. For the command-line interface, the linker commands are contained in the linker command file *project.ld.* Source 12.7 shows an example of a linker command file for building programs that run under D-Bug12. Notice that the linker command file is independent of the project source. It defines the configuration of the target environment. In this case it is for programs built to run under D-Bug12 on the M68HC912B32 EVB board.

Source 12.7 An Example of a Linker Command File

```
//******************************************************************
// wwudb12.ld - Linker command file for programs running under D-Bug12
//              on the 68HC912B32EVB. Both RAM and ROM sections use
//              on-chip RAM from $0800-$09FF.
//              Todd Morton, 4/2/98
//******************************************************************
//General configuration settings
set RAMORG = 0x0800;   /*Start of RAM*/
```

```
set RAMSIZE = 512;      /*User RAM size */
set REGORG = 0x0000;    /*Origin of 68hc11 registers */
set _chipregs = 0x0000;  /*Register location*/

//*********************************************************************
// Select F1Board memory locations
//*********************************************************************
group RAM rom origin RAMORG maxsize RAMSIZE;
section .start0 text;
section .start3 text;
section .startX text;
section .startZ text;
section .text text;
section .const data = .const, .strings;
section .init;
//*********************************************************************
section .data data copiedfrom .init = .data;
section .bss bss comms;
section .retm bss comm;
section .stack bss minsize 80;
group RAM;


//*********************************************************************
//      chip-specific sections:

section .chip bss origin _chipregs;

//*********************************************************************
// Import processor register set.
//*********************************************************************

import __hc912b32;

//***********************************************************************
// Start-up sequence: Not much here. D-Bug12 does most of the initialization.
//***********************************************************************
// .start0 - Label defined for reset start.
//          Do not add to this section.

import __start;

//***********************************************************************
// .start3 - Fundamental system initialization
//          Do not add to this section

import __bssinit;      // .start3 : zero out bss area (required by library)
import __datainit;     // .start3 : copy the initialized data area

//***********************************************************************
// .startx - Entry point definition:
```

```
//              Do not add to this section

set __entry = main;

//*************************************************************************
// .startz - Exit point if 'rts' at end of 'main' is reached.
//              Do not add to this section

import __exithalt;       // .startZ: program termination code (halts)

set __ramstart = startof(.bss);
set __ramend = endof(.bss);
set __stackstart = startof(.stack);
set __stackend = endof(.stack);
set __initstart = startof(.init);
set __initend = endof(.init);
set __datastart = startof(.data);
set __dataend = endof(.data);
set __fastend = endof(.start1);
```

It is not the intention at this point to cover the syntax of the linker command file. This is covered in the Introl-CODE manual. We can, however, look at Figure 12.7 to get an idea of what it contains.

The first part of the file uses the *set* command to set some important configuration constants. These are not required, but they help make the file more readable and easier to change.

The next part of the linker command file determines how code sections are mapped into memory blocks. In this case everything goes into the user RAM block on the EVB board.

Next the linker file contains *import* commands that are required for system initialization. The *import* command tells the linker to add the named code, typically from a library. Again this code is somewhat project independent. The initialization code added is based on the target requirements. It is more convenient to include them here instead of in the source code files, because it is dependent on the target, not the project code. So if the target is changed, the programmer should have to edit only the linker command file and then rebuild the project.

Using *make* to Automate the Build Process.   If we use the command-line interface, the program normally used to automate the execution of the build processes is called *make*. There are many different versions of the *make* program for both UNIX and Windows systems. Using *make,* the build process is specified in the *Makefile* and is executed by typing

```
make project
```

where *project* is the project name. Entering this command is essentially the same as pressing the *Build* button in the graphical interface.

Source 12.8 shows an example of a *Makefile* used to build the *demo1* project. It is easy to configure this *Makefile* for a specific project. Typically only the *PROJ, CSRCS,*

SRCS, and LD variables need to be set. The *PROJ* variable is set equal to the project name. The *CSRCS* variable is set equal to a space-separated list of all C source files in the project. The *SRCS* variable is set to a space-separated list of all assembly sources in the project. The *LD* variable is set equal to the linker command file name.

Source 12.8 An Example of a Makefile to Build demo1

```
#
# Makefile for GENERIC 68HC12 Introl-CODE project.
# Todd Morton, 1/19/99
#
# Enter the project name, C sources (CSRCS), assembly sources (SRCS),
# and the linker command file (LD). If there are multiple sources, list
# all sources separated by spaces.
#
PROJ = demo1
CSRCS = pulse2.c
SRCS = ms_delay.s12
#
# Use one of the following linker command files for a standard configuration
# or create a project linker command file specific for your project and enter
# it below on the 'LD = ' line.
#
#       wwudb12.ld - running under D-Bug12.(Everything at $0800-$09FF)
#       b32alone.ld - running stand-alone on the 68HC912B32 EVB.
#                       (ROM at $8000, RAM at $0800)
#
LD = wwudb12.ld
#
# In general you should not have to change anything from here on.
#
.SUFFIXES : .o12 .s12 .c
# Library and system header file paths.
#
LIBS = -lwwu912 -lgen -lc
ASLIBS = -yy=$(INTROL)/wwu/Libraries/wwu912
LDPATH = $(INTROL)/wwu/Libraries/wwu912
INCL = -i=. -i=$(INTROL)/wwu/include/wwu912
COBJS = $(CSRCS:.c=.o12)
OBJS = $(SRCS:.s12=.o12)
LSTS = $(SRCS:.s12=.lst) $(CSRCS:.c=.lst) $(PROJ).alst

$(PROJ): $(OBJS) $(COBJS)
        ild12 -g$(LD:.ld=) $(OBJS) $(COBJS) $(LIBS) -f$(PROJ).map -o$(PROJ)
        ihex $(PROJ)
        iadr $(PROJ) -o $(PROJ).alst

.s12.o12:
        as12 -l $(ASLIBS) $<
```

```
.c.o12:
        cc12 -l $(INCL) $<

clean:
        rm -f $(PROJ) $(OBJS) $(COBJS) $(PROJ).0 $(LSTS) $(PROJ).0 $(PROJ).map
```

Following is an example of the command lines that are executed when the *demo1* program is built with the *make* command shown previously:

```
> make demo1
as12 -l -yy=/usr/local/introl400b3/wwu/Libraries/wwu912 ms_delay.s12
cc12 -l -i=. -i=/usr/local/introl400b3/wwu/include/wwu912 -
i=/usr/local/introl400b3/wwu/include/mcx12 pulse2.c
ild12 -gwwudb12 ms_delay.o12 pulse2.o12 -lwwu912 -lgen -lc -fdemo1.map -odemo1
ihex demo1
iadr demo1 -o demo1.alst
```

As you can see, it would not be realistic to type these commands by hand every time you want to build the project. Using *make* allows us to build extremely complex development projects efficiently.

make also allows us to create some useful utilities. With the *Makefile* shown previously, we can key

```
make clean
```

to delete all the generated files. This allows us to archive the project with a minimal space requirement or to force a complete rebuild of the project.

Command-Line Interface Summary. If we use the command-line interface, we can have an efficient development system. In fact, for some parts of the process, using a command-line interface allows more flexibility and control. We must learn how to use *make* and how to create a linker command file, however. Both of these tasks are somewhat complicated. The primary advantage to using the graphical interface is that you do not have to deal with linker command files or *make* files. In this text we do not cover either of these tasks in detail. Detailed documentation for the linker command file can be found in the Introl-CODE manual. There are a few good books on *make;* one in particular is *Managing Projects with make,* from O'Reilly and Associates.

▶ 12.3 SOURCE-LEVEL DEBUGGING

Programming in C and using the modular build process complicates the debugging task. When we use C source, there is no longer a one-to-one correspondence between the source code and the CPU operations. We have to translate the code executed by the CPU back to the C source. Using C also adds another source of bugs during the translation from C to assembly. These errors may be due to compiler bugs, compiler inefficiencies, or a misinterpretation by the programmer of the C language.

In this section we cover a manual debugging process that is based on the debugging techniques covered in Chapter 3 and Chapter 10. We then introduce the Noral Flex Debugger and the BDM pod, a full-featured, source-level debugging system that uses the CPU12's background debug system.

12.3.1 Manual C Code Debugging

This process is based on the debugging systems we have already covered. It is an inexpensive but tedious method. To translate the CPU operations to the C source code, we make use of the generated listing and map files. The listing files are used to see the assembly code generated by the compiler, and the map file is used to find symbol values, including labels. Finally, to find the starting address for each line of C code, we use the absolute listing file.

Let's illustrate this process by debugging a portion of the *demo1* project. In this example we set a breakpoint and trace through the pulse-counting process and the decision to quit or loop back to generate a new pulse.

The first step is to find the critical piece of code for this example. Listing 12.3 shows the critical piece of code from the listing file for *pulse2.c* shown in Listing 12.1.

Listing 12.3 Critical Code from pulse2.lst

```
34                                        line      24
35 0000001f >730000                       dec       PulseCnt
36 00000022 >f60000         ?0.12         ldab      PulseCnt
37 00000025  26e6                          bne       ?0.7
```

Since this is from a relocatable listing, the address shown is not the actual address in the target system. We have to calculate the actual address by adding the offset shown to the value of the label *main*.

$$0x0838 + 0x1F = 0x0857$$

So we need to set the breakpoint to 0x0857 to stop execution before the code on line 35. The next piece of information we need is the address for the variable *PulseCnt*. We can find this in the map file shown in Figure 12.6 on page 468. *PulseCnt* is at 0x0865. Now we can set the breakpoint, run the program, and examine the value of *PulseCnt*. The following transcript shows the D-Bug12 commands and responses:

```
S>br 0857
Breakpoints: 0857
S>g 0800
R>
User Breakpoint Encountered

  PC    SP    X     Y     D = A:B   CCR = SXHI NZVC
 0857  09FE  0865  0866    00:00          1111 0100
 0857  730865         DEC   $0865
```

At this point the program reached the breakpoint. It is a good habit to look at the code shown and verify that it is the same as the code from the listing file. We can now examine *PulseCnt:*

```
S>md 0865

0860   56 01 20 FE - 3D 3C 00 00 - 00 00 00 00 - 00 00 00 00    V. .=<..........
```

As expected, *PulseCnt* is equal to 60 ($3C). Now let's trace through the next few instructions to watch the decision to generate another pulse or to quit:

```
S>t

PC     SP     X      Y      D = A:B    CCR = SXHI NZVC
085A   09FE   0865   0866     00:00          1111 0000
085A   F60865          LDAB  $0865
S>t

PC     SP     X      Y      D = A:B    CCR = SXHI NZVC
085D   09FE   0865   0866     00:3B          1111 0000
085D   26E6            BNE   $0845
S>t

PC     SP     X      Y      D = A:B    CCR = SXHI NZVC
0845   09FE   0865   0866     00:3B          1111 0000
0845   4C5601          BSET  $0056,#$01
```

As expected, the program looped back to generate a new pulse. We now proceed to get to the breakpoint again:

```
S>g
R>
User Breakpoint Encountered

PC     SP     X      Y      D = A:B    CCR = SXHI NZVC
0857   09FE   0865   0866     00:00          1111 0100
0857   730865          DEC   $0865
```

We can continue this process as long as necessary. As you can see, it is the same process we saw before, with the added task of calculating the address and looking up symbols.

12.3.2 Using a Source-Level Debugger

In general a source-level debugger refers to an application that can control program execution on a target system and correlate the program execution with the original source code. Many programmers are first introduced to this type of system while taking a programming course for writing PC applications that uses Microsoft's Visual Studio. When writing code for embedded microcontrollers, however, the target is not the same system that runs the debugger. The target may be a simulator, an emulator, or an MCU running under the control of a monitor program or background debugging system.

In traditional source level debugging systems, an emulator is used to control the CPU, collect data from the CPU memory, and load programs into memory. These are the most powerful debugging tools available for debugging high-level language programs. They are also the most expensive, however, because extensive hardware is required to emulate the processor.

As PCs have become more powerful, simulators have become more popular. No target hardware is required, so the cost is low and the source-level debugger has full control of the program execution. The Introl-CODE system includes simulators and debuggers for the processors supported. As we have discussed before, simulators are limited by their inability to accurately simulate the external circuitry connected to the system. They also cannot simulate real hardware-related effects on code execution. For example, if a program turns on a solenoid, a simulator can only indicate that the solenoid has been turned on. It cannot show a hardware problem like a power supply voltage drop due to the current required by the solenoid, which may in turn, reset the MCU. If we debug a program on a real target, we will catch software bugs and hardware-related problems.

To reduce the cost of the debugging system while running on the target hardware, we can use a source-level debugger that communicates with a resident monitor program. This helps with the task of mapping C source code to the target memory and execution. Since the program being tested must be stopped and control passed to the monitor, however, this system may be inadequate for real-time debugging.

The on-chip debugging features that are found on many new microcontrollers provide an ideal solution to this problem. Because the debugging hardware is integrated into the microcontroller, the CPU emulation circuitry is no longer required. The only hardware required is a communications pod to communicate with the on-chip debugging circuitry. This type of system provides us with most of the capabilities of a full emulator at a reduced cost.

In this section, we introduce the Noral Micrologics Flex BDM Debugger for the 68HC12 family of microcontrollers. It takes advantage of the 68HC12's Background Debug Module (BDM) to provide real-time source-level debugging while the code runs on the target hardware. The system consists of a source-level debugging application and a BDM pod connected between a PC parallel port and the target hardware.

This system is more expensive than using Motorola's 68HC12 EVB as a background debug pod but less expensive than a full emulator—if one existed. As a program developer, once you use a full-feature source-level debugger like this, you will never want to go back to the basic systems we have used in the rest of this book. In addition the cost savings realized by reduced development time and increased code-reliability may pay for this system many times over.

Figure 12.7 shows an example of the Noral 68HC12 BDM debugger's working environment while running the *demo1* code that we have used throughout this chapter. We cover a few of the capabilities of this debugger by going over each window shown in the figure. This includes the *Source* window, the *Registers* window, the *Monitor Points* window, the *Watch* window, and the *Call Tree* window.

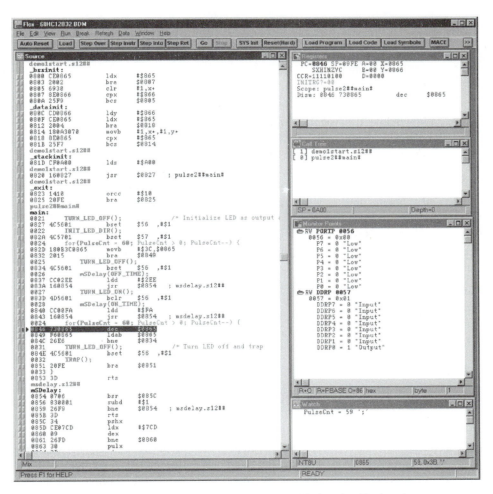

Figure 12.7 Noral 68HC12 BDM Source-Level Debugging Environment

***Source* Window.** The *Source* window shows the source code of the program being tested. In Figure 12.7 the *Source* window shows the source code, which is either C code or assembly, and the assembly code contained in memory. It also shows the module name and the routine name for the source of each function. Optionally we could also display the source in assembly only or C only, but when testing code for embedded microcontrollers, it is important to see both the source and assembly code, so we normally use the mixed mode as shown.

The highlighted line shows where the code execution was stopped with a breakpoint. The breakpoint was set by clicking on the breakpoint box directly to the left of the assembly code line. The buttons on the top of the display control program execution. We can run the program with the *Go* button or single-step to the next instruction with the *Step Instr* button. In addition, if we display C code only, we can step into a function or step over a function using the *Step Into* and *Step Over* buttons.

The debugger correlates the code from memory with the code in the source files by using information contained in an IEEE-695 standard debug file. To generate this file with the Introl-CODE system, the IEEE-695 check box is selected in the Output tab of the *Edit Project* dialog box. This generated the *demo1.695* file shown in Figure 12.5 on page 465.

Registers Window. The top window on the right-hand side of Figure 12.7 is the *Registers* window. It is roughly equivalent to the *RD* command in D-Bug12. It shows register contents at the breakpoint, the current module and routine, and the disassembled source code from memory. Register contents are normally updated when a breakpoint is reached or code execution is stopped because the 68HC12 BDM cannot access the registers without interrupting code execution. We can set the register display to be updated periodically, but the updates will affect the real-time execution of the program.

Call Tree Window. The *Call Tree* window displays the current source code context. It includes the module name followed by the function name for each nested function call. In the display it indicates that we started in the start-up module *demo1start.s12*. The program then called the function *main()* in the *pulse2.c* module.

Each nested function is indicated on a new line, or *frame,* in the *Call Tree*. The frame number indicates the number of returns required to return to that frame. Therefore the current context always has a frame number of zero.

The *Call Tree* is especially useful for large modular programs because it provides a clear indication of where we are in the program code.

Monitor Points Window. The *Monitor Points* window is used to display the current contents of a memory location. These points are absolute memory locations such as the MCU control and status registers. In Figure 12.7 we have the system set to monitor PORTP and DDRP. Specifically we are interested in bit 0 of these two registers, because this is the general purpose output on which the pulse will be generated. Since the 68HC12 BDM can read address contents without interrupting the program execution, we can monitor these locations in real time.

In D-Bug12 we would use the *MD* command to display this information. The update time for this window is configurable. If we set it to update at an appropriate rate, we would see *PORTP,* bit 0 generate a low pulse every second while the program runs. Since the *MD* command is a manual instruction, however, we could not look at the contents of a memory location at a fixed rate. With the Noral debugger, we can have the display updated at a fixed rate or have the display updated at designated points in the code execution called *Refresh Points. Refresh Points,* however, are implemented with breakpoints, so they can affect the real-time execution of the code.

Watch Window. The *Watch* window is essentially the same as the *Monitor Points* window except it allows us to monitor C data objects by name instead of first looking up the address of the object. Again, to do this with D-Bug12, we would use the *MD* command.

We have just touched on the main features of the Noral 68HC12 BDM debugger. There are many other features we have not covered, such as loading Flash EEPROM, monitoring local variables, and automating the debug process by creating debugging macros.

480

These are beyond the scope of this text. At this point, however, you should be able to see the benefits of using this type of system.

► SUMMARY

In this chapter we have looked at the contents of a C source program, covered the modular build process, and covered techniques for debugging C code. These are the basic tasks required for modular C code construction.

Throughout the chapter we used the *demo1* project as an example. Compared to the *demo1* program and development process described in Chapter 3, we went from three project files to 15. Indeed, it seems like we have made a simple program relatively complex. That is the first impression for many programmers who try to program in C for the first time. Notice, however, that the complexity is really due to the modular build process, not due to having a C source file.

The reason for adding this complexity is that it is scalable to very large programs. In this case the program is small and the benefits are negligible. We do not have to add much more complexity before the advantages of this development system are understood, however.

Debugging C code can be done inexpensively by using the EVB systems covered previously in this text. They are fine for hobbyists or for simple projects with limited funding, but for developing real products, it is most beneficial to use a full-featured, source-level debugger. Using modular C source code and a full-featured, source-level debugger, we can efficiently develop large and complex real-time embedded systems.

EXERCISES

1. The following program was found in the recycle bin. Reorganize it to meet readability standards.

```
void OCDelay
    (INT16U ms)
{INT16U term_cnt;term_cnt = OCmSCnt + ms;
    while(OCmSCnt
!=
term_cnt)
{}}
```

2. Find the hexadecimal value of the following constants:
   (a) `0x11`
   (b) `'\r'`
   (c) `010`
   (d) `0`
   (e) `'0'`
   (f) `'\020'`

3. The following line was written by mistake. If *b* is equal to 0x81 and *c* is equal to 0x99, what is *a* after the statement is executed?

```
a = b && c;
```

4. Rewrite the following statement using an assignment expression:

```
InVar = 0x80 ^ InVar;
```

5. Rewrite the following statement using a simple "=" assignment:

```
Port &= 0x80 | InVar;
```

Be careful to preserve the order of execution.

6. A program that uses the function *LibFun()* from a library generates errors when it is built. Describe the probable cause for the following error messages:
   (a) The error message is from the compiler and says something like the following:

```
"Call to nonprototyped function, LibFun()"
```

   (b) The error message is from the linker and says something like the following:

```
*** Undefined Symbols ***
In File: 'program.o12'
    LibFun
```

7. Using the map file in Figure 12.6
   (a) Find the starting address for the *msDelay* routine.
   (b) Sketch a memory map of the program space that shows the starting and ending address for each source module.
   (c) Determine how many total bytes of memory were used by the program. Do not include stack space.

8. Using the map file in Figure 12.6 and the listing in Listing 12.1, find the absolute address of the following listing line:

```
27 00000010  cc02ee                ldd #750
```

Creating and Accessing Data in C

In this chapter we focus on data types and the most common data structures used when programming embedded controllers in C. The chapter emphasize methods used for embedded systems that are not normally covered when programming computer systems. In fact some techniques we use here are actually seen as poor programming practice in the computer world.

First we analyze how the C compiler implements data objects. It is essential that the programmer understand how the various data types are actually implemented in order to understand the effect on storage space and execution speed. Using an incorrect data type or not understanding how the compiler handles type conversions can result in serious errors or inefficient code. This is especially important for the small 8-bit or 16-bit embedded systems described in this book. Some programming practices used by computer programmers would spell disaster if applied to these microcontrollers.

Second, when programming embedded microcontrollers, we never stray far from the hardware. This means having to operate on absolute locations for peripheral control or memory tests. Setting a pointer equal to an absolute location is rarely a good idea when writing computer applications, but we need to do this often. We also need to know the type of memory device in which the different parts of our code will reside. Parameters in EEPROM and stored constants in ROM are two examples of data objects that are not found in traditional computer programming where all data are stored in RAM.

13.1.1 Data Types in Assembly

Before we continue with our discussion on C, let's look at how we handled data in our assembly programs. In assembly language we create a data object using an assembler directive. The programmer is responsible for defining, accessing, and converting data objects. Some of the details can be missed easily, and since the assembler cannot perform type checking, this can result in unreliable code.

Recall the definition and implementation of a global variable in an assembly program. The variable was defined and memory was reserved using the *rmb* directive. For example

```
Var8        rmb 1        ; an 8-bit variable
Var16       rmb 2        ; a 16-bit variable
```

To use these variables, it was the programmer's responsibility to keep track of the data type. Following is an example of a type mismatch error in assembly:

```
        ldd Var8        ; data type error
```

This is an error because the destination is a 16-bit register and the source is an 8-bit variable. The assembler has no way of checking for this error, so it could be undetected for quite some time.

Another common data type error in assembly programming is caused by using the incorrect conditional branch instruction. It is easy to use a *bgt* instead of a *bhi* after two unsigned numbers are compared, for example. This is a data type error because *bgt* is used for signed numbers and *bhi* is used for unsigned numbers. Again this error would not generate an assembler error. The program may work during initial tests, but during normal operation a certain combination of events may cause the incorrect decision to be made.

13.1.2 Data Type Checking

When writing code in C, the programmer is required to explicitly define the data types so the compiler knows how to create and access the data. Once the compiler knows the data types, it can also perform some type checking or type conversion for us.

When a compiler detects a data type mismatch, it can stop compiling and generate an error message, or it can automatically convert one of the data objects and continue as if nothing has happened. A compiler that stops and generates an error is the most reliable, because it forces the programmer to look at the mismatch and explicitly define a conversion. The compiler that makes an automatic conversion is more convenient to use, because the programmer does not have to deal with the conversions. Unfortunately the automatic conversion that the compiler makes may not be the conversion expected. Because of this, data type errors may be missed.

C compilers are a compromise when it comes to data type conversions. They will disallow some mismatches, give an error message and stop, give warnings for other mismatches and continue, or make a conversion and continue without a warning or error message for other mismatches. This is not a good compromise, and it is the source of much criticism toward the C language. Some improvements have been made to ANSI-C, but it is still seen as a weakness.

▶ 13.2 ANSI-C DATA TYPES

A data object is one or more memory locations that contain data. Figure 13.1 shows a data object that is one byte. The name of the data corresponds to its address, and the value corresponds to the contents of that address.

A data object has three main attributes: its storage class, its scope, and its type. The storage class defines the memory type and persistence of the data object, the scope defines the region in the program where the data are accessible, and the type determines how the contents of the object are interpreted.

In C all data objects must be defined or declared in order to name and define the attributes of the data object. A declaration is used for external data objects, so no storage is allocated. A data definition allocates the storage for the data. For example the definition

```
unsigned char PulseCnt;        /*The number of pulses remaining*/
```

defines a single byte variable named *PulseCnt*. The data contents, the storage, and the scope of this data are determined by its data type and by its placement in the program. In this case the declared data type is an *unsigned char*. If we assume that this declaration was placed outside all function blocks, we know that *PulseCnt* is an 8-bit, unsigned, global variable. One byte of RAM space is permanently allocated for this variable.

The name of the object, in this case, *PulseCnt*, follows the data type. As shown in Figure 13.1, the name corresponds to the object's address. Data object names in C are the same as labels in assembly. For code readability the object's name should always be descriptive of the contents and its application. The name should also follow the naming conventions used for the project.

The comment contains the description of the data. It is not a requirement for the compiler, but it is considered good programming practice. In this case *PulseCnt* is the number of pulses remaining to be generated. Without this description, we can only look at the name and guess—is it equal to the number of pulses generated or the number of pulses remaining?

If we use a data object in a different file than the one in which it was defined, we must declare it. For example to declare *PulseCnt*, we would use

```
extern unsigned char PulseCnt;      /*The number of pulses remaining*/
```

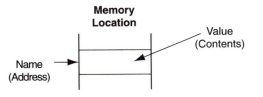

Figure 13.1   A Data Object

This declaration tells the compiler the characteristics of the data, but it does not allocate storage for the data because it is referring to an object that is defined in a different file.

13.2.1 Accessing Data

As described previously, the name of a data object corresponds to the object's address, and the value of the object corresponds to the contents of that location. Yet the name is used in a C program to access the object's value. For example

```
PulseCnt = TotalPulses;
```

PulseCnt and *TotalPulses* are the names of two data objects, so they are equal to the addresses of the objects. The contents are affected by this statement, however, not the addresses. In this case the contents of *TotalPulses* are copied into the contents of *PulseCnt*.

There are times when we need to refer directly to the address of the object. To do this in C we use the & operator. For example in the statement

```
CountPtr = &PulseCnt;
```

the contents of *CountPtr* will be set equal to the address of *PulseCnt*. In effect *CountPtr* will point to *PulseCnt*. When reading C code, we read the & operator as *the-address-of.*

13.2.2 Fundamental Data Types

The data type of a data object describes its storage class, scope, and type. First, let's look at the fundamental data types. All other types are derived from these fundamental types. Table 13.1 shows the fundamental data types for the CPU11 and the CPU12 using the Introl-CODE development system. Fundamental data types are dependent on the target processor and the compiler being used. This means that an *int* may be 16 bits for one processor and 32 bits for another. When working with small microcontrollers, it is especially important to understand the definitions of the fundamental data types.

For an ANSI-C compiler, you can count on two things: (1) The size of a *short* is always less than or equal to the size of an *int* and (2) the size of a *long* is always greater than or equal to the size of an *int*.

TABLE 13.1 FUNDAMENTAL DATA TYPES FOR THE CPU11 AND CPU12 USING INTROL-CODE

Type	Size in Bits	Signed Range	Unsigned Range
char	8	$-128 \ldots 127$	$0 \ldots 255$
short	16	$-32768 \ldots 32767$	$0 \ldots 65535$
int	16	$-32768 \ldots 32767$	$0 \ldots 65535$
long	32	$-2147483648 \ldots 2147483647$	$0 \ldots 429496729$
float	32	$\pm 1.2 \times 10^{-38} \ldots \pm 3.4 \times 10^{38}$	
double	32	$\pm 1.2 \times 10^{-38} \ldots \pm 3.4 \times 10^{38}$	
long double	64	$\pm 2.2 \times 10^{-308} \ldots \pm 1.8 \times 10^{307}$	

Type Modifiers. Type modifiers add additional type information to the fundamental type. Type modifiers are used to define the data's storage class and the data's scope. In addition the *signed* and *unsigned* modifiers are used to define whether the data are to be interpreted as signed or unsigned. For example

```
unsigned int k;
```

declares *k* as an unsigned 16-bit integer.

If the *signed* or *unsigned* modifiers are not used, the compiler will choose a default type. The Introl-CODE defaults are

```
char i;     /* i is an unsigned char */
int k;      /* k is a signed int */
short k;    /* k is a signed short */
long m;     /* m is a signed long */
```

Since these defaults are dependent on the compiler implementation, you should not rely on them and always should include the *signed* or *unsigned* modifier. Explicitly declaring the type can be cumbersome. Later we look at how we can use the *typedef* operator to derive new types that are much easier to work with.

13.2.3 Storage Class Modifiers

The storage class of a data object determines three attributes of the data object: the type of memory used for the data, the persistence of the data, and the method used to access the data. The storage class for a data object must be well understood, because it can have a big impact on the efficiency of the resulting code. In general when programming embedded systems declaring the storage classes is more complex than when programming computer systems. This is because of the different types of memory that are required in an embedded system such as RAM, ROM, EEPROM, and BBRAM. For most computer programs everything goes into RAM.

The *persistence* of a data object refers to how long the contents of that object are valid. Persistence ranges from no persistence for *volatile* variables to forever for stored constants. Both storage modifiers and scope modifiers affect the persistence of a data object.

When declaring data objects, two things determine the object's storage class—storage class modifiers and the placement of the definition. Table 13.2 shows the storage class modifiers. Notice that some of these modifiers have different meanings when the declaration is placed inside a function block as opposed to when it is placed outside all function blocks.

The *static* Modifier. When used inside a function block, the *static* modifier forces the compiler to create permanent storage so the data will persist outside the function. If the *static* modifier was not used, the variable would be an automatic variable, which uses the stack.

Using the *static* modifier inside a function block is normally not considered good programming practice. The programmer should first consider whether it makes sense to use a global variable or an automatic variable in the calling routine. The *static* modifier is more

TABLE 13.2 STORAGE CLASS MODIFIERS

Modifier	Placement	Memory Allocated	Comment
none	Outside all function blocks	Permanent, RAM	Variable. RAM space is allocated at compile time. Initialized variables also allocate ROM space for initial values.
none or `auto`	Within a function block	Temporary, Stack	Variable. No memory is allocated at compile time. Initialized variables also allocate ROM space for initial values.
`static`	Within a function block	Permanent, RAM	Variable. RAM space is allocated at compile time. Initialized variables also allocate ROM space for initial values.
`extern`	Outside all function blocks	None	A declaration for a public function or data that are defined in another module.
`const`	Outside all function blocks	Permanent, ROM	Stored constant. Must be initialized.
`register`	Within a function block	Temporary, CPU Register	Variable. The compiler attempts to use a CPU register. May revert to *auto* if the register is needed.
`volatile`	Outside all function blocks or within a function block with `static` modifier	None, qualifier only	Volatile variable. Data may change by an asynchronous event. Therefore it has no persistence.

commonly used with cooperative multitasking programs. *Beware:* It may result in a non-reentrant function.

The *extern* Modifier. The *extern* modifier indicates that the variable belongs to another module or library. It allows the compiler to access the variable without allocating space for it. The space is allocated in the file that defines the variable. The *extern* declaration is required so the compiler knows the data type.

The *const* Modifier. The *const* modifier is used to declare stored constants. It causes the compiler to store constant values into allocated space in ROM. When the *const* modifier is used, the data object must be initialized. If the programmer attempts to change the value of a stored constant, the compiler will give an error. This modifier is rarely used in computer applications because everything is in RAM. In embedded systems it may be used for lookup tables or strings.

The *register* Modifier. When the *register* modifier is used, the compiler tries to allocate a CPU register for the data. If it must give up the register for another operation, however, the data will revert to temporary storage on the stack. This class can reduce program size and execution time, but it is only practical for simple programs.

The *volatile* Modifier. The *volatile* modifier is used to indicate that the data contents can be changed as a result of some other asynchronous event. In effect it tells the compiler that the data have no persistence. Asynchronous events include external hardware events, interrupt driven events, and intertask events from a different task in a preemptive multitasking system. The most common example of a *volatile* data object is a general purpose input or general purpose I/O port. It forces the compiler to read the location every time the contents are required instead of optimizing the program by reading the data once and using a register for future accesses.

13.2.4 Scope Modifiers

The scope of a data object refers to the region in the program in which the object is valid and accessible. Data scope can range from only being accessible within a single function to being accessible everywhere.

The scope of a data object defined within a function block is limited to that function block. This makes sense because, as we saw in the last section, these objects do not persist beyond the function anyway.

In modular programming the terms *public* and *private* are used to describe the scope of a data object or a function across multiple files. The scope of a *public* data object or function includes all modules in the project. The scope of a *private* data object or function is limited to the module in which it is defined.

In Chapter 6 we used the term *global* to describe a data object that persists as long as power is applied. This is a common use of the term *global* when programming in assembly. When writing modular programs in C, however, we have to be more precise. Here we define a global variable as a public static variable. The term *global* implies the static storage class because the data object must persist as long as power is applied.

In C the scope of a data object is controlled by two things: the scope modifier and the placement of the declaration in the program code. The scope of a data object is shown in Table 13.3. Notice that the scope rules also apply to function definitions.

If the definition is placed outside all functions and has no scope modifier, the data or function is available everywhere. This is referred to as a *public data object*.

If the declaration is placed inside a function and has no modifier, the data are only available within the function. The scope of the data object is limited to the function even if the object persists outside the function. For example if a variable is declared inside a function with the *static* modifier, the scope is limited to that function, even though the contents of the variable persist beyond the function.

TABLE 13.3 SCOPE MODIFIERS

Modifier	Placement	Scope	Comment
None	Outside all function blocks	Everywhere	Public function or data definition
None	Within a function block	Within the function block	Scope only includes the function even if the data persist outside the function
static	Outside all function blocks	Within the module	Private function or data definition

The *static* Modifier. When used outside all function blocks, the *static* modifier limits the scope of an object to the module that contains the declaration. This is a definition for a private data object or function and is used for information hiding in modular programs.

Note _____

static versus static

Unfortunately there are two meanings for the *static* modifier in C. The term *static* is normally used to describe a storage class. The static storage class implies that the compiler allocates a permanent RAM location, and the data persist as long as power is applied. When we used the static modifier within a function, this is exactly what it means—use the static storage class for a variable that would otherwise use the stack.

If the static modifier is used outside all function blocks, it does not affect the storage class. Instead it affects the scope of the object. After all, every variable defined outside function blocks use the static storage class. The static modifier in this case tells the compiler to limit the scope to the file. It is used to declare a private data object or function.

Avoid the Global Mind-set. We could make all data available everywhere. That would certainly simplify accessibility, and for some simple programs, it may increase efficiency. We would lose some of the important benefits of using modular programming, however; we also would complicate programs that are in a multitasking environment. In general using global data for everything is lazy programming.

The scope of a data object can have serious implications on the program's reliability. In addition the scope of a data object may also have implications on its storage type. For example if all variables were made to be available everywhere, their contents must persist and therefore they must be stored in a permanent location. This can result in a large waste of RAM space. It also means that the name of every data object must be unique. By limiting the scope of an object, we can reuse common variable names such as loop counters and save precious RAM space. A good programming practice is to try to limit the scope of data objects to the smallest region practical.

13.2.5 Defining New Types

Defining a data object using the fundamental types and modifiers can be cumbersome. For example the following declaration is required to use the 68HC12 *PORTA* register:

```
extern volatile __mod1__ __mod2__ unsigned char _H12PORTA;
```

The *__mod1__ __mod2__* modifiers are special Introl-CODE modifiers that indicate that direct addressing should be used to access these registers because they are in the base page. Luckily we do not have to key this in every time we need to define or declare a register. These definitions are contained in the *hc912b32.h* header file provided by Introl.

Another problem with using the fundamental types and modifiers is that the fundamental type is CPU and compiler specific. If we changed to a different CPU with different fundamental type definitions, we would have to change the types throughout the whole program. It would be far more convenient to be able to define a new type based on the fundamental types in a header file. Then if we did have to change the type, we would only have to do it in one place.

In C the *typedef* modifier can be used to derive new data types. A *typedef* modifier is similar to a *#define* because it simply attaches a new name to a type. The new name can then be used throughout the file in which the type is defined. The syntax for the *typedef* instruction is

```
typedef type new_type_name;
```

For example the derived types used throughout this book and defined in the code conventions in Appendix A are

```
typedef unsigned char INT8U;
typedef signed char INT8S;
typedef unsigned short INT16U;
typedef signed short INT16S;
typedef unsigned long INT32U;
typedef signed long INT32S;

#define ISR void __interrupt
#define UBYTE INT8U
#define SBYTE INT8S
#define UWYDE INT16U
#define SWYDE INT16S
```

Using these new defined types, the definition

```
unsigned char PulseCnt;      /*The number of pulses remaining*/
```

can now be replaced with

```
UBYTE PulseCnt;              /*The number of pulses remaining*/
```

or

```
INT8U PulseCnt;              /*The number of pulses remaining*/
```

With either one of these new types, it is more obvious that *PulseCnt* is an unsigned 8-bit integer, because the type names directly correspond to the data size.

These type definitions not only make it more convenient, they also make it easier to port the code to a different CPU, compiler, or hardware resource. For example if we wanted to switch to a CPU with a *long* defined as 64 bits instead of 32 bits, we would only have to change the *typedef* lines

```
typedef unsigned long INT32U;
typedef signed long   INT32S;
```

to

```
typedef unsigned int INT32U;
typedef signed int   INT32S;
```

If we did not use a *typedef,* we would have to go through the complete program and change all *long*s to *int*s.

EXAMPLE 13.1

Using Defined Types for Portability

Let's take a look at how easy it can be to port a software module from one system to another by taking advantage of defined types. In this case we have a temperature acquisition module that must be ported from a system that uses 8-bit temperature values to one that uses 10-bit temperature values. Assume that the 10-bit system actually uses 16-bit data, which contain the sign extended 10-bit temperature values.

Following is some of the code contained in the module that is written without using defined types:

```
signed char CurrentTemp;

signed char ProcessTemp(signed char temp);

signed char ProcessTemp(signed char t) {
    signed char peak_temp;
    ...
}
```

To change this code to use 16-bit data for temperature values would require changing the variable definition, the function prototype, the function definition, and the local variable definition inside the function. You can imagine that if this were a large module, there would be a significant amount of revision to do.

If we use a defined type for the temperature data using a *typedef,* our code would look like this:

```
typedef signed char TEMP_T;

TEMP_T CurrentTemp;

TEMP_T ProcessTemp(TEMP_T temp);

TEMP_T ProcessTemp(TEMP_T t) {
    TEMP_T peak_temp;
    ...
}
```

Now to change the module to a 16-bit system we would only have to change the *typedef* line to

```
typedef signed short TEMP_T;
```

If this *typedef* is in a header file that is included in every project module, then this technique would work for all modules in the project.

13.2.6 Data Type Conversion

As mentioned previously, the C programming language is not a strongly typed language. This means that if you have a mismatch of data types in an expression, C may automatically perform a type conversion to one of the data objects. This is meant to be a convenience.

There are times when the automatic data type conversion can result in unexpected results, however. The rules for these automatic type conversions can be obscure and many are dependent on the compiler implementation. ANSI-C seems to be especially unclear with respect to integers smaller than an *int,* such as 8-bit integers.

It is best to avoid automatic type conversion and explicitly convert mismatched data types by using a *cast.* A data *cast* is an operator that changes the data type of a data object in that expression. The syntax is

```
(new type) data
```

A cast does not permanently change the data type for the object. It only makes a conversion in the expression containing the type cast.

Source 13.1 shows a typical example of an expression that involves type mismatches. In this example an 8-bit signed number, *n,* is multiplied by an 8-bit unsigned number, *m.* The result goes into a 16-bit signed integer, *p.* This would seem like a perfectly normal thing to do. The signed number may be data from a sensor. The unsigned number may be a scaling factor and the result is large enough to hold any result from the multiplication.

Source 13.1 Automatic Type Conversion versus Type Casting

```
unsigned char m;
signed char n;
signed int p;

...

    p = m * n;  /* doesn't work */

    p = (signed int)n * (signed int)m;  /* does work */
```

The first expression relies on the compiler's automatic type conversion. The Introl-CODE compiler promotes the two arguments to *unsigned int* and then performs a signed multiplication. This is probably not what we want. Promoting *n* to an *unsigned int* means it was converted from 8 bits to 16 bits without sign extension. If *n* was a negative number, the *unsigned int* version is incorrect.

The second line in *main* uses type casting to explicitly change the data types of *m* and *n* to *signed int.* Now the promotion of *n* from a *signed char* to a *signed int* is done with sign extension, and the result is correct. The type cast for *m* is not required, but it helps explicitly define the programmer's intention and it does not add any additional code.

When converting types you must make sure you understand what the compiler will do to make the conversion. In the previous example, all conversions were promotions. That is, smaller integers were converted to larger integers. There is no question about the results, because an *unsigned char* and a *signed char* will both fit into a *signed int*. We must be careful if we need to demote an integer to a smaller integer, however. Demotion is normally accomplished by truncating the higher-order bits. For example, when converting from an *unsigned int* to an *unsigned char*, the least-significant byte of the *unsigned int* is used for the *unsigned char*.

In general explicit type casts should be used for any mismatches. You should also check the code generated by the compiler to make sure it is doing what you expected. This is the only way you can make sure that the conversion you expected is the conversion the compiler produced.

▶ 13.3 VARIABLES AND STORED CONSTANTS

Now let's look at how the C compiler implements different data objects. In this section variables and stored constants are covered. The contents of a variable can be changed at run time. The contents of a stored constant can only be changed at build time.

13.3.1 Variables

Variables are data objects that can be directly changed at run time. "Directly changed" means that the contents of the memory location can be changed with a CPU write cycle. There is no intervening program required to change the location, which means variables are always stored in RAM. EEPROM locations can be changed at run time, but they require a program to perform an erase-and-write sequence. Therefore the contents of an EEPROM are not considered variables.

Peripheral registers are also considered variables. They are similar to RAM, but the contents may change as a result of the peripheral circuitry. For example a general purpose input port can be treated as a read-only variable. The contents of the variable change to reflect the logic state of the input pins. In C these registers are declared as *volatile* variables.

Variable Implementation. There are two implementations for variables: *static* and *automatic*. A static variable is implemented in permanent memory, which is allocated by the compiler when the variable is defined. The allocation is accomplished by the compiler by using an *rmb* or a *ds* directive to reserve memory locations for the variable.

If the definition is outside all function blocks, the variable is a static variable. For example if a variable is declared outside all function blocks as

```
INT16U PulseCnt;
```

then the compiler would generate the following assembly code:

```
PulseCnt    ds.w 1
```

which is the same implementation as a global variable defined in assembly language as described in Chapter 6. To access this variable, extended addressing would most likely be used. For example

```
CurrentCnt = PulseCnt;
```

would be implemented as

```
movw PulseCnt,CurrentCnt
```

if both *PulseCnt* and *CurrentCnt* are static variables.

If the definition is inside a function block and there is no modifier or an *auto* modifier, the data would be an *automatic* variable. Another name for an automatic variable is a *local* variable. Storage for an automatic variable is allocated on the stack at run time. Therefore the data only persist as long as program control is inside the function.

If the static modifier is added, the variable is implemented as a static variable as described earlier. The difference between a static variable declared inside a function with a *static* modifier and a static variable declared outside all function blocks is the scope. The scope of a static variable that is declared inside a function block is limited to that function even though the data persist as long as power is applied.

Following is an example of an automatic variable definition:

```
void CountPulses(void) {
    INT16U current_cnt;
    ...
}
```

This declaration causes the compiler to generate code that allocates space on the stack. Typically this would result is the following assembly code:

```
leas -2,sp
```

which decrements the stack pointer to save space for the variable.

To access the variable, indexed addressing with a constant offset is used. The name of the variable is associated with a constant offset that is added to an index register to generate the variable address. For example, for the following C statement within the function

```
current_cnt = PulseCnt;
```

the compiler would generate something like this:

```
movw PulseCnt,0,sp
```

It is important to understand the persistence and scope of a variable in embedded systems. This is especially the case when using a real-time, multitasking kernel.

Initialized Variables. Variable definitions that include an initial value require both extra storage and extra code. First the compiler must allocate additional storage in ROM for the initial value. Because a variable always uses RAM, its initial value is unknown. The compiler must also generate a start-up program to copy the initial value from ROM to RAM.

For example the following definition gives *PulseCnt* an initial value of 60.

```
INT16U PulseCnt = 60;
```

The resulting assembly code looks something like

```
            section .init
init1:      dc.w 60

            section .data
PulseCnt:   ds.w 1
```

In addition the compiler will generate start-up code that performs the equivalent of the following instruction:

```
        movw init1,PulseCnt
```

This copies the initial value into the variable location before any C code is executed.

13.3.2 Stored Constants

In Chapter 12 we saw how to create a defined constant using the *#define* pre-processor command. Most constants in our C programs will be defined constants because they do not require CPU memory. There are times, however, when we do want to store a constant value in memory. For example stored constants are commonly used for lookup tables, strings, and to aid the debugging process during program development.

In C a stored constant is created by using the *const* modifier in a definition. For example the definition

```
const unsigned int StartAddr = 0x8000;
```

will allocate two bytes in memory and attach the name *StartAddr*. Since we used the *const* modifier, the memory will be allocated in ROM instead of RAM. The assembly code generated by the compiler would be

```
StartAddr:  dc.w $8000
```

This is the same assembly code as for the stored constants described in Chapter 3. Notice that stored constants must always have an initial value in the definition. It would not make sense to create a stored constant without a value.

13.3.3 Data with Absolute Locations

One of most noticeable differences between programs written for an embedded microcontroller and programs written for a computer is the use of absolute memory locations. Data objects that have absolute memory locations are those objects that have a fixed address. Since computer programs must be relocatable, they do not access fixed locations unless they are peripheral driver routines. When programming embedded systems, we are always accessing absolute locations. Accessing the MCU control registers is the most common example. Other examples include accessing external peripherals and performing memory tests.

Because C was written as a computer language, it does not have a direct way to declare a data object with an absolute memory location. When we define a variable, for example, the compiler will allocate the next available space in the variable code section. We have no direct control over the address that is allocated.

There are two ways to declare data objects with absolute locations in C—the first one requires additional assembly code for the definition, and the second one uses type casts to "trick" the compiler into accessing an absolute location.

Some compilers also include compiler-specific keywords to define a data object with a fixed location. The problem with using these custom solutions is that it makes the code more difficult to port to another compiler. The techniques described here should work with any ANSI-C compiler.

Allocating Space with Assembly Directives.   This process uses an assembler directive to define the variable. Then we can use an *extern* modifier to declare the data in the C program. This is the method used by the Intro-CODE system to provide access to the MCU control registers. In an assembly code module the following code is used:

```
            section .chip
_H12PORTA   ds.b 1
_H12PORTB   ds.b 1
  ...
```

where the section *.chip* is defined in the memory map to start at 0x0000. This code defines the name of the data object and reserves the space for the data at a specific location.

In a C module we need to declare these data objects without allocating space, so the *extern* modifier is used. Following is the declaration for the first two 68HC12 registers, *PORTA* and *PORTB:*

```
extern volatile __mod1__ __mod2__ unsigned char _H12PORTA;
extern volatile __mod1__ __mod2__ unsigned char _H12PORTB;
```

The *extern* modifier tells the compiler not to allocate memory. The *volatile* modifier is used because the port contents can change based on the state of external pins, and the *__mod1__ __mod2__* modifiers are used to tell the compiler that it can use direct addressing. In the Intro-CODE system, these declarations are found in *hc912b32.h*.

The assembler directive created the name, which corresponds to a label, and allocates memory starting at address 0x0000. For the two registers shown, the memory allocated is at the addresses 0x0000 for *_H12PORTA* and 0x0001 for *_H12PORTB*. The C declarations then tell the compiler the type of data contained at these locations. Now to access either port, we can treat the names like any other variable name. For example

```
_H12PORTA = _H12PORTA | 0x01;
```

sets bit 0 of *PORTA*.

Accessing Locations by Type Casting the Address.   This second method does not allocate the space. Instead it tricks the compiler into accessing a location by using a type cast on the address. Since space is not allocated, care must be taken to make sure the linker does not place any other code at the location. This can be done by creating a memory map section at the location with nothing in it.

Remember that when we define an integer with the *#define* preprocessor command, it will by default cast that integer as an *int*. So after the command

```
#define ADDR 0x1800
```

the constant *ADDR* has the *int* data type. We can change the type of *ADDR* by type casting the number.

When we access a data object, remember that we use the name of the object, which is its address. The problem with *ADDR* is that it is seen by the compiler as a value, not an address. We can make it an address by using a type cast for a pointer. This is what we have to do to access absolute locations with a pointer. In this case we also need to dereference the pointer to refer to the contents of the address.

Following is the rather strange *#define* command to access two LCD registers mapped to addresses 0x1800 and 0x1801:

```
#define LCD_CNTL *(volatile UBYTE *)0x1800
#define LCD_DATA *(volatile UBYTE *)0x1801
```

In words this *#define* command says, "Set *LCD_CNTL* equal to the contents of address 0x1800, which is a pointer to a volatile *unsigned char*." Now to access the LCD control register, we can use normal C code. For example

```
LCD_CNTL = 0x01;
```

loads a 0x01 into the LCD control register. Note that after the compiler replaces the defined name *LCD_CNTL* this code would look like

```
*(volatile UBYTE *)0x1800 = 0x01;
```

The part in parentheses is a type cast that casts the number 0x1800 as a pointer to a *volatile unsigned char.* The asterisk at the beginning indicates that we are referring to the contents of the location pointed to. Later we also look at ways to access absolute locations by using pointers.

▶ 13.4 POINTERS

Pointers have to be one of the most widely discussed parts of the C language—for good reason. If used wisely pointers can result in very efficient code. It is one of the things that makes C a *low-level* high-level language and therefore a good language for embedded systems. If not used wisely, however, pointers can be the cause of some of the nastiest bugs and can seriously limit the portability of the program. The bottom line is to make sure you understand how to use pointers and how pointers are actually implemented. Let's start with a pointer definition:

```
UBYTE *CountPtr        /* Pointer to a unsigned byte */
```

The asterisk indicates that *CountPtr* is the name of a pointer. The type in the declaration indicates the type of data pointed to by *CountPtr*. So in this case *CountPtr* is a pointer to an unsigned byte.

When a pointer is defined, the compiler allocates space in RAM, which will contain the address pointed to by the pointer. The pointer size is CPU and compiler dependent. The Introl-CODE compiler uses 16-bit pointers for the CPU11 and CPU12. The CPU12 also has *far pointers,* which are 32-bit addresses that are used to point to functions that are contained in different pages in the memory map.

13.4.1 Working with Pointers

Initialization. This brings us to one of the most common pointer bugs—failing to initialize a pointer. The pointer in the definition given previously is not initialized. The definition allocated two bytes of RAM space for the pointer, but it did not put anything into that space. The contents of the pointer are unknown—they are whatever happens to be in memory when the system is powered up. If we use this pointer, no errors would be generated by the compiler. The pointer would be pointing to some unknown memory location, however. Changing the contents of an unknown location could result in a soft failure that may have catastrophic consequences.

Since pointer initialization is so important, it is good programming practice to initialize all pointers at the definition. Following we define a pointer called *CountPtr* that is initialized to point to the variable *PulseCnt:*

```
UBYTE PulseCnt;
 ...
UBYTE *CountPtr = &PulseCnt;
```

Notice that *PulseCnt* and *CountPtr* have the same data type except for the asterisk. Since we want *CountPtr* to point to *PulseCnt,* we initialize the pointer contents to the address of *PulseCnt.*

Now let's look at this pointer in more detail. Figure 13.2 shows the memory details for these two definitions. The hex values are shown for illustrative purposes. We normally do not know or care about the actual values unless we are dealing with absolute locations.

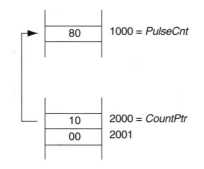

Figure 13.2 Memory Details of a Pointer

In this figure *PulseCnt* ended up at address 0x1000 and *CountPtr* ended up at 0x2000. Since the pointer was initialized, the contents of the pointer is the address of *PulseCnt*, 0x1000. Following is the syntax required to access the different addresses and contents of the pointer and the variable.

To Access…	Syntax	Hex Value from Figure 13.2
The address of the pointer	`&CountPtr`	0x2000
The contents of the pointer or the address of the variable	`CountPtr, &PulseCnt`	0x1000
The contents of the variable	`*CountPtr, PulseCnt`	0x80

Again the name of a data object, including pointers, is set equal to the address of that object. When the name is used in a program, however, it refers to the contents of that object, not the address.

In the last row of the table, two ways are shown to access the contents of the variable. To access the contents using the pointer, an asterisk is used. We read this asterisk as *the-contents-of*. So to access the contents of what *CountPtr* is pointing to, we use **CountPtr*.

Note
_____

Remember What * and & Mean

Programming with pointers becomes much easier once you memorize the meanings of the * and & operators:

```
*  -> 'The contents of'
&  -> 'The address of'
```

Pointer Operations.   The data type given in the pointer definition affects the operations involving pointers. Both incrementing a pointer and adding one to a pointer actually result in the pointer pointing to the next data object. For example

```
UWYDE *IntPtr;
…
IntPtr++;
IntPtr += 1;
```

In this case, since *IntPtr* points to 16-bit objects, both expressions actually add two to *IntPtr*. Because of this it is best to avoid the addition expression, because it is a bit deceptive. This seems strange now, but when we start working with arrays, it will make perfect sense. *Incrementing* a pointer also means to point to the next element in an array.

Generic Pointers.   There are times when we want to create a pointer that points to an object that may have different types. Essentially we would like to turn the compiler's data type checking off. To do this we can declare a generic pointer. To declare a generic pointer in ANSI-C, we use the *void* modifier. For example

```
void *pdata;
```

declares a pointer that points to data of an unknown type. Generic pointers are useful for passing unknown parameters to a generic function or to point to the space allocated by a memory allocation program.

Null Pointers.   A null pointer is a pointer that points to address zero. Null pointers are often used to indicate an error from a function that normally returns a pointer. To check for a null pointer, we should use the following definition:

```
#define NULL (void *)0
```

This allows *NULL* to be used regardless of the actual type of pointer we are checking.

In normal computer systems there is never a problem using a null pointer to indicate an error. When programming embedded microcontrollers, however, you must be careful when using null pointers because the address 0x0000 may be a valid address. For example, imagine if we had a function that returns a pointer to one of the 68HC12 GPIO ports. We cannot use the null pointer as an error indicator because address 0x0000, which is *PORTA,* is a valid return value.

Most of the time we do know by the context if a null pointer can be used as an error indicator. For example the *slicestr( )* function in Appendix B returns a null pointer if a token is not found. This is reasonable because we know that no strings will be located at address 0x0000. Source 13.2 shows an example program that uses *slicestr( )*. If *slicestr( )* returns a null pointer, an error handler routine is called.

Source 13.2 Checking for a Null Pointer to Detect an Error

```
sptr = SLICESTR(strgbuf,&arg1);
if(sptr == NULL){
    ErrorHandlr(5);
}
```

13.4.2 Pointers to Absolute Locations

When we are working with embedded systems, we often need to work with absolute locations on the memory map. In the last section we saw how we can create a variable that is at an absolute location. Here we look at two ways to point to absolute locations—the first uses a type cast on the address and the second creates a constant pointer in memory.

Pointing to Locations by Type Casting the Address.   This method is essentially the same as the method used to access variables at absolute locations. We use a type cast to cast the address to be a pointer. For example if we have an LCD module connected as described in the previous section, we can use pointers to access the LCD registers by using the following definitions:

```
#define LCD_CNTL_PTR (volatile UBYTE *)0x1800
#define LCD_DATA_PTR (volatile UBYTE *)0x1801
```

If we compare these to the definitions used earlier for variables, we can see that they are the same except for the missing asterisk before the type cast. These definitions cast the addresses to be pointers, not the contents of pointers. So to access the LCD control register we use pointer notation

```
*LCD_CNTL_PTR = 0x01;
```

For this application there is no real advantage to using this method instead of accessing these registers as variables. In fact they are equivalent and will result in the same code.

Source 13.3 shows an example that uses an absolute pointer definition to initialize a variable pointer. This is typically done for a memory test program. *START_ADDR* is the start of the block to be tested and *END_ADDR* is the end of the block to be tested. *AddrPtr* is used to point to each byte in the block.

Source 13.3 An Example of Using a Pointer to an Absolute Location

```
#define START_ADDR (UBYTE *)0x8000
#define END_ADDR (UBYTE *)0xFFFF

UBYTE *AddrPtr;

MemTest() {

    UBYTE dat;

    AddrPtr = START_ADDR;
    AddrPtr--;
    while(AddrPtr != END_ADDR){
        AddrPtr++;
        dat = *AddrPtr;
        /* Test dat*/
    }
}
```

Using Constant Pointers.   Another way to point to absolute locations is to declare a constant pointer. Let's first look a regular pointer definition that is initialized to an absolute location:

```
UBYTE *LcdCntl = (UBYTE *)0x1800;
```

This declaration allocates space in RAM for the pointer *LcdCntl* and initializes it to 0x1800. We could certainly use this, but there is a problem—it is a variable pointer. This means you could change the address to which it is pointing, which is not the intention. We want the pointer to be a constant so that it will be stored in ROM and cannot be changed.

A typical mistake in trying to create a constant pointer is to do the following:

```
const UBYTE *LcdCntl = (UBYTE *)0x1800;    /*Pointer to a constant*/
```

This definition does not declare a constant pointer. It declares a pointer to a constant byte. To define a constant pointer, we have to move the *const* modifier to the other side of the asterisk:

```
UBYTE *const LcdCntl = (UBYTE *)0x1800;    /*Constant pointer*/
```

This definition loads the address 0x1800 into ROM. To write to the LCD control register, we can use

```
*LcdCntl = 0x01;
```

In addition, since it is a constant pointer, the compiler will generate an error if we try to do something like this:

```
LcdCntl = 0x1900;     /* Error */
```

▶ 13.5 ARRAYS AND STRINGS

13.5.1 Arrays

Arrays are used for sets of related data objects with the same data type. Applications for arrays include strings of characters, jump tables, and lookup tables. In C, arrays and pointers are inexorably intertwined. This is because arrays are implemented in C with pointers. First let's look at two array definitions:

```
UBYTE VarArray[10];
const UBYTE
NumArray[] = {'0','1','2','3','4','5','6','7','8','9'};
```

The first definition creates a variable array. Ten bytes of RAM space are allocated for the array, and the array name, *VarArray,* is set equal to the address of the first byte in the array. If a variable array is initialized, the compiler allocates space for the array in RAM and allocates space for the initial values in ROM.

The second definition is for a constant array. Notice that you do not need to indicate the number of elements in the brackets. The number of elements is determined by the number of entries in the array list. In this case *NumArray* will store the ASCII values for 0 through 9 in 10 bytes of ROM space.

To access the array we can either use array notation or pointer notation. Array notation uses the array name and a bracket containing the element to be accessed. The first element in *VarArray* is

```
VarArray[0]
```

The C compiler also treats the array name as a pointer to the first element in the array. In a program, *VarArray* is the same as *&VarArray[0],* so we can also use pointer notation with *VarArray*. In fact the compiler will actually use pointers even if you use array

Pointer Notation	*NumArray*	Array Notation
*NumArray	30	NumArray[0]
*(NumArray+1)	31	NumArray[1]
*(NumArray+2)	32	NumArray[2]
*(NumArray+3)	33	NumArray[3]
*(NumArray+4)	34	NumArray[4]
*(NumArray+5)	35	NumArray[5]
*(NumArray+6)	36	NumArray[6]
*(NumArray+7)	37	NumArray[7]
*(NumArray+8)	38	NumArray[8]
*(NumArray+9)	39	NumArray[9]

Figure 13.3 Pointer and Array Notation for Accessing the Contents of an Array

TABLE 13.4 ARRAY IMPLEMENTATION

C Source		Resulting Assembly Code (Edited)
UBYTE *PassPtr;	→	PassPtr: ds.b 2
UBYTE PassArry[5];	→	PassArry: ds.b 5
...		
PassPtr = PassArry;	→	movw #PassArry,PassPtr

notation. So *VarArray[i]* is the same as *(VarArray+i),* where *i* is the *i*th element in the array. Figure 13.3 shows both notations for accessing each element in *NumArray.*

Array names and pointers are implemented differently. To see this difference, let's look at the actual implementation in more detail. Table 13.4 shows the C code and the resulting assembly code for a pointer definition, a variable array definition, and a statement to set the pointer equal to the first element in the array.

The first C definition is for a pointer, *PassPtr.* As we would expect, it allocates two bytes in RAM for the value of the pointer. The second definition is for an array, *PassArry.* This definition allocates five bytes of RAM space for the array. Notice that the name *PassPtr* is a label that is equal to the address of the pointer, and *PassArry* is a label that is equal to the first address of the array.

In the C program we can treat *PassArry* as a pointer, but it is actually implemented differently than a pointer. No RAM space was allocated for *PassArry* as it was for *PassPtr.* Therefore when we set *PassPtr* equal to *PassArry* in Table 13.4, it actually loads the contents of *PassPtr* with the address *PassArry.* Notice the different addressing modes in the *movw* instruction. The difference between an array name and a pointer means three things:

1. *PassArry* must be treated like a constant pointer. After all, it is really a label, not the contents of RAM space like a normal pointer. This means we cannot assign different values to *PassArry* such as the following:

```
PassArry++;                    /* Illegal operations */
PassArry = PassPtr;
```

2. When operating on the contents of an array with a pointer such as *PassPtr,* indirect addressing is used. When operating on the array with the array name such as *PassArry,* extended addressing is used. Typically using the array name is more efficient than using a pointer.

3. When debugging the system, *PassPtr* will not be equal to *PassArry,* even after the expression in Table 13.4 is executed. The contents of *PassPtr* (*PassPtr:PassPtr*+1) will be equal to *PassArry.*

Declaring a Pointer Does Not Create an Array. Another infamous pointer bug is to declare a pointer and then store things using the pointer as if it were an array name. Remember, when declaring a pointer, space is only allocated for the pointer value, not for the space pointed to by the pointer. We can set the pointer value to any address and then treat it like an array, but if space is not allocated by the compiler, then we may end up writing on top of other data or program code. For example we can define a pointer

```
UBYTE *PassPtr;
```

and then treat it as if it is an array name such as the following:

```
PassPtr[0] = '1';
PassPtr[1] = '2';
```

The compiler will allow you to do this, but it would result in a serious error. The reason is that no space was allocated for the array. And without initializing the pointer, we do not know where the *1* and *2* will end up.

Beware of Local Arrays. If we declare an array inside a function, it is a local array. It will use the automatic storage class, so stack space will be allocated for the array. If we have a large array, this is going to require a lot of stack space. If the function is also reentrant, then array space will be allocated for each instance of the function. As you can imagine, this can result in a large amount of required stack space. For small microcontrollers this may be unacceptable.

EXAMPLE 13.2

Seven-Segment Lookup Table

Let's try using an array as a lookup table to convert digits to seven-segment display codes. Lookup tables are a common application for arrays. They are especially useful when you have to map one variable to another and the mapping function is difficult to implement mathematically or logically.

Seven-segment displays are a classic example. To display each digit, you need a segment code that represents the set of segments to be activated. There is no simple mathematical or logical relationship between the digit and the segment code. Following is an illustration of the display, which shows the bit position corresponding to each segment:

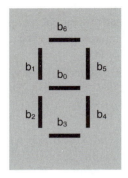

First, declare a constant array that contains the segment codes in order from zero through nine. If you assume that the segments are connected active-high, the segment lookup table should be

```
const UBYTE SegTable[] = {0x7e,0x30,0x6d,0x79,0x33,0x5b,0x5f,0x70,0x7f,0x7b};
```

Now to output the digit contained in *num* to a display connected to *PORTA,* we can use the following code:

```
_H12PORTA = SegTable[num];
```

Since the array index matches the digit to be displayed, the code is simple. This is certainly better than the alternative—a long case statement.

13.5.2 Strings

A C string is an array of ASCII characters followed by the ASCII NULL character, which indicates the end of the string. This is called a *null-terminated string.* The NULL character is different from the null pointer because it is a character, not a pointer. The data type of a NULL character is *UBYTE,* whereas the data type for a null pointer is *(void *).*

In C we can create strings by enclosing a string of characters in double quotation marks. The compiler creates an array of characters and adds the NULL automatically to the end of the array.

Source 13.4 shows two different ways to create C strings. Both of these strings are constant strings and will be stored in ROM. The first string, *Enter>,* is declared as a constant array called *Prompt.* The ASCII characters are stored in ROM with a terminating NULL. To use this string we reference the name of the array as shown in the first line of the *main()* program. In this example the *putstrg()* function has one parameter, a pointer to the string to be displayed. The first example can use *Prompt* because an array name can be used as a pointer to the first element of the array.

Source 13.4 Two Ways to Create Strings

```
const unsigned char Prompt[]  = "Enter>";
void putstrg(unsigned char *string);
  ...
main(){
    putstrg(Prompt);
    ...
    putstrg("Select Mode");
    ...
}
```

The second string, *Select Mode,* is actually shown as the argument of the function. There was no declaration required. In this case the C compiler will allocate and store the string into a constant array. Since it is a C string, a NULL character is placed at the end of the array. It then passes the address of the first element of the array to the *putstrg()* function.

The advantage of this method is that it is easier to read. The disadvantage, however, is that the string can only be used by that function call. If *Select Mode* is displayed in another part of the program in the same way, a duplicate array is created by the compiler. The first string in Source 13.4 does not have this problem. The same array can be used many times by referring to it by the array name, *Prompt.* If there is one thing to watch out for when using strings, it is memory usage. There are few ways to use ROM space faster than to fill it with strings. The second method shown in Source 13.4 makes this problem worse because the compiler will duplicate the string for each function call.

A solution to the duplication problem is to create a function that only displays the prompt. For example

```
void DisplayPrompt(void){
    putstrg("Enter>");
}
```

Here *DisplayPrompt()* is called each time a prompt is required. Since it uses the same *putstrg()* every time, only one copy of the constant string is generated by the compiler.

▶ 13.6 STRUCTURES

Structures are arrays of related objects that have different data types. Structures are useful to group sets of related data, especially when there may be many instances of the complete set. Structures can also be used in embedded systems to define and manipulate bit fields. Following is the syntax for a structure declaration:

```
struct struc_tag { structure list } struc_name;
```

where

> *struc_tag* = the type name for the structure
>
> *structure list* = the list of structure elements, including element type
>
> *struc_name* = the name of an instance of the structure

Both the *struc_tag* and the *struc_name* are optional. There can also be a list of *struc_name*s separated by commas if it is necessary to generate multiple instances. The elements of the structure are separated by semicolons.

It is important to understand the syntax of the structure definition so you know when memory is actually allocated. For example the following definition defines a structure type called *TcbInit:*

```
struct TcbInit {
    UBYTE state;
    void (*start)(void);
    void *stack;
    struct tcb *tcbp;
};
```

Because no structure names are given, it does not create an instance of the structure. An instance corresponds to separate memory that is allocated for the structure. The example shown only creates a structure type. It does not allocate any memory. To create an instance you must either add a structure name or add an additional definition that uses the structure type. For example to create a structure called *Task1TcbInit* of type *TcbInit,* we would use the following declaration:

```
struct TcbInit Task1TcbInit;
```

This line allocates memory, in this case seven bytes of RAM space—one byte for *state,* two bytes for a pointer to *start(),* two bytes for a generic pointer to a stack, and two bytes for a pointer to another structure, *tcbp.*

A member of the structure can be accessed by using the dot operator. For example to access the member *state* in the *Task1TcbInit* structure, we use

```
TaskState = Task1TcbInit.state;
```

You can also access structures through pointers. For example we can declare a pointer that points to the structure *Task1TcbInit* as follows:

```
struct TcbInit *Task1TcbPtr = &Task1TcbInit;
```

To access structure elements with a pointer, the -> notation is used. So to access the member *state* in the structure pointed to by *Task1TcbPtr,* we use

```
TaskState = Task1TcbPtr->state;
```

13.6.1 Using typedef to Define a Structure Type

It is preferable to use a *typedef* to create a structure type. For the example shown, we can define the structure type with

```
typedef struct {
    UBYTE state;
    void (*start)(void);
```

```
    void *stack;
    struct tcb *tcbp;
} TCB_INIT;
```

Now to create an instance of the structure, we use

```
TCB_INIT Task1TcbInit;
```

Using a *typedef* clearly indicates when you are defining a structure type as opposed to creating an instance of the structure. We will discuss structures used for bit fields in Section 13.8.

▶ 13.7 ENUMERATED TYPES

An enumerated type is a data type that is limited to a predefined set of values. For example if a state variable can only have values 1 through 4 during normal operation, it should be declared as an enumerated type.

The syntax is the same as a structure declaration in that it has an optional tag for the type name and an optional name for an instance. The following example is for a tape transport, which can be in only one of the predefined states listed as members:

```
enum TransportStates {STOP,PLAY,REV,FF,PAUSE};
```

This declaration defines the set of states that are possible, attaches a numerical value to each state, and declares the type with the name *TransportStates*. Again, because there was no name given in this declaration, it did not allocate memory; it only defined an enumerated type.

To create an instance we can declare a variable that uses the tag *TransportStates:*

```
enum TransportStates CurrentTransportState;
```

This line actually allocates one byte of RAM space. It can be accessed by using the name *CurrentTransportState.*

Again it is better to define an enumerated type by using a *typedef* than by using *enum* tags. We could replace the enumeration type declaration shown previously with

```
typedef enum {STOP,PLAY,REV,FF,PAUSE} TRANSPORT_STATES;
```

Then we create an instance with

```
TRANSPORT_STATES CurrentTransportState;
```

▶ 13.8 BIT OPERATIONS

In embedded systems we often need to manipulate or test individual bits or bit fields within a word. Examples include general purpose I/O, peripheral control and status, and bit representations for states in a state machine implementation. In this section we look at a variety of ways to operate on bits and bit fields using C. We also address the design issues involved, including readability, portability, and execution speed.

Because bit operations are so common, microcontroller assembly languages usually include special instructions for bit manipulation and testing. Examples are the 68HC12 *bit*, *bset*, and *brset* instructions described in Chapter 5. The C language, however, does not include specific instructions or data types for bit operations with the exception of structure bit fields. So when writing in C, all bit operations use the bitwise logic instructions.

We covered bit operations using assembly language in Chapter 5. Here we apply the same concepts to programs written in C, so you may want to review Section 5.5 first.

13.8.1 Bit Testing

Bit operations can be separated into two categories, reading bit values or *bit testing* and writing bit values or *bit manipulation*. We look at these for individual bits and sets of bits, which are called *bit fields*.

To find the status of a bit or bit field, we first need to mask the other bits in the word using the bitwise AND operation, &. The result of the AND will be all zeros except for any ones in the bit or bit field of interest. If the status of a single bit is needed, or if we are checking to see whether all bits in a bit field are cleared, we can test for a result of zero. Following is an example that tests bit 0 of *PORTP:*

```
/* Single bit status. BIT0 of PORTP */

if( (_H12PORTP & 0x01) == 0) {
        /* Do Something */
    }
```

We also can test for a bit combination of a bit field as follows:

```
#define RTR (UBYTE)0x07
#define LOW_RATE (UBYTE)0x02

    if( (_H12RTICTL & RTR) == LOW_RATE) {
        /* Do Something */
    }
```

In this case *RTR* defines the bit field by containing a one for each bit in the field. The byte tested is the *RTI* control register *_H12RTICTL*. This program checks to see whether the bit field *RTR* of *_H12RTICTL* is equal to *LOW_RATE,* which is defined as 0x02. When performing tests on bit fields, we must make sure that the fields we are comparing are aligned. In this case *LOW_RATE* is aligned with *RTR*. If *RTR* had been defined as 0xE0, for the most-significant three bits instead of the least-significant three bits, we would have to either redefine *LOW_RATE* or first perform a shift to align the bit fields.

13.8.2 Bit Manipulation

Bit manipulation is the process of writing to a bit or bit field while not affecting the other bits in the word. When performing bit manipulation, we must be aware of which bits we need to change and which bits we cannot change.

Direct Writes.   Sometimes we do want to change every bit in a word. This is usually the case during initialization of peripheral control registers. For example

```
_H12PORTP = BIT0;    /* Turn bit-0 off.( active-low) */
_H12DDRP = BIT0;     /* Make Bit0 an output all other bits inputs */
```

The first line initializes *PORTP*, bit 0 to a one to turn it off. The second line initializes the direction of bit 0 of *PORTP* as an output and all the other bits as inputs.

Direct writes to control registers can also be used to clear the 68HC12 timer flag bits. In order to clear a timer flag, we must write to the register with a one in the position of the flag to be cleared. Following is an example to clear the *RTI* flag:

```
#define RTIF (UBYTE)0x80

_H12RTIFLG = RTIF;            /* Clear the RTI flag */
```

Read-Modify-Writes.   Usually when we need to manipulate a bit or bit field, we only want to change that bit or bit field without changing the other bits in the word. To do this we need to perform a read-modify-write operation. First the word value is read, and then the bit or bit field is modified using bitwise logic instructions. The revised word is then written back to its location. Following are some examples for modifying individual bits:

```
_H12PORTP = _H12PORTP & ~0x01;    /* Clear Bit0 */
_H12PORTP &= ~0x01;               /* Clear Bit0 */
_H12PORTP = _H12PORTP | 0x01;     /* Set Bit0 */
_H12PORTP = _H12PORTP ^ 0x01;     /* Toggle Bit0 */
```

The first and second lines clear bit 0 of *PORTP*. The first line is the normal expression showing the read-modify-write. It reads *_H12PORTP*, modifies it by ANDing it with NOT bit 0, and then writes the results back to *_H12PORTP*. The second line performs the same task using the C assignment operator.

In order to clear bit 0 in the first two lines, we take advantage of the property of a bitwise AND function to clear individual bits. That is

$$X \cdot 1 = X \qquad X \cdot 0 = 0$$

Since BIT0 is equal to %00000001, ~BIT0 is %11111110. So the first two lines clear bit 0 of *PORTP* and do not change the rest of the bits.

The next line sets bit 0 of *PORTP* by using the bitwise OR function, |. Since

$$X + 1 = 1 \qquad X + 0 = X$$

bit 0 will be set, and all other bits in the register will remain the same.

The last line toggles, or inverts, bit 0 of *PORTP* by using the bitwise exclusive-OR function, ^. Since

$$X \oplus 1 = X' \qquad X \oplus 0 = X$$

bit 0 will be inverted, and the other bits will remain the same.

Bit Field Manipulation.   These ideas can be extended for manipulating bit fields. In the examples shown previously, we changed only the bit-0 position. To extend this technique to change bit fields, we simply include ones for each bit that needs to be changed. For example

```
_H12RTICTL = (_H12RTICTL & ~0x07);    /* Clear Bits 0, 1 and 2 of RTICTL */
_H12PORTP = _H12PORTP | (0x03);       /* Set Bits 0 and 1 of PORTP */
_H12PORTP = _H12PORTP ^ (0x06);       /* Toggle Bits 1 and 2 of PORTP */
```

The process gets more complicated when we need to set a bit field to a specific combination of ones and zeros. This is especially true when we need to set a bit field equal to the value of a variable. There are a few different options to set bit fields to variable values. The most efficient one is to first clear every bit in the bit field, and then OR the variable bit field to the result. Following are some examples using this method:

```
#define HI_NIBBLE (UBYTE)0xf0

...

_H12PORTP = (_H12PORTP & ~HI_NIBBLE) | (NewDigit & HI_NIBBLE);
_H12RTICTL = (_H12RTICTL & ~RTR) | (NewRtiRate & RTR);
```

In the first line the high-order nibble of *PORTP* is made equal to the high-order nibble of *NewDigit*. The right-hand side of the OR operator is *NewDigit* with the lower nibble cleared. The left-hand side of the OR operator is *PORTP* with its upper nibble cleared. These two values are then ORed together to generate the result, which is written back to *PORTP*. The second line uses the same process to set the *RTI* rate bits of *RTICTL* to *NewRtiRate*.

The expressions given only work if the bit fields of the words are aligned. If the bits are not aligned, we have to shift one of the arguments before the OR function. Following is an example that sets the high-order nibble of *PORTP* equal to the low-order nibble of *NewDigit*:

```
#define LOW_NIBBLE (UBYTE)0x0f

...

_H12PORTP = (_H12PORTP & ~HI_NIBBLE) | ((NewDigit & LOW_NIBBLE)<<4);
```

In this example we did not have to AND *NewDigit* with *LOW_NIBBLE* because the upper nibble is shifted out of the word anyway. This, however, is not true in general, so the AND operation is shown.

Extra Hardware Required for Bit Manipulation.   In order to perform bit manipulation operations, you must have hardware that supports a read-modify-write operation. This means that you must be able to read the current value of the port in addition to writing to it. This is the case for most of the HC11 and HC12 registers. But if we add a port to the bus system, we cannot simply use a single register or latch unless it is bidirectional.

13.8.3 Portability of Bit Operations

The code for the bit operations covered in the previous subsection illustrates the functional operation of the CPU. These examples are good for understanding how bit operations are implemented, but there are some practical problems with the programs as written. First they are not portable. If we want to switch microcontrollers, we have to rewrite the code. Also the programs are not very readable. In order to understand what is happening, we must remember the definitions of the logic instructions and mentally go through the expressions.

To solve these two problems, we need to add another layer of abstraction. We can do this by using *#define*s to define better symbol names and macros, or we can write functions to perform the operations.

Improve Portability by Defining Functional Names. Of course if an LED is connected to *PORTP,* bit 0, our code must access *PORTP,* bit 0. If we move the LED to another port or bit position, or if we change the microcontroller, we must change our code. There is no way around it.

We can make the code much easier to change by putting all hardware-related definitions in one file—typically a header file. Then if a hardware change is made, or the processor is changed, we only have to change that file to make the complete program portable. As an additional benefit, we can make our program easier to read by using more generalized names.

For example if we have an LED labeled *Busy* on the front panel, we can change the names from the low-level resource name like *_H12PORTP* and *BIT0* to functional names like *LED_PORT* and *BUSY_LED,* where *LED_PORT* is the general purpose I/O port and *BUSY_LED* is the bit position connected to the LED. Following is how the code would be changed to turn the LED on:

```
#define LED_PORT  _H12PORTP
#define BUSY_LED  0x01
...
    LED_PORT &= ~BUSY_LED;
```

This technique helps portability if the *#define*s are in a header file so they can easily be changed. The code is also easier for us to read and understand. Instead of having to look at a schematic and see what is connected to bit 0 of *PORTP,* we only have to read the code.

Abstraction by Creating Macros. Creating functional names helped with portability and helped a little with readability. We still have to work the expression out mentally while reading the code, however. In addition the code shown is not totally hardware independent, because it assumes that we have an active-low LED. The programmer should not have to be concerned with how the LED is connected.

The program can be improved by creating a macro for the operations to turn the LED on and off. For example, these are some macros for the *Busy* LED:

```
#define TURN_BUSY_LED_ON()   LED_PORT = LED_PORT & ~BUSY_LED
#define TURN_BUSY_LED_OFF()  LED_PORT = LED_PORT | BUSY_LED
#define INVERT_BUSY_LED()    LED_PORT = LED_PORT ^ BUSY_LED
```

Now the code to turn the *Busy* LED on is

```
TURN_BUSY_LED_ON();
```

The details of implementation and hardware configuration are hidden in the macro and the programmer can focus on the functional level of the code. In addition, because these are macros, this method does not require any more program space and will run at the same speed as the original program.

Abstraction Using Functions.   As an alternative to macros, we can create abstraction by using functions. Functions have the advantage over macros in that they can reduce the program code required. They will run slower, however, because a context change must be made. Following are examples of the programs to perform the operations in the macros above:

```
void TurnBusyLedOn(void) {
    LED_PORT = LED_PORT & ~BUSY_LED;
}

void TurnBusyLedOff(void) {
    LED_PORT = LED_PORT | BUSY_LED;
 }

void InvertBusyLed(void) {
    LED_PORT = LED_PORT ^ BUSY_LED;
 }
```

The program code to turn the LED on would be

```
TurnBusyLedOn();
```

Again the details are hidden in the function, so the programmer does not need to be concerned with them.

We can also set a bit field equal to a variable bit field by passing a parameter. For example, to set the *RTI* rate bits in *RTICTL,* you could create the following function:

```
void SetRtiRate(UBYTE rate) {
    _H12RTICTL = (_H12RTICTL & ~RTR) | (rate & RTR);
}
```

Now to set these bits to *NewRtiRate,* you use

```
SetRtiRate(NewRtiRate);
```

We can also pass a parameter to a macro if we need to squeeze every clock cycle out of the code.

13.8.4 Structure Bit Fields

Structure bit fields are yet another way to operate on individual bits or bit fields. When using structure bit fields we operate on the bit or bit field directly by name. We do not have to deal with the actual code to implement the operation—the compiler does it for us.

514   Chapter 13

Structure bit fields are useful, but as defined by the ANSI-C specification, their implementation is compiler dependent. This means structure bit fields are not as portable as the other methods for performing bit operations. This text covers the Introl-CODE implementation of structure bit fields.

Following is a type definition for a structure called *DEVICE_STATE*. It then defines an instance called *Device1State*. This structure is a state variable that contains bit fields.

```
typedef struct {
            UBYTE DevicePriority : 3;
            UBYTE Running        : 1;
            UBYTE OverloadError  : 1;
            UBYTE FatalError     : 1;
        } DEVICE_STATE;
...
DEVICE_STATE Device1State;
```

The number after the colon specifies the number of bits required for the structure member. For example in the definition above, *DevicePriority* is three bits wide.

In the Introl-CODE implementation, the structure type does not have to be an *int*. The member type will determine the smallest word size to be used for the structure. In this case the members use *UBYTE*, so the structure uses a single byte to hold all the members. If a member is defined as an *int* or the total number of bits exceeds eight, then the structure size will increase to 16 bits.

The actual bits used in a bit field are compiler dependent. For the Introl compiler, we can use command line options to select the order in which bits are assigned to the members. The default order is from bit 0 to bit 7. So in the declaration shown previously, *DevicePriority* will be assigned bits 0–2, *Running* will be assigned bit 3, *OverloadError* will be assigned bit 4, and *FatalError* will be assigned bit 5. Bits 6 and 7 are not used. In this example it does not really matter which actual bits are assigned to the members. The members are always accessed by name, using normal structure notation. Following is a program example that performs a test on *DevicePriority* and *OverloadError* and changes *DeviceNumber:*

```
if((DeviceState.DevicePriority == 7) || (DeviceState.OverloadError == 1)) {
    DeviceState.DevicePriority = 1;
}
```

As you can see, the programmer no longer needs to implement the details for alignment and the logic operations—the compiler takes care of it.

In this example, the actual bits assigned to the members did not matter. There are, however, some applications in which it does matter which bits are assigned to the members. Following is an example that manipulates individual hex characters in a byte:

```
typedef struct {
    char LDigit    : 4;
    char UDigit    : 4;
} HEX_BYTE;
...
HEX_BYTE InHex;
```

Now in order to access the 4-bit characters, we use the standard structure notation:

```
InHex.LDigit = 0xa;
InHex.UDigit = 0xb;
```

Notice that in the second line we can set the upper digit to a value without being concerned about alignment. Again the compiler will generate the alignment code for us.

If we know the bit assignments, we can also write code to access an MCU register. For example, to treat the *RTICTL* register in the 68HC12 as a structure bit field, we can use the following declarations:

```
typedef struct {
    UBYTE RTR    : 3;
    UBYTE RTBYP  : 1;
                 : 1;
    UBYTE RSBCK  : 1;
    UBYTE RSWAI  : 1;
    UBYTE RTIE   : 1;
} RTICTL_STRUC;
...
#define RTICTL *((RTICTL_STRUC *)0x0014)
```

In this example we needed to type cast an absolute address, 0x0014, to the structure type. Now to access individual fields within the register, we use the dot operator. For example to set the *RTR* bit field to 3, we use

```
RTICTL.RTR = 3;
```

If you are familiar with structure notation, this resulted in very readable code, and it did not require us to implement the details.

The last example illustrated the advantages of using structure bit fields—the compiler generates the code to implement the bit operation, and it results in a convenient and readable notation. However, these advantages are outweighed by the disadvantages of the code not being portable to other compilers and giving the compiler too much control. The code generated by the compiler may not be as efficient as our own implementation using bitwise logic and shifting operations.

▶ SUMMARY

In this chapter we saw how to create and use data objects in C. We also looked at the actual assembly code that is generated by the compiler. The emphasis was on portability and efficient use of memory.

Much of this material is quickly passed over in a normal C course. When you are designing C code for a small embedded microcontroller, you need to make sure that you understand all the details. It is a good idea to consistently look at the code produced by the compiler. In this way, you can learn more about the compiler, learn how to use it efficiently, and make sure an error was not introduced by the compiler.

EXERCISES

1. For each of the following lines of C code, indicate the memory type(s) that will be used and the number of bytes that will be allocated:
   - **(a)** `#define SIZE 4`
   - **(b)** `SBYTE convert(SBYTE);`
   - **(c)** `const UBYTE cvar = 0x40;`
   - **(d)** `INT16S var1;`
   - **(e)** `INT16U Table[SIZE];`
   - **(f)** `typedef enum{RST,STOP,GO} STATE;`
   - **(g)** `STATE newstate = RST;`
   - **(h)** `extern INT16S var1;`
   - **(i)** `UBYTE *aptr;`

2. What is the scope of a defined type?

3. A system has two external serial port peripherals. The data register for serial port 1 is mapped to 0x0300 and the data register for serial port 2 is mapped to 0x0400. Using type casting and *#defines*, define the two data ports as *SER1_DAT* and *SER2_DAT* so they can be accessed directly as data objects. Also show a statement that transfers the data from serial port 1 to serial port 2.

4. Define *Ser1Ptr* and *Ser2Ptr* as constant pointers that point to the two serial port registers described in Exercise 3. Show a statement that transfers the data from serial port 1 to serial port 2.

5. Using pointer notation, copy *RpmTbl[4]* to *RpmTbl[5]*, where *RpmTbl[]* is an array of 16-bit unsigned integers.

6. A project requires a time-of-day program.
   - **(a)** Define a structure type called *TIME* that has three members that are unsigned 8-bit integers and are called *hrs, mins,* and *secs*.
   - **(b)** Create an instance of the *TIME* structure called *TimeOfDay*.
   - **(c)** Write a code snippet that correctly increments the time given in *TimeOfDay; hrs* should roll over after 12.

7. A program that implements a state machine has the following state names: *STATE1, STATE2, STATE3,* and *RESET.*
   - **(a)** Create an enumerated type called *PSTATES* that defines these state names. *RESET* should have the value 0.
   - **(b)** Create an instance of *PSTATES* called *CurPState.*
   - **(c)** Set *CurPState* equal to *RESET.*

8. Write a code snippet that will call *Process()* if the bits in the 8-bit variable *Status* are as follows (do not change *Status* or use structure bit fields):
   - **(a)** $bit3 \oplus bit1 = 0$
   - **(b)** $bit2 \mid bit0 = 0$
   - **(c)** $bit2 \cdot (bit3 \oplus bit1) = 1$

9. Write a code snippet that will set bits 3, 4, and 5 of *PORTP* equal to bits 1, 2, and 3 of *PORTT,* respectively. Do not use structure bit fields.

10. Create a macro for the code snippet created in Exercise 9.

C Program Structures

In this chapter we cover C program structures, including control structures, functions, and modules. The focus is on program design for embedded microcontrollers, readability, portability, and reliability.

▶ 14.1 CONTROL STRUCTURES

In Chapter 6 we covered structured constructs for assembly language program design. In that chapter we used the C control constructs to design the equivalent assembly code. At this point it is a good idea to review that material. It is especially helpful to understand how the construct is represented as a flow diagram.

Here we look at the syntax and program design techniques we need when writing our programs in C. We also look at the *goto, continue,* and *break* instructions.

14.1.1 Conditional Constructs

Conditional constructs are used to conditionally execute a block of code. The conditional constructs in C are *if, if-else, if-else if-else,* and *case.*

The *if* Construct.   The *if* construct is used to conditionally execute a block of code. Figure 14.1 shows the *if* syntax and flow diagram. If the expression *expr* is true, then the code block contained in braces is executed. The braces are not required if the code block consists of a single statement. The code conventions followed in this text, however, require

518

C Syntax	Flow Diagram

`if(expr){`
`    /*True code block */`
`}`
	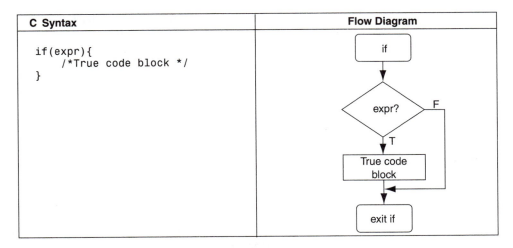

Figure 14.1 The *if* Construct

the braces regardless of the size of the code block. In fact we may show an empty code block in the braces to indicate that nothing is done for that condition.

Expressions.   All the control structures involve the conditional execution of a block of code. In C the condition is based on the evaluation of an expression. If the expression evaluates to true, the block of code is executed.

The condition is considered true if the expression evaluates to a nonzero result. If an expression evaluates to zero, the condition is considered false. Expressions may have a multivalued result or a binary result. For example the following expression has a multivalued result:

`(_H12PORTA & 0x0F)`

When the expression results are based on one of the following relational or logical operators, it has a binary result:

Relational Operators: `<, <=, ==, !=, >=, >`
Logical Operators: `&&, ||, !`

In C binary results are zero if the condition is false and one if the condition is true. Because of this, it is common to have the following *#define*s:

```
#define TRUE  1
#define FALSE 0
```

With these *#defines,* we can have an expression such as

```
/***********************************************************/
    if(state1 || state2 == TRUE){
        /* True code block */
    }

/***********************************************************/
```

TRUE and *FALSE* must be used only with expressions that have binary results, because other expressions are also seen as true, even if the result is not a one. The following bad example contains an expression that compares a multivalued result to *TRUE*:

```
/*************************************************************
 * A bad expression that compares a multivalue to a binary
 *************************************************************/
    if((port1 | port2) == TRUE){  /* Bad expression */
        /* True code block */
    }
/*************************************************************/
```

This expression performs a bitwise OR on *port1* and *port2*. The result is zero if both *port1* and *port2* are zero. But if either is nonzero, the result could be any nonzero value—not just one.

Because of this, we should design all expressions to end up with a binary result. This helps reduce errors and improves readability. To fix the last example, we need to compare the result of the bitwise OR to a multivalued integer. The result of that compare would then be a binary result.

```
/*************************************************************
 * A corrected version that compares two multivalues.
 *************************************************************/
if((port1 | port2) == 0x01){
    /* True code block */
}
/*************************************************************/
```

This is better because the result of a bitwise operation is a bit pattern, not a logical value.

Assignments in Conditional Expressions.   C allows assignments to be put almost anywhere, including in conditional expressions. This gives the programmer more freedom, but it is considered poor programming practice, because it hides errors. For example

```
/*************************************************************
 * A bad example. includes assignment in expression.
 *************************************************************/
    if(port1 = port2){
        /* True code block */
    }
/*************************************************************/
```

This expression will set *port1* equal to *port2*, and then if they are nonzero, it will be evaluated as true. Now is this what was intended? Normally an expression like this is a mistake. The "=" was accidentally used instead of "==". But since C allows assignments in conditional expressions, no error or warning message was given. This is one of the most common errors made by programmers—new and experienced.

Even if we did intend to use the assignment operator, it is considered by many to be poor programming practice. Using an assignment in a conditional expression is an example of code with a side effect, which should be avoided. The expression is evaluated, and as a side effect, the assignment is made. When we see this code, we have to figure out whether it is a mistake, and if it is not, what happens first, the assignment or the evaluation? To make this code acceptable, we need to separate the two operations. Following are two acceptable ways to write this code:

```
/************************************************************
* Taking assignment out of the expression onto a new line.
************************************************************/
port1 = port2;
if(port1 != 0x00){
    /* Do something */
}

/************************************************************
* Separating the assignment with the comma operator.
************************************************************/
if(port1 = port2, port1 != 0x00){
    /* True code block */
}

/************************************************************/
```

In the first example, the assignment is separated by taking it out of the expression and into another line. This is the most common solution. It is easy to read and simple.

The second example uses the comma operator to make the assignment first. When a comma operator is used, the two expressions are executed from left to right. The result of the last statement is used to decide if the code is to be executed. Using the comma operator is more appropriate in looping constructs, when both statements are part of the loop parameters.

The *if-else* Construct. The next conditional construct is the *if-else*. The *if-else* conditionally executes one of two blocks of code. The syntax and flow diagram are shown in Figure 14.2. If *expr* is nonzero, the first block of code is executed. If not, the second block is executed.

An *if-else* construct should always be considered before using an *if*. Many bugs can occur if the *else* condition is not considered. After consideration, if the *else* is not required, we can remove it. Or to explicitly document that it was considered, we can include the *else* but with an empty code block. For example

```
/************************************************************
* Explicitly show an empty else block
************************************************************/
    port1 = port2;
    if(port1 != 0x00){
        /* True code block */
    } else { /* Do nothing */
    }
/************************************************************/
```

C Syntax	Flow Diagram
| ```
if(expr){
    /*True code block*/
}else{
    /*False code block*/
}
``` | 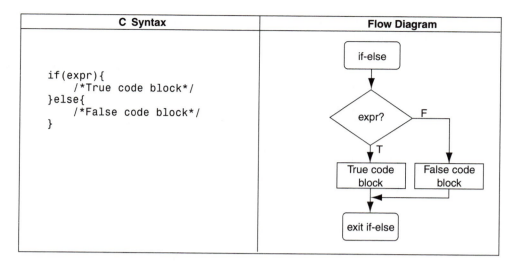 |

Figure 14.2 The *if-else* Construct

C Syntax	Flow Diagram
| ```
if(expr1){
    /*True code block1*/
}else if(expr2){
    /*True code block2*/
}else{
    /*False code block*/
}
``` | 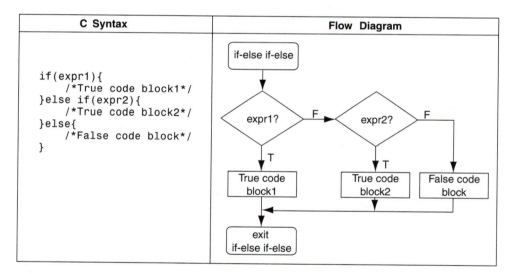 |

Figure 14.3 The *if-else if-else* Construct

The *if-else if-else* Construct. The *if-else if-else* construct is used to conditionally execute one of several mutually exclusive processes. The syntax and flow diagram are shown in Figure 14.3.

It contains multiple expressions that are evaluated in order until one is true. The code block following the first true expression is executed, and the *if-else if-else* construct is exited. Note that only the code block of the first true expression will be executed, even if more than one expression is true. If we do want to execute all blocks of code for every true expression, we would use multiple *if* constructs as shown below:

```
/**********************************************************
 * Using ifs to execute all code blocks with true expressions
 **********************************************************/
if(expr1){
    /* True code block1 */
}
if(expr2){
    /* True code block2 */
}
if(!(expr1 || expr2)) {
    /* False code block */
}
/**********************************************************/
```

The *case* Construct. The *case* construct is similar to the *if-else if-else* construct. As shown in Figure 14.4, the syntax for the *case* construct uses the *switch* statement followed by a list of cases.

The expression is evaluated one time and then compared to the constant expressions for each case. In a C *case* construct, a *break* is required to execute only one process. The *break* makes execution jump out of the *switch* construct. So if a case is true, that code block is executed, and the program exits the switch statement. If more than one case is true, only the code for the first true case is executed. There is also a default process defined with the *default* statement. It is equivalent to the last *else* in the *if-else if-else* construct.

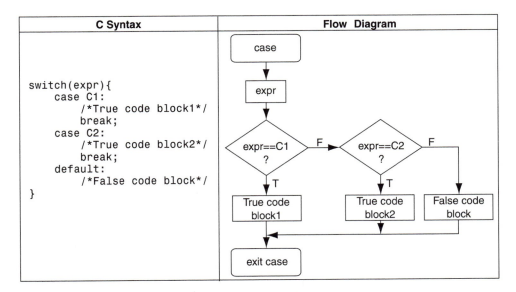

Figure 14.4 The *case* Construct

Logically the *case* construct is the same as the *if-else if-else* construct, but there are some subtle differences in the implementation. For a *case* construct, a single expression is evaluated one time. The result of the expression is then compared to constant expressions for each case. For the *if-else if-else,* an expression is evaluated at every *if* and *else if.* We can illustrate how this makes a difference with the following bad example:

```
/***********************************************************
* Bad example of using an if-else if-else
***********************************************************/
    if(getchar() == 'q') {
    /* quit */
    }else if(getchar() == '\r'){
        /* do something */
    }else{
        /* ignore character */
    }
/***********************************************************/
```

The intention of the code is to do something if a carriage return is received or to quit if 'q' is received. The problem is that since there are two expressions that are evaluated, *getchar()* is called both times. So if the program is at the first *if* expression and a carriage return is received, nothing will happen. Or if we are at the *else if* expression and a 'q' is received, nothing will happen. This is a common error for beginning programmers.

To fix the code, we can use a *case* construct or use a temporary variable to hold the character received. Following is the corrected code using the *case* construct:

```
/***********************************************************
* Corrected example using a case construct
***********************************************************/
    switch(getchar()){
    case('q'):
        /* quit */
        break;
    case('\r'):
        /* do something */
        break;
    default:
        /* do something else */
    }
/***********************************************************/
```

For this program *getchar()* is called only one time. The character received is then compared with the *case* expressions, and if one is equal, that code will be executed. The difference here is that the expression *getchar()* was executed one time instead of twice, as it was in the *if-else if* example.

Following is an example of the equivalent code using an *if-else if-else* construct and a temporary variable.

```
/************************************************************
* Corrected example of using an if-else if-else construct
************************************************************/
    c = getchar();
    if(c == 'q') {
        /* quit */
    }else if(c == '\r'){
        /* do something */
    }else{
        /* do something else */
    }
/************************************************************/
```

In this example *getchar()* was only called one time. The conditional expressions do not contain side effects. They are only based on the value of the character received.

Lookup Tables. If you have a requirement for a *case* construct with many cases, or if your code has to run very fast, you should consider using a lookup table. In Chapter 13 we saw how we could use a lookup table to display a number on a seven-segment display. The resulting code for the lookup table is

```
const UBYTE
SegTable[] = {0x7e,0x30,0x6d,0x79,0x33,0x5b,0x5f,0x70,0x7f,0x7b};
  …
_H12PORTA = SegTable[num];
```

The alternative to using this table would be to use the following *case* construct:

```
/************************************************************
* Example of using an equivalent case construct
************************************************************/
switch(num){
case(0):
    _H12PORTA = 0x7e;
    break;
case(1):
    _H12PORTA = 0x30;
    break;
case(2):
    _H12PORTA = 0x6d;
    break;
case(3):
    _H12PORTA = 0x79;
    break;
case(4):
    _H12PORTA = 0x33;
    break;
case(5):
```

```
        _H12PORTA = 0x5b;
        break;
    case(6):
        _H12PORTA = 0x5f;
        break;
    case(7):
        _H12PORTA = 0x70;
        break;
    case(8):
        _H12PORTA = 0x7f;
        break;
    case(9):
        _H12PORTA = 0x7b;
}
*************************************************************/
```

As you can see, the code using the *case* construct is much more cumbersome and would be much slower than the lookup table.

Nesting versus Complex Expressions.   We can implement a program by applying the conditional constructs in many different ways. Our goal is to find an implementation that is simple and easy to read. To do this, we try to avoid nesting and make the expression contain all the information required for a given condition. The problem with nesting is that it splits the condition for a given block of code into several expressions. Nesting can also make the code more difficult to read because of the indentation. For example

```
/*************************************************************
 * Poorly designed control structure
 *************************************************************/
if((_H12PORTA & 0x01) == 0x01){
    if((_H12PORTA & 0x10) == 0x10){
        if((_H12PORTP & 0x02) == 0x00){
            if((_H12PORTP & 0x20) == 0x00){
                StopPulse();
            }
        }
    }else{
        SampleFlag = 1;
    }
}
if((_H12PORTA & 0x01) == 0x01){
    StopPulse();
}
/*************************************************************/
```

This example has several *if* constructs, some nested and one in sequence. One of the *if* constructs also includes an *else,* but because it is separated from the *if,* it is difficult to determine what conditions will cause the *else* code to be executed.

We can implement this example with a single *if-else if* construct that uses more complex expressions. To convert nested *if* constructs into a complex expression, we combine the nested expressions with logical ANDs, &&. To convert sequenced *if* constructs into a complex expression, we combine the sequenced expressions with logical ORs, ||. The example shown previously now becomes

```
/*****************************************************************
* Better design with complex expressions
*****************************************************************/
if((((_H12PORTA & 0x01) == 0x01) &&
   ((_H12PORTA & 0x10) == 0x10) &&
   ((_H12PORTP & 0x02) == 0x00) &&
   ((_H12PORTP & 0x20) == 0x00) ||
   ((_H12PORTA & 0x01) == 0x01)){
     StopPulse();
}else if (((_H12PORTA & 0x01) == 0x01) &&
         ((_H12PORTA & 0x10) != 0x10)){
     SampleFlag = 1;
}
/*****************************************************************/
```

Now the condition to execute each code block is contained in a single expression and the nesting has been eliminated.

This code eliminated the nesting, but now we have complex expressions that span multiple lines. Let's look at a couple of methods that can be used to simplify these expressions.

Simplifying the Expression. We can remove the bit masking from the expression by using defines, macros, temporary variables, or functions. The following example uses simple macros for the bit mask operations:

```
/***********************************************************************
* Simplified expressions using defined macros
***********************************************************************/
#define PA0 (_H12PORTA & 0x01)
#define PA4 (_H12PORTA & 0x10)
#define PP1 (_H12PORTP & 0x02)
#define PP5 (_H12PORTP & 0x20)

...

if((PA0==0x01) && (PA4==0x10) && (PP1==0x00) && (PP5==0x00) || (PA0==0x01)){
    StopPulse();
}else if ((PA0 == 0x01) && (PA4 != 0x10)){
    SampleFlag = 1;
}


/***********************************************************************/
```

Another similar method is to simplify the expression by using a function or a more complex macro that performs the expression. For example we might implement the program shown with the following code:

```
/*****************************************************************
* Simplified expressions using a function
*****************************************************************/
if(PulseIsDone()){
    StopPulse();
}else if ((((_H12PORTA & 0x01) == 0x01) &&
        ((_H12PORTA & 0x10) != 0x10)){
    SampleFlag = 1;
}

/*****************************************************************/
UBYTE PulseIsDone(void){
    return ((((_H12PORTA & 0x01) == 0x01) &&
            ((_H12PORTA & 0x10) == 0x10) &&
            ((_H12PORTP & 0x02) == 0x00) &&
            ((_H12PORTP & 0x20) == 0x00) ||
            ((_H12PORTA & 0x01) == 0x01)))
}

/*****************************************************************/
```

The function *PulseIsDone()* returns a one or a zero based on the results of the expression. Of course, we could also do this using a macro.

Boolean Reduction of Complex Expressions. Another important method to reduce complex expressions is to use Boolean logic reduction. In the complex expression shown previously, we have several expressions combined with logical operators. We can use the same techniques we learned for digital design to reduce this expression. To illustrate this, let's first label the expressions in the example:

A: PA0 == 0x01
B: PA4 == 0x10
C: PP1 == 0x00
D: PP5 == 0x00

Now the logic for the first expression is

$$A \cdot B \cdot C \cdot D + A = A(B \cdot C \cdot D + 1) = A$$

The whole expression can be reduced to one simple expression. We can implement the first *if* construct with

```
if((_H12PORTA & 0x01) == 0x01) {
    StopPulse();
}
```

Of course this was an extreme case. A complex expression may not be reducible at all. Boolean reduction is just one way to simplify a complex expression or a deeply nested construct.

Another benefit in examining a complex expression with Boolean logic is to see what the results are for every combination of inputs. From a Boolean expression, we can generate a truth table and examine each case. This will avoid bugs caused by not considering every possible condition. Usually most of these bugs can be removed by simply adding an *else* to cover all other conditions.

Combining Bit Operations into Bit Fields. The second expression in the example shown earlier can be reduced in a different way. The expression is

```
((( _H12PORTA & 0x01) == 0x01) && (( _H12PORTA & 0x10) != 0x10))
```

The two bits we are testing are both in *PORTA,* so we can combine the two tests into a single test as follows:

```
(( _H12PORTA & 0x11) == 0x01)
```

Now the resulting code for the complete example has been reduced to

```
/**********************************************************/
if(( _H12PORTA & 0x01) == 0x01) {
    StopPulse();
}else if(( _H12PORTA & 0x11) == 0x01){
    SampleFlag = 1;
}
/**********************************************************/
```

14.1.2 Looping Constructs

A looping construct conditionally repeats a block of code. The C looping constructs are *do-while, while,* and *for.* It is especially important to understand the order of execution for looping constructs.

The *do-while* Loop. The *do-while* loop executes a block of code, and if an expression is true, it repeats the execution. The syntax and the flow diagram for the *do-while* are shown in Figure 14.5.

The distinguishing characteristic of a *do-while* loop is that it always runs the process at least one time because it evaluates the expression after it executes the code block. Consequently it is not as versatile as the *while* loop. It is also somewhat more difficult to read, because the looping expression is given at the end of the loop instead of the beginning. However, it can result in more efficient machine code than the *while* loop.

The *while* Loop. The *while* loop tests a condition, and if the condition is true, repeats a block of code. The syntax and the flow diagram for the *while* are shown in Figure 14.6.

The *while* loop is easier to read and more versatile than the *do-while*. Because of this it is preferred, unless you are trying to squeeze every byte out of the machine code. We can also write a *while* loop to run a process at least one time if that is what is required.

C Syntax	Flow Diagram

`do{`
`    /*Code block*/`
`}while(expr);`
	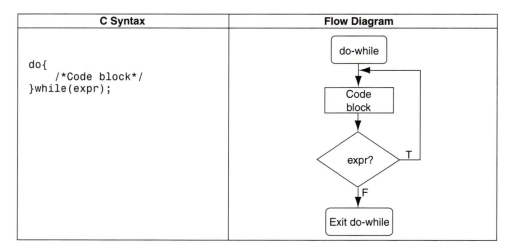

Figure 14.5 The *do-while* Construct

C Syntax	Flow Diagram

`while(expr){`
`    /*Code block*/`
`}`
	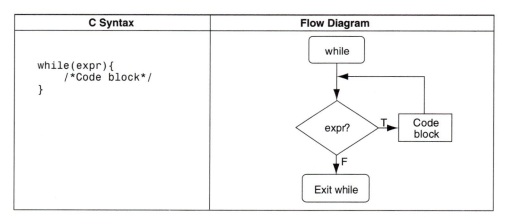

Figure 14.6 The *while* Construct

The *for* Loop. A *for* loop is a *while* loop with loop parameters and statements defined in an argument list. Because all the loop parameters are in the argument list, the *for* improves the program structure. The syntax and the flow diagram for the *for* loop are shown in Figure 14.7.

The first statement in the argument list, *S1,* is executed first. This statement is for initializing the looping parameter. Next the expression is evaluated, and if the expression is true, it executes the block of code in the braces. After the code block is executed, it executes the last statement in the *for* argument list, *S2.* This is the operation performed on the looping parameter every time through the loop. Typically it is an increment or decrement. Note that the first statement in the argument list, *S1,* is only executed one time. The last statement, *S2,* is executed each time through the loop after the code block is executed and before the expression is evaluated.

C Syntax	Flow Diagram

`for(S1;expr;S2){`
`    /*Code block*/`
`}`
	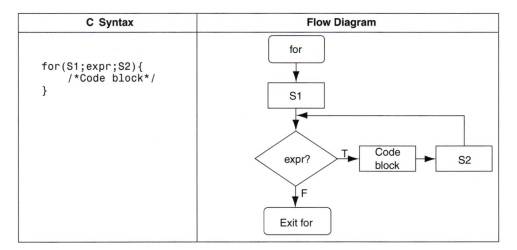

Figure 14.7 The *for* Construct

Terminal Count Bug. In Chapter 6 we saw that we have to be careful when designing a loop construct that may involve a terminal count. Let's look at this error again here and come up with some solutions. Source 14.1 shows a simple function intended to fill a block of memory with the specified fill character. It appears to be a good function, but if we pass an ending address of 0xFFFF, it will never stop.

Source 14.1 Memory Fill Function with Terminal Count Bug

```
/***************************************************************/
void BlockFill(UBYTE *startp, UBYTE *endp, UBYTE fillc){

    UBYTE *mptr;

    for(mptr = startp; mptr <= endp; mptr++){
        *mptr = fillc;
    }
}
/***************************************************************/
```

The reason the function fails is that the expression *mptr <= endp* can never be false if *endp* is 0xFFFF. Unfortunately the compiler will not give an error or warning message.

If we examine the program more closely, we can come up with several ways to fix this problem. First we need to change the "<=" operator to either "==" or "!=" in the expression. This presents a problem, because we need to include the last address in the fill process.

We can come up with several solutions. These solutions provide a solution to the terminal count bug, and they serve as illustrations in designing loops. To come up with these solutions, we have to understand how the loop constructs work.

Source 14.2 includes two solutions using a *while* loop. The first one reverses the order of the fill process and the increment. In this way the last location is filled before the address is compared. To avoid missing the first byte in the block, the pointer is initialized to point to the location before the first location to be filled. Then when the pointer is incremented the first time through the loop, it points to the first byte in the block.

Source 14.2 Two Solutions Using a while Loop

```
/*************************************************************/
void BlockFill(UBYTE *startp, UBYTE *endp, UBYTE fillc){

    UBYTE *mptr;

    mptr = (startp - 1);    /* point to byte before start */
    while(mptr != endp){    /*evaluate expression first*/
        mptr++;             /*point to next location */
        *mptr = fillc;      /*fill byte */
    }
}
/*************************************************************/

/*************************************************************/
void BlockFill(UBYTE *startp, UBYTE *endp, UBYTE fillc){

    UBYTE *mptr;

    mptr = startp;
    while(mptr != endp){    /*fill all but last address in loop*/
        *mptr = fillc;
        mptr++;
    }
    *mptr = fillc;          /*last address outside loop */
}
/*************************************************************/
```

The second example is a normal *while* loop that evaluates the expression, performs the fill process, and then increments the pointer. When this loop is finished, the last byte is not filled, so the fill process is executed one more time outside the loop. Note that upon exiting the loop, *mptr* is equal to *endp*.

Both of these solutions require some processing outside the loop. This is an invitation for bugs. If the programmer forgets to decrement the pointer in the first example or forgets to fill the last byte in the second example, the programs fail. The first example is preferable because the extra operations are performed before the loop. It is normal to perform initialization for a *while* loop before entering the loop. In the second example, the extra operation is at the end, so it may be easy to miss.

It is best to have everything contained in the loop for readability and to avoid errors when moving or reusing the loop. We can do this using a *for* loop as shown in Source 14.3.

Source 14.3 A Solution Using a for Loop

```
/*****************************************************************/
void BlockFill(UBYTE *startp, UBYTE *endp, UBYTE fillc){

    UBYTE *mptr;

    for(mptr = (startp - 1); mptr < endp; ){ /* no S2 */
        mptr++;    /* increment before fill */
        *mptr = fillc;
    }
}
/*****************************************************************/
```

This *for* loop is a little unconventional, but it does contain all the looping parameters within the loop structure. Because the pointer had to be incremented before the fill operation, we could not put the increment operation in the argument list. Because of this, a programmer may look at the argument list and think that the loop goes forever. An alternative would be to put everything in the argument list and use the comma operator as shown in Source 14.4. This is a bit cryptic, and only practical because the code block in the loop consists of only a single statement.

Source 14.4 Another Solution Using a for Loop

```
/*****************************************************************/
void BlockFill(UBYTE *startp, UBYTE *endp, UBYTE fillc){

    UBYTE *mptr;

    for(mptr = (startp - 1); mptr < endp; mptr++, *mptr = fillc){
    }
}
/*****************************************************************/
```

The *break, continue,* and *goto* Constructs.   We can write any program using only the control constructs already covered. As a good programming habit, we should first try to design our code only using the basic constructs. There are times when the code can be simplified by using the *break, continue,* or *goto* branching instructions, however. These commands can simplify a program, or if not used carefully, can ruin a program's structure.

The *break* command is used to exit a case or looping construct. We already saw how it was used with the *case* construct; now let's look at how it can be used with a *do-while, while,* or *for* loop.

Source 14.5 shows the *while* loop contained in the *getstrg()* function in the Basic I/O module. This loop inputs characters until a carriage return is received, or until the maximum string length is reached. Within the loop, if the character is a printable character, it is stored in the string buffer, *strg[]*. In order to avoid a buffer overflow, the number of printable characters are counted and compared with the maximum string length. If the string length is exceeded, it exits the loop. In this example this is done with the *break* command.

Source 14.5 The Use of break in getstrg() Loop

```
/***************************************************************/
    while(c = getchar(),c != '\r'){   /* receive until enter */
        if(' ' <= c <= '~'){          /* if printable */
            charnum++;                /* count character */
            if(charnum == (strlen)){  /* exit if strlen exceeded */
                break;
            }
            PUTCHAR(c);               /* otherwise store */
            *strg = c;
            strg++;
        }else if(c = '\b' && charnum > 0){
            PUTCHAR(c);
            PUTCHAR(' ');
            PUTCHAR(c);
            strg--;
            charnum--;
        }
    }

/***************************************************************/
```

The problem with the *break* command is that it is a bit sneaky. We cannot tell by reading the *while* expression that the loop exits when the maximum string length is reached. Because of this, the only time a *break* should be used is for an exception to normal operation. In the *getstrg()* example, the normal operation is to receive characters until a carriage return is received. Exceeding the maximum string length could be considered an exception to this normal operation, because it is treated as an error condition.

With a little more thought, we could have handled the maximum string length in the loop expression. Source 14.6 shows another version of the loop in which the string length is checked in the *while* expression. The expression is somewhat more complex, but it better documents the actual loop parameters.

Source 14.6 The getstrg() Loop Without a break

```
/***************************************************************/
    while(c = GETCHAR(),(c != '\r') && (charnum < (strlen-1))){

        if(' ' <= c <= '~'){          /*Echo and save printable chars */
            PUTCHAR(c);
            *strg = c;
            strg++;
            charnum++;
        }else if(c = '\b' && charnum > 0){   /* Handle backspace */
            PUTCHAR(c);
            PUTCHAR(' ');
            PUTCHAR(c);
            strg--;
            charnum--;
        }else{                /*Ignore all other nonprintable chars */
```

```
        }

    }
```

/**/

In general think before using a *break*. Try to design your loop without it, but if it really helps simplify the loop, go ahead and use it.

The *continue* control command is used to return to the start of a loop construct. As shown in Figure 14.8, for both the *while* and *do-while* constructs, the *continue* makes the program jump to evaluate the loop expression.

In the *for* loop, the *continue* command jumps to the second loop statement. This is shown in Figure 14.9.

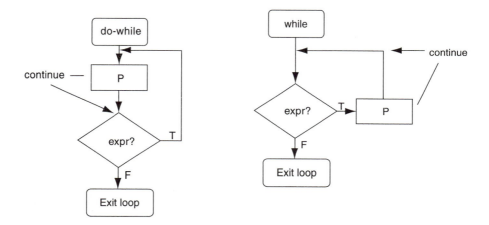

Figure 14.8   The Effects of a Continue on *while* and *do-while* Loops

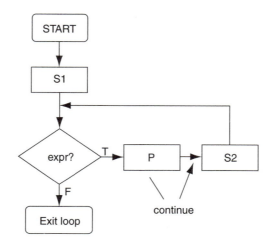

Figure 14.9   The Effects of a *continue* on a *for* Loop

The use of the *continue* command is like the *break;* if used carefully it can simplify the loop design. Source 14.7 shows a program that uses the *continue* command after error conditions. This program prompts for user input and then parses the input and outputs a string. Not all the code is shown, so we can focus on the *continue* usage.

Source 14.7 A Program That Effectively Uses continue

```
/***********************************************************************/
    while(1){
        OUTCRLF();
        PUTSTRG("Enter 2 decimal bytes:");
        if(GETSTRG(strgbuf,STRGLEN) == 0){
            sptr = SLICESTR(strgbuf,&arg1);
            if(sptr == NULL){    /* check for input error */
                ErrorHandlr(5); /* run handler and output new prompt */
                continue;
            }
            ...
            err = DECSTOB(arg1,&bin1);
            if(err == 1){
                ErrorHandlr(4);
                continue;
            }else if(err == 2){
                ErrorHandlr(3);
                continue;
            }
            ...
        }
        ...
    }
/***********************************************************************/
```

This loop is an endless loop that prompts for user input. Depending on the data received, there are several errors that could occur. If an error does occur, an error message should be sent, and then the program should return to the beginning of the loop to output a new prompt. The *ErrorHandlr()* function handles the error message. After the message is displayed, the *continue* command makes the program return to the start of the loop. Again the *continue* command is used to handle exceptions to the normal operation, and it makes the loop much simpler than it would be without using *continue*.

The *goto* command is rarely needed, and since using it can result in unstructured programs, it should not be used. It is equivalent to an assembly jump command, and like the jump command, it allows us to write spaghetti code. We do not use the *goto* command in this text. We could write structured programs using the *goto* command—it is just not necessary.

▶ 14.2 FUNCTIONS

In this section we cover the basic building blocks of a C program—*functions.* Well-designed functions are critical to the success of a project. We look at both design and implementation with an emphasis on programming small microcontrollers. Most of the implementation material is specific to the Introl-CODE development system and is documented in the Introl-CODE on-line manual.

All the C code we write is contained in a function. Even *main( )* is a function. For each C function, the compiler creates an assembly subroutine. So when we call a function, the compiler uses a *jsr* instruction, and to return from the function an *rts* instruction is used. At this point it may help to review the material in Chapter 6 on parameter passing and reentrancy.

14.2.1 The main() Function

main( ) is a special function that is in general the entry point to a C program. When writing C code for a computer system, a *main( )* function is required. To be consistent with this convention, all the code written for this text calls the entry point *main( )*.

When writing a C program for a computer system, we do not really care what the CPU does before it gets to *main( )*; we simply treat *main( )* as the start of the program. When writing C code for embedded systems, we do need to know what happens before *main( )*. There are a couple of reasons for this. First, most of our code must be ROMable, and we need control over what happens from the time the CPU is reset. Second we have limited memory resources. Even if the development system adds start-up code automatically, we need to know what was added and why. Figure 14.10 shows the code execution sequence for a C program in a embedded system.

If the program is in the development stage and is executed under the D-Bug12 monitor program, the start-up code is minimal. If the code is executed out of *RESET*, the start-up code includes all the MCU initialization and can be significant. We will look at this start-up code in more detail in Section 14.4.

At the end of the start-up sequence, the instruction

```
jsr main
```

passes control to *main( )*.

main( ) is also the only function that does not require a prototype. The compiler already knows about *main* and treats it as a

```
void main(void);
```

This makes sense for embedded systems, because in general we do not pass anything to *main( )* or return from *main( )*. In case our program does fall out of the end of *main( )*, the start-up code includes a trap to catch it.

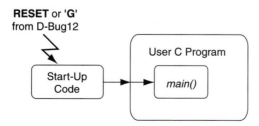

Figure 14.10 Execution Sequence for a C Program

14.2.2 Declaring and Defining Functions

When we use a function in C, there are three pieces of code required—the function prototype, the function definition, and the function call. Like an assembly subroutine, we write the function code, which is the definition, and then call the function by name. When we write ANSI-C code, we also have to declare a function with a function prototype. Source 14.8 shows these three pieces of code for a function called *AbsValue()*, which returns the absolute value of the parameter passed to it.

Source 14.8 Function Code Requirements

```
/********************************************************/
SBYTE AbsValue(SBYTE x);      /* function prototype */

/********************************************************/
void main(void) {

    SBYTE var;

    var = AbsValue(var);      /* function call */

}

/********************************************************/
SBYTE AbsValue(SBYTE x) {     /* function definition */
    x = (x >= 0) ? x : -x;
    return x;
}
/********************************************************/
```

The first line is the function prototype. The function prototype tells the compiler how many and what types of parameters are passed to the function, the type of data returned from the function, and the scope and type of function. Of course it must agree with the function definition.

In general a function prototype has the following form:

```
function_modifiers return_type function_name(p1_type x1,..,pn_type xn);
```

In Source 14.8 the *AbsValue()* function did not have a function modifier. The function modifiers are shown in Table 14.1. They specify the type of function, and they are always placed before the return type modifiers.

If there are no function modifiers, the function is a public function, so it can be called in other modules. If there are no functions modifiers, the compiler will also expect the function to be defined within the module containing the prototype. This is the case for the *AbsValue()* function.

The *static* modifier limits the scope of the function to the module that contains the function definition. It is used to create private functions.

TABLE 14.1 FUNCTION MODIFIERS

Modifier	Comment
None	Public function. The scope includes all modules.
static	Private function. The scope is limited to the module that contains the function definition.
extern	Used for function prototypes when the function is defined in another module. The function must be a public function.
_interrupt	The function is an interrupt service routine.

The *extern* modifier is used in a function prototype to indicate that the function definition is contained in another module. For example if we wanted to use the *AbsValue()* function in another module, that module must contain the following prototype:

```
extern SBYTE AbsValue(SBYTE x);        /* function prototype */
```

This prototype tells the compiler the types involved in the function so it can generate the correct calling code. It also tells the compiler to not look for the function definition, because it is defined in another module. This will be covered in more detail in Section 14.3.

The *_interrupt* modifier is the Introl-CODE modifier that tells the compiler that the function is to be implemented as an interrupt service routine. This modifier is compiler dependent. There is no official ANSI-C modifier for creating interrupt service routines. Consequently different compilers will use different methods for creating interrupt service routines; this will be covered later in this section.

Note that function prototypes are not strictly required. Typically a compiler will only give a warning message when a function is defined or used without a prototype. In fact the original, non-ANSI C did not have function prototypes. The problem with creating programs without function prototypes is that the compiler assumes that the parameters are *int*s. This has little effect on a typical computer program, but on a small microcontroller, it can result in very inefficient code.

The fourth line of Source 14.8 contains a call to the *AbsValue()* function. This call will replace *var* with the absolute value of *var*. A function call must include parentheses, even if there are no parameters. The parentheses tell the compiler that the name refers to a function.

The last part of Source 14.8 is the function definition. It starts with the function name and types, which must agree with the function prototype, and is followed by the block of code that performs the function. In this case the *AbsValue()* function is a simple function, which is contained in a single line. If an argument is returned, the return statement should be used to explicitly specify the value returned. In the case of *AbsValue()*, the value of *x* is returned to the calling function.

Since a function does something, we should try to use a verb for the name of a function. For example

```
GetStrg()
Delay()
```

With this in mind, is there a better name for *AbsValue()?* Functions that return a Boolean TRUE or FALSE should be named in a way that poses a question. Examples include the following:

```
isdigit()
IsTaskDone()
```

14.2.3 Passing Parameters

A function performs some kind of action. Usually it performs an action on one or more data objects. When writing functions, it is important to make it clear which data are affected by the function. Otherwise a programmer may use a function not realizing it changes a critical piece of data.

To make it clear which data are affected by a function, we should always pass all affected data as parameters. Passing parameters to functions will also make the function more general and therefore reusable for other programs. At this point it might be a good idea to review Section 6.5, which covers passing parameters in assembly language programs.

In the C language parameters are always passed by value. It is important to understand what this actually means. When a parameter is passed to a function, the value of the argument is copied into temporary storage to be passed to the function. This means that the function can only change the copy, not the original argument. The reason parameters are passed this way is to make the functions reentrant.

For example in the *AbsValue()* program in Source 14.8, the value of *var* is copied by the main program and passed to *AbsValue()*. *AbsValue()* then creates storage for the parameter passed and replaces it with its absolute value. The function then returns this new value back to *main()*, which assigns it to *var*.

Programmers who do not understand this concept often make the mistake shown in Source 14.9. This is another version of the absolute value function that would appear at first glance to work.

Source 14.9 Absolute Value Program with a Parameter Error

```
/********************************************************************/
void AbsValue1(SBYTE x);
SBYTE Var;

/********************************************************************/
void main(void) {

    AbsValue1(Var);

}

/********************************************************************/
void AbsValue1(SBYTE x) {    /* Bad */
    x = (x>=0) ? x : -x;
}

/********************************************************************/
```

Again we want to replace *Var* with its absolute value. It looks like this function would work until you realize that *Var* is not passed to the function. To help us understand this error better, let's look at the resulting assembly code in Listing 14.1. In the main program, *Var* is copied into ACCB to be passed to the function. *AbsValue1()* then creates storage on the stack to store the parameter. The function then takes the absolute value and stores the result on the stack. The new value is then immediately destroyed, because the stack space is deallocated before the function returns to *main()*. *AbsValue1()* changed the copy of *Var* passed to it. It did not change *Var* itself.

Listing 14.1 Assembly Code Showing Parameter Error

```
****************************************************************
main            ldab    Var
                bsr     AbsValue1
                rts
****************************************************************
AbsValue1:      pshb
                ldab    0,sp
                bge     ?2.7
                negb
                stab    0,sp
?2.7            leas    1,sp
                rts
****************************************************************
```

The *AbsValue()* function in Source 14.8 solves this problem by returning the new value back to *main()* with a *return* statement. *main()* then assigns this new value to *var* to replace the value of *var* itself.

Another way to design the *AbsValue()* function is to pass the parameter by reference. Instead of passing the value of the argument, we pass a pointer to the argument. The *AbsValue()* function can then change the argument by using the pointer. Source 14.10 shows a function that has its parameter passed by reference.

Source 14.10 Passing a Parameter by Reference

```
/**************************************************************/
void AbsValue2(SBYTE *ptr);
SBYTE Var;

/**************************************************************/
main(void) {

    AbsValue2(&Var);

}

/**************************************************************/
void AbsValue2(SBYTE *ptr) {
    *ptr = (*ptr >= 0) ? *ptr : -*ptr;
}
/**************************************************************/
```

Instead of passing the value of *Var*, we pass the address of *Var* to the function. *AbsValue2()* then assigns the contents of the location pointed to by the parameter to its absolute value. This technique also is used when the data object to be changed is too large to pass to the function.

14.2.4 Functions versus Macros

In Chapter 12 we saw how to create macros with the *#define* preprocessor command. The difference between macros and functions is that the code for the macro is inserted into the program every time the macro is used, whereas a function is implemented as a subroutine, so the code is only added one time. With respect to code size, a function is more efficient. The advantage of a macro is that a subroutine call and return are not required. The code is inserted in line. With respect to speed, a macro is more efficient.

Typically, as the code becomes more complex, it is better to implement it as a function. The absolute value function shown in the previous section is simple, so it is a good candidate for a macro. Source 14.11 shows the absolute value function implemented as a macro.

Source 14.11 Absolute Value Macro

```
/**************************************************************/
#define ABS_MAC(t) (((t) >= 0) ? (t) : -(t))

SBYTE Val1;

/**************************************************************/
main(void) {

    Val1 = ABS_MAC(Val1);

}
/**************************************************************/
```

The *#define* preprocessor command defines a macro called *ABS_MAC()*. This is an example of a macro that contains an argument. When using an argument in a macro, use a lot of parentheses. This is because we do not know where this macro will be placed.

When there is a macro argument, it is also passed by value. So this macro is equivalent to the *AbsValue()* shown in Source 14.8. To see the difference in the resulting assembly code, Listing 14.2 shows the code for both implementations.

Listing 14.2 Assembly Code Comparison of AbsValue() and ABS_MAC()

```
*********************************************************************
* Assembly code using AbsValue() function
*********************************************************************
main:       ldab      Var
            bsr       AbsValue
            stab      Var
            rts
```

```
*******************************************************************
AbsValue:     pshb
              ldab    0,sp
              bge     ?1.7
              negb
              stab    0,sp
?1.7          ldab    0,sp
              tfr     b,d
              leas    1,sp
              rts


*******************************************************************
* Assembly code using ABS_MAC() macro
*******************************************************************
main:         ldab    Val1
              bge     ?0.7
              negb
              stab    Val1
?0.7          rts


*******************************************************************
```

As you can see, there is significantly more code for the function *AbsValue()* than there is for the macro *ABS_MAC()*. Not only are the *bsr* and *rts* instructions added in the *AbsValue()* code, there is also additional code for temporary storage of the parameter. This is not actually required, but remember that this code was generated by a compiler, not a programmer. There is no way for it to know that the parameter value would not be needed later in the function.

Again macros are preferred if the code is small and there is less than one parameter. When the code is larger or has several parameters, it is preferable to use a function.

14.2.5 Assembly Functions

Sometimes we need to call an assembly subroutine from a C program. This may be because the function should be written in assembly, like the *msDelay()* routine in Chapter 12, or because we are using a routine that has already been written in assembly. We have seen that a function is implemented as a subroutine, so we can call the subroutine by treating it as a function. Since the assembly function is always in a different file, the function prototype will include the *extern* modifier. We also need to consider how the parameters, if any, are passed to and from the routine.

Note _____

The Introl 68HC11 compiler uses a different technique to deallocate the parameter stack space. It expects the function to deallocate the space instead of the calling routine. You can have it use the method described here by declaring the function with the *_cdecl* modifier.

_____

The Basic I/O module described in Appendix B is an example of module made up of assembly routines that can be used in a C program. First let's look at a simple subroutine example that does not have any parameters. Source 14.12 shows a C program that calls the *OUTCRLF()* routine, which sends a carriage return and line feed out the serial port. As you can see, the C program looks identical to the code that would be used if *OUTCRLF()* was a C function from another file.

Source 14.12 Calling a Simple Assembly Routine from C

```
/*************************************************************/
extern void OUTCRLF(void);

/*************************************************************/
void main(void){

    OUTCRLF();

}
/*************************************************************/
```

C Calling Conventions.   The program in Source 14.12 did not require any parameters. When parameters need to be passed to or from the assembly routine, we need to know the C calling conventions for our compiler. The C calling conventions define how the compiler implements the parameter passing. There is not a standard for calling conventions, so they are dependent on the compiler and the CPU. Calling conventions may be different for different compilers, but they are typically very similar. The calling conventions for the Introl-CODE 68HC12 compiler are as follows:

Parameters:	The first parameter in the parameter list is passed as follows:
	8-bit: ACCB
	16-bit: ACCD
	32-bit: IY:ACCD
	>32-bit: On the stack
	The remaining parameters are pushed onto the stack in the opposite order as the order of the parameters in the list.
Return Values:	8-bit or 16-bit: ACCD
	32-bit: IY:ACCD

If we follow these conventions, we can write an assembly routine that can be used with an Introl-CODE C program. Figure 14.11 shows how each parameter is passed for an example function.

Let's look at another example from the Basic I/O module now. Source 14.13 shows a C code example that calls the *getstrg()* function. There are two parameters passed to the function and one 8-bit return value.

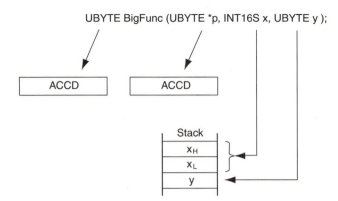

UBYTE BigFunc (UBYTE *p, INT16S x, UBYTE y);

Figure 14.11   Introl-CODE Parameter Passing Example

Source 14.13 Calling an Assembly Routine with Parameters

```
/****************************************************************/
#define STRLEN 5
extern UBYTE GETSTRG(UBYTE *strg, UBYTE strlen);

UBYTE StrgBuffer[STRLEN];

/****************************************************************/
void main(void){

    UBYTE err;

    err = GETSTRG(StrgBuffer,STRLEN);

}
/****************************************************************/
```

Again from the C side, we call the routine like any other C function from another module. Now let's look at the resulting assembly code. Listing 14.3 shows both the assembly code generated by the compiler and parts of the *getstrg()* subroutine.

In the calling routine, the compiler first places *strlen* on the stack. It then loads the pointer to *StrgBuffer* into ACCD. This agrees with the conventions defined previously for the Introl-CODE compiler.

Listing 14.3 Assembly Code That Shows the
Parameter Passing to getstrg()

```
********************************************************************
* Assembly code in calling routine. Generated by compiler
*    - comments added
********************************************************************
```

```
main:            leas -1,sp          ; reserve stack space for err
                 ldab #5             ; pass strlen on stack
                 pshb
                 ldd #StrgBuffer     ; pass pointer to buffer in ACCD
                 jsr getstrg         ; call function
                 leas 1,sp           ; reallocate strlen stack space
                 stab 0,sp           ; save return value in err
                 leas 1,sp           ; reallocate err stack space
                 rts

*******************************************************************
* Assembly code in getstrg. Generated by programmer.
*******************************************************************
getstrg          pshx                ;preserve IX
                 tfr d,x             ;strlen->4,sp;   strg->IX
                 ...
                 cmpa 4,sp           ;compare with strlen
                 ...
gs_ovrfl         ldab #1             ;return 1 if overflow
gs_rtn           pulx                ;recover IX
                 rts
*******************************************************************
```

Now the *getstrg()* routine must correctly access the parameters. In this case, upon entry into the subroutine, the pointer to *StrgBuffer* is in ACCD and *strlen* is on the stack just below the program counter. To preserve the IX register, the subroutine first pushes IX onto the stack. It then transfers the pointer passed in ACCD to IX so index addressing can be used. From now on, the pointer to *StrgBuffer* is in IX and *strglen* can be accessed at $(sp + 4)$. For example the *cmpa* line accesses the *strlen* using indexed addressing.

One line that affects the return value is also shown. Before returning from the subroutine, ACCB may be loaded with zero or one for the return value. Listing 14.3 shows the line that sets ACCB to one. The routine then recovers IX and returns. Once the program returns to the calling routine, the stack space for *strlen* is deallocated from the stack. The return value, which was passed in ACCB, is then stored into *err*.

There is another example of using a function written in assembly in the *demo1* project described in Chapter 12. In that example a 16-bit parameter is passed to *msdelay()* in ACCD.

Following the C calling conventions, we can also write assembly programs to call C functions. For example we can use the C standard math functions from an assembly program.

14.2.6 Interrupt Service Routines

We have seen that when we create a C function, the compiler implements that function as a subroutine. Now how do we create an interrupt service routine (ISR)? There is nothing in the ANSI-C standard that addresses interrupt service routines, and they are not covered in a typical first course in C. This is because we do not need to write ISRs when writing computer applications. When writing embedded microcontroller code, however, just about every project will require an interrupt service routine. We know we can write an ISR in assembly code, but what about in C?

First let's go over the differences between a normal subroutine and an ISR:

1. An ISR ends with an *rti* instead of an *rts*.
2. An ISR is never called. The only way to get to an ISR is through the interrupt mechanism.
3. There are no parameters passed to or returned from an ISR. Parameters are not possible, because we do not know when the interrupt will occur.

As far as the compiler is concerned, it can implement a function as an ISR by replacing the *rts* with a *rti* instruction. The other two differences are handled by the programmer. The second difference simply means that we should never actually call the function, and the third difference means the parameter and return types are always *void*.

To implement a function as an ISR, the Introl-CODE compiler uses a special modifier, *_interrupt*. To define an ISR in Introl-CODE, we would use the following syntax:

```
_interrupt void IrqIsr(void);
```

The *_interrupt* modifier tells the compiler to use an *rti* instead of an *rts*. Since the modifier *_interrupt* is compiler dependent, we should use a #*define* to make the code easier to port to another system. In this text we will use the following code:

```
#define ISR _interrupt void
```

Now the function prototype becomes

```
ISR IrqIsr(void);
```

This is easier to read and easier to change if another compiler is used. Source 14.14 shows an example of an ISR that makes the output compare for *Timer Channel 0* generate a 1ms periodic interrupt. As you can see, the ISR is declared and defined in the same way as a normal function, except the *ISR* modifier is used.

Source 14.14 An Interrupt Service Routine for Output Compare 0

```
/**************************************************************
* ISR function prototype
**************************************************************/
ISR OC0Isr(void);


/**************************************************************
* OC0 Service Routine
*    MCU: 68HC912B32, E = 8MHz
*    - Requires TC0 be set for an output compare (OC0Init()).
*    - setup for a 1ms periodic interrupt.
```

```
********************************************************/
ISR OC0Isr(void){
    _H12TC0 = _H12TC0 + 8000;
    _H12TFLG1 = C0F;
    OCmSCnt++;
}


/*******************************************************/
```

▶ 14.3 MODULES

As we learned in Chapter 12, a *module* is a source file that contains a collection of related routines. Modular programming can greatly enhance the portability of a large program. By designing program modules, we can create collections of routines that can easily be used in future projects. Creating a module is also the first step to creating a library.

Libraries are precompiled. This means we do not have to rebuild them when we build a project. But it also means we cannot change the configuration of the code, because the configuration is determined at build time. If we use a source-level module, we can configure the code for the specific project.

Every programmer or programming group should have an ongoing commitment to developing a collection of application-oriented modules. A new project can then be designed by adapting the existing modules, which will significantly reduce the development time of the project.

In this section we look at some of the design concepts and organization issues involved in writing program modules and using them in a project. Let's first look at some goals we are working toward when designing modular programs.

14.3.1 Portability

The number one reason for creating program modules is to organize our code in a way that allows us to easily adapt the code to other projects, other processors, and other peripheral hardware. To be successful at this, our modules should be as independent as possible; that is, each module should be able to perform its function without the haphazard use of code from other modules. Modules should also be independent of the hardware unless it is a driver module. Driver modules are used to interface a hardware resource with the rest of the program.

To use a module in a project, we should only have to add the module to the project, include the module header file in the master header file, and set any global configuration constants required for the module.

14.3.2 Reliability

Reliability can be enhanced in two ways by using program modules. It allows us to reuse pretested code, and it allows us to use information hiding.

Information hiding is the technique of allowing access to a limited number of module resources. In order to use a module, we should only need to know the specifications of the public resources. We do not need to dig into the complete module to determine which function to use and to figure out how to use it. Also we can keep other programmers away from resources that require a detailed and specific access process. For example a shared data object that requires a semaphore to control access can be kept private so a programmer cannot directly access it without using the semaphore. A public function then can be provided for accessing the data object correctly.

14.3.3 File Organization

Figure 14.12 shows the overall file organization for a modular program. The figure shows three modules. Additional modules would be added in the same way. Each module consists of a single C file. Associated with the module is a header file, which is used by other modules to access public resources within that module.

One of the most confusing issues concerning modular programs is how to use header files. Header files are used in several different ways: They are used to isolate configuration data; they are used for general definitions used by all programs; and they are used for external declarations of public resources in a module.

The system in Figure 14.12 shows the organization we use in this text. There is a single master header file for each project. This header file includes all definitions that are

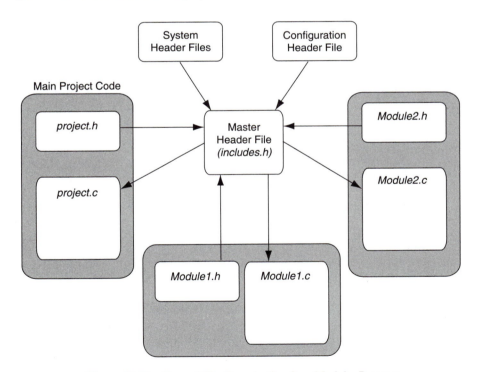

Figure 14.12   General File Organization for a Modular Program

required for every module, such as type definitions. The master header file also includes all other required header files. So all other header files are included through the master header file, and each C module contains only one *#include* command. For example the *project.c* program will not include *project.h*. It will include *includes.h,* which in turn includes *project.h* along with all the other header files required for the project.

System header files are those files that are used to access standard libraries and other code that comes with the development software. The configuration header file is optional. It contains project-wide configuration definitions. It may also include definitions for the specific MCU, so if you wanted to change MCUs, you only need to change the configuration header file. If there is only a small amount of configuration information, it can be placed in the master header file instead of creating a separate configuration file.

Because each header file may be included in several files, we need to make sure the contents of a header file do not cause a redefinition error. For example imagine that the *module2.h* file contains the following definition:

```
UBYTE Var;
```

This definition creates storage for an 8-bit variable and defines *Var* as the address of that variable. Since the definition is outside all function blocks, the scope of *Var* includes all modules in the project. From Figure 14.12 we can see that *module2.h* is included in the master include file, *includes.h,* which is then included in all the modules for the project. This means the program attempts to define *Var* in each file, but since its scope includes all files, we will get a redefinition error from the linker.

We should restrict items in a header file to those items that have a scope limited to the current file or to items that can be defined multiple times, such as *extern* declarations. This includes

> *#define*s
> *typedef*s
> *extern* declarations

If we need to place a definition that can only be made one time in a header file, we can use the conditional preprocessor commands. For the example shown previously, we can use

```
#ifndef VAR
#define VAR
UBYTE Var;
#endif
```

The first time the header file is compiled, *Var* will be defined. After that the definition will be ignored.

14.3.4 demo2 Project Example

Let's look at an example project *demo2*. Figure 14.13 shows the program organization for *demo2,* which is a new version of the *demo1* program that was introduced in

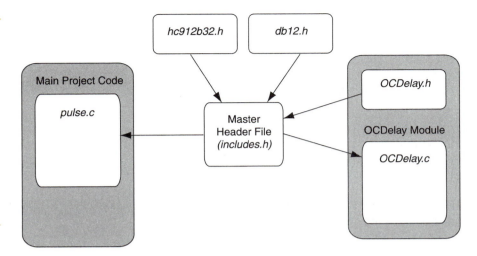

Figure 14.13 File Organization for the *demo2* Project

Chapter 12. This version uses a delay routine that is based on an output compare interrupt. It consists of two modules, the main project module and the *OCDelay* module.

Let's start with the *includes.h* header file. In this example it is used as the master header file and as the configuration header file. Source 14.15 shows the contents of *includes.h*.

It starts with one general configuration constant, *DBUG12*. This constant is used to determine whether the program is to be built to run under the D-Bug12 monitor or as a stand-alone program. If it is to run under D-Bug12, this constant is defined. If any module contains code that is specific to D-Bug12, it can be conditionally included by using this constant.

Source 14.15 demo2 Project Master Header File

```
/**********************************************************************
* includes.h - Master header file for demo2 project.
*
* Todd Morton 11/12/99
***********************************************************************
* Config Constants
**********************************************************************/
#define DBUG12              /* Define if program runs under D-Bug12   */


/**********************************************************************
* WWU Project Type Definitions
**********************************************************************/
typedef unsigned char    INT8U;
typedef signed char      INT8S;
typedef unsigned short   INT16U;
typedef signed short     INT16S;
typedef unsigned long    INT32U;
typedef signed long      INT32S;
```

```c
#define UBYTE INT8U
#define SBYTE INT8S
#define UWYDE INT16U
#define SWYDE INT16S

#define ISR _interrupt void

/************************************************************************
* General Defined Constants
************************************************************************/
#define FALSE    0
#define TRUE     1


/************************************************************************
* General Defined Macros
************************************************************************/
#define FOREVER()         while(1)
#define TRAP()            while(1){}


/************************************************************************
* MCU-Specific Configurations
************************************************************************/
#include "hc912b32.h"              /* CODE 68HC912b32 register defines    */

#define SWI()         asm("\tswi\n")
#define ENABLE_INT()  asm("\tcli\n")
#define DISABLE_INT() asm("\tsei\n")
#define LED_PORT_DIR  _H12DDRP  /* LED Port Definitions               */
#define LED_PORT      _H12PORTP
#define LED_BIT       0x01

/************************************************************************
* Project Constant and Macro Definitions
************************************************************************/
#define ON_TIME  250
#define OFF_TIME 750
#define TURN_LED_ON()  LED_PORT &= ~LED_BIT
#define TURN_LED_OFF() LED_PORT |= LED_BIT
#define INIT_LED_DIR() LED_PORT_DIR |= LED_BIT

/************************************************************************
* OCDelay Definitions
************************************************************************/
#define E_PER_MS 8000               /* E clock cycles per 1ms              */

/************************************************************************
* System Header Files
************************************************************************/
#ifdef DBUG12
#include "db12.h"                   /* D-Bug12 function prototypes         */
```

```
#endif

/*************************************************************************
* Project Module Header Files
*************************************************************************/
#include "OCDelay.h"              /* OCDelay module                   */

/*************************************************************************/
```

Next in *includes.h* are the type definitions. Since the type definitions are required in all modules, they should come before any other header file is included. After the type definitions, there are several general constant and macro definitions. These definitions may be used by any module.

The MCU-specific definitions come next. These definitions are used to isolate MCU-dependent definitions to one place. It is common to have these in a separate header. If we need to change MCUs, we only need to change these definitions and any driver modules that rely on a specific MCU.

The next code sets configuration definitions. These definitions should be made before the module header files are included, because there may be constants in those headers that depend on these configuration constants. The configuration constants should be divided into module groups. Not all the definitions for a module should be included. Only the constants that are intended to be public and are MCU or hardware dependent are placed here. The other definitions that are specific to the module and not dependent on the hardware configuration should be kept in the module itself.

Now that all the configuration definitions have been made, the system and module header files can be included. The system header file included in this case is *db12.h*. The *db12.h* header file contains function prototypes for some of the D-Bug12 routines. Note that *db12.h* is conditionally included only if the DBUG12 constant is defined.

After the system header files, the project module header files are included. There should be one header file for each module, except for the main program module.

The next file we will look at is the source file for the project's main program module, *pulse.c,* which is shown in Source 14.16. Notice that the only header file that is included is the master header file.

The *pulse.c* module should look familiar. It is essentially the same as the *pulse2.c* file from *demo1*. The only changes made are to use the interrupt-based delay routine. Instead of calling *mSDelay()*, this program calls *OCDelay()*. *OCDelay()* also requires additional initialization, *OCDlyInit(),* and it requires the interrupts be enabled, *ENABLE_INT().*

Source 14.16 The Main Project Code for demo2

```
/*************************************************************************
* pulse.c - A simple demonstration program.
* It generates 60 active-low pulses at a rate of 1Hz out of
* LED. This version uses a time-delay routine based on the
* Output compare 1.
*
*************************************************************************
/* Include Project Master Header File
```

```
***************************************************************/
#include "includes.h"

/*****************************************************************
* Global Variable Definitions
***************************************************************/
UBYTE PulseCnt;                /* The number of pulses remaining    */

/***************************************************************/
void main(void) {

    OCDlyInit();
    INIT_LED_DIR();            /* Initialize LED as output and turn off*/
    TURN_LED_OFF();
    ENABLE_INT();              /* Enable interrupts to use OCDelay()   */

    for(PulseCnt = 60; PulseCnt > 0; PulseCnt--) {
        TURN_LED_OFF();
        OCDelay(OFF_TIME);
        TURN_LED_ON();
        OCDelay(ON_TIME);
    }

    TURN_LED_OFF();            /* Turn LED off and trap                */
    TRAP();
}


/***************************************************************/
```

Now let's look at the *OCDelay* module. This module includes all the program code required to implement a delay based on an interrupt from *Timer Channel 0. Timer Channel 0* is configured as an output compare that generates a 1ms periodic interrupt.

Let's first look at the module header file *OCDelay.h,* shown in Source 14.17. This header file has one sole purpose—it provides *extern* declarations for the public resources in the *OCDelay.c* module.

Source 14.17 OCDelay Module Header File

```
/*****************************************************************
* OCDelay.h - Project Header file for the Output Compare - based
*             delay routine.
*
*****************************************************************
* Public Function Prototypes
***************************************************************/
extern void OCDelay(INT16U ms); /* Blocking delay routine.         */
                                /* The parameter is the number of  */
                                /* ms to delay.                    */
```

```
extern void OCDlyInit(void);        /* OCDelay initialization. This   */
                                    /* function must be called before */
                                    /* OCDelay() can be called.       */

extern INT16U GetmsCnt(void);       /* Read the current ms count.     */

/*******************************************************************/
```

In this case there are three public functions available, *OCDelay(), OCDlyInit(),* and *GetmsCnt()*. The module header file should also provide some level of documentation. So if we want to use the *OCDelay* module, we only have to include the header file in the master header file and read the descriptions contained in the header file. The module itself is either added to the project to be built along with the main code or added to the project as a precompiled library. Remember, if you use the module as a library, the header file provides the only documentation the programmer may have.

The code for the *OCDelay* module is all contained in *OCDelay.c,* which is shown in Source 14.18. First the master header file *includes.h* is included. This provides project configuration constants, system header files, and the type definitions used for the project. This is the first item in all modules.

Source 14.18 OCDelay Module

```
/*******************************************************************
* OCDelay.c - A delay module based on Output Compare 0
*
*******************************************************************
* Project Master Header File
*******************************************************************/
#include "includes.h"

/*******************************************************************
* Public Resources
*******************************************************************/
void OCDlyInit(void);               /* OCDelay Initialization Routine */
void OCDelay(INT16U ms);            /* OCDelay Function               */
INT16U GetmsCnt(void);              /* Read the current ms count      */

/*******************************************************************
* Private Resources
*******************************************************************/
ISR OC0Isr(void);                   /* OC0 interrupt service routine  */
static volatile INT16U OCmsCnt;     /* 1ms counter variable           */

/*******************************************************************
* Module Defines
*******************************************************************/
#define C0F 0x01
#define OL0 0x01
```

C Program Structures

555

```
#define C0 0x01
#define TEN 0x80
#define TC0_NUM 23

/********************************************************************
* OCDelay Function
*      - Public
*      - Delays 'ms' milliseconds
*      - Accuracy +/- 1ms
********************************************************************/
void OCDelay(INT16U ms){
    INT16U term_cnt;
    term_cnt = OCmSCnt + ms;
    while(OCmSCnt != term_cnt){}   /* wait for terminal count       */
}


/********************************************************************
* OCDlyInit() - Initialization routine for OCDelay()
*     MCU: 68HC912B32, E = 8MHz
*      - Sets TC0 for an output compare.
*      - Call SetUserVec() to run interrupt under D-Bug12
********************************************************************/
void OCDlyInit(void){

#ifdef DBUG12
    SetUserVector(TC0_NUM, &OC0Isr);        /* Init. D-Bug12 vector     */
#endif

    _H12TSCR |= TEN;                        /* Enable timer             */
    _H12TIOS |= C0;                         /* Set Channel 0 to OC      */
    _H12TCTL2 = OL0|(_H12TCTL2&0xFC);       /*Toggle OC0 pin for debug  */
    _H12TMSK1 |= C0F;                       /*Enable OC0 interrupt      */
    _H12TFLG1 = C0F;                        /*Clear Channel 0 flag      */

}


/********************************************************************
* GetmsCnt() - Read the current ms counter value. (Public)
********************************************************************/
INT16U GetmsCnt(void) {
    return OCmsCnt;
}


/********************************************************************
* OC0Isr() - OC0 Service Routine. (Private)
*     MCU: 68HC912B32, E = 8MHz
*      - Requires TC0 be set for an output compare (OC0Init()).
*      - setup for a 1ms periodic interrupt.
********************************************************************/
```

```
ISR OC0Isr(void){
    _H12TC0 = _H12TC0 + E_PER_MS;    /* Interrupt in 1ms later  */
    _H12TFLG1 = C0F;                 /* Clear Channel 0 flag     */
    OCmSCnt++;                       /* Increment 1ms counter    */
}
```

/***/

After the master header file is included, we declare the functions and data objects that are part of the module. Public and private resources should be clearly documented here. The public resources match those found in the module header file. The private resources are those data objects and functions that should not be accessed from another module. In this case the private resources include the ISR, *OC0Isr()*, and a static variable, *OCmSCnt*. Of course *OC0Isr()* is private. It should never be called at all.

In this case we also have made *OCmsCnt* a private variable. This was done to keep future programmers from breaking the *OCDelay* function. If *OCmsCnt* were a public variable, a programmer could change it from another module. If the *OCDelay()* function were in the middle of a delay, changing *OCmsCnt* would cause the delay time to be different than specified. For the code in this module to be accurate, the only time *OCmsCnt* is changed is by *OC0Isr()*.

On the other hand it would be useful to have access to a counter that counts milliseconds. For example we could use a counter like *OCmsCnt* to create a periodic time slice for executing tasks at a fixed rate. In this case, however, we only need read access. To provide read access to *OCmsCnt,* the *OCDelay* module provides the function *GetmsCnt()*, which simply returns the current value of *OCmsCnt*. This is an example of providing controlled access to a shared resource. We keep the resource private and provide a public function for controlled access.

If we compare the files used for the *demo1* program in Chapter 12 with the files used here for *demo2,* we can see how modules enhance reusability. We replaced *msdelay.s12* with the *OCDelay* module and made minor changes to the main program and the master include file.

▶ 14.4 START-UP AND INITIALIZATION

In Section 14.2 we saw that a program does not actually start at *main()*. Before we get to *main()*, start-up tasks must be executed for MCU configuration and program initialization. In this section we look at these start-up tasks in more detail. We cover the start-up code required when running a C program from the D-Bug12 monitor when D-Bug12 is resident on the target board. We then look at the start-up code required for stand-alone applications and applications that are executed using D-Bug12 from an EVB in pod mode.

Start-up and initialization code can be executed before jumping to *main()* or after entering *main()*. We use the term *start-up code* for the code that is executed before *main()*. Start-up code is always written in assembly code, because it includes initialization that is required before C code can be executed. The initialization code in *main()* may be written in C or assembly.

14.4.1 Start-Up Tasks

Microcontroller Configuration.　The first start-up code that must be executed out of a system *RESET* is the MCU configuration code. This includes setting MCU modes, initializing external chip selects, and other register initialization required before code can be executed. Detailed descriptions of the MCU configuration can be found in Chapter 10.

For the M68HC912B32, the most important register initialization is the *Computer Operating Properly* control register, *COPCTL*. Out of *RESET,* the COP system is enabled, so if it is not disabled within the COP timer rate, a COP reset will occur. For an E-clock frequency of 8MHz, the M68HC912B32's timer rate is 1.024ms out of *RESET,* and the 68HC812A4's timer rate is 1.04 seconds out of *RESET*.

Also several 68HC12 MCU registers are write-once registers. These registers control critical MCU hardware, which means that if an inadvertent write is made to one of these registers, a catastrophic error may occur. To avoid inadvertent writes, we should write to these registers in the start-up code to lock out any writes during normal operation.

For the 68HC11 family of microcontrollers, several registers must be initialized within the first 64 E-clock cycles out of *RESET*. After 64 E-clock cycles, they cannot be changed, so these registers should be initialized first.

C Program Requirements.　Before executing a C program, we have to initialize initialized variables, clear uninitialized variables, and initialize the stack pointer. In addition we have to include program control code for *main()*. If we are using the standard C I/O library, it also needs to be initialized before calling *main()*.

Before a C program can be executed, all variables must be initialized. There are two types of variables in C, initialized variables and uninitialized variables. To initialize the initialized variables, the initial value stored in ROM is copied to the RAM variable location. For example the following is a declaration for an initialized variable:

```
UBYTE InitVar = 0x80;
```

When this declaration is compiled, two bytes are allocated. One byte in ROM is loaded with the initial value 0x80, and another byte is allocated in RAM to hold the current value of the variable. *InitVar* will be assigned the address of the allocated RAM byte. When the start-up code is executed, the initial value in ROM is copied into the allocated RAM byte. Now when we start the C code, the variable contains the initial value, 0x80.

Some ANSI-C standard libraries also expect uninitialized variables be initialized to zero. So before the program jumps to *main()*, all uninitialized variables must be cleared. Also, because *main()* is a function, the stack pointer must be initialized first.

Remember that *main()* is a function, so it must be called and there must be code to trap the program if there is a return from *main()*. The last section of any start-up code is

```
              jsr     main         ; Call the program
_exit::       sei                  ; mask interrupts
loop          bra     loop         ; and loop forever
```

The *jsr* to *main* is the point at which program control is transferred to *main()*. For normal operation, the program should never return, because *main()* should contain an endless loop.

If the program does reach the end of *main()*, however, it will return to *_exit*, interrupts will be disabled, and the program will enter an endless trap.

We can change the name of our main program by changing the *jsr* instruction in this start-up code. We can also add additional code in the return trap that reports error messages. The code conventions for this text always call for the C entry function be called *main()*.

Interrupt Vectors. Before an interrupt can be used, the interrupt vector must be initialized. For stand-alone systems, the interrupt vectors are loaded when the ROM is programmed, so there is no run-time initialization code required. If we are using D-Bug12 resident in the target EVB ROM, however, we cannot change the vectors. In this case D-Bug12, along with most other monitor programs, has an interrupt jump table that can be changed at run time. To initialize the interrupt vectors, we should use D-Bug12's *SetUserVector()* function. *SetUserVector()* must be called before interrupts are enabled and before a computer operating properly or clock monitor reset occurs.

Program Initialization. Several program initialization tasks must be completed before normal operation. Most program initialization should be called by *main()*, however, not placed in the start-up code. The decision to place initialization code in the start-up code or in *main()* is a design decision. Some start-up tasks, such as those mentioned previously, must be completed before we can jump to *main()*. Most of the remaining initialization tasks should be placed in *main()*. This allows you to write the code in C and ensures that these initialization tasks are part of the application code. Since the start-up code may be generated by the development system, it is not part of the application unless it is saved as a file and made part of the project.

Sometimes the interrupts are also enabled in the start-up code. Interrupts are normally enabled in *main()*, however, because it is usually necessary to complete the remaining program initialization code before enabling interrupts. Also, if *SetUserVector()* is used to initialize the D-Bug12 interrupt jump table, the interrupts cannot be enabled until after *SetUserVector()* is called.

14.4.2 Start-Up Code for Programs Executed under D-Bug12

When the D-Bug12 monitor is resident on the target board, programs that are executed with the *G* command do not require much start-up code. This is because most of the initialization is already completed by the D-Bug12 monitor. The only remaining start-up code that is required is initialization for a C program.

Let's look at the start-up code and initialization for the *demo2* module shown in Section 14.3. First we will build this module to be executed out of D-Bug12 resident in the target ROM.

The Introl-CODE start-up configuration is set in the *Edit Project → Configure Environment* dialog box shown in Figure 14.14. Because this version of *demo2* is being compiled for D-Bug12, we do not have to initialize the device registers. The stack pointer is initialized by D-Bug12, but there are conditions when the program can be restarted without D-Bug12 initializing the stack pointer. It is included here to make sure the stack pointer is always initialized. After setting up this dialog box, a start-up file will be generated when the project is built. The resulting start-up code for this example is shown in Source 14.19.

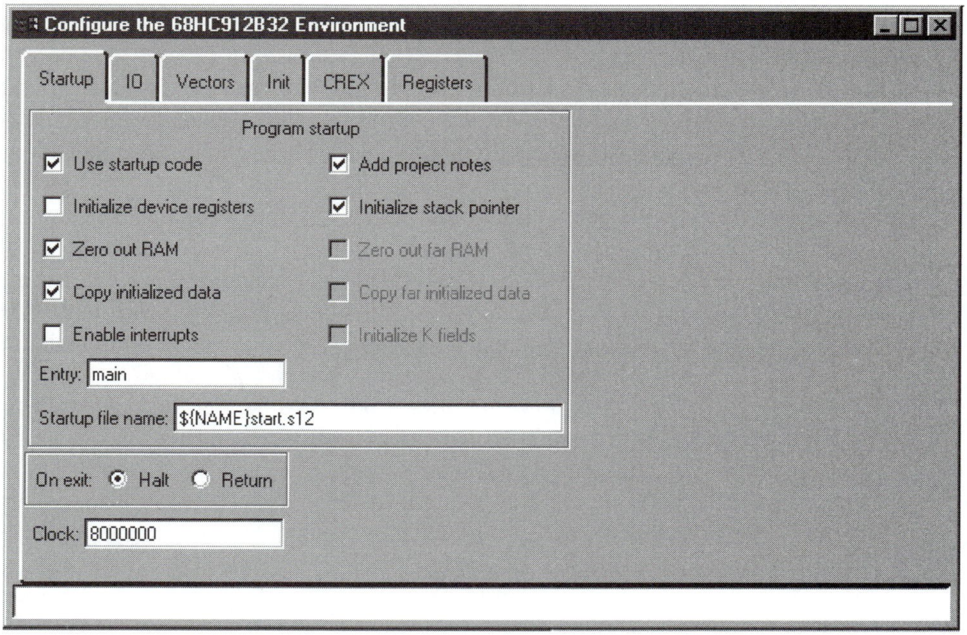

Figure 14.14   Introl-CODE Start-Up Configuration Dialog Box

Source 14.19 Generated Start-Up Code for demo2

```
;
; demo2start.s12 - startup file for the project demo2.
; This file was automatically generated by Introl-CODE. DO NOT EDIT.
;

          section    .start0
; The program entry point (at RESET)
_start:
;
;               stackinit - initialize the stack pointer
;

                  import     _stackend

                  section    .start4
_stackinit:  lds   #_stackend          ; point to the system stack

;
;               bssinit - initialize bss area
;
;               Clear the uninitialized RAM area.  This is required
;               so that uninitialized variables have a value of 0.
;
```

```
                    section   .start3
_bssinit:

;
;            Symbols used (defined in the linker command file):
;                 _ramstart - low address of the RAM area
;                 _ramend - address following the RAM area
;
             import    _ramstart,_ramend

             ldx       #_ramstart    ; point to start of data
             bra       1l
2            clr       1,x+
1            cpx       #_ramend      ; compare to end address
             blo       2s            ; continue if not finished


;
;            datainit - copy non-constant initialized data to RAM

             section   .start3
_datainit:

;
;            Symbols used (defined in the linker command file):
;
;            _datastart - destination address
;            _initstart - source address
;            _initend - address following the source block
;
             import    _datastart,_initstart,_initend

             ldy       #_datastart   ; target of the copy in RAM
             ldx       #_initstart   ; source of the copy in RAM
             bra       1l
2            movb      1,x+,1,y+
1            cpx       #_initend     ; at the end?
             blo       2s            ; jump if not

             section   .startX
             import    main
             jsr       main          ; Call the program
;
;            exithalt.s - program termination point
;
;
; Entry points:
;            _exit     - the program termination point
;

                    section   .startZ
```

```
_exithalt:

_exit::                sei                ; mask interrupts
loop                   bra      loop        ; and loop forever

                       end
```

Note that this source code does not reflect the actual order of execution. In the Introl-CODE system, there are start-up sections labeled *.start0* through *.start9* and *.startX* and *.startZ*. These sections are placed in order by the linker. The actual order of execution is shown in Figure 14.15. The program initializes the C variables, initializes the stack pointer, and jumps to *main()*. Again, for normal programs, the control will never return from *main()*, but if it does, it will return to the endless trap.

As shown in Source 14.16 on page 553 the rest of the program initialization is done in *main()*. The following code initializes the output compare module and the LED port, and enables the interrupts:

```
OCDlyInit();
INIT_LED_DIR();          /* Initialize LED as output and turn off*/
TURN_LED_OFF();
ENABLE_INT();            /* Enable interrupts to use OCDelay()   */
```

demo2 uses an output compare interrupt, so the interrupt vector must be set up. Because it

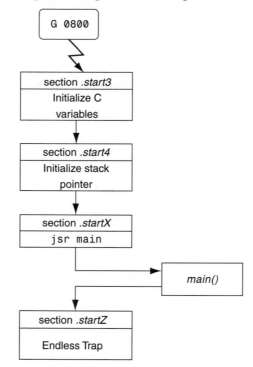

Figure 14.15 Start-Up Sequence for *demo2* with D-Bug12

was built to run under D-Bug12, the vector is not set up in the *Edit Project → Configure Environment → Vectors* dialog box. The *Timer Channel 0* interrupt vector is set up by *SetUserVector()*, which is called by *OCDlyInit()*.

14.4.3 Stand-Alone Programs

When we build a stand-alone program, we must include all the required initialization tasks. Stand-alone programs include those programs that are executed out of *RESET* or from D-Bug12 when D-Bug12 is resident on another EVB in pod mode.

To build *demo2* as a stand-alone program, we need to include additional start-up code and comment out the *#define DBUG12* line in *includes.h*. We also need to designate the interrupt vectors and change the memory map to reflect the new environment. In this subsection we will look at the start-up code and setting the interrupt vectors.

First, to add register configuration to the start-up tasks, we need to select the *Initialize device registers* check box shown in Figure 14.14. The rest of the settings in this dialog box remain the same.

Now we designate the initial values for the registers we need to configure in the *Registers* tab. Figure 14.16 shows some of the register settings used for *demo2*. The most important is the *COPCTL* register. Clearing this register disables the COP reset.

The other registers with designated values are write-once registers. At start-up they are written in order to lock out any accidental changes during normal operation. Notice that the I/O port settings for *LED_PORT* and *LED_PORT_DIR* are not made here. These are application-specific settings and should be configured by the application code. The only time that an application's register configuration should be contained in the start-up code is

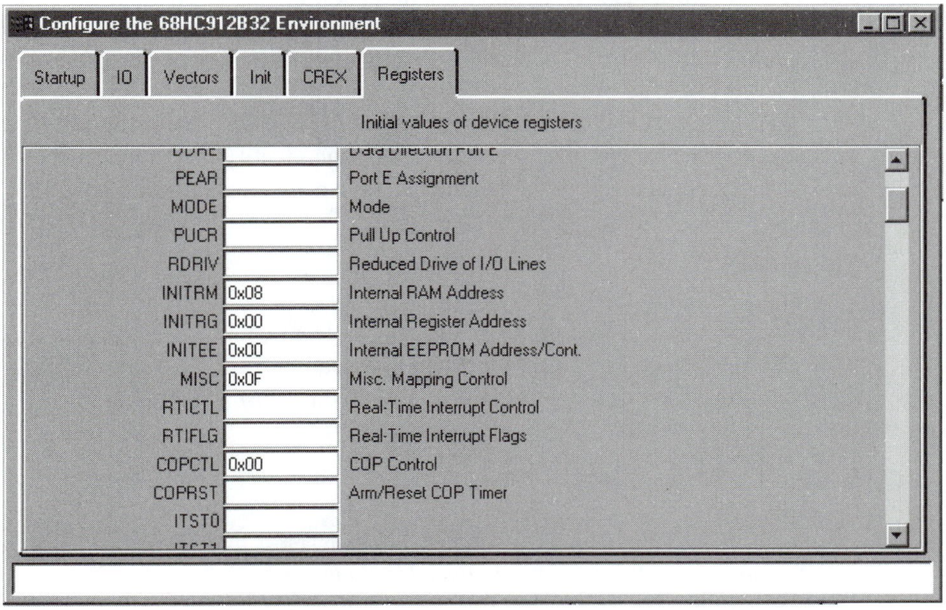

Figure 14.16 Introl-CODE Start-Up Register Configuration

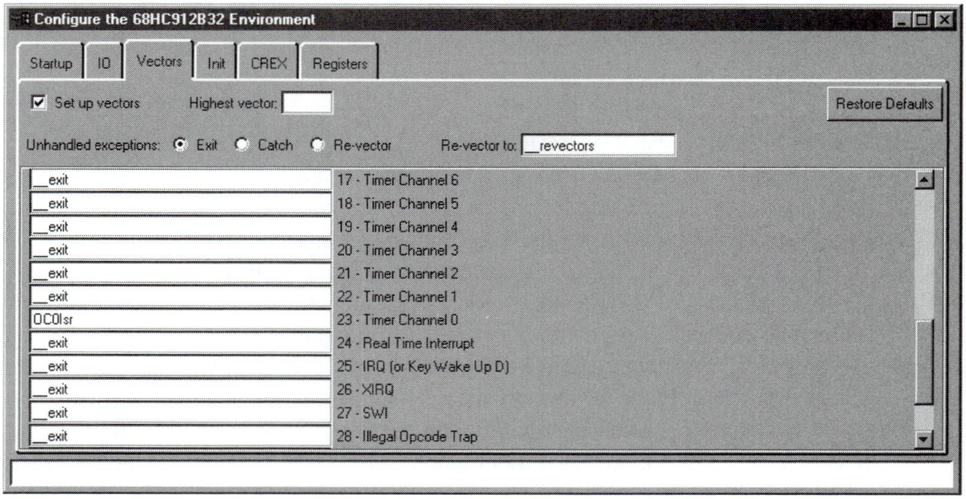

Figure 14.17 Introl-CODE Interrupt Vector Dialog Box

when it is required immediately out of *RESET*.

To build *demo2* as a stand-alone application, we also have to set up the interrupt vectors. In Introl-CODE, this is handled in the *Vectors* dialog box shown in Figure 14.17. In the figure the interrupt vector for *Timer Channel 0* is set to the interrupt service routine *OC0Isr( )*.

The default value for unused interrupt vectors is *_exit*. This is the trap created in the start-up code for an exit from *main( )*. This is a minimal solution. If the program stops working and we find that it is trapped in *_exit,* we only know that either *main( )* was exited or an unexpected interrupt occurred. We cannot tell which interrupt was the source of the error.

To help detect and debug interrupt errors, the unused interrupts could be directed to an interrupt error handler that generates a message identifying the interrupt that occurred unexpectedly. An interrupt error handler can save many hours of debugging time.

▶ SUMMARY

In this chapter we covered program structures—from control constructs to modules. We were also introduced to writing functions in assembly for a C program, writing interrupt service routines, and creating the start-up code required for a C program.

EXERCISES

1. Write a code snippet that calls *Func1( )* if *Num* is less than 10 but not zero, calls *Func2* if *Num* is equal to 10, and calls *Func3* if *Num* is greater than 10.

2. Convert the following nested *if* sequence into a single complex expression:

```
/******************************************************************/
        if(Num1 == 3){
            if(Num2 <= 4){
                ProcNum();
            }
        }
        if(Num4 == 5){
            ProcNum();
        }
/******************************************************************/
```

3. Write a function that converts the temperature from degrees Celsius to degrees Fahrenheit. The Celsius value is passed to the function and the Fahrenheit value is returned. Both temperature parameters are signed 8-bit integers. Include prototype.

4. Show a single statement that calls the conversion program in Exercise 3 to convert the variable *Temp* from Celsius to Fahrenheit.

5. Repeat Exercises 3 and 4 by designing a function that uses a pointer to the variable to be converted instead of passing the variable as a parameter. This function should return −1 if the result would be greater than 255 degrees Fahrenheit.

6. Write an assembly code snippet that calls the function in Exercise 3 to convert the value in the global variable *Temp*. The results should be stored in *Temp*.

Real-Time Multitasking in C

As we have seen throughout this text, real-time programming is an essential part of embedded system design. In Chapter 8 we looked at the basic analysis and design techniques used for real-time I/O and simple multitasking in assembly. In Chapter 9 we looked at the on-chip I/O resources on the M68HC912B32. In this chapter we review some of the material from Chapter 8 using C, introduce real-time multitasking kernels, and look at some design techniques for constructing our own simple kernels. In Chapter 16 we continue this topic by covering some design techniques for constructing programs that use a full-featured preemptive kernel, MicroC/OS-II.

▶ 15.1 REAL-TIME PROGRAMMING REVIEW

In Chapter 8 we saw that endless loops are required to sample and respond to real-time signals. These loops are called event loops. Figure 15.1 shows the C code and the block diagram for a basic event loop. If the event loop does not include the event detection, it is considered a continuous unconditional I/O loop.

Virtually every task in an embedded system is designed as an endless loop. This makes sense if we think about what an embedded system does. An embedded system, after power-up initialization, endlessly performs tasks until it is turned off.

Flow Diagram	C Code
| Event Loop → Event? → No / Yes → Service Event or Set Flag | ```
while(1){
    if(Event){
        /*Service Event*/
    }
}
``` |

Figure 15.1   Basic Event Loop

15.1.1 Stand-Alone Tasks

Figure 15.2 illustrates the overall program flow for a system that performs a single task. After initialization is complete, the task is called, and since the task contains an endless loop, it never returns. In the flow diagram the task is shown as an endless process.

In the C code the *main()* function ends abruptly when *task()* is called. At first it appears that the program flow will reach the end of *main()* and return. This is something that we would never want to happen during normal operation. Since *task()* is an endless process, however, program control will never return to *main()* from *task()*.

The task itself is shown in Figure 15.3. Upon entering the task, the initialization required for the task is executed. Then the task code, which is contained in an endless loop, is executed forever. The task shown in this figure is called an *independent task* because it appears to take over the CPU forever. It is also called a *free-running task* because a pass through the task code is not dependent on a timer event.

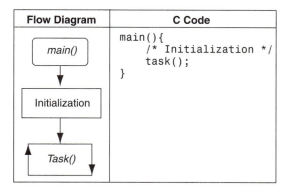

Flow Diagram	C Code
| main() → Initialization → Task() | ```
main(){
    /* Initialization */
    task();
}
``` |

Figure 15.2   The Program Flow for a System with a Single Task

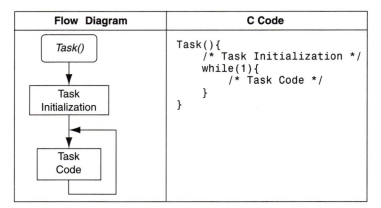

Flow Diagram	C Code
| Task() → Task Initialization → Task Code (loop) | ```
Task(){
    /* Task Initialization */
    while(1){
        /* Task Code */
    }
}
``` |

Figure 15.3   An Independent, Free-Running Task

There are two fundamental timing characteristics for the task, the task execution period, T_p, and the task execution time, T_T. From these two parameters, the load placed on the CPU by running the task is

$$L = \frac{T_T}{T_p}$$

For a free-running task such as that shown in Figure 15.3, the CPU load is always one because the task period is equal to the task execution time. As soon as the task is completed, it is executed again. With a CPU load of one, there is no opportunity to run other tasks.

Another problem with a free-running task is that it does not have a deterministic task period. The task period for a free-running task loop is equal to the task execution time, which may be unknown for a given pass through the loop. Many applications require a deterministic or even a constant task period.

A Timed Task Loop.   From the CPU load equation shown, we can see that if we increase the task period T_p, the CPU load would go down. This in turn would allow the CPU to execute other tasks.

So if we increase the task execution period, the CPU load will be reduced. If we increase the task period too much, however, it may be possible to miss events or have an unacceptable event response time. Therefore the maximum task period is

$$T_p \leq T_{et}$$

where

$$T_{et} = \text{the event valid time}$$

To control the task period, we can make the execution of the task code dependent on a periodic timer event. This is called a *timed task loop*. As shown in Figure 15.4, each pass through the task code in a timed task loop is conditional on a timer event. The timer event is a periodic event set to the desired task period.
```

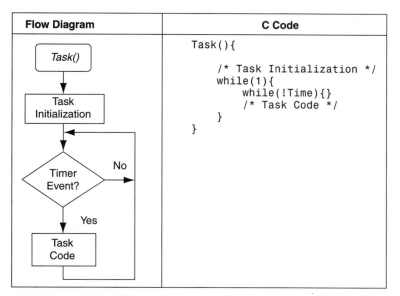

Flow Diagram	C Code
Task() Task Initialization Timer Event? — No Yes Task Code	```Task(){

    /* Task Initialization */
    while(1){
        while(!Time){}
        /* Task Code */
    }
}``` |

Figure 15.4 Timed Task Loop

The timed task loop allows us to execute the task at a constant rate. Once we can control the task period, we can lower the CPU load by reducing the task period. This is the basic concept that allows for multitasking. By running a task only as often as required, we can free up the CPU to run other tasks.

15.1.2 Event Response Time

Another important timing parameter for an event-driven task is the event response time. The event response time is the time between the occurrence of the event and the time that the response is made. In general the event response time for a single task is

$$T_R \leq T_p + T_{srv} + T_{CIR}$$

where

$$T_p = \text{the task period}$$

$$T_{srv} = \text{the service time}$$

$$T_{CIR} = \text{the circuit delays}$$

The event service time T_{srv} is the time it takes for the program to respond once the event has been detected. It can be divided into the task switch time, the task delay, and the sum of all time spent servicing interrupts as follows:

$$T_{srv} \leq T_{TSW} + T_{TD} + \sum T_{int, \, T_{srv}}$$

where

$$T_{TSW} = \text{the task switch time—the time it takes to get to the servicing task}$$

T_{TD} = the task delay—the time it takes to respond once the servicing task is entered

$\sum T_{int, T_{srv}}$ = the total time spent servicing interrupts during the service time T_{srv}

Since the task period for a free-running task loop is equal to the task execution time, the response time for a free-running event loop becomes

$$T_R \leq T_T + T_{srv} + T_{CIR}$$

At first it appears that the free-running task is more responsive because the timed loop period, T_p, must always be greater than the task execution time, T_T. Most tasks must be run at a periodic rate for multitasking or for a timed response, so a timed loop is required anyway.

If we require a faster response time, then we can use an interrupt-based response. The response time for an interrupt-based response is much faster because the response does not have to wait for the signal to be sampled. The event is detected by the interrupt hardware, which then interrupts execution to run the interrupt service routine. The resulting response time is

$$T_R \leq T_I + T_{TD} + T_{CIR}$$

where

$$T_I = \text{the interrupt latency}$$

Notice that in this case, the task switch time is the interrupt latency T_I. The interrupt latency depends on time spend in other interrupt tasks. If we assume that interrupts are disabled during the service routine, however, the task delay time is not extended due to other interrupts.

The interrupt-based response times look enticing, but remember that when an interrupt-based response is used, the worst-case response times for all other tasks increase because of the increased time spent in interrupt service routines. Interrupt-based responses can also have a negative effect by adding jitter to the task period in timed loops.

It is important to carefully evaluate the response time requirements for a design. We should only use an interrupt-based response if the normal background task loops cannot respond fast enough. The interrupt service routines should also be as short as possible to minimize the time spent in the foreground. Remember that a foreground task can interrupt a background task but no task can interrupt a foreground task.

▶ 15.2 REAL-TIME KERNEL OVERVIEW

So far we have looked at systems that perform a single task. Most embedded microcontroller programs must perform several tasks concurrently, however. Since the CPU is a sequential system that can only execute one program at a time, we need to design a system that uses the time spent waiting for one task to execute another task.

To coordinate the process of sharing the CPU between multiple tasks, we use a program called a *kernel*. A kernel is the part of an operating system that schedules and dispatches tasks. There are many types of kernels ranging from simple task loops to full-featured preemptive kernels. In this section we introduce real-time kernel concepts. Then in the next section, we look at designing our own simple cooperative kernels. In Chapter 16 we cover MicroC/OS-II, which is a preemptive kernel.

15.2.1 Tasks and Kernels

A *task* is a routine that is called by the kernel to do something. Tasks are different from normal routines because we must treat them as being asynchronous to all other tasks. For example a regular C function is always called by another function, so they are synchronized in program time. A task, however, is never called by another task. It is only called by the kernel's scheduler. We will always treat a task as if it were a stand-alone program running asynchronously to the rest of the program.

The process of designing a complete multitasking system involves breaking the system down into tasks, designing each task, and using a kernel to run the tasks. The tasks are normally made up of a set of device-driving tasks and a system control task. For most embedded systems we need a real-time kernel because deterministic response times are required.

Scheduling of multiple tasks can be done in two ways, preemptively or cooperatively. The design and structure of a task depends on the type of scheduling used by the kernel.

Preemptive Kernels. When using a preemptive kernel, each task is designed as an independent task that has complete control over the CPU. To share the CPU, the kernel can preempt a task to execute another higher priority task. Figure 15.5 shows the flow diagram of a preemptive multitasking kernel with three tasks. Each task is normally designed as a stand-alone timed loop or event loop. The kernel always executes the highest priority task that is ready to run.

Figure 15.6 shows the program context for a preemptive kernel. In the figure *Task1* has the highest priority and *Task3* has the lowest priority. A task is either running, waiting because it is not ready to run, or waiting because it is ready but a higher-priority task is running. When no tasks are ready to run, the kernel runs an *Idle* task, which is a loop that does nothing. The *Idle* task can be considered the lowest-priority task, but it is really part of the kernel, so we normally do not have to include it in our code.

When *Task3* becomes ready to run, the *Idle* task is preempted and the CPU starts executing *Task3*. Then when *Task1* becomes ready, *Task3* is preempted, because *Task1* has a higher priority. When *Task2* becomes ready, *Task1* is not preempted, because it is still the highest-priority task ready to run. At this point both *Task2* and *Task3* are ready to run but must wait for *Task1*. When *Task1* is complete, it puts itself into the waiting state until it

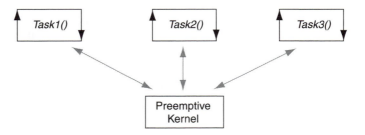

Figure 15.5   Preemptive Multitasking Program Flows

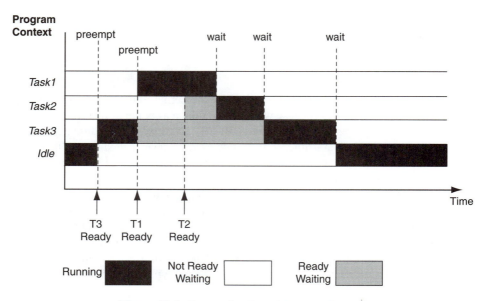

Figure 15.6 Preemptive Kernel Program Context

becomes ready to run again. Control is then passed to *Task2* because it is the highest-priority task ready to run. Finally when *Task2* is complete, control is passed back to complete *Task3*.

Notice that the highest-priority task never has to wait for other tasks to finish. Because of this, the higher-priority tasks in a preemptive kernel can have better response times.

The primary benefit to using a preemptive kernel is that the tasks can be designed as independent tasks. This is because the programmer does not have to be concerned with exiting the task to get back to the kernel so it can execute other tasks. The only concern is to make sure that the CPU load for a given task is low enough to allow other lower-priority tasks to run. This makes the tasks easier to design and more portable to other designs using the same type of kernel.

On the other hand a preemptive kernel is more complex and requires more memory resources than a cooperative kernel. So if the design is simple and the memory resources are scarce, a simple cooperative kernel should be used. If the design is complex and there is adequate RAM and ROM space, a preemptive kernel should be used.

For most of us it would be a waste of time to design our own complete preemptive kernel. This would be analogous to writing an operating system on a PC when an application is all that is required. Because of this we do not cover the design of preemptive kernels in this text. Instead it is assumed that an off-the-shelf product will be used when a preemptive kernel is required.

Cooperative Kernels. In a nonpreemptive or cooperative kernel, each task must give up the CPU before another task can run. Therefore the tasks used in a cooperative kernel must cooperate by giving up the CPU in a timely manner. These tasks are called *cooperative tasks*.

A cooperative kernel and a cooperative task are shown in Figure 15.7. The cooperative kernel shown is a cyclic scheduler, which is a loop that sequentially executes each

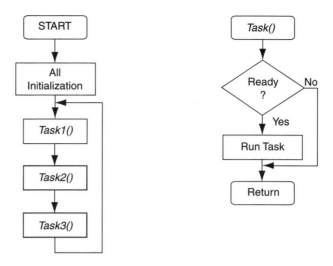

Figure 15.7 Cooperative Multitasking Loop and Cooperative Task

task. A cyclic scheduler is the most common and simplest cooperative scheduler. We can think of it as one big task loop that runs multiple tasks.

In order to use this type of kernel, the tasks can no longer be endless loops. You can imagine if *Task1* in Figure 15.7 were implemented as an endless loop. The scheduler would never get to *Task2* or *Task3*. Each task must cooperate by running only one time and then returning to the scheduler loop.

Figure 15.8 shows the program context for a cooperative scheduler. In this case the highest-priority task has to wait for the lowest-priority task to complete before it can run.

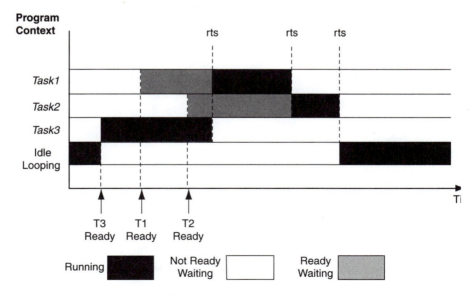

Figure 15.8 Cooperative Kernel Program Context

Once *Task3* is completed, control is passed to *Task1* because it has a higher priority than *Task2*. This is only the case for some cooperative schedulers. Many use a round-robin technique, which results in the same response times for all the tasks.

So cooperative kernels have slower response times for higher-priority tasks than preemptive kernels. In addition cooperative tasks are more difficult to write, and the tasks are not as portable as those written for a preemptive kernel. On the other hand cooperative kernels can be very simple to design and require less memory. They are the appropriate solution to most simple real-time designs using small microcontrollers with limited memory resources. We cover the design of cooperative kernels in Section 15.3.

15.2.2 CPU Load for Multiple Tasks

In all systems, preemptive or cooperative, we need to find the CPU load to determine if a design is realizable. For systems that require *n* tasks, the load is

$$L = \frac{T_{T1}}{T_{p1}} + \frac{T_{T2}}{T_{p2}} + \cdots + \frac{T_{Tn}}{T_{pn}}$$

where

$$T_T = \text{the task execution time}$$

$$T_p = \text{the task execution period}$$

A system cannot possibly work if the load is greater than one. So a necessary condition is

$$L \leq 1$$

This is not a sufficient condition for the design to work, however. This is especially true as the number of tasks increase. We normally try to keep the CPU load less than 0.7.

Again it is important to note that the CPU load does not guarantee that the specific real-time requirements for the design will be met. Because the actual timing is dependent on the implementation, we still need to analyze the timing parameters after the program is completed. The CPU load just tells us if the design might be possible.

▶ 15.3 COOPERATIVE KERNEL DESIGN

Many projects based on embedded microcontrollers are simple designs in which it may be more cost effective to design our own simple cooperative kernel. In this section we look at some simple, cooperative kernel designs. We focus on a basic kernel that is simple to design, analyze, and debug. We then look at some techniques to deal with tasks that require long execution times.

The cooperative kernel designs described in this section are often referred to as *foreground/background* systems. *Foreground* refers to the tasks that are executed in an interrupt service routine, and *background* refers to the tasks that are executed by the main program loop. The systems we look at in this chapter are primarily background systems that only use foreground processing when necessary to meet response time requirements. When it is required, we limit the processing done in the foreground to simple things such as setting flags, incrementing counters, or quickly storing data. Everything else is accomplished by background tasks.

15.3.1 Free-Running Cyclic Scheduler

We will start with the free-running cyclic scheduler shown in Figure 15.9. This is a cooperative kernel and therefore requires cooperative tasks. The scheduler shown is called a *round-robin scheduler* because each task is given an opportunity to run in the order that they appear in the loop.

With this type of scheduler, all the tasks in effect have the same priority, because the event response times are the same for every task. The scheduler does execute the tasks in order, but since an event is asynchronous to the program loop, we cannot predict where the event will occur.

The response time for every task in the round-robin scheduler is

$$T_R \leq T_{srv} + T_{CIR}$$

where

$$T_{srv} = \text{the service time}$$

$$T_{CIR} = \text{the circuit delays}$$

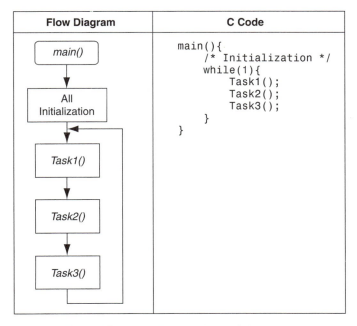

Flow Diagram	C Code
| main() | ```
main(){
    /* Initialization */
    while(1){
        Task1();
        Task2();
        Task3();
    }
}
``` |

Figure 15.9 Free-Running Cyclic Scheduler
```

The event service time T_{srv} is the time it takes the program to respond once the event has been detected. It can be divided into the task switch time, the task delay, and the sum of all time spent servicing interrupts as follows:

$$T_{srv} \leq T_{TSW} + T_{TD} + \sum T_{int,\ T_{srv}}$$

where

T_{TSW} = the task switch time—the time it takes to get to the servicing task

T_{TD} = the task delay—the time it takes to respond once the servicing task is entered

$\sum T_{int,\ T_{srv}}$ = the total time spent servicing interrupts during the service time T_{srv}

For a free-running round-robin scheduler with N tasks, the task switch time is the sum of all task execution times, so the event response time is

$$T_{Rn} \leq \sum_N T_{Ti} + T_{TDn} + T_{CIR} + \sum T_{int,\ T_{srv}}$$

This means that the worst-case response time can never be better than the sum of all task execution times. Therefore, if we have a task that requires a faster response time, we would have to use an interrupt-based response. A solution to this problem is to give some tasks priority over others by using a priority-based scheduler.

In addition many tasks need to run periodically at a constant rate. We can do this by making the task dependent on a timing event, but since the timing event is asynchronous to the scheduler, the response time is increased to

$$T_{Rn} \leq T_{pn} + \sum_N T_{Ti} + T_{TDn} + T_{CIR} + \sum T_{int,\ T_R}$$

where

T_{pn} = the task execution period

If a task must be timed, the total response time is better if we use a time-slice scheduler.

Priority-Based Scheduler.   A priority-based cyclical scheduler is shown in Figure 15.10. Notice that the event detection required to determine whether a task is ready must be performed in the scheduling function, not in the task itself.

The task scheduler shown uses an *if-else if* construct in the *main()* program to dispatch the highest-priority task. This could also be accomplished using a separate scheduler function and a lookup table.

The advantage of using the priority-based scheduler is that it can improve the response times for higher-priority tasks. In general the response time for *Taskn* is

$$T_{Rn} \leq \sum_{i<n} N_i T_{Ti} + max[T_{Tn}, T_{Tn+1}, \ldots] + T_{TDn} + T_{CIR} + \sum T_{int,\ T_{Rn}}$$

The first two terms make up the task switch time. The first term is the sum of execution times for all higher-priority tasks multiplied by the number of times those tasks need to run before reaching *Taskn*. The second term represents the longest execution time out of the

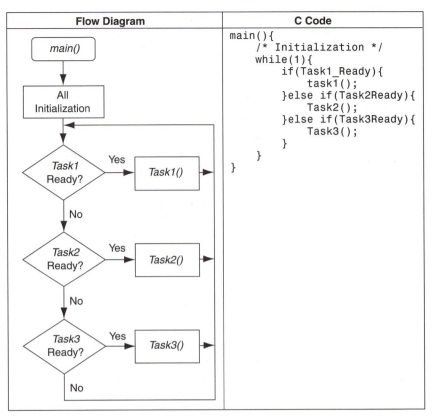

Flow Diagram	C Code

```
main(){
    /* Initialization */
    while(1){
        if(Task1_Ready){
            task1();
        }else if(Task2Ready){
            Task2();
        }else if(Task3Ready){
            Task3();
        }
    }
}
```

Figure 15.10 Priority-Based Free-Running Cyclic Scheduler

task itself and all lower-priority tasks. This term is needed because the program may be in any of these tasks when the event occurs, and this system cannot preempt a task.

Task response time for high-priority tasks will be less than the response time for a round-robin scheduler. The response time for lower priority tasks will be greater, however. This is normally an acceptable trade-off.

For example, for the system shown in Figure 15.10, the response time for the highest-priority task becomes

$$T_{R1} \leq max[T_{T1}, T_{T2}, T_{T3}] + T_{TD1} + T_{CIR} + \sum T_{int, T_{R1}}$$

This is almost as fast as a system with only one task. The rest of the tasks response times do not look so good, however. The response time for *Task3* is

$$T_{R3} \leq N_1 T_{T1} + N_2 T_{T2} + T_{T3} + T_{TD3} + T_{CIR} + \sum T_{int, T_{R3}}$$

where

N_1 = the number of times *Task1* runs during the task switch time

N_2 = the number of times *Task2* runs during the task switch time

Real-Time Multitasking in C

Again, if a task must be executed periodically, we must make it dependent on a timer event. When a task is based on a periodic timing event, the response time is

$$T_{Rn} \leq T_p + \sum_{i<n} N_i T_{Ti} + max\left[T_{Tn}, T_{Tn+1}, \ldots\right] + T_{TDn} + T_{CIR} + \sum T_{int, \, T_{Rn}}$$

15.3.2 Time-Slice Cyclic Scheduler

The time-slice cyclic scheduler was introduced in Chapter 8. It is considered the easiest cooperative scheduler to design, analyze, and debug. It allows the tasks to be executed at deterministic periods, and it has some prioritization built into it because the tasks that are executed first have the lowest response times.

The flow diagram for a time-slice scheduler is shown in Figure 15.11. Each time through the loop, the scheduler starts by waiting for a timer event. In this case the timer event detection is performed in a function, *WaitForTimer()*.

Each pass through the loop is called a *time slice*. During each time slice, each task has an opportunity to run. So the execution period for each task can easily be set to an integer multiple of the time-slice period.

Figure 15.12 shows the program context for a time-slice cyclic scheduler that runs three tasks. As we can see from the figure, the tasks must be complete by the time the next timing event occurs. If not, the response to the timing event may be delayed, or the event may be missed. If the response to the timing event is delayed, an error is introduced to the short-term timing characteristics of the kernel. In effect jitter is introduced to the time-slice period. If a timing event is missed, however, a long-term timing error occurs.

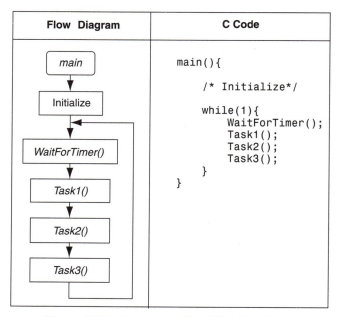

Figure 15.11 Background Time-Slice Scheduler

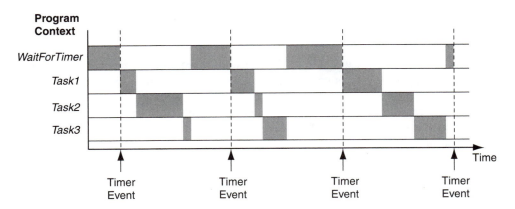

Figure 15.12 Program Context for the Time-Slice Cyclic Scheduler

In order to avoid jitter in the time-slice period or missed timing events, the time-slice period must be greater than the sum of the all task execution times. This includes both foreground and background tasks. So

$$T_{slice} > \sum_n T_{Ti, \, max} + \sum T_{int, \, T_{slice}}$$

The next step is to determine the required execution periods for each task. In a time-slice scheduler, we try to make all task execution periods integer multiples of the time-slice period. In this way we can make the execution for all the tasks synchronous to the time-slice period, which in turn reduces the system's response time. Therefore the time-slice period, T_{slice}, must be a common divisor to all the task periods. The largest acceptable time-slice period is the greatest common divisor of the task periods:

$$T_{slice} \leq gcd\big(T_{p1}, \ldots, T_{pn}\big)$$

where

$$gcd() = \text{the greatest common divisor}$$

This means that the time-slice period must be greater than the sum of all task execution times but cannot be larger than the greatest common divisor of the task periods. Many times it is difficult to meet both of these requirements.

The time-slice period also affects the response time for each task. The response time for *Taskn* in a time-slice scheduler is

$$T_{Rn} \leq x_n T_{slice} + \sum_{i<n} T_{Ti} + T_{TDn} + T_{CIR} + \sum T_{int, \, T_{SW}}$$

where

$$x_n T_{slice} = \text{the execution period for } Taskn \text{ (i.e., } Taskn \text{ runs every } x_n \text{ time slices)}$$

Real-Time Multitasking in C

579

$$\sum T_{int, T_{SW}} = \text{the time spent in interrupts during } \sum_{i<n} T_{Ti} + T_{TDn}$$

This assumes that all the task periods are based on the time-slice period. If a task period is based on a separate timing event, the task period must be added to the response time shown.

At first it appears that the free-running schedulers are more responsive. Once we consider that most tasks must run at a deterministic rate, however, the time-slice scheduler results in the best possible response time. That is, it will result in the best possible response time as long as we keep the T_{slice} as small as possible.

The problem is that we cannot have a time-slice period less than the total execution time of all tasks, or there will be timing jitter or even long-term timing errors. This means we need to reduce the execution time for each task as much as possible. Let's look at some simple methods to reduce the time-slice period.

15.3.3 Mutual Exclusion

In general not all tasks in a time-slice scheduler are executed every slice. We can reduce the total task execution time within one time-slice period by spreading these tasks out to mutually exclusive time slices.

For example let's say we have a system with three tasks, and *Task2* and *Task3* need to be executed every second slice. We can execute *Task2* during even-numbered slices and run *Task3* during odd-numbered slices. In this way *Task2* and *Task3* would never execute during the same slice.

The total task execution time for a given time slice has been reduced because we know that *Task2* and *Task3* will never run during the same slice. For this situation the time-slice period can be reduced to

$$T_{slice} > T_{T1, \, max} + max\left[T_{T2, \, max}, T_{T3, \, max}\right] + \sum T_{int, \, T_{Rn}}$$

15.3.4 Task Decomposition

When the execution time for one task is greater than the required period for another task, mutual exclusion will not help. The only way to make the system work in a cooperative scheduler is to break the long task down into shorter parts. This is referred to as *task decomposition.*

In this subsection we look at two ways to break down tasks so they have shorter execution times—*coroutines* and *state decomposition.*

Note that these methods do not reduce the total execution time of a task. What we are doing is breaking down a task into parts that have shorter executions times. We then can run one part during one slice and return to the scheduler so it can run other tasks. In this way we can reduce the response time of the other tasks at the cost of increasing the response time of the task that is decomposed. This is normally acceptable, because the tasks that have long execution times tend to be the tasks that do not require short response times.

Coroutines. One way to reduce the execution time of a long task is to break it down into shorter coroutines. Coroutines are routines that when combined perform a complete task. In general they cannot be executed alone, because individually they do not complete a task.

Once we break down a task into coroutines, each coroutine can be placed in a time-slice scheduler that has minor cycles within the major time-slice cycle. For example, in the first program shown in Source 15.1, the minimum time slice period is

$$T_{slice} > T_{T2} + T_{T1} + \sum T_{int, \, T_{Rn}}$$

In this case the system will not work, because the execution time for *Task1* is too long, so it prevents *Task2* from being executed often enough.

Source 15.1 Time-Slice Scheduler Using Coroutines

```
/*************************************************************
 * A cyclic scheduler that doesn't work because Task1 is
 * too long.
 *************************************************************/
main(){
    while(1){
        WaitForTimer();
        Task1();   /* Task1 is too long */
        Task2();
    }
}


/*************************************************************
 * A major/minor cyclic scheduler that uses coroutines for
 * Task1.
 *************************************************************/
main(){
    while(1){
        WaitForTimer();
        Task1a();
        Task2();
        WaitForTimer();
        Task1b();
        Task2();
    }
}

/*************************************************************/
```

In the second program shown in Source 15.1, *Task1* has been broken into two coroutines, *Task1a* and *Task1b*. The time slice is then broken down into two minor cycles. The first minor cycle executes the first coroutine and then executes *Task2*. The second minor cycle executes the second coroutine and then executes *Task2* again. Each minor cycle is a separate time slice. The time-slice period can now be smaller, because we are only executing part of *Task1* during each time slice. The minimum time-slice period is

$$T_{slice} > T_{T2} + max\left[T_{T1a}, T_{T1b}\right] + \sum T_{int, \, T_{Rn}}$$

There are some problems with coroutines. The most important problem is that a single task is no longer contained within a single function. Because of this, the variables used for the task must have a scope that includes all the coroutines for that task, and they must persist from one coroutine to another. This limits us to using global variables, which now introduces another separate piece that is required for the complete task.

This can cause problems when revising or reusing the code. Coroutines must be well documented so that it is obvious that one will not work without the other. The best way to organize the coroutines and the global data that they use is to create a separate module for the task. The module should contain all the coroutines and variables. The variables can then be declared as *static* variables outside all the coroutines. This limits the scope of the variables to the coroutines, but allows them to persist throughout the program.

Coroutines also require that the program be decomposed sequentially. We cannot change the order in which the coroutines are executed, and we cannot jump from one coroutine to another without returning to the scheduler first.

State Decomposition. The most popular and useful method for breaking down a long task into shorter pieces is to use state decomposition. *State decomposition* involves designing a task as a state machine. Each time the task is called, it performs a different part of the task based on the current state. The advantage of using state decomposition instead of coroutines is that all the code for a given task is contained in a single function.

Source 15.2 shows a task broken down into a set of sequential states. This example is functionally equivalent to the coroutine example shown in Source 15.1.

Source 15.2 Sequential State Decomposition

```
/************************************************************************
* A simple example of a task that is broken down into two
* sequential states.
************************************************************************/
typedef enum{STATE1,STATE2} T1_STATES;

void main(void) {
    /*A typical time-slice scheduler*/
    while(1) {
        WaitForTimer();
        Task1();
        Task2();
    }
}


/************************************************************************
* Task1 - Decomposed into two sequential states
************************************************************************/
```

```
void Task1(void) {

    static T1_STATES CurT1State = 0; /* Current Task1 state */

    switch (CurT1State) { /*Determine the current state*/
    case STATE1:
        /*Execute State1 Code*/
        CurT1State = STATE2; /*Change to next state*/
        break;
    case STATE2:
        /*Execute State2 Code*/
        CurT1State = STATE1;
        break;
    }
}

/*****************************************************************/
```

The sequence of execution for two time slices is now

$$Task1_{S1} \rightarrow Task2 \rightarrow Task1_{S2} \rightarrow Task2 \rightarrow \text{Repeat}$$

So the minimum time slice period can be reduced to

$$T_{slice} > T_{T2} + max\left[T_{T1,\,S1}, T_{T1,\,S2}\right] + \sum T_{int,\,T_{Rn}}$$

Notice that the current state variable, *CurT1State,* must be static so it will persist between calls to *Task1*. We can use either a global variable or a static local variable defined inside the function *Task1*. The static local variable is preferable, because the scope is limited to *Task1*.

15.3.5 Stopwatch Example

In this section we look at a stopwatch program that uses a cooperative multitasking kernel. It is an enhanced version of the stopwatch program shown in Chapter 8 that displays minutes, seconds, and tenths of seconds as follows:

```
99:59.9
```

As shown in Source 15.3, this version has one time-slice function, *WaitForSlice()*, and three tasks, *UpdateTime()*, *ScanSw()*, and *OutTime()*. The time-slice period, which is set to 10ms, is based on the *GetmsCnt()* function in the *OCDelay* module that was introduced in Chapter 14.

Source 15.3 Source Code for Stopwatch Example

```
/**************************************************************************
* StpWtchTSC - A stopwatch demonstration to demonstrate a time-slice
*              scheduler and task decomposition.
*
```

```
 * Modules: OCDelay, EEBIO
 *
 ******************************************************************************
 /* Include Project Master Header File
 ******************************************************************************/
 #include "includes.h"


 /******************************************************************************
 * Local Function Prototypes
 ******************************************************************************/
 void WaitForSlice(void);      /* Time slicer */
 void ScanSw(void);            /* Switch scanning and debouncing task */
 void UpdateTime(void);        /* Timer update task */
 void OutTime(void);           /* Timer output task */
 void ChangeTimMode(void);     /* Stopwatch mode function */


 /******************************************************************************
 * Global Variables
 ******************************************************************************/
 UBYTE  SliceCnt;              /* Slice counter for UpdateTime() and OutTime() */
 UWYDE  SliceTime              /* Next slice time*/
 UBYTE  Hms;                   /* Elapsed time, hundred milliseconds */
 UBYTE  Secs;                  /* Elapsed time, seconds */
 UBYTE  Mins;                  /* Elapsed time, minutes */
 TMODES TimMode;               /* Current stopwatch mode */


 /******************************************************************************
 *
 ******************************************************************************/
 void main(void) {

     SCIOPEN();                /* Init. SCI for EEBIO module */
     OCDlyInit();              /* Init. OCDelay module */
     SliceCnt = 0;             /* Init. all globals */
     SliceTime = GetmsCnt();
     Hms = 0;
     Secs = 0;
     Mins = 0;
     TimMode = CLEAR;
     ENABLE_INT();             /* Enable interrupts to use OCDelay()  */

     FOREVER(){                /* Endless time-slice cyclic scheduler  */
         WaitForSlice();       /* Wait for periodic time-slice event   */
         UpdateTime();         /* Run elapsed time update task         */
         ScanSw();             /* Run switch scan and debounce task     */
         OutTime();            /* Run timer display task                */
     }
 }
```

```
/***********************************************************************
* WaitForSlice() - Time slicer. Uses OCDelay module for a time-slice
*                  period of 10ms (SLICE_PER = 10). Also increments
*                  SliceCnt for UpdateTime() and OutputTime() tasks.
* Modules: OCDelay
* Member: GetmsCnt()
***********************************************************************/
void WaitForSlice(void){

    while(GetmsCnt() != slicetime){}    /* wait for next time slice */
    SliceTime += SLICE_PER;             /* set up for next time slice */
    SliceCnt++;                         /* Increment slice counter */
}


/***********************************************************************
* ScanSw() - Detects and debounces an active-low switch on the SW_BIT bit
*            on the SW_PORT port.
*            - With noise immunity
*            - Must be executed every 10ms for a switch bounce time < 20ms.
***********************************************************************/
void ScanSw(void){

    UBYTE cur_sw;                       /* Current switch position */
    static UBYTE SwState = SW_OFF;      /* Current switch debounce state */

    cur_sw = SW_PORT & SW_BIT;          /* Get current switch position */
    if(SwState == SW_OFF){              /* wait for switch edge */
        if(cur_sw == PRESSED){
            SwState = SW_EDGE;
        }
    }else if(SwState == SW_EDGE){       /* Verify switch press */
        if(cur_sw == PRESSED){
            SwState = SW_VERF;          /* Switch press verified */
            ChangeTimMode();            /* Change stopwatch mode */
        }else{                          /* False switch press, ignore */
            SwState = SW_OFF;
        }
    }else if(SwState == SW_VERF){       /* Wait for release  */
        if(cur_sw == RELEASED){
            SwState = SW_OFF;
        }
    }
}


/***********************************************************************
  ChangeTimMode() - Stopwatch mode state machine.
                    - Cycles: CLEAR->COUNT->STOP->CLEAR...
                    - Called every valid button push.
```

```
*****************************************************************/

void ChangeTimMode(void) {

    if(TimMode == CLEAR){              /* If CLEAR change to COUNT */
        TimMode = COUNT;
        SliceCnt = 0;                  /* Reinitialize SliceCnt */
    }else if(TimMode == COUNT){        /* If COUNT change to STOP */
        TimMode = STOP;
    }else{                             /* If STOP change to CLEAR */
        TimMode = CLEAR;
        Hms = 0;                       /* Clear all time variables */
        Secs = 0;
        Mins = 0;
        SliceCnt = 0;                  /* Clear SliceCnt for display update*/
    }
}


/*****************************************************************
UpdateTime() - This task increments the time every tenth of a second
               when stopwatch is in the COUNT mode.
             - It produces time in Mins:Secs.Hms
             - SliceCnt is used to update the time every tenth second.
*****************************************************************/
void UpdateTime(void){

    if(TimMode == COUNT && SliceCnt == 10){   /* Ready to update? */
        SliceCnt = 0;
        Hms++;                         /* Increment tenths */
        if(Hms == 10){                 /* Update seconds */
            Secs++;
            Hms = 0;
            if(Secs == 60){            /* Update minutes */
                Mins++;
                Secs = 0;
                if(Mins == 100){       /* Overflow, roll to back to zero */
                    Mins = 0;
                }
            }
        }
    }
}


/*****************************************************************
OutTime() - Displays the current time when SliceCnt is 1. This occurs
            every ten slices except when in the CLEAR or STOP mode. Then
            it updates the display every 256 slices because SliceCnt is
            free running.
            Note: Execution time is ~6.3ms when SCI is set for 9,600bps
```

```
*************************************************************************/
void OutTime(void) {

    UBYTE ones;                             /* Temporary display vars */
    UBYTE tens;

if(SliceCnt == 1){                  /* Output on slice 1 */
        PUTCHAR('\r');                      /* return to start of line */
        ones = (Mins % 10) + '0';   /* Calc. minute digits */
        tens = (Mins / 10) + '0';

        PUTCHAR(tens);                      /* Display minutes */
        PUTCHAR(ones);
        PUTCHAR(':');

        ones = (Secs % 10) + '0';   /* Calc. seconds digits */
        tens = (Secs / 10) + '0';

        PUTCHAR(tens);                      /* Display seconds */
        PUTCHAR(ones);
        PUTCHAR('.');
        PUTCHAR(Hms + '0');                 /* Display tenths */
    }
}

/************************************************************************/
```

The only task that is decomposed into states is the switch scanning and debouncing task, *ScanSw()*. The design of a switch debouncing task is covered Chapter 8. It is an example of a task that must be decomposed into states because of the required timing for noise immunity and edge-detection.

Timing Analysis. Figure 15.13 is a captured set of debugging signals that show several time slices for the program in Source 15.3. If the signal is high, the CPU is executing that task. Notice that the time-slice period is 10ms, and the timer output is sent every 10 slices, which is every 100ms. During the nine slices when the output is not sent, most of the time is spent in the *WaitForSlice()* routine, which is shown as *TSLICE* in the figure. The *OutTime()* task, shown as *OTTASK* in the figure, is an example of a task that has a long maximum execution time.

The measured worst-case execution times are

Task	Maximum Execution Time	Execution Period
UpdateTime()	6.1μs	100ms
SwScan()	4.5μs	10ms
OutTime()	6.3ms	100ms
OCDelay ISR	10.25μs	1ms

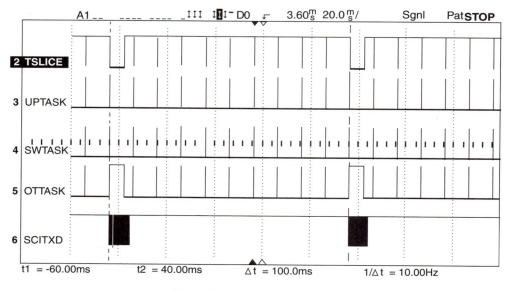

Figure 15.13 Stopwatch Timing

So the nominal CPU load is

$$L = \frac{6.1\mu s}{100ms} + \frac{4.5\mu s}{10ms} + \frac{6.3ms}{100ms} + \frac{10.25\mu s}{1ms} = 0.074$$

So theoretically the system should be easy to design.

The potential problem is that the *OutTime()* task has a large execution time relative to the time-slice period. It is designed for the Basic I/O module running on the SCI at 9,600bps. For every display update, it must output eight characters: a carriage return, two minutes digits, a colon, two seconds digits, a period, and one tenth-of-second digit. At 9,600bps, each character takes ~1ms to send.

Figure 15.14 shows the detailed timing for the *OutTime()* task and the serial bit stream from the SCI. Notice that the task actually ends ~2ms before the bit stream ends. This is because the SCI allows two characters to be buffered in the SCI itself, one in the SCI shift register and one in the SCI transmit register. So after the *OutTime()* task writes the eighth character to the SCI, the SCI still must send the last two characters.

You can also see in the figure that the *OutTime()* task is finished less than 4ms before the next time slice. So, although the CPU load is small, this design is close to violating the rule that the sum of all task execution times be less than the time-slice period. In fact, if we were required to send data out at 4,800bps or less, this design would not work because the *OutTime()* task execution time would be longer than the 10ms time-slice period.

Decompose *OutTime()* into States.   Now let's look at a redesign that will break the *Out-Time()* task down into five sequential states so the output will work with a 1,200bps modem.

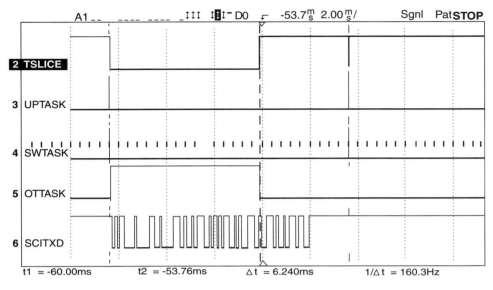

2 TSLICE

3 UPTASK

4 SWTASK

5 OTTASK

6 SCITXD

t1 = -60.00ms   t2 = -53.76ms   Δt = 6.240ms   1/Δt = 160.3Hz

Figure 15.14 *OutTime()* Timing Detail

First we need to make sure that the system load is acceptable. Since most of the load is due to the output task, the load placed on the system is slightly higher than

$$L = \frac{8\dfrac{\text{ms}}{char} \times 8char}{100\text{ms}} = \frac{66.67\text{ms}}{100\text{ms}} = 0.667$$

In this case the system load is close to the limit, but the design may be possible.

If we used the *OutTime()* task shown in Source 15.3, it would take ~50ms to execute, which is far longer than the maximum of 10ms. We can solve this problem by breaking the *OutTime()* task down into multiple states so all the characters are not sent during a single time slice.

Source 15.4 shows the *OutTime()* task decomposed into five sequential states, *OUT1–OUT5*. It is designed so that only one or two characters are transmitted during odd-numbered time slices. Since there are five states that are executed every two slices, it takes the full 100ms to update the display.

Source 15.4 OutTime() Decomposed into Five Sequential States for Slow Output Device

```
/**********************************************************************
OutTime() - Displays the current time when TimeSlice is odd. This version
            is decomposed into 5 sequential states for a 1,200bps SCI.
**********************************************************************/
void OutTime(void) {
```

```c
static OUTSTATES CurOutState = OUT1;
UBYTE ones;                              /* Temporary display vars */
UBYTE tens;

if(SliceCnt & 0x01 == 0x01){             /* Handle output on odd slices */
    if(CurOutState == OUT1){             /* Output carriage return */
        CurOutState = OUT2;
        PUTCHAR('\r');                   /* Return to start of line */

    }else if(CurOutState == OUT2){       /* Output minute chars */
        CurOutState = OUT3;
        ones = (Mins % 10) + '0';        /* Calc. minute digits */
        tens = (Mins / 10) + '0';
        PUTCHAR(tens);                   /* Display minutes */
        PUTCHAR(ones);

    }else if(CurOutState == OUT3){       /* Output colon */
        CurOutState = OUT4;
        PUTCHAR(':');

    }else if(CurOutState == OUT4){       /* Output second chars */
        CurOutState = OUT5;
        ones = (Secs % 10) + '0';        /* Calc. seconds digits */
        tens = (Secs / 10) + '0';
        PUTCHAR(tens);                   /* Display seconds */
        PUTCHAR(ones);

    }else if(CurOutState == OUT5){       /* Output point and tenths */
        CurOutState = OUT1;
        PUTCHAR('.');
        PUTCHAR(Hms + '0');              /* Display tenths */
    }else{
        CurOutState = OUT1;              /*In case CurOutState gets lost */
    }
    }
}

/**************************************************************************/
```

Figure 15.15 shows the new timing for this system. Notice that the SCI is sending data almost continuously. This design has distributed the SCI transmission over different time slices instead of trying to send them all during a single slice.

Another thing to note about this design is the dramatically reduced task execution time for *OutTime()*. The number of characters sent each time was purposely limited to two. These two characters can be loaded immediately into the SCI as long as the last two characters are finished being sent. Therefore the *OutTime()* task never has to wait for a character to be transmitted and can immediately return to the scheduling loop.

Because the *OutTime()* task does not have to wait for the SCI, the CPU load has been reduced to almost zero. Be careful, though. We can add a large amount of CPU load only as long as it does not involve sending more characters. If many more characters were required,

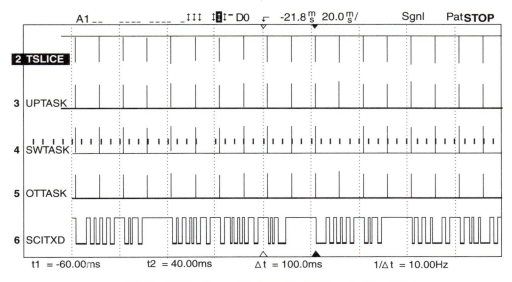

Figure 15.15 System Timing with the New *OutTime()* Task

the SCI would not be able to send them in the 100ms period. We can see this by recalculating the system load. We can also see it in the implementation. If we add more characters per time slice, the transmission time may exceed two time slice periods. If we add more output states, the display cannot be updated in the required 100ms.

Note

Another thing to be careful of is the temporary variables used in the *OutTime()* task. Remember that if a variable is used across more than one state, it must persist beyond the end of the function. The two temporary variables, *tens* and *ones,* are used in two different states, but they do not have to persist from one state to the other. Perhaps it would be safer to define these variables as *static* so if that other programmers revise this code, they will not create a bug by accidentally using these variables across more than one state. The cost for making these variables *static* would be only two more permanently allocated RAM locations.

This example illustrates some of the methods required to use state decomposition in a cooperative kernel. It is typical to have to do this in tasks that must communicate with a slower external device such as the SCI or a display module. If we are spending a lot of time trying to make a cooperative kernel work and the tasks are becoming complicated and unwieldy, however, it may be time to look at using a full preemptive kernel like MicroC/OS-II.

▶ SUMMARY

In this chapter we reviewed and expanded on the real-time programming material covered in Chapter 8. All the code was written in C and we looked at more techniques to implement a cooperative scheduler.

We first saw that most real-times tasks consist of an endless task loop. If we use a pre-emptive kernel, we can design the tasks as if they have complete control over the CPU. If we use a cooperative kernel, the tasks must be designed as cooperative tasks. Cooperative tasks must return to the scheduler in a timely manner so other tasks can run. This complicates the design of the tasks, but it allows us to use a cooperative kernel, which is simpler and does not require as much memory. In both cases the system will only work if the load placed on the CPU is not too high.

We then looked at different cooperative kernel designs, including a free-running cyclic scheduler and a time-slice cyclic scheduler. In both of these designs, the minimum task period must be greater than the sum of all execution times. This turned out to be a serious limitation to cooperative schedulers.

In the last two sections, we looked at techniques that allow us to decrease the total execution time during one pass through the scheduler loop. These techniques can be used to improve the response time and the minimum task period.

At some point the design of the cooperative tasks is too complicated, and it becomes more cost effective to use a preemptive kernel. In the next chapter, we will look at designing a system using a preemptive kernel, MicroC/OS II.

EXERCISES

1. Write a C program that implements a free-running event loop that samples *PORTP*, bit 0, and if it is zero, sets bit 0 of *InFlag*.

2. Repeat Exercise 1 using a timed event loop that runs the task code every 5ms. Use the *GetmsCnt( )* function from the *OCDelay* module for timing. Show only the event loop.

3. The task service time, T_{srv}, for a task is calculated to be 10ms. The calculation did not include the total time spent servicing interrupts, however. If there is one periodic interrupt that runs every 1ms and takes 10μs to complete, what is the actual service time?

4. Show the program context versus time given the three tasks and a priority-based, non-preemptive scheduler. Assume that all tasks are ready to run at the start of your diagram. The diagram should show at least 10 ms. Following is a description of each task:

Task	Priority	Execution Time	Execution Period
Task1	1	1ms	4ms
Task2	2	2ms	6ms
Task3	3	3ms	10ms

5. Repeat Exercise 4 using a priority-based, preemptive scheduler.

6. Calculate the CPU load for the tasks given in Exercise 4.

7. For the following three tasks, calculate the required CPU load. Based on the CPU load, is it possible to run these three tasks concurrently?

Task	Priority	Execution Time	Execution Period
Task1	1	10μs	4ms
Task2	2	200μs	6ms
Task3	3	1.5ms	10ms

8. Is it possible to design a free-running round-robin scheduler that can meet the task requirements shown in Exercise 7? If so, find the event response time for each task. Assume that the task delays are equal to the task execution times and there are no interrupts or circuit delays.

9. Repeat Exercise 8 using a priority-based, free-running scheduler.

10. If a time-slice scheduler is used to run the tasks given in Exercise 7, what is the optimum time-slice period? During which slices should each task be executed?

11. Find the event response times for the system in Exercise 10.

12. Can a time-slice scheduler be used to run the tasks given in Exercise 4 without task decomposition? Explain your answer.

13. The marketing department for the company developing the stopwatch described in this chapter has requested that the display include a 10ms digit. Is it possible to add this digit and update the display every 10ms with the program shown in Source 15.3? Give the approximate CPU load and the maximum execution time during one slice. Assume the extra time that is required to calculate the 10ms digit is insignificant, so the only added execution time is the time required to display the extra digit.

Using the MicroC/OS-II Preemptive Kernel

In this chapter we look at design and analysis of software written for MicroC/OS-II, a real-time preemptive kernel. We will cover the use of MicroC/OS-II, not the design of the kernel itself. For a detailed description of the kernel, refer to *MicroC/OS-II, The Real-Time Kernel* by Jean J. Labrosse, which also includes the source code for the kernel.[1]

Note _____

All of the examples and descriptions in this text are based on MicroC/OS-II. For simplicity, however, from this point on the kernel will simply be referred to as μC/OS.

All the programs used in this chapter use μC/OS as a precompiled library. This makes the build process and file organization of our projects simpler so we can focus on the application code. By using a precompiled library, we do not have control over the μC/OS build configuration, so our programs may not be as efficient as possible.

Appendix C covers reference material for the μC/OS kernel and some of the customization done for this text. Included in Appendix C are the μC/OS configuration settings for the precompiled library used in this text and descriptions of the μC/OS service functions.

_____

[1]Jean J. Labrosse, *MicroC/OS-II, The Real-Time Kernel* (Lawrence, KS: R&D Books, 1999).

594

▶ 16.1 OVERVIEW

µC/OS is a real-time preemptive multitasking kernel. It is written in C with the exception of a small CPU-specific assembly module. This means µC/OS can be ported easily to different processors. In fact it has been ported to a large number of processors ranging from simple 8-bit controllers like the 8051 to large 32-bit CPUs. The disadvantage to having the kernel written in C is a small loss in speed and code-size efficiency. However, this is a small price to pay for the portability and versatility. The kernel can be configured to be small and efficient for an 8-bit CPU with limited memory and processor speed, or it can be configured for large systems with essentially unlimited resources.

There are many real-time kernels available for 8-bit or 16-bit microcontrollers like the 68HC11 and 68HC12. µC/OS has a good balance of features versus required resources. It is widely available because the source, which has been available since 1992, is well written and it is free for noncommercial, peaceful use. It is also well documented in the book *MicroC/OS-II, The Real-Time Kernel* by Jean J. Labrosse.

Recall from last chapter that when we use a preemptive kernel, we can write the tasks as if they where independent tasks that have complete control over the CPU. Figure 16.1 shows the overall program flow for a system that uses a preemptive kernel. The tasks include task-related initialization and task code that is contained in an endless loop. As long as the CPU load for the task is not too high, the kernel can preempt one task to run another higher-priority task.

A preemptive kernel not only makes it easier for us to design, construct, and reuse our tasks, it also allows the higher priority tasks to have a much better event response time. To illustrate, let's look at the response to an event that is detected through an interrupt but serviced by a task.

Figure 16.2 shows the response time for the time-slice cyclic scheduler described in Chapter 15. The event is first detected by the interrupt circuitry. Then the interrupt service routine sets a flag that is sampled by *Task1*. If the flag is set, *Task1* responds to the event. In the figure, the event occurs at the worse possible time—right after *Task1* samples the event flag. Therefore the flag is not detected until the flag is sampled by *Task1* during the next time slice, and the response time is

$$T_R \approx T_{slice}$$

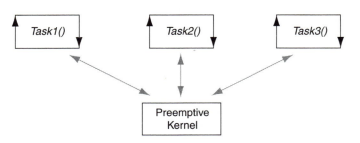

Figure 16.1 Preemptive Multitasking Program Flow

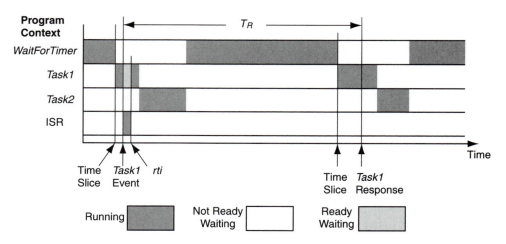

Figure 16.2 Event Response in a Time-Slice Cyclic Scheduler

The only way to improve the response with a time-slice scheduler is to put the event servicing code in the interrupt service routine. But as we have seen, this then slows the response for all other tasks and ISRs—including tasks and ISRs with higher priorities.

Figure 16.3 shows the response from a preemptive kernel. When the ISR is complete, control is not necessarily returned to the interrupted task, which in this case was *Task2*. Instead the kernel switches control to the highest-priority task that is ready to run. Since the event flag set by the ISR made *Task1* ready, control is switched to *Task1* instead of *Task2*. Therefore the response time is reduced dramatically.

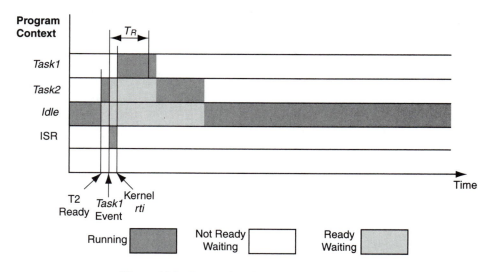

Figure 16.3 Preemptive Kernel Program Context

When designing tasks for a preemptive kernel, we have to keep in mind that the tasks run asynchronously to each other and that any task can be preempted by another higher-priority task. Because of this, we need to learn different techniques for intertask communications and protecting shared resources. A full-featured kernel will include services for intertask communications such as semaphores, messages, and queues. Semaphores are also used as keys to provide exclusive access to shared resources.

16.1.1 Program Organization

In this subsection we look at the basic organization used in this text for programs that use the µC/OS kernel. To demonstrate we will use an Introl-CODE project called *UcosDemo*.

The file organization for the project's program modules is the same as described in Section 14.3 in Chapter 14. Because we are using µC/OS as a precompiled library, using the kernel does not require any additional modules in the project. We only need to include the precompiled library and its header file.

If we did need to compile the kernel with our application code, the µC/OS modules would fit easily into the project, because they are designed for essentially the same module and header file organization.

Figure 16.4 shows the Introl-CODE directory window for the *UcosDemo* project before the project is built. The application code is contained in a single module, *UcosDemo.c,* and a master header file, *includes.h.* Notice that there is also a library included in the project called *libucosii.a12.* This is the precompiled kernel library.

The Main Program Module. When writing a program for µC/OS, the main program module contains the *main()* function, a start-up task, and, some or all the other tasks. Additional tasks and support functions may be contained in separate modules.

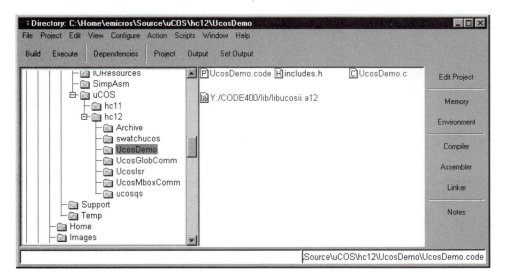

Figure 16.4 MicroC/OS Demonstration Project

The main module for the *UcosDemo* project, *UcosDemo.c,* is shown in Source 16.1. In this case it is the only module in the project, so it contains all the tasks. Let's quickly go over each part of this source code.

Like all the C modules described in this text, *UcosDemo.c* starts by including the master header file *includes.h.* We will look at the contents of this header file later.

Next the program defines the µC/OS events and task functions contained in the module. In this case there is a single event, *SecFlag,* and three tasks, *StartTask, Task1,* and *Task2.* We cover µC/OS events in Section 16.5 and tasks in Section 16.2. Notice that the task functions are declared as private functions. This is because task functions should never actually be called by another function. They are executed by the kernel, which knows about each task through the *OSTaskCreate()* service routine. The event objects are handled like normal C resources. They may be private if only used in the current module, or they must be public if used by a task in another module.

In a preemptive kernel each task uses its own stack space. So after the task functions are declared, the stack space for each task is allocated. The size of each stack is defined in the *includes.h* header file. The stack space should always be allocated in the same module as the task, so it too can be declared as a private resource.

Source 16.1 MicroC/OS-II Demonstration and Test Program

```
/**********************************************************************
 * A simple demo program for µC/OS-II.
 * 01/19/00, Todd Morton
 **********************************************************************/
#include "includes.h"


/**********************************************************************
 * Public Event Definitions
 **********************************************************************/
OS_EVENT *SecFlag;                    /* A one-second flag semaphore      */


/**********************************************************************
 * Task Function Prototypes.
 *    - Private if in the same module as start-up task. Otherwise public.
 **********************************************************************/
static void   StartTask(void *pdata);
static void   Task1(void *pdata);
static void   Task2(void *pdata);


/**********************************************************************
 * Allocate task stack space. (Must be public)
 *    - Private if in the same module as start-up task. Otherwise public.
 **********************************************************************/
static OS_STK   StartTaskStk[STARTTASK_STK_SIZE];
static OS_STK   Task1Stk[TASK1_STK_SIZE];
static OS_STK   Task2Stk[TASK2_STK_SIZE];


/**********************************************************************
```

```
 * main()
 **************************************************************************/
void main(void) {

    OSInit();                          /* Initialize µC/OS-II              */

    OSTaskCreate(StartTask,            /* Create start-up task             */
                 (void *)0,
                 (void *)&StartTaskStk[STARTTASK_STK_SIZE],
                 STARTTASK_PRIO);

    SecFlag = OSSemCreate(0);          /* Create a semaphore flag          */

    OSStart();                         /* Start multitasking               */
}

/***************************************************************************
 * START-UP TASK
 **************************************************************************/
static void StartTask(void *pdata) {

    OSTickInit();                            /* Initialize the µC/OS ticker */

    DBUG_PORT_DIR = 0xff;                     /* Initialize debug port       */
    DBUG_PORT = 0xff;

    OSTaskCreate(Task1,                      /* Create all other tasks      */
         (void *)0,
         (void *)&Task1Stk[TASK1_STK_SIZE],
         TASK1_PRIO);

    OSTaskCreate(Task2,
         (void *)0,
         (void *)&Task2Stk[TASK2_STK_SIZE],
         TASK2_PRIO);

    FOREVER() {                              /* Start-up task ending trap   */
        DBUG_PORT ^= DBUG_STSK;
        OSTaskSuspend(STARTTASK_PRIO);
    }
}

/***************************************************************************
 * TASK #1
 **************************************************************************/
static void Task1(void *pdata){

    INT8U TimCntr = 0;                       /* Counter for one-second flag*/
```

Using the MicroC/OS-II Preemptive Kernel   599

```
    FOREVER() {

        OSTimeDly(10);                        /* Task period = 10ms        */
        DBUG_PORT ^= DBUG_TSK1;               /* Toggle task 1 debug bit   */
        TimCntr++;
        if(TimCntr == 100){
            OSSemPost(SecFlag);               /* Signal one second         */
            TimCntr = 0;
        }
    }
}


/**********************************************************************
 * TASK #2
 **********************************************************************/
static void Task2(void *pdata){

    INT8U err;                                /* Storage for error codes   */

    FOREVER() {
        OSSemPend(SecFlag, 0, &err);          /* Wait for one-second event */
        DBUG_PORT ^= DBUG_TSK2;               /* Toggle task 2 debug bit   */
    }
}

/**********************************************************************/
```

The next item in the module is the *main()* function. Recall that *main()* is the function called after the system is reset and the start-up code is completed. Therefore it is the entry point for our application code. When writing a program for μC/OS, the *main()* function should initialize the kernel with *OSInit()*, create the start-up task with *OSTaskCreate()*, create any kernel event services, and start the kernel with *OSStart()*. Once *OSStart()* is called, the kernel is running and program control never returns back to *main()*.

There is rarely anything else in the *main()* function. All the hardware and system initialization is completed in the start-up code before reaching *main()*. The application-wide initialization is completed in the start-up task and the task-specific initialization is completed in the tasks themselves.

In the *UcosDemo* project, there are three tasks in the main module, *StartTask, Task1,* and *Task2. StartTask()*is the start-up task, which is required in all programs using μC/OS. When we started the kernel in *main()*, the start-up task was the only task that had been created. Therefore it is guaranteed that the kernel will run the start-up task first.

In the start-up task, we need to initialize the μC/OS timer service, create other tasks, and initialize application-wide resources that were not already initialized in the start-up code before *main()*. Because the start-up task creates other tasks, it must have the highest priority or the kernel may switch to another task before the initialization in the start-up task is complete.

In Source 16.1, the µC/OS timer is initialized using the service *OSTickInit()*. The timer must be initialized before any timer services can be used but only after the kernel has been started. Therefore we could not move this to *main()*.

The start-up task is unique because it is meant to run only one time. As we have seen throughout this text, most tasks are endless loops. To make sure the start-up task is executed only one time, it ends with a trap that calls the *OSTaskSuspend()* service. *OSTaskSuspend()* does stop the task, but just in case another task accidentally starts it up again with *OS-TaskResume()*, it is contained in an endless trap.

After the start-up task, there are two simple tasks, *Task1* and *Task2*. Notice that the two tasks appear never to exit. They are designed as endless independent tasks. *Task1* is a timed loop with a loop period of 10ms. When it first enters the loop, it waits 10ms by calling the *OSTimeDly()* service. When the delay is completed, the task code is executed. The task code toggles a general purpose output bit and increments a counter, *TimCntr*. If *TimCntr* reaches 100, *Task1* signals an event flag to communicate to *Task2* that one second has expired. The event flag in this case is implemented as a µC/OS event service called a *semaphore*.

Task2 is configured as an event loop. It waits for a *TimCntr* event from *Task1* and then runs the task code. The task code toggles another general purpose output bit. Therefore the output bit is toggled every second. One significant difference between this event loop and the event loops described previously in this text is that the event signal does not have to be polled. The task is placed in a waiting state when it calls *OSSemPend()*, and then when the event occurs, the kernel makes the task ready to run.

Master Header File.   Source 16.2 shows the master header file used for the *UcosDemo* project. As you can see, it is the same as the master header file described in Chapter 14 except it includes some definitions for the µC/OS.

Source 16.2 Master Header File for UcosDemo Project

```
/************************************************************************
* includes.h - Master header file for UcosDemo project.
*
* 01/18/00, Todd Morton
*************************************************************************
* WWU Project type definitions
*************************************************************************/
typedef unsigned char    INT8U;
typedef signed   char    INT8S;
typedef unsigned short   INT16U;
typedef signed   short   INT16S;
typedef unsigned long    INT32U;
typedef signed   long    INT32S;

#define UBYTE INT8U
#define SBYTE INT8S
#define UWYDE INT16U
#define SWYDE INT16S
```

```
#define ISR __interrupt void

/**************************************************************************
 * General Defined Constants
 **************************************************************************/
#define FALSE     0
#define TRUE      1


/**************************************************************************
 * General Defined Macros
 **************************************************************************/
#define FOREVER()        while(1)
#define TRAP()           while(1){}
#define SWI()            asm("\tswi\n")
#define ENABLE_INT()     asm("\tcli\n")
#define DISABLE_INT()    asm("\tsei\n")


/**************************************************************************
 * MCU-Specific Configurations
 **************************************************************************/
#define DBUG_PORT_DIR _H12DDRP          /* Debug port definitions       */
#define DBUG_PORT _H12PORTP
#define DBUG_STSK 0x01
#define DBUG_TSK1 0x02
#define DBUG_TSK2 0x04


/**************************************************************************
 * MicroC/OS Task Configurations
 **************************************************************************
 * Task Priorities
 **************************************************************************/
#define STARTTASK_PRIO  4             /* Priority for StartTask()        */
#define TASK1_PRIO      6             /* Priority for Task1()            */
#define TASK2_PRIO      10            /* Priority for Task2()            */


/**************************************************************************
 * Define Task Stack Sizes
 **************************************************************************/
#define STARTTASK_STK_SIZE 64         /* Stack size for StartTask()      */
#define TASK1_STK_SIZE     64         /* Stack size for Task1()          */
#define TASK2_STK_SIZE     64         /* Stack size for Task2()          */


/**************************************************************************
 * System Header Files
 **************************************************************************/
#include "hc912b32.h"                 /* CODE 68HC912b32 register defines*/
#include "ucos12.h"                   /* µC/OS-II library definitions    */


/**************************************************************************/
```

The µC/OS definitions that should be included in the master header file are the task priorities, the task stack sizes, and the message queue sizes. These definitions are placed in this header file because in general tasks may be defined in multiple modules. By putting these definitions here, we can look in one place to see what tasks are used in the project, what priority each task has, and how much RAM space is used for the stacks.

The µC/OS definitions could also be placed in a separate header file that in turn would be included in the master header file. A separate header file may be preferable for some designers, especially for larger projects.

***UcosDemo* Operation.**   When the *UcosDemo* program is executed, the *Task1* output bit should contain a 50Hz square wave and the *Task2* output bit should contain a 0.5Hz square wave. By using an oscilloscope, we can verify that the kernel is operating correctly. If an oscilloscope is not readily available, a qualitative test can be made by connecting an LED to the *Task2* bit.

This demonstration program can be used to test basic task switching, the kernel timer, and semaphore events. It is not a complete test, but it is a good starting point. Once this program works, additional tests can be performed to verify the other kernel services.

▶ 16.2 TASKS AND TASK SWITCHING

In this section we look at how task switching is accomplished in µC/OS and some basic task design techniques. Again the emphasis is on the design of the application, not on the design of the kernel itself.

16.2.1 Task Switching

In Figure 16.1 on page 595, we saw that the system consists of a set of tasks and the kernel. The tasks are designed as independent tasks. That is, they consist of an endless task loop, and it appears when looking at the task code that each task always has control of the CPU. Contrast this with the cooperative tasks that we used when using a cooperative kernel. The cooperative tasks had to cooperate by returning to the scheduler in a timely manner so other tasks could run.

When using a preemptive kernel like µC/OS, the task switching is accomplished by the kernel. When the kernel detects that another higher-priority task is ready to run, it will preempt the current task and switch to the higher-priority task. In Figure 16.1, this process is represented by the gray arrows between the kernel and the tasks.

This concept is similar to the way that interrupts can interrupt the background program to run an interrupt service routine. A preemptive kernel, however, is very different than an interrupt. An interrupt uses hardware in the CPU to detect an interrupt event and make the context switch. The kernel is another program, and if an endless task is running, how can the kernel be running at the same time? Of course, it is not running.

To understand how task switching works in µC/OS, refer to Figure 16.5. This figure shows a functional state diagram for µC/OS tasks.

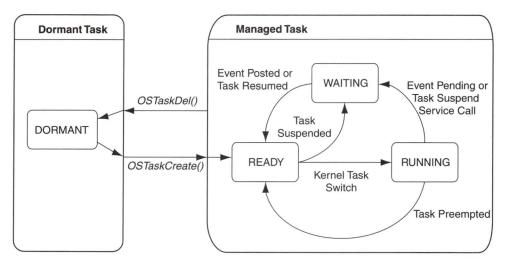

Figure 16.5 Task State Diagram

All tasks start out in the *DORMANT* state. While in the *DORMANT* state, the task code exists and its stack space is allocated, but the task is not managed by the kernel. This means that a task in the *DORMANT* state will not even be considered during the kernel's scheduling process.

To exit the *DORMANT* state, the task is created with the kernel service function *OS-TaskCreate( ),* which moves the task to the *READY* state. A task can also be placed back into the *DORMANT* state from any other state by an *OSTaskDel( )* service call.

Once a task is created, it is managed by the kernel and can be in one of three states, *READY, RUNNING,* or *WAITING.*

Tasks can be moved from one state to another by the kernel only when a kernel service call is made. When a service call is made by the task that is running or by an interrupt service routine, the kernel's scheduler is executed. The scheduler always switches CPU control to the highest-priority task that is in the *READY* state. This is called a *task switch.*

To perform a task switch, the kernel preempts the currently running task by moving it to the *READY* state or the *WAITING* state. Then it moves the highest-priority task in the *READY* state into the *RUNNING* state. Let's look at three scenarios in which a task switch may occur.

Preemption by an Interrupt Event. One way that the running task can be preempted is by an interrupt that causes a higher-priority task to be ready to run. If the interrupt service routine uses the *OSIntExit( )* service call to exit, and the ISR has made a higher-priority task ready to run, the kernel will perform the task switch immediately out of the ISR. So instead of returning to the task that was interrupted, a task switch is made and the higher-priority task is run. The task that was running is moved to the *READY* state and will be resumed when it is the highest-priority task that is ready to run.

The most important example of an interrupt causing a task switch is the kernel timer tick. The kernel timer tick is a periodic interrupt that is part of the kernel. When the timer interrupt occurs, the kernel timer is incremented, the task states are updated based on the new timer value, and a task switch is made. If the timer tick caused a higher-priority task to be ready to run, it is moved from the *WAITING* state to the *READY* state, and a task switch is then made to the higher-priority task.

Preemption by the Running Task Making a Service Call.   Every time a task makes a kernel service call, the kernel runs the scheduler. So if the task that is running makes a service call and the service call makes a higher-priority task ready to run, a task switch will be made to the higher-priority task.

For example suppose a high-priority task is in the *WAITING* state because it is waiting for a semaphore event. The running task then signals the semaphore by calling *OSSemPost()*. This call made the higher-priority task ready, so instead of the kernel returning to the running task, a task switch is made to the higher-priority task that is now ready to run.

Running Task Makes a Pending Service Call.   The third scenario for a task switch is caused by the running task removing itself from the *RUNNING* state. When the running task must wait for an event, it makes a pending service call, which moves the task to the *WAITING* state. The kernel scheduler is then run, and since the running task is no longer ready, a task switch will be made to the highest-priority task that is ready. Service pending calls include *OSMboxPend(), OSQPend(), OSSemPend(), OSTimeDly(),* and *OSTimeDly-HMSM()*. All of these calls move the task to the *WAITING* state until the event that the task is waiting for occurs.

Making a pending service call is not only important for synchronizing a task to an event, it also allows a task to cooperate by reducing its CPU load and allowing the kernel to run other tasks. Because the scheduler always runs the highest-priority task in the *READY* state, the low-priority tasks will only be run when all the higher-priority tasks are in the *WAITING* state.

So, a μC/OS task switch can be initiated by an interrupt service routine or by the running task calling a kernel service call. Recall from Chapter 15 that the event response time is primarily determined by the task switch time. The more times the kernel runs its task scheduler, the better the response time for the system. In addition, in order for lower-priority tasks to have an opportunity to run, the higher-priority tasks must cooperate by periodically moving into the *WAITING* state.

16.2.2 Task Design

As shown in Figure 16.1 on page 595, independent tasks are used when designing for a preemptive kernel like μC/OS. Figure 16.6 shows a free-running independent task. It first executes any task-related initialization and then enters an endless loop that unconditionally executes the task code.

Source 16.3 shows an example of a task that consists of a free-running task loop. This is a free-running independent task, because it is always either polling the event signal or servicing the event. Because it is a free-running loop, it requires a CPU load of one.

When a task like this is run in a preemptive kernel like μC/OS, it does not allow any task with a lower priority to run. Higher-priority tasks will still run because a task switch

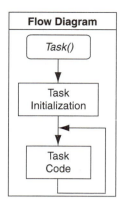

Figure 16.6 An Independent Free-Running Task

caused by the kernel tick or an interrupt service routine still can occur. Because this task is always ready to run, however, no task with a lower priority can ever preempt it. If this task happened to be the highest-priority task, no other task would ever run.

Source 16.3 A Free-Running Task with a CPU Load of One

```
/**********************************************************************
* Free-running independent task
**********************************************************************/
static void Task1(void *pdata){
    FOREVER() {
        if(Event){
            /* Service Event */
        }
    }
}

/**********************************************************************/
```

This illustrates the need for some level of cooperation required by tasks that run in a preemptive multitasking kernel. The tasks must cooperate by keeping their CPU load down to a level that allows all other tasks to run. To do this the task must periodically make an event-pending service call to implement a timed loop or an event loop based on a kernel event.

A Timed Loop.   Recall that a timed loop is one that waits for a timer event before making a pass through the task code. To implement a timed loop, we first need to analyze the timing of the event signal. We can then increase the polling period as much as possible without missing an event.

This in turn will reduce the CPU load and give other tasks an opportunity to run. Source 16.4 shows a simple way to implement a timed loop. The *OSTimeDly()* service is used to move the task to the *WAITING* state for 10ms before running the task code. During that 10ms period, lower-priority tasks have the opportunity to run.

Source 16.4 A Task That Reduces Its CPU Load by Running a Timed Loop

```
/**********************************************************************
* A better task
**********************************************************************/
static void Task1(void *pdata){
    FOREVER() {
        OsTimeDly(10)              /* Poll event every 10ms */
        if(Event){
            /* Service Event */
        }
    }
}


/*********************************************************************/
```

An Event Loop. A better way to reduce the CPU load for this task is to implement the event as a kernel event called a semaphore. The task can then wait for the event semaphore using the *OSSemPend()*, as shown in Source 16.5. This reduces the CPU load even further, because the only time the task code is executed is when the event occurs. The rest of the time it is in the *WAITING* state, and lower-priority tasks can run.

Source 16.5 An Event-Driven Task Based on a Kernel Semaphore

```
/**********************************************************************
* A better task
**********************************************************************/
static void Task1(void *pdata){
    INT8U err;                              /* Storage for error codes   */

    FOREVER() {
        OSSemPend(EventFlg, 0, &err);       /* Wait for event            */
        /* Service Event */
    }
}


/*********************************************************************/
```

This task almost acts like an interrupt service routine because no polling was required. The task is placed in the *WAITING* state until the event occurs. When the event occurs, the kernel moves the task to the *READY* state. So the next time the kernel's scheduler is run, if the task is the highest-priority task in the *READY* state, it will be run by the kernel.

The most important point here is to make sure that no task is a free-running task by having at least one pending service call each time through the loop.

16.2.3 Task Stack

Each task that runs under the μC/OS kernel requires its own stack space. The stack space is permanently allocated, as shown in Source 16.1 on page 598. Because each task must have its own task space, a preemptive kernel requires more RAM than the cooperative schedulers shown in Chapter 15, which use a single shared stack space.

Because of the limited amount of RAM in most embedded microcontrollers, we have to be careful about the amount of stack space we allocate. The minimum amount of stack space can be determined by adding the stack space required while running the task plus a copy of the CPU registers. The amount of space required while running the task includes the space required for nested interrupts, the space required for nested function calls, and the space required for all local variables.

In the *UcosDemo* program shown in Section 16.1, 64 bytes were allocated for each task's stack. This was more than required, but there is also plenty of RAM available because the program only has three simple tasks. It is a good idea to keep an eye on the task stacks. If a BDM debugger is used, we can look at the stack space while the program is running. Since the stack space is cleared during initialization, we can tell how much stack space is actually used by finding the last nonzero entry in each stack. μC/OS has a service that will check for this if needed.

16.2.4 Task Variables

Recall that when we designed tasks for a cooperative kernel, if the task was decomposed into states, we had to define the task's variables as static so they would persist from one task state to another. This is because in a cooperative system, the task function exits to return to the scheduler. Automatic variables do not persist after the task function is exited, so they will not be valid the next time the task is run.

When using a preemptive kernel like μC/OS, the task function is never exited, and each task has its own stack. Therefore we can use local (automatic) variables in our tasks, and those variables will persist when the task is preempted. In effect each task's local variables become static because the task function is never exited, so the byte allocated on the stack is permanently allocated.

16.2.5 Task Priorities

Whenever we create a new task, we need to decide on its priority. Task priorities can make or break a real-time system design. In μC/OS each task must be assigned a unique priority value. The lower the value assigned to a task, the higher the priority that task has.

To decide the priority to assign to a task, we first need to consider some μC/OS requirements and limitations. When μC/OS is compiled, it uses a constant called *OS_LOWEST_PRIO* to determine the lowest priority any task can have. In addition μC/OS uses the constant called *OS_MAX_TASKS* to determine the maximum number of tasks that can be created. The smaller these two constants are, the less RAM is used by the kernel.

μC/OS requests that you reserve several priorities for its own use and for future expansion. These include the four highest priorities and the four lowest priorities. Because of this, *OS_LOWEST_PRIO* should be greater than *OS_MAX_TASKS* + 7. Otherwise there would not be enough priority values to give each task a unique priority.

For example the μC/OS library used in this text was compiled with *OS_MAX_TASKS* equal to 10 and *OS_LOWEST_PRIO* equal to 17. Seventeen is the minimum value that

OS_LOWEST_PRIO can have to actually use the 10 tasks. When assigning priorities to the application tasks, we are limited to priorities 4 through 13.

Another requirement when designing for µC/OS is that the start-up task must have the highest priority of the application tasks. Otherwise the kernel may switch to another task before all the required start-up code is complete. In the examples used in this text, the start-up task is assigned priority 4 because it is the highest available.

Now we need to assign priorities to the rest of the tasks in our application. In general, when using a preemptive kernel, the optimum priority scheme is to assign the highest priorities to the tasks that execute at the highest rate. This rule is called *Rate Monotonic Scheduling (RMS)*.

For example the two tasks shown in Source 16.1 have the correct relative priorities for RMS. *Task1* runs every 10ms and *Task2* runs every second. Therefore, *Task1* should have the highest priority. The priority assignments are given in the master header file shown in Source 16.2 on page 601. The start-up task is given a priority of 4, *Task1* is assigned priority 6, and *Task2* is assigned priority 10.

Sometimes we have to make exceptions to this rule for application and implementation reasons. For example, as we will see in Section 16.4, in order for a task to run at an exact periodic rate using the *OSTimeDly()* function, the total task execution time must be less than one clock tick. And since the execution time includes the time spent in all higher-priority tasks and ISRs, we have to assign these tasks the highest priority. The stopwatch program in Section 16.6 provides a good example for assigning task priorities.

▶ 16.3 INTERRUPT SERVICE ROUTINES

As we learned in the last section, one way to preempt a task is with an interrupt that makes a higher-priority task ready to run. In order to use interrupts in this way, we need to include some code in the ISR to notify µC/OS when it is entering and leaving the ISR. This allows interrupts to be nested and a task switch to occur at the end of the ISR.

Source 16.6 shows the general form for most ISRs written to run with µC/OS on a small microcontroller. First the ISR must be declared as an ISR so Introl-CODE will end the function with an *rti* instead of an *rts*.

The first thing that should be done upon entering the ISR is to increment the µC/OS interrupt nesting level, *OSIntNesting*. We also could use *OSIntEnter()* to do this, but incrementing it directly is more efficient, which is very important for an ISR.

Source 16.6 General Form of an ISR Running with µC/OS

```
/*****************************************************************************
* General form of an ISR running with µC/OS.
*****************************************************************************/
ISR IsrName(void){
    OSIntNesting++;                          /* Inform µC/OS                 */

    /* Interrupt Service Routine */
    ...
    OSIntExit();                             /* µC/OS ISR exit routine       */
}


/*****************************************************************************/
```

After *OSIntNesting* is incremented, the interrupts could be enabled if we want to have nested interrupts. Normally if we keep the ISR short, interrupt nesting is not required. This is especially the case for the 68HC11 and the 68HC12 because, if we enable interrupts in an ISR, a lower-priority interrupt can interrupt a higher-priority interrupt.

Once the ISR code is complete, *OSIntExit()* must be called. This kernel service call decrements the interrupt nesting level and allows the kernel to make a task switch directly from the interrupt service routine.

Source 16.7 shows an example of a periodic timer ISR that runs under µC/OS. It uses an output compare on *Timer Channel 0* to generate an interrupt every 1ms. After 10 interrupts have occurred (10ms), the service routine signals a semaphore used as an event flag. This flag can then be used by a task to do something every 10ms.

Source 16.7 Example of a Periodic Timer ISR for a µC/OS Application

```
/********************************************************************
* OC0Isr() - OC0 Service Routine.
*    MCU: 68HC912B32, E = 8MHz
*    - Requires TC0 be set for an output compare (OC0Init()).
*    - set up for a 1ms periodic interrupt.
********************************************************************/
ISR OC0Isr(void){
    OSIntNesting++;                    /* Notify µC/OS              */
    _H12TC0 = _H12TC0 + E_PER_MS;      /* Interrupt in 1ms later    */
    _H12TFLG1 = C0F;                   /* Clear Channel 0 flag      */
    OCmsCnt++;                         /* Increment 1ms counter     */
    if(OCmsCnt == 10){                 /* Signal flag every 10ms    */
        OCmsCnt = 0;
        OSSemPost(TenmSFlag);
    }
    OSIntExit();                       /* µC/OS exit ISR routine    */
}


/********************************************************************/
```

▶ 16.4 TIMERS

In this section we look at different ways to generate and use timing events with µC/OS. We have seen throughout this text that timing events are important for tasks that require a timed response and for timed loops used to reduce the CPU load required by a task.

16.4.1 µC/OS Timer Services

µC/OS includes a periodic timer service called the kernel timer or the kernel *tick*. Because the kernel relies on this timer for timeouts, it is always running and available for application code through kernel service functions. The timer involves a periodic interrupt (the tick) and a free-running 32-bit counter that is incremented every tick.

TABLE 16.1 µC/OS TIMER SERVICE FUNCTIONS

Timer Service Functions	Description
OSTimeDly()	Delay task for *N* timer ticks.
OSTimeDlyHMSM()	Delay task in hours, minutes, seconds, and millisecond units.
OSTimeDlyResume()	Resume another task that is currently waiting for the timer.
OSTimeGet()	Get the current value of the µC/OS timer.
OSTimeSet()	Set the current value of the µC/OS timer.

The µC/OS timer period, or tick, must be configured at build time. This means that once µC/OS is compiled, the timer period cannot be changed. Typically the tick period is set in the range from 1ms to 100ms. A smaller tick period provides better timing resolution and response times, but because it runs more often, it also places a higher load on the CPU.

The tick period in the µC/OS library that is used in this text is 1ms. This small tick period is reasonable for simple applications that require fast response times. If the application is more complex with longer task execution times, this tick period may be too small. One millisecond is also small for µC/OS running on the slower 68HC11 MCU.

The µC/OS service functions for the timer are shown in Table 16.1. For a more detailed description of these functions, refer to Appendix C.

Notice that we could set the value of the timer's counter using *OSTimeSet()*. This is not recommended, however, because many programs that use the timer rely on it being a free-running counter. So if *OSTimeSet()* is used to change the counter value, these programs will no longer work correctly.

The most important timer functions are *OSTimeDly()* and *OSTimeDlyHMSM()*. These kernel service functions allow us to delay a task for an integral number of timer ticks. The parameter for *OSTimeDly()* is the number of ticks to delay. *OSTimeDlyHMSM()* allows us to specify the delay in normal time units—hours, minutes, seconds, and milliseconds. The resolution of both functions is one timer tick.

To see how *OSTimeDly()* works, let's look at *Task1* from the *UcosDemo* project. The source code for *Task1* is shown again in Source 16.8. It is a timed loop that uses *OSTimeDly()* to set the loop period to 10ms. When *OSTimeDly()* is called, this task is moved to the *WAITING* state, which allows other lower-priority tasks to run. After 10 timer ticks occur, the task is moved back to the *READY* state and is executed when it becomes the highest-priority task that is ready to run.

Source 16.8 An Example of a Timed Task Loop

```
/**********************************************************************
* TASK #1
**********************************************************************/
static void Task1(void *pdata){

    INT8U TimCntr = 0;                      /* Counter for one-second flag*/

    FOREVER() {
```

```
    OSTimeDly(10);                      /* Task period = 10ms        */
    DBUG_PORT ^= DBUG_TSK1;             /* Toggle task 1 debug bit   */
    TimCntr++;
    if(TimCntr == 100){
        OSSemPost(SecFlag);            /* Signal one second          */
        TimCntr = 0;
    }
  }
}
```

The function name implies a delay, and at first glance it appears that the task period would actually be greater than 10ms, because if the task waits 10ms and then executes the task code, it appears that the task period would be the sum of these two times. This would result in an inconsistent task period, as shown in Figure 16.7. This is not the case, however.

Let's look at how *OSTimeDly()* works in more detail. When *OSTimeDly(10)* is called, it tells the kernel to move the task to the *WAITING* state for 10 clock ticks. After the tenth clock tick, the kernel moves the task back to the *READY* state. Now, since the clock ticks are coming from a free-running periodic source, the actual delay will be between 9ms and 10ms because *OSTimeDly()* is asynchronous to the timer source and will land somewhere between two ticks.

Figure 16.8 shows the timing for a task that is delayed two ticks with *OSTimeDly(2)*. We changed the delay from 10 to 2 to simplify the figure. Notice that the first time *OSTimeDly(2)* is called, it falls right after a clock tick. In this case the actual delay is almost 2ms. The second call falls right before a clock tick. In this case the delay is only slightly more than 1ms.

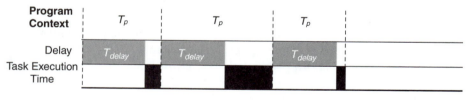

Figure 16.7   Inconsistent Loop Times Caused by Using a Timer Delay Routine

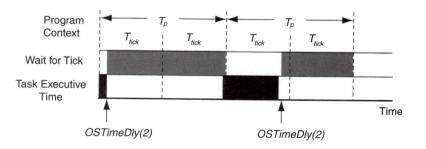

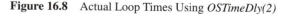

Figure 16.8   Actual Loop Times Using *OSTimeDly(2)*

Now look at the task period. Notice that it is always two timer periods wide because the difference in the actual delays is made up by the task execution time. This is what we want if we need to have an exact task period.

However, the timing shown in Figure 16.8 results in an exact task period only if the task execution time is less than one tick period. In addition the task shown in the figure is the highest-priority task. If it was not the highest-priority task, other tasks might have had to run first, delaying the execution of the task shown. Also not shown is the interrupt time required for the tick ISR. All of these things must be done before the next timer tick to have an exact task period. Otherwise the next tick will be missed and we lose short-term and long-term accuracy.

16.4.2 User-Designed Timer Events

If we cannot meet timing requirements with the kernel timer service, then we may need to create our own timer source. For example the timing in Figure 16.8 depends on the task being completed before the next tick. If it is not, both the short-term and long-term accuracy suffers.

We can improve the long-term accuracy if we use the periodic timer interrupt shown in Source 16.7. This is because it uses a counting semaphore, *TenmSFlag,* to signal an elapsed time of 10ms. Source 16.9 shows how *Task1* is changed so it will execute the task loop every time the semaphore is signaled. If a timer event is missed because a higher-priority task is running, the semaphore will be incremented again. Then when the CPU becomes available, *Task1* will be executed until the semaphore counter reaches zero again.

Source 16.9 Task1 Revised to Use a Counting Semaphore

```
/*******************************************************************
* TASK #1
*******************************************************************/
static void Task1(void *pdata){

    INT8U TimCntr = 0;                      /* Counter for one-second flag*/

    FOREVER() {

        OSSemPend(TenmSFlag);               /* Wait for 10ms flag        */
        DBUG_PORT ^= DBUG_TSK1;             /* Toggle task 1 debug bit   */
        TimCntr++;
        if(TimCntr == 100){
            OSSemPost(SecFlag);             /* Signal one second         */
            TimCntr = 0;
        }
    }
}
```

This timer may have short-term errors if timer events are missed, but long-term it will be accurate. Of course in order for this system to work, we are assuming that the average CPU load is less than one. Otherwise *Task1* might never get the opportunity to catch up.

Because tasks running in a preemptive kernel are running asynchronously to each other, there is no way to directly pass parameters from one task to another. We saw this before when trying to communicate between an interrupt service routine and the program running in the background.

Like communicating between an interrupt service routine and the background, we can use global variables to communicate between tasks. Using global variables alone is a simple and fast method to communicate, but because they are a shared resource, they are limited to simple message passing from one task to another. To implement intertask communication that is more complex, μC/OS includes three intertask communication services—semaphores, message mailboxes, and message queues. These services can be used alone or in some cases must be used in combination with global variables used as buffers.

Let's quickly review the use of a global variable and the problems associated with using a shared resource in a preemptive multitasking environment. Then we will look at each of the μC/OS intertask communication services.

16.5.1 Global Variables

A global variable, or more precisely, a static variable, uses permanently allocated RAM space. Therefore it persists until power is removed from the system and can have a scope that includes more than one task. This means that tasks can communicate by having one task, the sender, write to the variable and another task, the receiver, read it. The data flow diagram for this technique is shown in Figure 16.9.

Source 16.10 shows an example that illustrates the use of a global variable to pass data from one task to another. In this example *Task1* is a simple task that counts 100ms periods. It then passes the count, *TimCntr,* to a display buffer, *DispBuf,* in order to send this new value to the display task. *Task2* is a display task that reads the contents of the display buffer, converts it to decimal, and sends it out the SCI port to be displayed on a terminal.

This simple application is well suited for using a global variable because the display buffer is only one byte and there is only one sender task, *Task1,* and one receiver task, *Task2.*

Source 16.10 Two Tasks That Use a Global Variable to Communicate

```
/*****************************************************************************
* TASK #1
*****************************************************************************/
static void Task1(void *pdata){

    INT8U TimCntr = 0;                      /* 100ms Counter              */

    FOREVER() {

        OSTimeDly(100);                     /* Task period = 100ms        */
        TimCntr++;                          /* Increment counter          */
```

```
        DispBuf = TimCntr;                      /* Copy to display buffer    */
    }
}

/**********************************************************************
* TASK #2
***********************************************************************/
static void Task2(void *pdata){

    PUTCHAR('\r');                              /* Initialize to Column 1    */

    FOREVER() {
        OSTimeDly(100);                         /* Update display every 100ms */
        PUTCHAR('\r');
        PUTSTRG("    ");                        /* Clear previous display    */
        PUTCHAR('\r');
        OUTDECB(&DispBuf);
    }
}

/**********************************************************************/
```

Note _____

Notice that *Task2* does not run at an accurate 100ms period. The display output requires ~6ms, which is longer than the kernel timer tick. As discussed in Section 16.4, if the task execution time is longer than one timer tick, the task period will be longer, because one or more ticks are missed. However, since the *TimCntr* is actually updated in *Task1,* the long-term accuracy of the display is good. Occasionally the display routine will miss a timer value and the display will skip a count.

Problems with global variables start to show up as soon as the messages are longer than 16 bits or when the receiver cannot keep up with the messages sent.

When the message is more than 16 bits, it requires multiple assembly instructions for the sender to write the message to the buffer. If the sender is preempted in the middle of the write process, the buffer is caught in an intermediate state. Part of it is updated, while another part is old. If the preemption then causes the display task to run, or if it causes another sender to write to the display buffer, the displayed message will be corrupted.

The write process in this case is an example of a *critical region*. To avoid its being preempted, interrupts must be disabled while the write process is running. Another way to protect a shared resource is to use a semaphore as a key to limit access to one task at a time. We will cover semaphores in the next section.

Figure 16.9 Communication between Two Tasks via a Global Variable *DispBuf*

Global Variables as Event Signals. Another problem with global variables is that they do not make efficient software event signals in a μC/OS application. We saw in Chapter 6 that variables are one way to realize software events between asynchronous routines. In order to detect an event, the servicing routine is required to poll the variable with an event loop.

Our μC/OS tasks can poll a global variable to detect an event, but the polling process requires some CPU load. If the event loop is a free-running loop, then the task requires a CPU load of one and no lower-priority task will run. Therefore we would have to use a timed event loop to poll the variable at a lower rate.

We can get rid of the whole polling loop by using a μC/OS semaphore as an event flag. Until the semaphore flag is signaled, the servicing task is in the *WAITING* state and requires no CPU load.

Global variables by themselves are good for *as-long-as* applications such as communicating operating states or modes. These applications require a task to do something based on the current value of a variable and do not need to detect a change in the variable. Again the global variable must be less than 16 bits. There are a few examples of this type of global variable in the stopwatch program in Section 16.6.

16.5.2 Semaphores

Semaphores are the most important kernel service for intertask communications. Any full-featured preemptive kernel should include counting semaphores that allow multiple tasks to signal or wait for a semaphore.

Just about all the μC/OS code examples shown so far use a semaphore. This illustrates how widely used and important semaphores are in a preemptive kernel. In fact, any desired intertask communication can be implemented with a combination of a semaphore and a global variable. In this subsection we look at μC/OS semaphores in more detail.

Semaphore Service Functions. Table 16.2 shows the μC/OS semaphore service functions. For a more complete description, refer to Appendix C. Generally, most applications use *OSSemCreate()*, *OSSemPost()*, and *OSSemPend()*. Occasionally *OSSemAccept()* is used in an application, whereas *OSSemQuery()* is normally used only for debugging an application.

In order to use a semaphore, it must be created first by using the *OSSemCreate()* function. To create a semaphore, there must be an event control block available. An event control block is a data structure used by the kernel for semaphores, message mailboxes, and queues. The number of event control blocks is determined when μC/OS is built.

TABLE 16.2 μC/OS SEMAPHORE SERVICE FUNCTIONS

Semaphore Service Functions	Description
OSSemAccept()	Check the status of a semaphore.
OSSemCreate()	Create a semaphore.
OSSemPend()	Wait for a semaphore.
OSSemPost()	Signal a semaphore.
OSSemQuery()	Obtain information about a semaphore.

Following are the two lines from Source 16.1 used to create the semaphore called *SecFlag:*

```
OS_EVENT *SecFlag;              /* A one-second flag semaphore   */

...

SecFlag = OSSemCreate(0);       /* Create a semaphore flag       */
```

The first line defines a pointer that will point to the event control block used for a semaphore called *SecFlag.* The event control block is a structure of type *OS_EVENT.* The second line configures one of the event control blocks as a semaphore and returns the address of that block. Therefore this line makes *SecFlag* point to the event control block used for this semaphore.

The argument in the *OSSemCreate()* service call is used to initialize the value of the semaphore. The initial value of a semaphore depends on the application. If the semaphore is used as an event flag, the initial value should be zero. If the semaphore is used as a resource key, the initial value should be 1 or more.

OSSemPost() is used to *signal* the semaphore. It first checks to see whether there are tasks waiting for the semaphore. If there are, *OSSemPost()* moves the highest-priority task that is waiting to the *READY* state and returns. All other tasks waiting for that semaphone remain in the *WAITING* state. If there are no tasks waiting for the semaphore, *OSSemPost()* increments the semaphore value and returns.

When the semaphore is used as an event flag, *OSSemPost()* is used to tell the semaphore that the event occurred. If the semaphore is used as a resource key, *OSSemPost()* is used to return the key so that another task can access the shared resource.

OSSemPend() makes a task wait until the semaphore is signaled. When *OSSemPend()* is called, it checks the value of the semaphore. If it is zero, *OSSemPend()* moves the task to the *WAITING* state to wait for the semaphore to be signaled. If the semaphore is greater than zero, it decrements the semaphore value and returns to run the task.

When the semaphore is used as an event flag, *OSSemPend()* is used to make the task wait until the event occurs. When the semaphore is used as a resource key, *OSSemPend()* is used to wait for the resource key. When the key is available, *OSSemPend()* takes the key by decrementing the semaphore and returns the calling task to the *READY* state so it can proceed.

Semaphores as Event Flags.   One of the primary applications for a semaphore in real-time systems is to implement intertask event flags. When an event is generated or detected by one task or an ISR, a semaphore can be used to signal other tasks that may be waiting to service the event. In addition event flags can be used to synchronize two tasks with each other.

For example the *UcosDemo* program uses a semaphore as a periodic event flag. *Task1* signals the semaphore, *SecFlag,* every second. *Task2* waits for *SecFlg* to be signaled in order to execute its task code. The data flow diagram for the event flag is shown in Figure 16.10.

To better understand the operation of the *SecFlag* semaphore in *UcosDemo,* let's look at the context diagram shown in Figure 16.11. This diagram shows the state for *Task1* and *Task2* during the time that the one-second event occurs. The diagram starts out

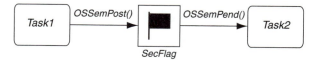

Figure 16.10 Data Flow Diagram for the Semaphore Event Flag Used in *UcosDemo*

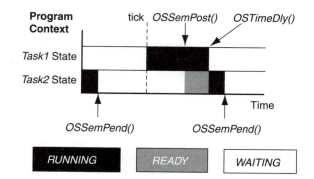

Figure 16.11 Context Diagram for the Semaphore in *UcosDemo*

showing *Task1* waiting for the next timer tick and *Task2* running. *Task2* calls *OSSemPend()* to wait for the one-second event. At this point, we assume that the semaphore is zero, which means that there are no pending events.

Because the semaphore is zero, *Task2* goes to the *WAITING* state to wait for *SecFlag* to be signaled. Then when the next timer tick occurs, *Task1* starts running. *Task1* determines that one second has expired, so it calls *OSSemPost()* to signal *SecFlag*. *OSSemPost()* moves *Task2* to the *READY* state, and then, since *Task1* is still the highest-priority task, it returns to *Task 1*. When *Task1* is finished, it executes an *OSTimeDly()* and goes back to the *WAITING* state. *Task2* is now the only task ready, so it is executed to service the semaphore.

Semaphores as Resource Keys. When a resource is shared in a preemptive kernel, we need to make sure that only one task has access to the resource at a time. A shared resource may be a global data object, a peripheral device, or a function that accesses a data object or peripheral. To illustrate this, Source 16.11 shows two tasks that need to write to an LCD display.

The first task is an ADC processing task that needs to display a processed ADC sample. The sample is to be displayed on the first row and first column of the LCD. First *LcdMoveCursor()* is called to move the cursor to the correct position. Then *LcdDispDecByte()* is called to display the sample in decimal.

Source 16.11 Two Tasks That Write to a Shared LCD Display without a Key

```
/****************************************************************************/
static void AtoDProcTask(void *pdata){

    ...
```

```
    FOREVER() {

        ...

        LcdMoveCursor(1, 1);
        LcdDispDecByte((UBYTE)disval);              /* Display sample    */
    }
}
/*******************************************************************/
static void ETimeTask(void *pdata){

    ...

    FOREVER() {

        ...

        LcdMoveCursor(2, 1);                        /* Display HH:MM:SS      */
        LcdDispTime(hours,minutes,seconds);
    }
}

/*******************************************************************/
```

The second task is used to display the elapsed time on the first column of the second row. Again it calls *LcdMoveCursor()* to position the cursor and then calls *LcdDispTime()* to format and display the time. Figure 16.12 shows the correct display format for these two tasks.

Since these tasks are running in a preemptive kernel, it is possible that one task preempts the other while it is in the process of displaying data. Figure 16.13 shows the result of such an error. In this case the elapsed time task was just finished displaying **00:5** when it was preempted by the ADC task. The ADC task then moved the cursor up the first row and displayed its sample, **123**. When the ADC task was finished, the elapsed time task continued, but since there is no change in cursor position, the rest of the time characters are displayed after the ADC sample.

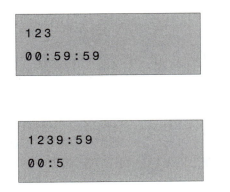

Figure 16.12   Correct LCD Display

Figure 16.13   LCD with a Preempted Display Task

Actually this is a rather mild error compared to what could happen if the display process was preempted at a more critical moment. The result could have been a display with unreadable flickering characters.

In order to avoid this type of error, we need to make sure that only one task writes to the LCD at one time. The normal way to do this is with a resource key implemented by a semaphore.

Source 16.12 shows how we can rewrite the two tasks shown previously using a resource key. Before either task writes to the display, it waits for a semaphore called *LcdKey* with an *OSSemPend()*. Once the task finishes with the display code, it signals the semaphore with an *OSSemPost()*.

Source 16.12 A Semaphore Key Used to Protect a Shared LCD Display

```
/**********************************************************************/
static void AtoDProcTask(void *pdata){

    …

    FOREVER() {

        …

        OSSemPend(LcdKey, 0, &err);        /* Get LCD display key      */
        LcdMoveCursor(1, 1);
        LcdDispDecByte(disval);            /* Display sample    */
        OSSemPost(LcdKey);                 /* Return LCD display key   */
    }
}

/**********************************************************************/
static void ETimeTask(void *pdata){

    …

    FOREVER() {

        …

        OSSemPend(LcdKey, 0, &err);        /* Get LCD display key      */
        LcdMoveCursor(2, 1);               /* Display HH:MM:SS         */
        LcdDispTime(hours,minutes,seconds);
        OSSemPost(LcdKey);                 /* Return LCD display key   */
    }
}

/**********************************************************************/
```

The data flow diagram for this system is shown in Figure 16.14. To send data using the LCD routines, the two tasks must first acquire the semaphore key, *LcdKey*. The LCD routines then write to the LCD.

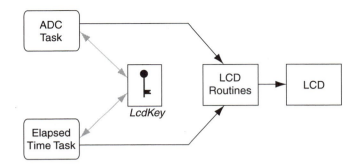

Figure 16.14 Semaphore Key Data Flow

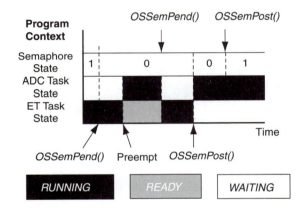

Figure 16.15 Context Diagram for the Resource Key Used in Source 16.12

Let's look at what would happen if the ADC task tried to display while the elapsed time task was in the process of displaying. Figure 16.15 shows a context diagram for this discussion.

First a semaphore that is used as a resource key is always initialized with a value of 1. When the semaphore is one, we can think of the key as being available in the semaphore.

At the start of the diagram, the elapsed time task is running and needs to display the time to the LCD. So it calls *OSSemPend()* to get the *LcdKey*. *OSSemPend()* decrements the semaphore to zero, which in effect is taking the key. Since the semaphore was nonzero, control is passed back to the elapsed time task and it continues with the display routines.

After the elapsed time task displays its first four characters, it is preempted by the ADC task. The ADC task then calls *OSSemPend()* to display its sample. This time the semaphore is zero, so the task is moved to the *WAITING* state. In effect the key was not in the semaphore, so it needs to wait for it. Once it goes to the *WAITING* state, the elapsed time task can run again and proceed with its display.

After the elapsed time task finishes the display routines, it calls *OSSemPost()* to return the key to the semaphore. But since there is a task waiting for the key, the semaphore remains zero and the ADC task gets the key and runs its display routines. Once it is finished with the display, it calls *OSSemPost( )* to return the key. This time there is no task waiting, so the semaphore is incremented, which means the key is now available for another task.

Added Protection through Information Hiding.   The problem with the code shown in Source 16.12 is that it relies on the programmer to include the semaphore code before and after running a display routine. There is nothing that restricts the programmer from writing to the display without using the key. This leaves too much room for error.

A better way to do this is to limit access to the shared resource to routines that handle the key themselves. The first step to do this is to put all the LCD routines into a separate module. This is the best way to organize the routines anyway. Then limit the public routines to those that use the key. All other LCD routines would be declared as private, so no other task could access them.

For the example given, we could write a routine called *LcdDispSample( )* and another called *LcdDispETime( )*. These functions would wait for the LCD key before trying to update the display. The programmer only has to call one function and all the resource protection is provided by that function. Source 16.13 shows what these functions might look like and how they are called. Notice that the programmer writing the task code does not have to remember to wait for the resource key.

Source 16.13 Access to a Shared Resource with Semaphore-Aware Functions

```
/*********************************************************************
 * In the main module, ADC module, or ETime module
 *********************************************************************/
static void AtoDProcTask(void *pdata){

    ...

    FOREVER() {

        ...

        LcdDispSample(disval);
    }
}


/*********************************************************************/
static void ETimeTask(void *pdata){

    ...

    FOREVER() {

        ...

        LcdDispETime(hours,minutes,seconds);
    }
}
```

```
/*****************************************************************************
* In the LCD module:
******************************************************************************
* Public Resources
*****************************************************************************/
void LcdDispSamp(UBYTE disval);
void LcdDispETime(UBYTE hours, UBYTE minutes, UBYTE seconds);


/*****************************************************************************
* Private Resources
*****************************************************************************/
static void LcdMoveCursor(UBYTE row, UBYTE col);
static void LcdDispTime(UBYTE hours, UBYTE minutes, UBYTE seconds);
static void LcdDispDecByte(UBYTE data);
 ...


/*****************************************************************************/
void LcdDispSamp(UBYTE disval){
    UBYTE err;

    OSSemPend(LcdKey, 0, &err);              /* Get LCD display key        */
    LcdMoveCursor(1, 1);
    LcdDispDecByte(disval);                  /* Display sample    */
    OSSemPost(LcdKey);                       /* Return LCD display key    */
}

/*****************************************************************************/
void LcdDispETime(UBYTE hours, UBYTE minutes, UBYTE seconds){
    UBYTE err;

    OSSemPend(LcdKey, 0, &err);              /* Get LCD display key        */
    LcdMoveCursor(2, 1);                     /* Display HH:MM:SS           */
    LcdDispTime(hours,minutes,seconds);
    OSSemPost(LcdKey);                       /* Return LCD display key    */
}

/*****************************************************************************/
```

Two important characteristics of the µC/OS semaphores that we have not covered are timeouts and error handling. Notice that the *OSSemPend()* has two additional parameters besides the semaphore name. The second parameter is a timeout in ticks. If it is zero, there is no timeout. This is used for fault protection, in case the source of a semaphore fails. Without a timeout, a failed event source can cause a task to block indefinitely. Timeouts can also be used to make a task periodically execute some code, while waiting for a semaphore.

The third parameter in *OSSemPend()* is a reference to an error code location. If an error is encountered in the *OSSemPend()* function, it will place an error code into this location. For more information about the error codes, refer to Appendix C.

Because we have not used these two parameters in our examples does not mean that they are not important. They are important features that help us to use defensive programming techniques. The error handling code and timeouts are not included in the examples so we can focus on the normal operation of the semaphores themselves.

16.5.3 Sending Messages with a Semaphore and Global Variables

In the earlier subsection on global variables we saw two problems with using a global variable to communicate between tasks: (1) the receiver task had to poll the variable with an event loop and (2) the global variable had to be limited to 16 bits or it might end up with corrupted data. In this subsection we look at using semaphores with the global data to solve these problems.

Simple Task Synchronization.　First we can use a semaphore to signal the receiver that the global data have been changed. Figure 16.16 shows the data flow diagram that uses a semaphore to signal the receiver when new data are ready. With this design, the receiver task does not have to poll the global, it only has to wait for the semaphore. Now the receiver task has a CPU load of zero until new data are actually ready.

This system solves the shared resource problem if these are the only two tasks that access the global data, and the receiver is guaranteed to be finished with the data before the sender task changes them again. This may require that the global data be a copy of the actual data being processed by the sender task. For this reason we normally refer to the global data as a *buffer.*

Full-Task Synchronization.　For a more general solution that will solve the shared resource access problem for multiple sender tasks, we need to add another semaphore to implement full-task synchronization. Figure 16.17 shows the data flow diagram for using two semaphores to implement full-task synchronization.

First a sender task waits for the buffer to become available by pending on the data acknowledge flag. Once the buffer becomes available, the sender task will copy its message into the message buffer. When it is done loading the message into the buffer, it signals the data ready flag so the receiver task knows there are new data.

The receiver task waits for new data by pending on the data ready flag. Once new data are ready and a sender task signals the data ready flag, the receiver task processes the data in the buffer. Once it is finished, it signals the data acknowledge flag to indicate to the sender

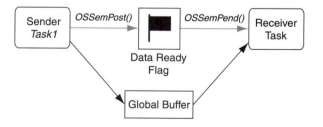

Figure 16.16　Sending a Message Using a Semaphore for Simple Task Synchronization

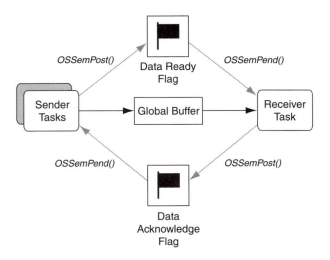

Figure 16.17 Message Sending Using Two Semaphores for Full-Task Synchronization

tasks that it is ready for new data. At this point, if there are more than one sender tasks pending on the data acknowledge flag, the highest-priority task waiting will be made ready to run. The others will have to wait for the receiver to signal the data acknowledge flag again.

Source 16.14 shows a program that demonstrates this method. There are four sender tasks and one receiver task that displays the messages sent by the sender tasks on an LCD. Only one sender task, *SendATask()*, is shown. The other three sender tasks are identical except that they send a unique message header and they are executed at different periods. The message header is used by the display routine to place the message at different locations on the LCD. The sender task periods are different, so there will be times when all four sender tasks want to send data to the receiver at the same time. To keep the display readable, the sender task periods should be greater than 500ms. An example would be to set the sender task A period to 653ms, the sender task B period to 710ms, the sender task C period to 597ms, and the sender task D period to 800ms.

Source 16.14 Message Communication Using Full-Task Synchronization

```
/**************************************************************************
*Sender Task A
**************************************************************************/
static Void SendATask(void *pdata){
    UBYTE i, Cntr;
    UBYTE err;
    FOREVER() {
        OSTimeDly(653);                     /* Task period                 */
        DBUG_PORT | = DBUG_TSK1;            /* Set debug bit               */
        OSSemPend(MsgAckFlag, 0, &err);    /* Wait for free display buffer */
        Cntr++;                             /* Fill buffer with string     */
        if(Cntr > 9){
            Cntr = 0;
        }
```

```
        MsgBuf[0] = 'A';
        for(i = 1; i <= 6; i++){
            MsgBuf[i] = Cntr + 0;
        }
        MsgBuf[7] = 0;
        OSSemPost(MsgRdyFlag);              /* Signal message ready       */
        DBUG_PORT &= ~DBUG_TSK1;            /* Clear debug bit            */
    }
}
/**************************************************************************/
...
// Sender tasks B, C, and D
...
/**************************************************************************
* RECEIVER TASK
**************************************************************************/
Static void RcvTask(void *pdata){
    UBYTE err;

    FOREVER() {
        OSSemPend(MsgRdyFlag, 0, &err);
        DBUG_PORT |= DBUG_TSK5;            /* Set debug bit              */
        OSSemPend(LcdKey, 0, %err);         /* Get LCD resource key       */
        switch(MsgBuf[0]) {                 /* Get message header and . . .*/
        case('A'):                          /* . . .set position          */
            LcdMoveCursor(1,1);
            break;
        case('B'):
            LcdMoveCursor(1,9);
            break;
        case('C'):
            LcdMoveCursor(2,1);
            break;
        case('D'):
            LcdMoveCursor(2,9);
        }
        LcdDispStr(MsgBuf);                 /* Display message string     */
        OSSemPost(LcdKey);                  /* Return LCD resource key     */
        DBUG_PORT &= ~DBUG_TSK5;            /* Clear debug bit            */
        OSSemPost(MsgAckFlag);             /* Signal free display buffer */
    }
}


/**************************************************************************/
```

In this program the data ready flag semaphore is called *MsgRdyFlag* and the data acknowledge semaphore is called *MsgAckFlag*.

This full-synchronization method is a versatile technique. It can be used to send messages of any size, it works with volatile data as long as the global buffer contains a copy of

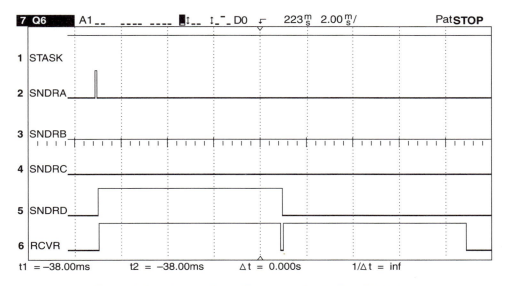

Figure 16.18   A Scope Trace That Shows Sender Task D Blocked

the data and is not the original, and it works for multiple sender tasks. The primary problem with it is that it is a blocking technique. All sender tasks are blocked until the buffer becomes available again. And since the receiver task involves sending data to a slow display, a sender task may be blocked for a significant amount of time.

For example consider a case in which we required a sender task that executed at an exact period. This system would not work because the sender task may be blocked for several milliseconds. If it is blocked longer than one timer tick, its execution period is no longer accurate.

To illustrate this Figure 16.18 shows a mixed-signal oscilloscope trace at a time when both sender task A *(SNDRA)* and sender task D *(SNDRD)* need to send a message to the receiver task to display a message. The sender trace is high when it is ready to send a message and while it is copying the message into the buffer. In this case *SNDRA* has the highest priority, so it runs and copies a message into the display buffer. *SNDRD* has the next-highest priority, so it is made ready to run, but since the receiver task *(RCVR)* has not processed the *SNDRA* message, it is blocked at the pending on *MsgAckFlag* service call. This puts *SNDRD* into the *WAITING* state so the receiver task can run. Once the receiver task has completed the display update for *SNDRA,* it signals the *MsgAckFlag* and *SNDRD* copies data into the display buffer. The receiver task then updates the display with the message from *SNDRD*.

In this example *SNDRD* was blocked for about 8ms. If it was supposed to be doing other things during this time, this system would not work.

Nonblocking Message Passing.   To implement a nonblocking message passing scheme that can have multiple sender tasks and/or sender tasks that transmit messages in bursts, we need to remove the wait for the data acknowledge flag. But if this wait is removed, the message buffer is no longer protected from being accessed by multiple tasks at one time, and we are back in a situation in which buffer data can be corrupted. So to implement a

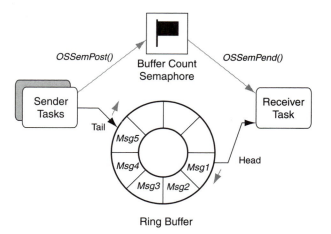

Figure 16.19   Nonblocking Message Passing with a Ring Buffer Message Queue

nonblocking message passing scheme, we also need to have multiple message buffers. Let's look at implementing a nonblocking message passing technique that uses a counting semaphore and a ring buffer configured as a message queue. Figure 16.19 shows the data flow diagram for a system that uses a ring buffer and a counting semaphore.

In this system there is an array of message buffers. A sender task copies its message into the next available message buffer in the array and signals the buffer count semaphore. The receiver task then pends on the buffer count semaphore, which indicates the number of messages to be processed. The messages are processed in the order sent by having the ring buffer head follow the ring buffer tail. Note that when the buffer count semaphore is equal to zero, the head is pointing to the same message buffer as the tail. The size of the ring buffer must be large enough so that the tail never raps around and reaches the head. Otherwise the oldest data will be destroyed by the newest data.

Source 16.15 shows one way to implement a ring buffer message queue. This system is designed specifically for the mesages used in Source 16.14. First a structure is defined for a single message, *MSG*. This structure shows that the message includes a header, *hdr*, and a 6-byte message, *msg[6]*. By defining the message structure, it is easier to define a ring buffer designed for this type of message.

The ring buffer is also defined in a structure, *BUFFER*. Its members include the write (tail) index, *wrindx*, a read (head) index, *rdindx*, a pointer to a counting semaphore, **cntr*, a pointer to a semaphore used as a resource key for the buffer, **key*, and the buffer itself, *msgbuf[4]*, which is an array of four *MSG* structures. One instance of the ring is then defined as *DispBuffer.*

Source 16.15 Ring Buffer Definition and Access Routines

```
/*****************************************************************************
* Message and ring buffer definitions
*****************************************************************************/
typedef struct {
    UBYTE hdr;
```

```
    UBYTE msg[6];
} MSG
typedef struct {
    UBYTE wrindx;
    UBYTE rdindx;
    OS_EVENT *cntr;
    OS_EVENT *key;
    MSG msgbuf[4];
} BUFFER;                        */ Static display message buffer        */

BUFFER DispBuffer;

/*********************************************************************
* Ring Buffer write function
*********************************************************************/
UBYTE BufWrite(MSG *msg, BUFFER *buff){
    UBYTE err;

    OSSemPend(buff->key, 0, &err);
    buff->wrindx++;
    if(buff->wrindx > 3){
        buff->wrindx = 0
    }
    if(buff->wrindx == buff->rdindx){
        return 1;
    }
    (*buff).msgbuf[buff->wrindx] = *msg;
    OSSemPost(buff->cntr);
    OSSemPost(buff->key);
    return 0;
}

/*********************************************************************
* Ring Buffer read function
*********************************************************************/
void BufRead(MSG *msg, BUFFER *buff){
    UBYTE err;

    OSSemPend(buff->key, 0, &err);
    *msg = (*buff).msgbuf[buff->rdindx];
    buff->rdindx++;
    if(buff->rdindx > 3){
        buff->rdindx = 0;
    }
    OSSemPost(buff->key);
}
/*********************************************************************/
```

To access the ring buffer the sender tasks use the function *BufWrite()*, which copies a message into the ring buffer. A pointer to the message to be copied along with a pointer to the ring buffer are passed to *BufWrite()*. Accessing the ring buffer is itself a critical region. We would not want the process of accessing the buffer to be preempted by another task that is trying to access the buffer. So a resource key must be acquired before reading or writing to the ring buffer. Once the key is acquired, the *BufWrite()* function increments the ring buffer index to the next available message buffer. If the index is too large for the array, it is cleared so it will wrap around to the first element in the array. This is how the ring is implemented. The write index is also compared to the read index. If they are equal the buffer is full and the function returns a one to indicate an error. Once the write index has been set, the *BufWrite()* function copies the contents of the message into the ring buffer. Once the message is copied, the function signals the ring buffer counting semaphore and returns the ring buffer resource key.

The receiver task uses the *BufRead()* function to copy messages from the ring buffer to the display message buffer. A pointer to the message to be copied along with a pointer to the ring buffer are passed to *BufRead()*. It is assumed that the *BufRead()* function is only called after the ring buffer counting semaphore has been signaled (i.e., there is a message in the buffer). *BufRead()* first waits for the ring buffer resource key and then copies the message from the ring buffer into a message buffer. It then increments the read index and wraps it around to zero if required. It then returns the ring buffer resource key.

Source 16.16 shows a sender task to demonstrate the complete technique. The resulting display will be identical to the display from Source 16.14.

Source 16.16 A Demonstration Program for Nonblocking Message Passing

```
/*********************************************************************
* SENDER TASK A
*********************************************************************/
static void SendATask(void *pdata){
    UBYTE i, Cntr;
    UBYTE msg[6];
    MSG newmsg;                          /* Sender message buffer          */

    FOREVER() {
        OSTimeDly(653);                  /* Task period                    */
        DBUG_PORT |= DBUG_TSK1;          /* Set debug bit                  */
        Cntr++;                          /* Fill buffer with string        */
        if(Cntr > 9){
            Cntr = 0;
        }
        newmsg.hdr = 'A';
        for(i = 0; i <= 5; i++){
            newmsg.msg[i] = Cntr + '0';
        }
        newmsg.msg[6] = 0
        BufWrite(&newmsg, &DispBuffer);
        DBUG_PORT &= ~DBUG_TSK1;         /* Clear debug bit                */
    }
}
...
// Other sender tasks
...
/*********************************************************************
* RECEIVER TASK
```

```
***********************************************************************/
static void RcvTask(void *pdata){
    UBYTE err;
    MSG dmsg;                               /* Display buffer             */

    FOREVER() {
        OSSemPend(DispBuffer.cntr, 0, &err);
        DBUG_PORT |= DBUG_TSK5;             /* Set debug bit              */
        BufRead(&dmsg, &DispBuffer);
        OSSemPend(LcdKey, 0, &err);         /*Get LCD resource key        */
        switch(dmsg.hdr){                   /*Get message header and. . . */
        case('A'):                          /*. . . set position          */
            LcdMoveCursor(1,1);
            break;
        case('B'):
            LcdMoveCursor(1,9);
            break;
        case('C'):
            LcdMoveCursor(2,1);
            break;
        case('D'):
            LcdMoveCursor(2,9);
        }
        LcdDispStrg(&dmsg);                 /* Display message string     */
        OSSemPost(LcdKey);                  /* Return LCD resource key     */
        DBUG_PORT &= ~DBUG_TSK5;            /* Clear debug bit            */
    }
}
```

The difference between the program in Source 16.16 and the one shown in Source 16.14 is that the new program is nonblocking. The sending tasks can return to do other things after they copy the message into the buffer. Figure 16.20 shows a mixed-signal oscilloscope trace showing three sender tasks trying to send a message close to the same time. Using full-task synchronization, *SNDRC* and *SNDRA* would have been blocked while the output displayed previous messages.

In this subsection we have developed solutions for passing messages in simple systems to complex systems. The methods shown can be used for any message passing requirement.

16.5.4 Message Mailboxes

In the subsection on semaphores, we saw how to use a semaphore to communicate between two or more tasks. When using a semaphore, the message sent is binary—the semaphore is zero or nonzero. As we saw, binary messages are useful for signaling events or for implementing a resource key. They can also be useful for passing messages when used in conjunction with a global data buffer. A message mailbox is essentially the same as a noncounting semaphore that contains a pointer to a message. They are convenient for sending one of many nonvolatile messages, because the message does not have to be copied into a buffer first.

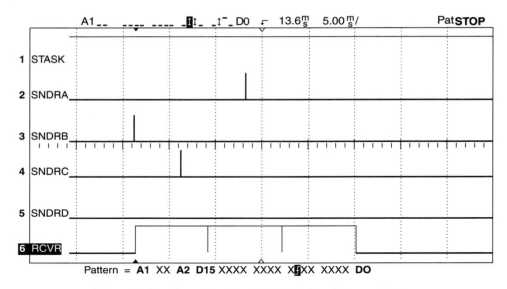

Figure 16.20   Traces Showing the Nonblocking Message Passing

Note

When we say "a message is placed in a mailbox," the message itself is not actually placed in the mailbox. Instead a pointer to the message is placed in the mailbox. This makes the mailbox flexible enough to handle messages of any size.

In addition the message pointer parameter in *OSMboxPost()* and the return value for *OSMboxPend()* and *OSMboxAccept()* are declared as generic pointers *(void *)*. This enables us to use a pointer to data of any type. By using a generic pointer, a mailbox is flexible enough to handle any type or size of message.

Message Mailbox Service Functions.   The µC/OS service functions used for message mailboxes are shown in Table 16.3. Notice the mailbox functions are essentially the same as the semaphore service functions.

To use a mailbox, we must first create it using *OSMboxCreate()*. Since *OSMbox-Create()* will configure an event control block as a mailbox, this requires a spare event control block. *OSMboxCreate()* also initializes the contents of the mailbox. In µC/OS a

TABLE 16.3   µC/OS MAILBOX SERVICE FUNCTIONS

Mailbox Service Functions	Description
OSMboxAccept()	Check for a message without waiting.
OSMboxCreate()	Create a message mailbox.
OSMboxPend()	Wait for a message.
OSMboxPost()	Send a message through a mailbox.
OSMboxQuery()	Obtain information about a mailbox.

mailbox is considered empty if it contains a null pointer and considered full if it contains anything else. Typically, a mailbox will be initialized with a null pointer to start the application with an empty mailbox.

OSMboxPost() is used by the sender task to send a message to another task through a mailbox. If no tasks are waiting for the message, the message pointer is placed in the mailbox. At this point the mailbox is considered full.

The receiving task can use *OSMboxAccept()* or *OSMboxPend()* to receive the message. *OSMboxAccept()* checks the mailbox and returns immediately with the pointer contained in the mailbox. If the mailbox was full, it returns the pointer to the message and, if the mailbox was empty, it returns a null pointer. When *OSMboxPend()* is called, the task is moved to the *WAITING* state until a message is sent to the mailbox. When a message is sent to the mailbox, the task is moved from the *WAITING* state to the *READY* state and the message pointer is returned by *OSMboxPend()*.

Message Server with Full Synchronization.   The µC/OS message mailbox and message queues are convenient alternatives to semaphores and global buffers when you want to send one of many nonvolatile messages. For example, to display one of many messages with a semaphore and global buffer, you would have to copy the message into the buffer and then signal the data-ready semaphore. When using a message mailbox in place of the data-ready semaphore, you can send the address of the message to the receiver task. Figure 16.21 shows the data flow diagram for mailbox-based message sending with full-task synchronization. This system is very similar to the system shown in Figure 16.17. Instead of placing a message into a buffer, this method simply sends a pointer to the message to the receiver task. Like the system in Figure 16.17, this system can block tasks while they wait for the acknowledge flag.

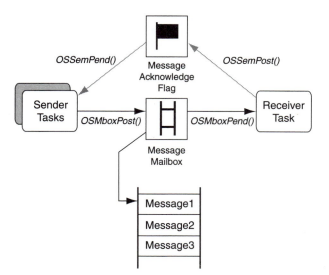

Figure 16.21   Mailbox-Based Message Server

Source 16.17 shows a demonstration program that uses this method. Two sender tasks send one of three constant messages. After waiting for the acknowledge semaphore *DispAck*, the sender task sends the address of one of three messages to the receiver task via a message mailbox, *DispMbox*.

The receiver task pends on the mailbox. When a message is received, it accesses the message using the pointer passed in the mailbox. Once it is finished displaying the message, it signals the acknowledge semaphore to indicate that it is ready for a new message.

Source 16.17 Mailbox-Based Message Passing with Full-Task Synchronization

```
/*****************************************************************************
* SENDER TASK A
*****************************************************************************/
static void SendrATask(void *pdata){
    UBYTE msgcnt = 0;
    UBYTE err;

    FOREVER() {
        OSTimeDly(603);                  /* Sender A period           */
        DBUG_PORT |= DBUG_TSK1;
        OSSemPend(DispAck, 0, &err); /*Wait for Mbox access           */
        msgcnt++;                        /* Choose and post message   */
        switch(msgcnt){
            case 1:
                OSMboxPost(DispMbox, &MsgA1[0]);
                break;
            case 2:
                OSMboxPost(DispMbox, &MsgA2[0]);
                break;
            case 3:
                OSMboxPost(DispMbox, &MsgA3[0]);
                msgcnt = 0;
        }
        DBUG_PORT &= ~DBUG_TSK1;
    }
}
/*****************************************************************************
* SENDER TASK B
*****************************************************************************/
static void SendrBTask(void *pdata){
    UBYTE msgcnt = 0;
    UBYTE err;

    FOREVER() {
        OSTimeDly(750);                          /* Sender B period       */
        DBUG_PORT |= DBUG_TSK2;
        OSSemPend(DispAck, 0, &err);             /* Wait for Mbox access  */
        ...
        /* Choose and post message */
        ...
        DBUG_PORT &= ~DBUG_TSK2;
    }
}
/*****************************************************************************
```

```
* RECEIVER TASK
********************************************************************************/
static void RcvrTask(void *pdata){
    UBYTE err;
    UBYTE *dptr;

    FOREVER() {
        dptr = OSMboxPend(DispMbox, 0, &err);
        DBUG_PORT |= DBUG_TSK3;                 /* Set debug bit              */
        OSSemPend(LcdKey, 0, &err);             /* Get LCD resource key       */
        switch (*dptr){                         /* Get message header and  ...*/
        case ('A'):                             /* ... set postion            */
            LcdMoveCursor(1,1);
            break;
        case ('B'):
            LcdMoveCursor(1,9);
        }
        LcdDispStrg(dptr);                      /* Display message string     */
        OSSemPost(LcdKey);                      /* Return LCD resource key     */
        DBUG_PORT &= ~DBUG_TSK3;                /* Clear debug bit            */
        OSSemPost(DispAck);                     /* Signal free display buffer */
    }
}
/********************************************************************************/
```

There are a couple of potential problems with this mailbox system. First, since the address of the original data is sent to the receiver, the data must be nonvolatile. If the data are shared variables that can be accessed by other tasks or by the sender task, they must also be protected by a resource key or corrupted data may be displayed. If we send the address of a copy of the volatile data, we might as well use the system described previously using a semaphore and global buffer.

The second potential problem is that it is a blocking system because of the full-task synchronization. To solve this problem we will need to use a message queue.

16.5.5 Message Queues

Message queues are an array of mailboxes that can be used as a FIFO (first-in-first-out) buffer or LIFO (last-in-first-out) buffer for sending multiple nonvolatile messages. Normally they are used as FIFO buffers so the messages are received in the same order as they were sent.

Message queues are important for systems that have one or more sender tasks that cannot be blocked. Like message mailboxes, queues by themselves should only be used for nonvolatile messages. When combined with dynamically allocated memory, they are also important for sending volatile messages. A message queue combined with a dynamic memory buffer is equivalent to using a counting semaphore and a ring buffer message queue.

Message Queue Service Functions. The service functions used for the μC/OS queue service are shown in Table 16.4. Again, for a more detailed description of these functions, refer to Appendix C.

TABLE 16.4 µC/OS MESSAGE QUEUE SERVICE FUNCTIONS

Queue Service Functions	Description
OSQCreate()	Create a message queue.
OSQFlush()	Empty the contents of a queue.
OSQPend()	Wait for a message in a queue.
OSQPost()	Send a message through a queue (FIFO).
OSQPostFront()	Send a message through a queue (LIFO).
OSQQuery()	Obtain information about a queue.

In order to create a queue with *OSQCreate(),* we need an event control block and memory allocated for the queue itself. Unlike semaphores and message mailboxes in which everything is contained in the event control block, queues require allocated space for an array of pointers. This is where the message pointers go. The event control block controls access to the array based on the service function called.

Source 16.18 shows a program with two tasks that demonstrates the use of a µC/OS message queue. First in the *Event Definitions,* an event pointer is allocated and the pointer array is allocated. The size of the allocated pointer array, *OutStrgPtrs[],* is *OUTSTRGQ_SIZE,* which determines the queue size. In the main function, *OSQCreate()* is used to configure and initialize an event control block for the queue. The first argument in *OSQCreate()* is the first address of the pointer array and the second argument is the size of the queue. Because the size of the queue must agree with the size designated in the definition, the defined constant, *OUTSTRGQ_SIZE,* is used.

Source 16.18 A µC/OS Queue Demonstration Program

```
/********************************************************************
* Event Definitions
********************************************************************/
OS_EVENT *OutStrgQ;              /* Serial Output message queue    */
void *OutStrgPtrs[OUTSTRGQ_SIZE]; /* Queue pointer storage array    */

    ...

/********************************************************************
* main()
*
********************************************************************/

void main(void) {

    OSInit();                    /* Initialize µC/OS-II             */

       ...

    OutStrgQ = OSQCreate(&OutStrgPtrs[0], OUTSTRGQ_SIZE);

    OSStart();                   /* Start multitasking              */
}
```

```
/*******************************************************************/

    ...

/********************************************************************
* Sendr1Task - A task that sends message to the serial output server
********************************************************************/
static void Sendr1Task(void *pdata){

    FOREVER() {
        OSQPost(OutStrgQ, (void *)"FIFO");
        OSTimeDly(10);
        OSQPost(OutStrgQ, (void *)"FMsg 1");
        OSQPost(OutStrgQ, (void *)"FMsg 2");
        OSQPost(OutStrgQ, (void *)"FMsg 3");
        OSTimeDly(2000);
        OSQPost(OutStrgQ, (void *)"LIFO");
        OSTimeDly(10);
        OSQPostFront(OutStrgQ, (void *)"LMsg 1");
        OSQPostFront(OutStrgQ, (void *)"LMsg 2");
        OSQPostFront(OutStrgQ, (void *)"LMsg 3");
        OSTaskSuspend(SENDR1_PRIO);
    }
}

/********************************************************************
* SoServrTask - Serial output server
********************************************************************/
static void OutStrgSrvrTask(void *pdata){

    UBYTE err;
    UBYTE *msgp;

    SCIOPEN();
    FOREVER() {

        msgp = OSQPend(OutStrgQ, 0, &err);
        OUTCRLF();
        PUTSTRG(msgp);
    }
}

/********************************************************************
```

The first task, *Sendr1Task()*, sends two bursts of messages to the second task, *OutStrgSrvrTask()*, through *OutStrgQ*. The first burst sends three messages with *OSQPost()*. This places the messages in FIFO order. The second burst sends three messages with *OSQPost-Front()*, which places the messages in LIFO order in the queue. This task has the highest priority, so it sends the message bursts before the queue can be read by *OutStrgSrvrTask()*.

The second task, *OutStrgSrvrTask()*, is a server task that outputs strings that are sent through the *OutStrgQ* queue. Because this task has a lower priority than *Sendr1Task()*, the messages will be sent before any of them are displayed.

The resulting display is

```
FIFO
FMsg 1
FMsg 2
FMsg 3
LIFO
LMsg 1
LMsg 3
LMsg 2
```

The messages sent in the first burst are displayed in the order that they were sent. This is the normal usage of a message queue. The messages in the second burst are displayed in a different order. When the first message, *LMsg1,* was sent, the *OutStrgSrvr-Task( )* was waiting for the queue, so the message pointer was passed directly to the waiting task. After that the other two messages were placed in the queue. Since the *OSQPost-Front( )* function was used, they were placed in the queue in reverse order. Thus *LMsg3* was displayed before *LMsg2.*

OSQPostFront( ) is normally used for exceptions to normal operation. For example if a fault was detected somewhere in the system, we might want to send a message to the display before something bad happens. If we use *OSQPost( ),* the message would have to wait for the messages in front of it in the queue to be displayed. By then it might be too late. By using *OSQPostFront( ),* the message would be placed in front of all other messages and would be the next message displayed.

► 16.6 μC/OS-BASED STOPWATCH PROGRAM

Throughout this text we have seen several versions of a stopwatch program to demonstrate real-time programming concepts. In this section we will look at how we might design the stopwatch using μC/OS. Recall that the stopwatch has a single button that cycles the watch through three modes:

$$\text{CLEAR} \rightarrow \text{COUNT} \rightarrow \text{STOP} \rightarrow \text{CLEAR}\dots$$

The elapsed time is displayed in minutes, seconds, and tenths-of-seconds.

Source 16.19 shows the complete source for the main stopwatch module. By now most of the pieces in the program should look familiar because we have used them in other examples.

16.6.1 Stopwatch Tasks

Let's start with a summary of the tasks that are used in the program. There are five tasks for the complete program, including the start-up task. The priorities assigned to each task follows the rate monotonic scheduling rule. Following are the task execution rates and the assigned priorities:

Task	Task Period	Priority
StartTask( )	One time only	4[*]
UpdateTimeTsk( )	1ms	6
ScanSwTsk( )	10ms	8
DispTimeTsk( )	100ms	10
TimerModeTsk( )	1/keypress	12

[*]Required to be the highest priority

Start-up Task, *StartTask().* The start-up task should look familiar. It starts by initializing the kernel timer with *OSTickInit( )*. It then initializes the LCD and creates the rest of the tasks. Once the rest of the tasks are complete, the start-up task suspends itself indefinitely.

Source 16.19 µC/OS Stopwatch Program

```
/*********************************************************************
* SWatchµCOS.c - A stopwatch program written to demonstrate and test
* µCOS-II. This version uses a counting semaphore and ring buffer for
* sending the time to the display task.
*
* Todd Morton, 05/29/00
*********************************************************************/
#include "includes.h"


/*********************************************************************
* Global Variables
*********************************************************************/
UBYTE MsCntr;                      /* ms counter for UpdateTimeTsk    */
TMODES TimMode;                    /* Stopwatch mode                  */
ETIME ETime;                       /* Elapsed time structure          */
DISPBUFF DispBuffer;


/*********************************************************************
*Ring Buffer Function Prototypes
*********************************************************************/
void BufRead(ETIME *et, DISPBUFF *buff);
UBYTE BufWrite(ETIME *et, DISPBUFF *buff);
void BufFlush(DISPBUFF *buff);


/*********************************************************************
* Event Definitions
*********************************************************************/
OS_EVENT *LcdKey;
```

```
OS_EVENT *SwFlag;

/***************************************************************************
 * Task Function Prototypes
 ***************************************************************************/

static void  StartTask(void *pdata);       /* Start-up task          */
static void  UpdateTimeTsk(void *pdata);   /* Timer update task      */
static void  ScanSwTsk(void *pdata);       /* Switch scan/debounce task */
static void  TimerModeTsk(void *pdata);    /* Timer mode state machine */
static void  DispTimeTsk(void *pdata);     /* Timer display routine, SCI */

/***************************************************************************
 * Allocate task stack space.
 ***************************************************************************/
OS_STK   StartTaskStk[START_STK_SIZE];
OS_STK   UpdateTimeTskStk[UPDATETIME_STK_SIZE];
OS_STK   ScanSwTskStk[SCANSW_STK_SIZE];
OS_STK   TimerModeTskStk[TIMERMODE_STK_SIZE];
OS_STK   DispTimeTskStk[DISPTIME_STK_SIZE];

/***************************************************************************
 * main()
 ***************************************************************************/
void main(void) {
    OSInit();                              /* Initialize µC/OS-II     */

    OSTaskCreate(StartTask, (void *)0,
                 (void *)&StartTaskStk[START_STK_SIZE],
                 START_PRIO);

    LcdKey = OSSemCreate(1);               /* LCD key                 */
    SwFlag = OSSemCreate(0);               /* Switch flag semaphore   */
    DispBuffer.cntr = OSSemCreate(0);
    DispBuffer.key = OSSemCreate(1);       /* Message acknowledge flag */

    OSStart();                             /* Start multitasking      */
}

/***************************************************************************
 * STARTUP TASK
 ***************************************************************************/
static void StartTask(void *pdata) {
    OSTickInit();                          /* Initialize the µcos ticker */
    LcdInit();                             /* Initialize LCD module     */

    OSTaskCreate(UpdateTimeTsk, (void *)0,
                 (void *)&UpdateTimeTskStk[UPDATETIME_STK_SIZE],
                 UPDATETIME_PRIO);
```

```
    OSTaskCreate(ScanSwTsk, (void *)0,
                 (void *)&ScanSwTskStk[SCANSW_STK_SIZE],
                 SCANSW_PRIO);
    OSTaskCreate(TimerModeTsk, (void *)0,
                 (void *)&TimerModeTskStk[TIMERMODE_STK_SIZE],
                 TIMERMODE_PRIO);
    OSTaskCreate(DispTimeTsk, (void *)0,
                 (void *)&DispTimeTskStk[DISPTIME_STK_SIZE],
                 DISPTIME_PRIO);

    FOREVER() {                                   /* Forever suspend           */
        OSTaskSuspend(START_PRIO);
    }
}

/***************************************************************************
 * UpdateTimeTask() - This task increments the time every tenth of a second
 *                    when stopwatch is in the COUNT mode.
 *                  - For an accuracy of 1ms, it runs with a period of 1ms
 *                    and increments the ms counter, msCntr.
 *                  - It updates ETime, a structure with Mins:Secs.Hms
 ***************************************************************************/
static void UpdateTimeTsk(void *pdata){

    ETime.hms = 0;                      /* Initialize elapsed time display  */
    ETime.secs = 0;
    ETime.mins = 0;
    BufWrite(&ETime, &DispBuffer);
    msCntr = 0;

    FOREVER() {
        OSTimeDly(1);                                /* 1ms period            */
        if(TimMode == COUNT){                        /* COUNT mode?           */
            msCntr++;                                /* Count ms              */
            if(msCntr == 100){                       /* Update ETime every 100ms */
                msCntr = 0;                          /* Update tenths         */
                ETime.hms++;
                if(ETime.hms == 10){                 /* Update seconds        */
                    ETime.secs++;
                    ETime.hms = 0;
                    if(ETime.secs == 60){            /* Update minutes        */
                        ETime.mins++;
                        ETime.secs = 0;
                        if(ETime.mins == 100){       /* Overflow to zero      */
                            ETime.mins = 0;
                        }
                    }
                }
                BufWrite(&ETime, &DispBuffer);       /* Display new ETime     */
```

```
                }
            }
        }
    }
}

/*************************************************************************
* ScanSwTsk()  - Detects and debounces an active-low switch on the SW_BIT
*                  bit on the SW_PORT port.
*                - With noise immunity
*                - With an execution period of 10ms it works with debounce
*                  times < 20ms.
*                - Posts SwFlag semaphore when a keypress is accepted
*************************************************************************/
static void ScanSwTsk(void *pdata){

    UBYTE cur_sw;                          /* Current switch position     */
    UBYTE SwState = SW_OFF;                /* Current switch debounce state*/

    FOREVER() {
        OSTimeDly(10);
        cur_sw = SW_PORT & SW_BIT;         /* Get current switch position */
        if(SwState == SW_OFF){             /* Wait for switch edge        */
            if(cur_sw == PRESSED){
                SwState = SW_EDGE;
            }
        }else if(SwState == SW_EDGE){      /* Verify switch press         */
            if(cur_sw == PRESSED){
                SwState = SW_VERF;         /* Switch press verified       */
                OSSemPost(SwFlag);         /* Signal valid switch press   */
            }else{                         /* False switch press, ignore  */
                SwState = SW_OFF;
            }
        }else if(SwState == SW_VERF){      /* Wait for release            */
            if(cur_sw == RELEASED){
                SwState = SW_OFF;
            }
        }
    }
}

/*************************************************************************
* TimerModeTsk() - Stopwatch mode state machine.
*                 - Cycles: CLEAR->COUNT->STOP->CLEAR...
*                 - Runs every valid keypress by waiting for SwFlag
*                   semaphore.
*************************************************************************/
static void TimerModeTsk(void *pdata) {
    UBYTE err;
```

```
    FOREVER() {
        OSSemPend(SwFlag, 0, &err);          /* Wait for keypress          */
        if(TimMode == CLEAR){                /* If CLEAR change to COUNT    */
            msCntr = 0;                      /* Restart ms count for accuracy*/
            TimMode = COUNT;
        }else if(TimMode == COUNT){          /* If COUNT change to STOP     */
            TimMode = STOP;
        }else{                               /* If STOP change to CLEAR     */
            TimMode = CLEAR;
            ETime.hms = 0;                   /* Clear ETime                 */
            ETime.secs = 0;
            ETime.mins = 0;
            BufWrite(&ETime, &DispBuffer;       /* Display new ETime        */
            BufWrite(&ETime, &DispBuffer;       /* and clear buffer         */
        }
    }
}


/************************************************************************
* DispTimeTsk() - Displays the ETIME structure received through DispBuffer.
************************************************************************/
static void DispTimeTsk(void *pdata) {
    UBYTE err;                               /* returned error codes       */
    ETIME *det;
    UBYTE ones;                              /* Temporary display vars     */
    UBYTE tens;

    FOREVER() {
        OSSemPend(DispBuffer.cntr, 0, %err);
        BufRead(&det, &DispBuffer);
        OSSemPEnd(LcdKey, 0, %err);          /* Get LCD resource key       */
        LcdMoveCursor(1,1);
        ones = (det.mins % 10) + '0';        /* Calc. minutes digits       */
        tens = (det.mins / 10) + '0';
        LcdDispChar(tens);                   /* Display minutes            */
        LcdDispChar(ones);
        LcdDispChar('.');                    /* mins:secs separator        */
        ones = (det.secs % 10) + '0';        /* Calc. seconds digits       */
        tens = (det.secs / 10) + '0';
        LcdDispChar(tens);                   /* Display seconds            */
        LcdDispChar(ones);
        LcdDispChar('.');                    /* secs:tenths separator      */
        LcdDispChar(det.hms + '0');          /* Display tenths             */
        OSemPost(LcdKey);                    /* Return LCD resource key    */
    }
}


/************************************************************************
```

```
* Ring Buffer write function
********************************************************************/
UBYTE BufWrite (ETIME *et, DISPBUFF *buff){
    UBYTE err;

    OSSemPend(buff->key, 0, &err);
    buff->wrindx++;
    if(buff->wrindx > 1){
        buff->wrindx = 0;
    }
    if(buff->wrindx == buff->rdindx){
        return 1;
    }
    (*buff).etbuf[buff->wrindx] = *et;
    OSSemPost(buff->cntr);
    OSSemPost(buff->key);
    return 0;
}
/********************************************************************
* Ring Buffer read function
********************************************************************/
void BufRead(ETIME *et, DISPBUFF *buff){
    UBYTE err;

    OSSemPend(buff->key, 0, &err);
    *et = (*buff).etbuf[buff->rdindx];
    buff->rdindx++;
    if(buff->rdindx > 1){
        buff->rdindx = 0;
    }
    OSSemPost(buff->key);
}
/********************************************************************/
```

Update Time Task, *UpdateTimeTsk().* This is the primary time-keeping task. It has the highest priority to keep the stopwatch accuracy within 1ms. At first it seems more complicated than it needs to be. For example we could give this task a period of 100ms and update the *ETime* structure each time through the task code. The result would be an error as great as 100ms, or one least-significant display digit, which may be fine for some applications. The reason for this error is that the stopwatch is started asynchronously to the execution of this task. So it may occur immediately before the 100ms delay expires or immediately after.

The task increments a global variable called *msCntr* every millisecond. When the stopwatch is started, the timer mode task resets this counter to zero, so the first time that *ETime* is updated is between 99 and 100ms after the button is pressed. The long-term error is good as long as the execution time of this task is less than 1ms. For this reason we could not include the display code in this task. So to initiate a display update, *ETime* is copied to the display ring buffer with *BufWrite().* This is the same method used in Source 16.15 and Source 16.16 on pages 628 and 630, respectively.

Switch-Scan and Debounce Task, *ScanSw()*.   The next task is the switch-scanning and debouncing task. We have seen this task many times throughout this text. The main requirement is that it has to run with a period that is at least one-half the switch bounce time. Since the task period is 10ms, it is designed for switch bounce times less than 20ms. It also rejects noise pulses up to 10ms wide. Notice that the task period does not have to be exactly 10ms. It can vary as much as 20% without causing significant errors.

When a valid keypress is accepted, *ScanSw()* signals a semaphore event flag, *SwFlag*. This flag can then be used by other tasks to service a keypress. In this application the timer mode task changes the mode each time the key is pressed.

Timer Mode State Machine Task, *TimerModeTsk()*.   This task is a simple state machine that controls the mode of the stopwatch. Each time a key is pressed, the *SwFlag* semaphore is signaled by the switch-scanning task. When *SwFlag* is signaled, this task makes a state transition and some actions based on the state change. For example when the state is changed from *CLEAR* to *COUNT,* the *msCntr* is cleared to restart the millisecond counter in the time update task. When the *CLEAR* mode is entered, the display must be cleared one time at the transition so it is done by this task. Notice that the buffer must be written to twice to clear old buffer contents.

Timer Display Task, *DispTimeTsk()*.   This display task displays the current elapsed time by waiting for a value to be written to the display ring buffer. It then uses *BufRead()* to copy the time value stored in the ring buffer into a local display buffer. By using the ring buffer technique, the other tasks will not be blocked to wait for the display.

▶ ## SUMMARY

In this chapter the preemptive kernel µC/OS was introduced. We looked at some basic techniques for designing programs that run under µC/OS, including using the timer and intertask communications services.

To simplify the examples, µC/OS was included as a precompiled library, and the timeout and error handling was not included. In this way we could focus on the new concepts covered in this chapter, such as task design for preemptive kernels, semaphores, message mailboxes, and message queues.

Using a full-featured kernel like µC/OS greatly simplifies the design of the tasks, especially when revising tasks for reuse in another project. The cost for this design simplicity is the requirement for more memory resources. When using a single-chip microcontroller, RAM is especially limited. Since the source code is normally available, however, we have the flexibility to build µC/OS specifically for use in systems with limited resources. With this in mind, using a preemptive kernel like µC/OS can significantly reduce the development time required to build a complex real-time system.

We did not cover all the services and capabilities of the µC/OS kernel here.[2] For example µC/OS has a service for dynamic memory allocation. This is an important addition,

---

[2]For more information on µC/OS, the book *MicroC/OS-II, The Real-Time Kernel* by Jean J. Labrosse (Lawrence, KS: R&D Books, 1999) is highly recommended. For application examples that use µC/OS, refer to *Embedded Systems Building Blocks* by Jean J. Labrosse (Lawrence, KS: R&D Books, 2000).

because the standard C function, *malloc()*, was not designed for a real-time preemptive kernel. The service functions for memory allocation are included in Appendix C. The configuration for the precompiled library used in this text is also shown in Appendix C, along with a more detailed description of the µC/OS service functions.

EXERCISES

1. What are the three scenarios in which a running task can be preempted?

2. Although we do not use *cooperative tasks* for a preemptive kernel, we still need to design tasks that cooperate to some extent. What is the "cooperation" required by the tasks designed for a preemptive kernel? Which kernel services can be used to realize this cooperation?

3. A project uses µC/OS for six tasks with priorities 4 through 9 available. Given the following tasks and task execution rates, assign priority values based on the rate monotonic scheduling rule:

Task	Task Period
StartTask()	One time only
TaskA()	1ms
TaskB()	20ms
TaskC()	500ms
TaskD()	10ms
TaskE()	1/keypress

4. After the system in Exercise 3 was tested, it was determined that *TaskD* had to have an exact period of 10ms, and with the current priorities, it was longer than 10ms. Readjust the priorities so *TaskD* will have an exact period of 10ms. Assume its execution time is just under one tick period.

5. Replace the *OSTimeDly()* call in Source 16.8 with the equivalent *OSTimeDlyHMSM()* call.

6. An on-chip input capture is to be used to count pulses on *Timer Channel 1*. Design the input capture's interrupt service routine so it will run under µC/OS and will signal the semaphore, *NewPulse,* every time a pulse is received. Include the code required to create the semaphone.

7. Design a task that increments a 16-bit variable, *PulseCnt,* to count the pulses from Exercise 6. Use the *NewPulse* semaphore for intertask communications.

8. Expand on the task in Exercise 7 to include a timeout so that the function *PulseError()* will be called if a pulse is not received within two seconds.

9. The following definition is used for a message to be displayed by a µC/OS program:

```
UBYTE HelloMsg[] = {"Hello Happy User"};
```

(a) Show the code required in a sender task and a receiver task to send a pointer to this message as a global variable. The receiver task waits for the pointer and displays it using the Basic I/O routine, *PUTSTRG()*. Include the code required to define and initialize the variable.

(b) Show the code required in the two tasks to send the message through a μC/OS mailbox. Include the code required to create and initialize the mailbox.

(c) Show the code required in the two tasks to send the message through a μC/OS queue. Include the code required to create and initialize a queue that can hold eight messages.

(d) Revise the code in Exercise 9a so the tasks use a semaphore as a resource key to access the global variable. Include the code required to create and initialize the semaphore.

10. The following two tasks send messages to the *OutStrgSrvrTask()* in Source 16.15. Assuming *Sendr1Tsk()* has priority 5, *Sendr2Tsk()* has priority 6, and *OutStrgSrvrTask()* has a priority of 10, what should the resulting display look like? (Assume *OutStrgQ* can hold eight messages.)

```
/*****************************************************************
* Sendr1Task - A task that sends a message to the serial output server
*****************************************************************/
static void Sendr1Task(void *pdata){
    FOREVER() {
        OSQPost(OutStrgQ, (void *)"MESSAGES");
        OSTimeDly(10);
        OSSemPost(OutFlag);
        OSQPost(OutStrgQ, (void *)"Sendr1, Msg1");
        OSQPost(OutStrgQ, (void *)"Sendr1, Msg2");
        OSQPost(OutStrgQ, (void *)"Sendr1, Msg3");
        OSTaskSuspend(SENDR1_PRIO);
    }
}
/*****************************************************************
* Sendr2Task - A task that sends a message to the serial output server
*****************************************************************/
static void Sendr2Task(void *pdata){
    UBYTE err;
    FOREVER() {
        OSSemPend(OutFlag, 0, &err);
        OSQPost(OutStrgQ, (void *)"Sendr2, Msg1");
        OSQPostFront(OutStrgQ, (void *)"Sendr2, Msg2");
        OSQPost(OutStrgQ, (void *)"Sendr2, Msg3");
        OSTaskSuspend(SENDR2_PRIO);
    }
}
/*****************************************************************/
```

11. A new design has the following requirements: 5 tasks, 4 semaphores, and 2 message queues. What are the minimum values for the following configuration constants to build • C/OS specifically for this project? Assume that the four highest priorities and four lowest priorities are reserved for the kernel.

```
OS_MAX_EVENTS
OS_MAX_QS
OS_MAX_TASKS
OS_LOWEST_PRIO
```

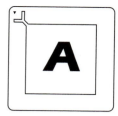

Programming Conventions

This appendix includes some basic conventions for writing assembly and C source code. This set of conventions is not meant to be the "best and final" coding rules. It is simply a place to start when creating your own or your company's conventions. Remember that the ultimate goal of coding conventions is the productivity of reliable programs. This means the code must be readable, revisable, and understandable. All of this starts with writing simple, straightforward programs instead of trying to be tricky and obscure.

▶ A.1 SOURCE FILES

1. Limit source lines to 80 columns. This line length is the longest that is compatible with most terminals and printer programs.
2. Avoid tabs. Use spaces only. A good editor will have tab-to-space conversion. This is important for portability. Tabs are set differently on just about every program and printer. In addition word processor tabs are set by length, not by number of characters, so tabs cause problems when inserting code into a document.

▶ A.2 CONVENTIONS FOR ASSEMBLY CODE

Following are some basic conventions for assembly programs:

Organization

1. Absolute/Single File Source. Use organization described in Chapter 3 for monitor-run programs. Use organization described in Chapter 10 for stand-alone programs.
2. All assembly sources must be at least source relocatable, as described in Chapter 5.

Comments

1. Comments should make the code easier to read and understand. Therefore they should not add clutter to the source and describe the source at a higher level.
2. Comments should not be a redundant description of a statement or instruction.
3. Comment headers are required at the start of a module, for each routine or function.

Naming Conventions

1.

Symbol Type	Convention
Equated constants	All caps with underscores to separate words.
Stored constants, and global variables	Mixed case. Multiple words are combined into one word with the first letter of each word capitalized.
Subroutine names	Mixed case. Multiple words are combined into one word with the first letter of each word capitalized.
All other labels	All lowercase with or without underscores to separate words.
Section names	All start with a period.

2. Data object names should be nouns, whereas routines should describe an action.

▶ A.3 CONVENTIONS FOR C CODE

Following are the basic conventions for C programs:

Naming Conventions

1.

Symbol Type	Convention
#define constants	All caps with underscores to separate words.
#define macros	All caps with underscores to separate words followed by ().
Stored constants, and global variables	Mixed case. Multiple words are combined into one word with the first letter of each word capitalized.
Function names	Mixed case. Multiple words are combined into one word with the first letter of each word capitalized.
Local variables	All lowercase with or without underscores to separate words.
typedef names	All caps.

2. Data object names should be nouns, whereas routines should describe an action.

Standard Type Definitions

```
typedef unsigned char INT8U;
typedef signed char INT8S;
typedef unsigned short INT16U;
typedef signed short INT16S;
```

```
typedef unsigned long INT32U;
typedef signed long INT32S;

#define ISR void __interrupt
#define UBYTE INT8U
#define SBYTE INT8S
#define UWYDE INT16U
#define SWYDE INT16S
```

Control Constructs

1. Indent three to six characters (four preferred).
2. Use comma operator instead of using assignments in conditional expressions.
3. Braces are always used for control structures, even if the code block consists of a single statement.

Function Names

1. Entry function is always called *main()*.
2. Function names should start with module name or abbreviation.
3. Task functions should always include the word *task* in the name.

Header Files

1. Master header file organization is described in Chapters 12 and 14.
2. Header files should contain items that are used across different modules.
3. Avoid declarations that allocate memory. These should *belong* to a specific module. The header file will then contain an *extern* declaration for public resources.

Comments

1. Comments should make the code easier to read and understand.
2. Comments should be indented with the code so programs blocks remain visual.
3. Comments should not be a redundant description of a statement or instruction.
4. Comment headers are required at the start of a module and for each routine or function.

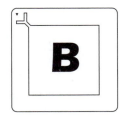

Basic I/O

$00	0	$01	1	$02	2	$03	3	$04	4	$05	5	$06	6	$07	7
NUL		SOH		STX		ETX		EOT		ENQ		ACK		BEL	
$08	8	$09	9	$0A	10	$0B	11	$0C	12	$0D	13	$0E	14	$0F	15
BS		HT		LF		VT		FF		CR		SO		SI	
$10	16	$11	17	$12	18	$13	19	$14	20	$15	21	$16	22	$17	23
DLE		DC1		DC2		DC3		DC4		NAK		SYN		ETB	
$18	24	$19	25	$1A	26	$1B	27	$1C	28	$1D	29	$1E	30	$1F	31
CAN		EM		SUB		ESC		FS		GS		RS		US	
$20	32	$21	33	$22	34	$23	35	$24	36	$25	37	$26	38	$27	39
SP		!		"		#		$		%		&		'	
$28	40	$29	41	$2A	42	$2B	43	$2C	44	$2D	45	$2E	46	$2F	47
(		)		*		+		,		−		.		/	
$30	48	$31	49	$32	50	$33	51	$34	52	$35	53	$36	54	$37	55
0		1		2		3		4		5		6		7	
$38	56	$39	57	$3A	58	$3B	59	$3C	60	$3D	61	$3E	62	$3F	63
8		9		:		;		<		=		>		?	
$40	64	$41	65	$42	66	$43	67	$44	68	$45	69	$46	70	$47	71
@		A		B		C		D		E		F		G	
$48	72	$49	73	$4A	74	$4B	75	$4C	76	$4D	77	$4E	78	$4F	79
H		I		J		K		L		M		N		O	
$50	80	$51	81	$52	82	$53	83	$54	84	$55	85	$56	86	$57	87
P		Q		R		S		T		U		V		W	
$58	88	$59	89	$5A	90	$5B	91	$5C	92	$5D	93	$5E	94	$5F	95
X		Y		Z		[		\		]		^		_	
$60	96	$61	97	$62	98	$63	99	$64	100	$65	101	$66	102	$67	103
`		a		b		c		d		e		f		g	
$68	104	$69	105	$6A	106	$6B	107	$6C	108	$6D	109	$6E	110	$6F	111
h		i		j		k		l		m		n		o	
$70	112	$71	113	$72	114	$73	115	$74	116	$75	117	$76	118	$77	119
p		q		r		s		t		u		v		w	
$78	120	$79	121	$7A	122	$7B	123	$7C	124	$7D	125	$7E	126	$7F	127
x		y		z		{		|		}		~		DEL	

▶ B.2 BASIC I/O ASSEMBLY MODULE

B.2.1 Function Descriptions

Following is a description for each Basic I/O function. If the Basic I/O module is built for the byte-erasable EEPROM (EEBIO) as shown in the source code that follows, these functions can be called by using the addresses shown. As a convention, when accessing these functions this way, capital letters should be used for the function name.

For example to call *btod()*, the following code is used:

```
*****************************************************************
    BTOD      equ $0fd1
              ...
              jsr BTOD
*****************************************************************
```

btod()	Description:	One binary byte is converted to three unpacked digits in *bcd[]* (*bcd2:bcd1:bcd0*).
	Parameters:	Binary byte is passed in ACCB.
		Address of *bcd[0]* is passed on the stack.
	Return:	None
	Registers:	All registers preserved except ACCB, CCR.
	Address:	$0FD1
	C Prototype:	`void BTOD(UBYTE bin, UBYTE *bcd);`

bwtod()	Description:	16-bit word is converted to five unpacked digits in *bcd[]* (*bcd4:bcd3:bcd2:bcd1:bcd0*).
	Parameters:	Binary word is passed in ACCD.
		Address of *bcd[0]* is passed on the stack.
	Return:	None
	Registers:	All registers preserved except ACCD, CCR.
	Address:	$0FD4
	C Prototype:	`void BWTOD(INT16U bin, UBYTE *bcd);`

decstob()	Description:	A routine that converts a string of decimal characters to a binary byte until a CR, white space, or a NULL is encountered
	Parameters:	Pointer to the string is passed in ACCD.
		Pointer to binary byte is passed on the stack.
	Return:	Error code is returned in ACCB:
		0 -> No error
		1 -> Nondecimal character
		2 -> Too large
	Registers:	All registers preserved except ACCD, CCR.
	Address:	$0FDA
	C Prototype:	`UBYTE DECSTOB(UBYTE *strg, UBYTE *bin)`

getchar()	Description:	Read 68HC912B32 SCI until character is received.
	Parameters:	None
	Return:	ACCB contains ASCII character received. Blocks until a character is received.
	Registers:	All registers preserved except ACCB, CCR.
	Address:	$0FB3
	C Prototype:	`UBYTE GETCHAR(void);`

getstrg()	Description:	A routine that inputs a character string to an array until a carriage return is received or the maximum string length is exceeded. Only printable characters are recognized except carriage return and backspace. Backspace erases displayed character and array character. A NULL is always placed at the end of the string. Only printable characters are echoed.
	Parameters:	Pointer to the array is passed in ACCD. The maximum string length is passed on the stack. The string length includes the carriage return.
	Return:	Return value is passed in ACCB: 0 is returned if the string ended with a carriage return. 1 is returned if the maximum string length was exceeded.
	Registers:	All registers preserved except ACCD, CCR.
	Address:	$0FDD
	C Prototype:	`UBYTE GETSTRG(UBYTE *strg, UBYTE strlen);`

hexstobw()	Description:	A routine that converts a string of hex characters to a 16-bit word until white space or a NULL is encountered.
	Parameters:	Pointer to the string is passed in ACCD. Pointer to binary word is passed on the stack.
	Return:	Error code is returned in ACCB: 0 -> No error 1 -> Too large 2 -> Nonhex character
	Registers:	All registers preserved except ACCD, CCR.
	Address:	$0FD7
	C Prototype:	`UBYTE HEXSTOBW(UBYTE *strg,INT16U *bin);`

htoa()	Description:	Converts one binary nibble to uppercase ASCII.
	Parameters:	Nibble passed in least-significant nibble of ACCB. Most-significant nibble of ACCB must be zero.
	Return:	ASCII character returned in ACCB
	Registers:	All registers preserved except ACCB, CCR.
	Address:	$0FCB
	C Prototype:	`UBYTE HTOA(UBYTE bin);`

htob(c)	Description:	A routine converts an ASCII hex character to binary.
		Assumes a legal hex character [run *ishex()* first].
	Parameters:	The ASCII digit is passed in ACCB.
	Return:	Returns the binary nibble in ACCB.
	Registers:	All registers preserved except ACCB, CCR.
	Address:	$0FCE
	C Prototype:	`UBYTE HTOB(UBYTE c);`

INPUT(),	Description:	Read 68HC912B32 SCI.
sci_read()	Parameters:	None
	Return:	ACCB contains character received or 0 (NULL) if no
		character is received.
	Registers:	All registers preserved except ACCB, CCR.
	Address:	$0FB0
	C Prototype:	`UBYTE INPUT(void);`

isdigit(c)	Description:	A routine that checks if an ASCII character is a digit (0–9).
	Parameters:	Character is passed in ACCB.
	Return:	Return value is passed in ACCB and Z bit of CCR.
		Returns TRUE if character is digit.
	Registers:	All registers preserved except ACCB, CCR.
	Address:	$0FE6
	C Prototype:	`UBYTE ISDIGIT(UBYTE c);`

ishex(c)	Description:	A routine that checks if an ASCII character is a hexadecimal
		character (0..9), (a..f), or (A..F).
	Parameters:	Character is passed in ACCB.
	Return:	Return value is passed in ACCB and Z bit of CCR.
		Returns TRUE if character is hex.
	Registers:	All registers preserved except ACCB, CCR.
	Address:	$0FE3
	C Prototype:	`UBYTE ISHEX(UBYTE c);`

outcrlf()	Description:	Outputs a carriage return and line feed.
	Parameters:	None
	Return:	None
	Registers:	All registers preserved except CCR.
	Address:	$0FC8
	C Prototype:	`void OUTCRLF(void);`

outdecb()	Description:	A routine that outputs the decimal value of one byte.
	Parameters:	Pointer to byte is passed in ACCD.
	Return:	None
	Registers:	All registers preserved except ACCD, CCR.
	Address:	$0FC2
	C Prototype:	`void OUTDECB(UBYTE *bin);`

outdecw()	Description:	A routine that outputs the decimal value of a 16-bit word.
	Parameters:	Pointer to the 16-bit word is passed in ACCD.
	Return:	None
	Registers:	All registers preserved except ACCD, CCR.
	Address:	$0FC5
	C Prototype:	`void OUTDECW(INT16U *bin);`

outhexb()	Description:	A routine that outputs the hex value of one byte.
	Parameters:	ACCD contains pointer to byte to be sent.
	Return:	Returns with ACCD pointing to the next byte in memory.
	Registers:	All registers preserved except ACCD, CCR.
	Address:	$0FBC
	C Prototype:	`UBYTE * OUTHEXB(UBYTE *bin);`

outhexw()	Description:	A routine that outputs the hex value of a 16-bit word.
	Parameters:	ACCD contains pointer to the word to be sent.
	Return:	Returns with ACCD pointing to the next byte in memory.
	Registers:	All registers preserved except ACCD, CCR.
	Address:	$0FBF
	C Prototype:	`INT16U * OUTHEXW(INT16U *bin);`

PUTCHAR(), sci_write()	Description:	Write to 68HC912B32 SCI.
	Parameters:	ACCB contains the ASCII character to be transmitted.
	Return:	None. Blocks as much as one character at a time.
	Registers:	All registers preserved except ACCB, CCR.
	Address:	$0FB6
	C Prototype:	`UBYTE PUTCHAR(UBYTE c);`

putstrg()	Description:	Outputs a NULL terminated string.
	Parameters:	ACCD contains pointer to the null terminated string.
	Return:	Returns with ACCD pointing to the byte following the NULL.
	Registers:	All registers preserved except ACCD, CCR.
	Address:	$0FB9
	C Prototype:	`UBYTE * PUTSTRG(UBYTE *str);`

sci_open()	Description:	Initializes the 68HC912B32 SCI. Normal 8-bit mode, 9600bps, no interrupts.
	Parameters:	None
	Return:	None
	Requirements:	Assumes 8MHz E-clock
	Registers:	All registers preserved except CCR.
	Address:	$0FEC
	C Prototype:	`void SCIOPEN(void);`

slicestr()	Description:	A routine that slices a delimited token from a string to make it a substring. White space (tabs and space) are used as delimiters. The first white space character after a token is replaced by a NULL.
	Parameters:	Pointer to string to be sliced is passed in ACCD. Pointer to the pointer to be set to point to the substring is passed on the stack.
	Return:	Return value is a pointer to the character following the inserted NULL, or zero (a null pointer) if no token is found before a NULL.
	Registers:	All registers preserved except ACCD, CCR.
	Address:	$0FE9
	C Prototype:	UBYTE * SLICESTR(UBYTE *str, UBYTE **substr)

toupper(c)	Description:	Converts lowercase alpha characters to uppercase. All other characters remain unchanged.
	Parameters:	Character is passed in ACCB.
	Return:	Return character is passed in ACCB.
	Registers:	All registers preserved except ACCB, CCR.
	Address:	$0FE0
	C Prototype:	UBYTE TOUPPER(UBYTE c);

B.2.2 Basic I/O Source Code

```
*****************************************************************
*BasicIO - Basic I/O Utility Routines.
*
* The origin directives in this module are set for the
* byte-erasable EEPROM ($0D00-$0FFFF). When built in this form,
* the module is called EEBIO and the addresses shown in the
* function descriptions are valid.
*****************************************************************
NULL            equ 0
CR              equ $0d
LF              equ $0a
BS              equ $08
SPACE           equ $20
TAB             equ $09
TDRE            equ $80
RDRF            equ $20


                import _H12SC0BD,_H12SC0CR1,_H12SC0CR2,_H12SC0SR1
                import _H12SC0DRL
*****************************************************************
                org $0800
*****************************************************************
* getchar() - Read input until character is received.
*             Returns character in ACCB.
```

```
*Pseudo-C:
* ubyte getchar(){
*     while(1){
*         c = sci_read();
*         if(c != 0){
*             return(c);
*         }
*     }
* }
*******************************************************************
getchar     jsr sci_read        ;wait for characater
            cmpb #0
            beq getchar
            rts


*******************************************************************
* putstrg() - Output string to sci.
*     ACCD contains pointer to null terminated string
*     Differs from standard C puts() because it does not send '\n'
*     Returns with ACCD pointing to the byte following the NULL
*     All other registers but CCR preserved
* Pseudo-C:
* ubyte *putstrg(ubyte *strg){
*     while(*strg != NULL){
*         PUTCHAR(*strg);
*         strg++;
*     }
* }
*******************************************************************
putstrg     pshx                ;preserve IX
            tfr d,x             ;strg pointer->IX
ps_nxt      ldab 1,x+
            cmpb #NULL          ;put until NULL
            beq puts_rtn
            jsr PUTCHAR
            bra ps_nxt
puts_rtn    tfr x,d
            pulx                ;recover IX
            rts


*******************************************************************
* outhexb() - Output one hex byte.
*     ACCD contains pointer to byte to be sent.
*     Returns with ACCD pointing to the next byte in memory
*     All other registers but CCR preserved
* Pseudo-C:
* ubyte *outhexb(ubyte *bin){
*     PUTCHAR(htoa(*bin>>4));
*     PUTCHAR(htoa(*bin&0x0f));
* }
*******************************************************************
```

```
outhexb        pshx                 ;preserve IX
               tfr d,x              ;pointer to bin->IX
               ldab 0,x             ;put most sig. nibble
               lsrb
               lsrb
               lsrb
               lsrb
               jsr htoa             ;convert to ASCII
               jsr PUTCHAR
               ldab 0,x             ;put least sig. nibble
               andb #$0f
               jsr htoa
               jsr PUTCHAR
               inx                  ;point to next byte
               tfr x,d
               pulx                 ;recover IX
               rts


*****************************************************************
* outhexw() - Outputs two hex bytes.
*     ACCD contains pointer to displayed word.
*     Returns with ACCD pointing to the next byte in memory.
*     All other registers but CCR preserved
*Pseudo-C:
* uwdye *outhexw(uwyde *bin){
*     outhexb(bin[0]);
*     outhexb(bin[1]);
* }
*****************************************************************
outhexw        jsr outhexb
               jsr outhexb
               rts


*****************************************************************
* htoa()
* Description: Converts one binary nibble to uppercase ASCII.
* Arguments: Nibble passed in ACCB (LCN)
*            ASCII character returned in ACCB
* Pseudo-C:
* ubyte htoa(ubyte bin){
*     if(bin <= 9){
*         return(bin + 0x30);
*     }else{
*         return(bin + 0x37);
*     }
* }
*****************************************************************
htoa           cmpb #9
               bhi ha_alpha
               addb #$30
               bra ha_rtn
```

```
ha_alpha      addb #$37
ha_rtn        rts


*****************************************************************
* outcrlf() - Output a carriage return and line feed.
*     All registers but CCR preserved
* Pseudo-C:
* outcrlf(){
*     PUTCHAR('\r');
*     PUTCHAR('\n');
* }
*****************************************************************
outcrlf       pshb
              ldab #CR
              jsr PUTCHAR
              ldab #LF
              jsr PUTCHAR
              pulb
              rts
*************************************************************************
* toupper(c)
*
* Description: A routine that converts lowercase alpha characters to
*              uppercase. All other characters remain unchanged.
* Arguments: Character is passed in ACCB
*              Return value is passed in ACCB.
* Pseudo-C:
* ubyte toupper(c){
*     if('a'<= c <= 'z'){
*         c = c - 0x20;
*     }
* }
*************************************************************************
toupper       cmpb #'a'
              blo tou_rtn
              cmpb #'z'
              bhi tou_rtn
              subb #$20
tou_rtn       rts


*************************************************************************
* getstrg()
*
* Description: A routine that inputs a character string to an array
*              until a carriage return is received or strglen is exceeded.
*              Only printable characters are recognized except carriage
*              return and backspace.
*              Backspace erases displayed character and array character.
*              A NULL is always placed at the end of the string.
*              All printable characters are echoed.
*              Return value:
```

Basic I/O

```
*                    0 -> ended with CR
*                    1 -> if strglen exceeded.
* Arguments: Pointer to array is passed in ACCD
*            strglen is passed on the stack. strglen includes CR/NULL.
*            Return value is passed in ACCB.
* Pseudo-C:
* ubyte getstrg(ubyte *strg, ubyte strglen){
*     charnum = 0;
*     while(c = getchar(),c != '\r'){
*         if(' ' <= c <= '~'){
*             charnum++;
*             if(charnum == (strlen)){
*                 break;
*             }
*             PUTCHAR(c);
*             *strg = c;
*             strg++;
*         }else if(c = '\b' && charnum > 0){
*             PUTCHAR(c);
*             PUTCHAR(' ');
*             PUTCHAR(c);
*             strg--;
*             charnum--;
*         }
*     }
*     outcrlf();
*     *strg = NULL;
*     if(c=='\r'){
*         return(0);
*     }else{
*         return(1);
*     }
* }
****************************************************************************
getstrg     pshx               ;preserve IX
            tfr d,x            ;strlen->4,sp;  strg->IX
            clra               ;charnum->ACCA; c->ACCB
gs_nxt      jsr getchar        ;Character input
            cmpb #CR           ;done if CR
            beq gs_finish
            cmpb #' '          ;Check if printable
            blo gs_chkbs
            cmpb #'~'
            bhi gs_nxt
            inca               ;increment charnum
            cmpa 4,sp          ;break out if too many chars
            beq gs_finish
            jsr PUTCHAR        ;echo char
            stab 1,x+          ;store char and increment strg
            bra gs_nxt         ;get next character
gs_chkbs    cmpb #BS           ;Check for backspace
```

```
            bne gs_nxt
            tsta            ;ignore if no chars yet
            beq gs_nxt
            jsr PUTCHAR     ;erase displayed char
            ldab #SPACE
            jsr PUTCHAR
            ldab #BS
            jsr PUTCHAR
            dex             ;decrement strg and charnum
            deca
            bra gs_nxt      ;get next character
gs_finish   jsr outcrlf     ;echo CR
            clr 0,x         ;place NULL at end of array
            cmpb #CR        ;determine return value
            bne gs_ovrfl
            clrb            ;return 0 if CR
            bra gs_rtn
gs_ovrfl    ldab #1         ;return 1 if overflow
gs_rtn      pulx            ;recover IX
            rts


**************************************************************************
* btod()
*
* Description: One binary byte is converted to three unpacked digits
*              in bcd[],(bcd2:bcd1:bcd0).
* Arguments: Binary byte is passed in ACCB.
*            Address of bcd[0] is passed on the stack.
* Pseudo-C:
* btod(ubyte bin, ubyte *bcd){
*     bcd[2] = bin%10;
*     bin = bin/10;
*     bcd[1] = bin%10;
*     bcd[0] = bin/10;
* }
**************************************************************************
btod        pshy            ;preserve IY, IX, and ACCA
            pshx
            psha
            ldy 7,sp        ;Array pointer -> IY
            clra
            ldx #10         ;divide bin by 10
            idiv
            stab 2,y        ;remainder = bcd0
            tfr x,d         ;divide quotient by 10
            ldx #10
            idiv
            stab 1,y        ;remainder = bcd1
            tfr x,d
            stab 0,y        ;quotient = bcd2
            pula            ;recover IY, IX, and ACCA
```

```
            pulx
            puly
            rts


*********************************************************************
* bwtod()
*
* Description: 16-bit word is converted to five unpacked digits
*              in bcd[],(bcd4:bcd3:bcd2:bcd1:bcd0).
* Arguments: Binary word is passed in ACCD.
*              Address of bcd[0] is passed on the stack.
* Pseudo-C:
* bwtod(uwyde bin, ubyte *bcd){
*     bcd[4] = bin%10;
*     bin = bin/10;
*     bcd[3] = bin%10;
*     bin = bin/10;
*     bcd[2] = bin%10;
*     bin = bin/10;
*     bcd[1] = bin%10;
*     bcd[0] = bin/10;
* }
*********************************************************************
bwtod       pshy          ;preserve IY and IX
            pshx
            ldy #0
            ldx #10       ;divide bin by 10
            ediv
            ldx 6,sp
            stab 4,x      ;remainder = bcd0
            tfr y,d       ;divide quotient by 10
            ldy #0
            ldx #10
            ediv
            ldx 6,sp      ;remainder = bcd1
            stab 3,x
            tfr y,d       ;divide quotient by 10
            ldy #0
            ldx #10
            ediv
            ldx 6,sp      ;remainder = bcd2
            stab 2,x
            tfr y,d       ;divide quotient by 10
            ldx #10
            idiv
            ldy 6,sp      ;remainder = bcd3
            stab 1,y
            tfr x,d
            stab 0,y      ;quotient = bcd4
            pulx          ;recover IY and IX
            puly
            rts
```

```
***************************************************************************
* outdecb()
*
* Description: A routine that outputs the decimal value of one byte.
* Arguments: Pointer to byte is passed in ACCD
*
* Pseudo-C:
* void outdecb(ubyte *bin){
*     bcd[] = btod(*bin);
*     i = 0;
*     while(bcd[i] == 0 && i<2){
*         i++;
*     }
*     while(i<3){
*         outchar(bcd[i] + 0x30);
*         i++;
*     }
* }
***************************************************************************
outdecb     pshx             ;preserve IX
            leas -3,sp       ;allocate space for BCD digits
            tfr d,x          ;get binary byte
            ldab 0,x
            tfr sp,x         ;convert bin to BCD
            pshx
            jsr btod
            leas 2,sp
            ldaa #0
odb_lp1     cmpa #2          ;ignore leading zeros
            bhs odb_lp2
            ldab A,sp
            bne odb_lp2
            inca
            bra odb_lp1
odb_lp2     cmpa #3          ;convert bcd[i] to ASCII and put
            bhs odb_finlp
            ldab A,sp
            addb #$30
            jsr PUTCHAR
            inca
            bra odb_lp2
odb_finlp   leas 3,sp        ;deallocate BCD space
            pulx             ;recover IX
            rts


***************************************************************************
* outdecw()
*
* Description: A routine that outputs the decimal value of a 16-bit word.
* Arguments: Pointer to word is passed in ACCD
*
```

```
* Pseudo-C:
* void outdecw(uwyde *bin){
*     bcd[] = bwtod(*bin);
*     i = 0;
*     while(bcd[i] == 0 && i < 4){
*         i++;
*     }
*     while(i<5){
*         outchar(bcd[i] + 0x30);
*         i++;
*     }
* }
******************************************************************
outdecw     pshx                ;preserve IX
            leas -5,sp          ;allocate space for BCD digits
            tfr d,x             ;get binary word
            ldd 0,x
            tfr sp,x            ;convert bin to BCD
            pshx
            jsr bwtod
            leas 2,sp
            ldaa #0
odw_lp1     cmpa #4             ;ignore leading zeros
            bhs odw_lp2
            ldab A,sp
            bne odw_lp2
            inca
            bra odw_lp1
odw_lp2     cmpa #5             ;convert bcd[i] to ASCII and put
            bhs odw_finlp
            ldab A,sp
            addb #$30
            jsr PUTCHAR
            inca
            bra odw_lp2
odw_finlp   leas 5,sp           ;deallocate BCD space
            pulx                ;recover IX
            rts


*************************************************************************
* ishex(c)
*
* Description: A routine to check if an ASCII character is an
*              hexadecimal character (0..9),(a..f) or (A..F).
* Arguments: Character is passed in ACCB
*              Return value is passed in ACCB and Z bit of CCR.
*
* Pseudo-C:
* ubyte ishex(c){
*     if(('0'<=c<='9') || ('a'<=c<='f') || ('A'<=c<='F')){
*         return(1);
```

```
*       else
*           return(0);
* }
*
**************************************************************************
ishex           cmpb #'0'               ;Check if between '0' and '9'
                blo isnoth
                cmpb #'9'
                bls ish
                cmpb #'A'               ;check if between 'A' and 'F'
                blo isnoth
                cmpb #'F'
                bls ish
                cmpb #'a'               ;check if between 'a' and 'f'
                blo isnoth
                cmpb #'f'
                bls ish
isnoth          clrb                    ;Not hex, return 0
                bra ish_rtn
ish             ldab #1                 ;is hex, return 1
ish_rtn         rts

**************************************************************************
* ubyte htob(ubyte c)
*
* Description: A routine converts an ASCII hex character to binary.
*              Assumes a legal hex character.(run ishex() first)
* Arguments: The ASCII digit is passed in ACCB.
*              Returns the binary nibble in ACCB.
* Pseudo-C:
* ubyte htob(ubyte c){
*     if('0'<=c<='9') {
*         bin = c - '0';
*     }else if('a'<=c<='f'){
*         bin = c - 'a';
*     }else if('A'<=c<='F'){
*         bin = c - 'A';
*     }
*     return(bin);
* }
**************************************************************************
htob            cmpb #'0'       ;is digit?
                blo hb_rtn
                cmpb #'9'
                bhi hb_lw
                subb #'0'       ;subtract $30
                bra hb_rtn
hb_lw           cmpb #'A'       ;is uppercase hex?
                blo hb_rtn
                cmpb #'F'
                bhi hb_up
```

```
            subb #$37          ;subtract $37
            bra hb_rtn
hb_up       cmpb #'a'          ;is lowercase hex?
            blo hb_rtn
            cmpb #'f'
            bhi hb_rtn
            subb #$57          ;subtract $57
hb_rtn      rts

*************************************************************************
* hexstobw()
*
* Description: A routine that converts a string to a hex 16-bit word
*              until white space or NULL is encountered.
* Arguments: Pointer to the string is passed in ACCD
*            Pointer to binary word is passed on the stack
*            Error code is returned in ACCB
*               0 -> No error
*               1 -> Too large
*               2 -> Nonhex character
* Pseudo-C:
* ubyte hexstobw(ubyte *strg,uwyde *bin){
*      cnt = 0;
*      *bin = 0;
*      while(*strg != NULL || *strg != ' ' || *strg != '\t'){
*          if(ishex(*strg)){
*              *bin = (*bin<<4) | htob(*strg);
*          }else{
*              return(2);
*          }
*          strg++;
*          cnt++;
*          if(cnt > 4){
*              return(1);
*          }
*      }
*      return(0);
* }
*************************************************************************
hexstobw    pshx               ;preserve IX
            tfr d,x            ;string pointer -> IX
            leas -2,sp         ;cnt->1,sp; tmpc->0,sp
            clr 1,sp           ;clear cnt
            ldd #0
            std [6,sp]         ;clr binary number
sh_lp       ldab 1,x+          ;get next character
            cmpb #NULL         ;finish if NULL, space, or tab.
            beq sh_finlp
            cmpb #' '
            beq sh_finlp
            cmpb #TAB
```

```
                beq sh_finlp
                jsr ishex          ;check for legal hex digit
                beq err2           ;if not return error 2
                jsr htob           ;convert ASCII hex to binary
                stab 0,sp
                ldd [6,sp]         ;pack binary nibble into bin
                lsld
                lsld
                lsld
                lsld
                orab 0,sp
                std [6,sp]
                inc 1,sp           ;check if four digits done
                ldab 1,sp
                cmpb #4
                bhi err1           ;error 1 if too many chars
                bra sh_lp
sh_finlp        clrb               ;return 0
                bra sh_rtn
err1            ldab #1
                bra sh_rtn
err2            ldab #2
sh_rtn          leas 2,sp          ;reallocate stack
                pulx               ;recover IX
                rts
*****************************************************************
* decstob()
*
* Description: A routine that converts a string of decimal
*              characters to a binary byte until a CR, white space,
*              or a NULL is encountered.
* Arguments: Pointer to the string is passed in ACCD
*            Pointer to binary byte is passed on the stack
*            Error code is returned in ACCB
*               0 -> No error
*               1 -> Nondecimal character
*               2 -> Too large
*
* Pseudo-C:
* ubyte decstob(ubyte *strg,ubyte *bin){
*     if(isdigit(*strg)){
*         *bin = (*strg-0x30);
*     }else if(*strg == NULL){
*         return(0);
*     }else{
*         return(1);
*     }
*     strg++;
*     while(*strg != NULL || *strg != ' ' || *strg != '\t'){
*         if(isdigit(*strg)){
*             *bin = *bin * 10 + (*strg-0x30);
```

```
*                if(*bin>255){
*                     return(2);
*                }
*          }else{
*               return(1);
*          }
*          strg++;
*      }
*      return(0);
* }
**********************************************************************
decstob    pshx             ;preserve IX
           tfr d,x          ;strg pointer->IX
           clr [4,sp]       ;clear bin
           ldab 0,x         ;get first digit
           cmpb #NULL       ;if NULL done
           beq dsb_finlp
           jsr isdigit      ;if not digit error 1
           cmpb #0
           beq dsb_err1
           ldab 0,x
           subb #$30        ;convert char to binary
           stab [4,sp]
           inx              ;point to next char
dsb_lp     ldab 0,x
           cmpb #NULL       ;finished if NULL or white space
           beq dsb_finlp
           cmpb #' '
           beq dsb_finlp
           cmpb #TAB
           beq dsb_finlp
           jsr isdigit      ;error if not digit
           beq dsb_err1
           ldab [4,sp]      ;bin*10+bcdx
           ldaa #10
           mul
           tsta             ;check for overflow
           bne dsb_err2
           ldaa 0,x
           suba #$30        ;convert net digit to bcd
           aba
           bcs dsb_err2     ;check for overflow
           staa [4,sp]
           inx              ;point to next digit
           bra dsb_lp
dsb_finlp  clrb             ;done no error
           bra dsb_rtn
dsb_err1   ldab #1          ;nondigit error
           bra dsb_rtn
dsb_err2   ldab #2          ;too-large error
dsb_rtn    pulx             ;recover IX
           rts
```

```
*****************************************************************
* isdigit(c)
*
* Description: A routine to check if an ASCII character is a
*              digit (0-9).
* Arguments: Character is passed in ACCB
*            Return value is passed in ACCB and Z bit of CCR.
* Pseudo-C:
* ubyte isdigit(c){
*    if('0'<= c <= '9'){
*        return(1);
*    else
*        return(0);
* }
*****************************************************************
isdigit      cmpb #'0'
             blo isd_not
isd_9        cmpb #'9'
             bhi isd_not
             ldab #1
             bra isd_rtn
isd_not      clrb
isd_rtn      rts


*****************************************************************
* slicestr()
*
* Description: A routine that slices a delimited token from a
*              string to make it a substring. White space
*              (tabs and space) are used as delimiters. The first
*              white-space character after a token is replaced by a
*              NULL.
* Arguments: Pointer to string to be sliced is passed in ACCD
*            Pointer to be set to point to the substring is passed
*            on the stack
*            Return value is a pointer to the character following
*            the inserted NULL or zero (a NULL pointer) if no
*            token is found before a NULL.
* Pseudo-C:
* ubyte *slicestr(ubyte *str, ubyte **substr){
*     while(*str == '\t' || *str == ' '){
*         if(*str == 'NULL'){
*             return(*0);
*         }
*         str++;
*     }
*     *substr = str;
*     str++;
*     while(*str!='\t' && *str!=' ' && *str!=NULL){
*         str++
*     }
```

Basic I/O   669

```
*       *str = NULL;
*       str++;
*       return(str);
* }
*******************************************************************
slicestr    pshx                ;preserve IX
            tfr d,x             ;str->IX
ss_wslp     ldab 0,x            ;ignore leading white space
            cmpb #TAB
            beq ss_nc
            cmpb #' '
            beq ss_nc
            cmpb #NULL
            bne ss_tk           ;token found
            ldd #0              ;no token return 0
            bra ss_rtn
ss_nc       inx                 ;check next char
            bra ss_wslp
ss_tk       stx [4,sp]          ;save pointer to token in substr
            inx                 ;find end of token (white space)
ss_tklp     ldab 0,x
            cmpb #TAB
            beq ss_endtk
            cmpb #' '
            beq ss_endtk
            cmpb #NULL
            beq ss_endtk
            inx
            bra ss_tklp
ss_endtk    clrb                ;end token with NULL (substring)
            stab 0,x
            inx                 ;point to next char after NULL
            tfr x,d             ;return current string pointer
ss_rtn      pulx                ;recover IX
            rts


*******************************************************************
* sci_open - Initializes SCI.
*    Normal 8-bit mode, 9600bps,no interrupts
*       -Assumes 8MHz Eclk
*    All registers preserved except CCR.
*******************************************************************
sci_open    movw #52,_H12SC0BD      ;9600bps @ 8MHz Eclk
            movb #$00,_H12SC0CR1    ;Normal 8-bit Mode
            movb #$0c,_H12SC0CR2    ;No ints, no parity
            rts


*******************************************************************
* sci_read() - Read sci.
*    Returns ACCB=char or 0 if no character received.
*******************************************************************
```

```
sci_read          clrb
                  brclr _H12SC0SR1,RDRF,sci_read_rtn  ; get status
                  ldab _H12SC0DRL                     ; get data
sci_read_rtn      rts
****************************************************************
* sci_write() - write to sci.
*    Outputs the value passed in ACCB.
*    Blocks as long as 1 character at a time.
****************************************************************
sci_write         brclr _H12SC0SR1,TDRE,sci_write  ; wait for TDRE
                  stab _H12SC0DRL                   ; send data
                  rts
****************************************************************
* Function Jump Table
****************************************************************
                  org $0fb0
****************************************************************
INPUT:            jmp sci_read        ;Alias for sci_read
GETCHAR:          jmp getchar
PUTCHAR:          jmp sci_write       ;Alias for sci_write
PUTSTRG:          jmp putstrg
OUTHEXB:          jmp outhexb
OUTHEXW:          jmp outhexw
OUTDECB:          jmp outdecb
OUTDECW:          jmp outdecw
OUTCRLF:          jmp outcrlf
HTOA:             jmp htoa
HTOB:             jmp htob
BTOD:             jmp btod
BWTOD:            jmp bwtod
HEXSTOBW:         jmp hexstobw
DECSTOB:          jmp decstob
GETSTRG:          jmp getstrg
TOUPPER:          jmp toupper
ISHEX:            jmp ishex
ISDIGIT:          jmp isdigit
SLICESTR:         jmp slicestr
SCIOPEN:          jmp sci_open
****************************************************************
```

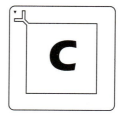

µC/OS Reference

▶ C.1 CONFIGURATION

The µC/OS kernel consists of several files, including C source files, an assembly source file, and several header files. In this section we look at which files need to be changed in order to port µC/OS to the configuration used in this text. Table C.1 shows the files that are used to build the µC/OS library used in this text. They can be grouped into the CPU-independent files and the files that are designed for a specific CPU, MCU, or MCU family.

TABLE C.1 µC/OS SOURCE FILES

µC/OS Kernel—CPU/MCU-Independent Core and Service Source Files

File Name	Description
os_core.c	Core kernel code
os_mem.c	Memory partition service code
os_mbox.c	Message mailbox service code
os_q.c	Message queue service code
os_sem.c	Semaphore service code
os_task.c	Task management code
os_time.c	Kernel CPU/MCU independent timer code
ucos-ii.c	Single source file that includes all core and service source files
ucos-ii.h	Header file for all µC/OS CPU/MCU independent code

µC/OS Kernel—CPU/MCU-Dependent Files and Configuration Files

File Name	Description
os_cpu_c.c	CPU/MCU specific C code
os_cpu_a.s12	CPU/MCU specific assembly code
os_tick.c	CPU/MCU specific timer code
os_cfg.h	Kernel build configuration header file
os_cpu.h	CPU/MCU specific definitions
includes.h	Master header file

672

Normally when building μC/OS, the CPU-independent files never have to be changed. These files make up the core or the kernel, and the kernel services make up the primary distribution for μC/OS. A minor change to some of these files has been made for the configuration used in this text. These changes will be discussed later. Normally these files are kept in a separate subdirectory, and when a new version of μC/OS is released, these are the only files that need to be replaced.

When building μC/OS for a project or library, we should restrict any changes to the CPU-dependent and configuration files. These files can be used to customize μC/OS to a particular system. Let's quickly go over each file.

os_cpu_c.c. This module contains the CPU-specific C code. Functions included in this file are *OSTaskStkInit()* and the kernel hook functions. *OSTaskStkInit()* is of course dependent on the CPU and how the CPU stacks its registers during a context switch. The hook functions are user-created functions that are useful for debugging the kernel.

os_cpu_a.s12. This module is the only assembly module required for the kernel. It consists of functions that cannot practically be written in C. The functions in this file are small task-switching functions that must be very fast. They are also dependent on the CPU because they directly manipulate the stack.

os_tick.c. This module contains the CPU/MCU-dependent kernel timer code, including *OSTickInit()* and *OSTickISR()*. *OSTickInit()* is always called immediately after the kernel is started. It should initialize and enable the periodic interrupt source used for the kernel tick. *OSTickISR()* is the actual timer interrupt service routine. It includes the code required to inform the kernel that it has been entered. The periodic timer source used in the configuration for this text is *Timer Channel 6* configured as an output compare.

os_cpu.h. This header file includes all the CPU-specific definitions with the exception of the MCU register names. The registers are defined in the Introl-CODE header, *hc912b32.h,* which is included through the master header file. *os_cpu.c* contains all the standard type definitions, some CPU-specific parameters, and CPU-specific macros. It also includes some defines used in the *os_tick.c* module, which of course is also CPU/MCU-dependent.

includes.h. This is the master *includes* file. It includes all other header files only. These include, in order, *hc912b32.h, os_cfg.h, os_cpu.h,* and *ucos-ii.h.*

os_cfg.h. This is the most important file for configuring the μC/OS kernel for your application. The configuration header file used for the text is shown in Source C.1. This file is used to configure the kernel for a particular environment. In our case we are using the M68HC912B32 microcontroller, which only has 1K-byte of RAM space. This is limiting for using a preemptive kernel. Because of the configuration parameters available in *os_cfg.h,* however, we can build the kernel for limited RAM space.

Source C.1 μC/OS Configuration for This Text

```
/***************************************************************************
* File : os_cfg.h
* By   : Jean J. Labrosse
*      : Revised by Todd Morton
```

```
/**************************************************************************
* μC/OS-II CONFIGURATION
*       - This is the configuration for ETec454 and Embedded Microcontrollers
*       - Last Revised: 01/31/2000
**************************************************************************/
#define OS_MAX_EVENTS     10    /* Ten ECBs available for application       */
#define OS_MAX_MEM_PART   4     /* Four memory partitions available for app. */
#define OS_MAX_QS         4     /* Four queues available for application      */
#define OS_MAX_TASKS      10    /* Ten tasks available for application        */
#define OS_LOWEST_PRIO    17    /* Priorities available for tasks: 4..13      */

#define OS_TASK_IDLE_STK_SIZE 64     /* Idle task stack size                  */

#define OS_TASK_STAT_EN 0    /* Disable(0) the statistics task               */

#define OS_TASK_STAT_STK_SIZE 64     /* Statistics task stack size            */

#define OS_CPU_HOOKS_EN 1    /* μC/OS-II hooks are found in cpu.c            */

#define OS_MBOX_EN   1       /* Include code for MAILBOXES                   */
#define OS_MEM_EN    1       /* Include code for MEMORY MANAGER             */
#define OS_Q_EN      1       /* Include code for QUEUES                      */
#define OS_SEM_EN    1       /* Include code for SEMAPHORES                  */

#define OS_TASK_CHANGE_PRIO_EN  1 /*Include code for OSTaskChangePrio()     */
#define OS_TASK_CREATE_EN       1 /* Include code for OSTaskCreate()         */
#define OS_TASK_CREATE_EXT_EN   0 /* No code for OSTaskCreateExt()           */
#define OS_TASK_DEL_EN          1 /* Include code for OSTaskDel()            */
/* Include code for OSTaskSuspend() and OSTaskResume()                      */
#define OS_TASK_SUSPEND_EN      1

#define OS_TICKS_PER_SEC    1000  /* The number of ticks in one second      */

/**************************************************************************/
```

We can predetermine the maximum number of kernel events, memory partitions, queues, and tasks. The lower we keep these numbers, the less memory is required. Because we intended this configuration to be built as a general-purpose library, some of the numbers may be higher than we would use for a specific application. For example if we built the kernel specifically for the stopwatch example in Chapter 16, we could get by with two kernel events (one queue and one semaphore), no memory partitions, one queue, and five tasks. We also did not use any kernel hook functions, mailboxes, or memory partitions, so we could remove all the code for these from the build.

Imagine, if the source code were not available, the kernel would have to be used in the form it was received. In the application-specific world of embedded microcontrollers, this would be a serious limitation.

μC/OS Interrupt Vectors.   When building a μC/OS project, it is required to load two interrupt vectors for μC/OS. First a vector must be loaded for the kernel timer. In the precompiled library used in this text, the timer uses Timer Channel 6 and the interrupt service routine is called *OSTickISR*. Therefore *OSTickISR* must be added to the Timer Channel

6 entry in the Introl-CODE vectors dialog box. Second the software interrupt (SWI) is used by the kernel for context switching. Therefore _OSCtxSw must be added to the SWI entry in the Introl-CODE vectors dialog box. Without these vectors µC/OS will not run.

▶ C.2 µC/OS SERVICE FUNCTIONS

C.2.1 Task Functions

OSTaskCreate()

Description:	This function starts creates a µC/OS task. It must be called before a task can be used. A task can be created before *OSStart()* or by another task.
C Prototype:	`INT8U OSTaskCreate(void(*task)(void *pd),void *pdata,` `                   OS_STK *ptos,INT8U prio);`
Parameters:	*task* is a pointer to the task's code. *pdata* is a pointer to optional data to be passed to the task (normally this is a NULL pointer). *ptos* is a pointer to the top of the task's stack. *prio* is the task's priority.

Return:	*OSTaskCreate()* returns the following error codes:
	    OS_NO_ERR if the function was successful.
	    OS_PRIO_EXIST if the requested priority already exists.
	    OS_PRIO_INVALID if *prio* is higher than *OS_LOWEST_PRIO*.
	    OS_NO_MORE_TCB if there are no more task control blocks available.

Notes:	*OSTaskCreate()* must be called before a task will run. Do not use priorities 0, 1, 2, 3, or the lowest three priorities. So if *OS_LOWEST_PRIO* is 16, priorities 4 through 13 are available for the user application.

OSTaskResume()

Description:	This function resumes a task that was suspended by *OSTaskSuspend()*.
C Prototype:	`INT8U OSTaskResume(INT8U prio);`
Parameters:	*prio* is the priority of the task to be resumed.

Return:	*OSTaskResume()* returns the following error codes:
	    OS_NO_ERR if the function was successful.
	    OS_TASK_RESUME_PRIO if the task does not exist.
	    OS_TASK_NOT_SUSPENDED if the task was not suspended.
	    OS_PRIO_INVALID if *prio* is higher than *OS_LOWEST_PRIO*.

OSTaskSuspend()

Description:	This function suspends execution of a task unconditionally. A task can suspend itself using either the task priority or *OS_PRIO_SELF*.
C Prototype:	`INT8U OSTaskSuspend(INT8U prio);`
Parameters:	*prio* is the priority of the task to be suspended.

Return:	*OSTaskSuspend()* returns the following error codes:
	    OS_NO_ERR if the function was successful.
	    OS_TASK_SUSPEND_PRIO if the task does not exist.
	    OS_TASK_SUSPEND_IDLE if you try to suspend the idle task, which is not allowed.
	    OS_PRIO_INVALID if *prio* is higher than *OS_LOWEST_PRIO*.

C.2.2 Time Functions

OSTimeDly()

Description:	This function allows a task to delay itself for a number of clock ticks.
C Prototype:	`void OSTimeDly(INT16U ticks);`
Parameters:	*ticks* is the number of clock ticks to delay the current task.
Return:	None
Notes:	To delay a task for *at least N* ticks, you should set *ticks* to *N+1*.

OSTimeDlyHMSM()

Description:	This function allows a task to delay itself for a specified amount of time in hours, minutes, seconds, and milliseconds.
C Prototype:	`void OSTimeDlyHMSM(INT8U hours, INT8U minutes,` `                   INT8U seconds, INT16U milli);`
Parameters:	*hours* is the number of hours to delay. Range: 0–55. *minutes* is the number of minutes to delay. Range 0–59. *seconds* is the number of second to delay. Range 0–59. *milli* is the number of milliseconds to delay. Range: 0–999. Note *milli* is rounded off to the nearest number of ticks.

Return:	*OSTimeDlyHMSM()* returns one of the following error codes:
	    OS_NO_ERR if the arguments are valid and the call was successful.
	    OS_TIME_INVALID_MINUTES if minutes was greater than 59.
	    OS_TIME_INVALID_SECONDS if seconds was greater than 59.
	    OS_TIME_INVALID_MILLI if *milli* was greater than 999.
	    OS_TIME_ZERO_DLY if all four arguments are zero.

Notes:	If the total delay is greater than 65,535 clock ticks, you cannot resume the task with *ODTimeDlyResume()*.

OSTimeDlyResume()

Description:	This function resumes a task that has been delayed by *OSTimeDly()* or *OSTimeDlyHMSM()*.
C Prototype:	`INT8U OSTimeDlyResume(INT8U prio);`
Parameters:	*prio* is the priority of the task you want to resume.

Return:	*OSTimeDlyResume()* returns one of the following error codes:
	    OS_NO_ERR if the call was successful.
	    OS_PRIO_INVALID if *prio* was greater than *OS_LOWEST_PRIO*.
	    OS_TIME_NOT_DLY if the task was not delayed.
	    OS_TASK_NOT_EXIST if the task has not been created.

Notes:	If you use this with a task that is waiting for a timeout, it will appear that a timeout occurred. You cannot resume a task that was delayed by *OSTimeDlyHMSM()* if the total delay time was greater than 65,535 clock ticks.

OSTimeGet()

Description:	This function returns the current value of the system clock. The system clock contains the number of clock ticks since reset or since the clock was last set.
C Prototype:	`INT32U OSTimeGet(void);`
Parameters:	None
Return:	The current value of the system clock.

OSTimeSet()

Description:	This function sets the value of the system clock.
C Prototype:	`void OSTimeSet(INT32U ticks);`
Parameters:	*ticks* the desired value of the system clock.
Return:	None

C.2.3 Semaphore Functions

OSSemAccept()

Description:	This function checks to see if a semaphore is pending (>0). Unlike *OSSemPend()* it does not suspend the block task until the semaphore is pending. It returns immediately.
C Prototype:	`INT16U OSSemAccept (OS_EVENT *pevent)`
Parameters:	*pevent* is a pointer to the event control block associated with the semaphore.
Return:	When *OSSemAccept()* is called and the semaphore value is greater than 0, the semaphore value is decremented and the value of the semaphore before the decrement is returned. If the semaphore value is 0, the semaphore is not pending and 0 is returned.

OSSemCreate()

Description:	This function creates and initializes a semaphore.
C Prototype:	`OS_EVENT OSSemCreate (INT16U value)`
Parameters:	*value* is the initial value of the semaphore. It can range from 0 to 65,535. A semaphore used as a flag is normally initialized to 0 (not pending),whereas a semaphore used as a key is normally initialized to 1 (resource available).
Return:	*OSSemCreate()* returns a pointer to the event control block allocated to the semaphore. If no event control block is available, it returns a NULL pointer.

OOSSemPend()

Description:	This function checks to see if a semaphore is pending (>0). If not, *OSSemPend()* will suspend the task until the semaphore is pending. If the semaphore is greater than 0, it will decrement the semaphore and return.
C Prototype:	`void OSSemPend(OS_EVENT *pevent, INT16U timeout, INT8U *err)`
Parameters:	*pevent* is a pointer to the event control block associated with the semaphore. *timeout* allows the task to resume if the semaphore is not pending within the specified number of clock ticks. *err* is a pointer to a variable to hold an error code.
Return:	The error codes for *OSSemPend()* are as follows: *OS_NO_ERR* if the semaphore was available. *OS_TIMEOUT* if the semaphore was not signaled by the specified timeout. *OS_ERR_PEND_ISR* if you called this function from an ISR. *OS_ERR_EVENT_TYPE* if *pevent* is not pointing to a semaphore.

OSSemPost()

Description:	This function signals a semaphore. If the semaphore is 0 or more, it is incremented and *OSSemPost()* returns. If there are tasks waiting for the semaphore, those tasks will be made ready to run.
C Prototype:	`INT8U OSSemPost (OS_EVENT *pevent)`
Parameters:	*pevent* is a pointer to the event control block associated with the semaphore.
Return:	*OSSemPost()* returns one of three error codes. The error codes are as follows: *OS_NO_ERR* if the semaphore was signaled successfully. *OS_SEM_OVF* if the semaphore count overflowed. *OS_ERR_EVENT_TYPE* if *pevent* is not pointing to a semaphore.

OSSemQuery()

Description:	This function can be used to obtain information about a semaphore. You must allocate an *OS_SEM_DATA* data structure to receive the data from the semaphore event control block.
C Prototype:	`INT8U OSSemQuery (OS_EVENT *pevent, OS_SEM_DATA *pdata)`
Parameters:	*pevent* is a pointer to the event control block associated with the semaphore. *pdata* is a pointer to an *OS_SEM_DATA* structure.
Return:	*OSSemQuery()* returns one of two error codes. The error codes are: *OS_NO_ERR* if the call was successfully. *OS_ERR_EVENT_TYPE* if *pevent* is not pointing to a semaphore.

C.2.4 Mailbox Functions

OSMboxAccept()

Description:	This function checks the mailbox to see if a message is available. *OSMboxAccept()* does not suspend the calling task if a message is not available.
C Prototype:	`void *OSMboxAccept (OS_EVENT *pevent)`
Parameters:	*pevent* is a pointer to the event control block
Return:	A pointer to the message in the mailbox if one is available. NULL if the mailbox is empty or if you did not pass the proper event pointer.

OSMboxCreate()

Description:	This function creates and initializes a mailbox.
C Prototype:	`OS_EVENT *OSMboxCreate (void *msg)`
Parameters:	*msg* is used to initialize the mailbox. To create an empty mailbox, *msg* is a NULL pointer.
Return:	A pointer to the event control block allocated to the mailbox. If no event control block is available, a NULL pointer is returned.

OSMboxPend()

Description:	This function is used when a task expects to receive a message. When called, it checks the mailbox. If there is a message, it is returned and the mailbox is emptied. If there is no message, it waits until a message is received or until the timeout occurs.
C Prototype:	`void *OSMboxPend(OS_EVENT *pevent, INT16U timeout, INT8U *err)`
Parameters:	*pevent* is a pointer to the event control block associated with the mailbox. *timeout* allows the task to resume if a message is not received within the specified number of clock ticks. *err* is a pointer to a variable to hold an error code.
Return:	A pointer to the message in the mailbox if one is available. NULL if the mailbox is empty or if you did not pass the proper event pointer. *err* is set to the appropriate error code. The error codes from *OSMbox Pend()* are as follows: OS_NO_ERR if the message was received. OS_TIMEOUT if the message was not received by the specified timeout. OS_ERR_PEND_ISR if you called this function from an ISR. OS_ERR_EVENT_TYPE if *pevent* is not pointing to a mailbox.

OSMboxPost()

Description:	This function sends a message to a task through a mailbox. The message is a pointer-sized variable. If the mailbox is full, *OSMboxPost()* immediately returns without placing the message in the mailbox. Once the message is placed in the mailbox, a context switch may occur.
C Prototype:	`INT8U OSMboxPost(OS_EVENT *pevent, void *msg)`
Parameters:	*pevent* is a pointer to the event control block associated with the mailbox. *msg* is the message (or a pointer to the message) to be sent to the mailbox. It should never be a NULL pointer because that indicates an empty mailbox.
Return:	*OSMboxPost* returns one of three error codes. The error codes are as follows: OS_NO_ERR if the message was placed in the mailbox. OS_MBOX_FULL if the mailbox already contained a message. OS_ERR_EVENT_TYPE if *pevent* is not pointing to a mailbox.

OSMboxQuery()

Description:	This function obtains information about the mailbox and places it in a *OS_MBOX_DATA* structure.
C Prototype:	`INT8U OSMboxQuery(OS_EVENT *pevent, OS_MBOX_DATA *pdata)`
Parameters:	*pevent* is a pointer to the event control block associated with the mailbox. *pdata* is a pointer to a *OS_MBOX_DATA* structure.
Return:	*OSMboxQueryt* returns one of two error codes. The error codes are as follows: *OS_NO_ERR* if the message was placed in the mailbox. *OS_ERR_EVENT_TYPE* if *pevent* is not pointing to a mailbox.

C.2.5 Queue Functions

OSQCreate()

Description:	This function creates and initializes a message queue.
C Prototype:	`OS_EVENT OSQCreate (void **start, INT8U size)`
Parameters:	*start* is the base address of the message storage area. A message storage area is an array of pointers to *voids*. *size* is the size (in number of messages) of the message storage area.
Return:	*OSSemCreate()* returns a pointer to the event control block allocated to the queue. If no event control block is available, it returns a NULL pointer.

OSQFlush()

Description:	This function empties the contents of the message queue and eliminates all the messages sent to the queue.
C Prototype:	`INT8U *OSQFlush(OS_EVENT *pevent)`
Parameters:	*pevent* is a pointer to the event control block associated with the queue.
Return:	*OSSemPend()* returns one of the following error codes: *OS_NO_ERR* if the message queue was flushed. *OS_ERR_EVENT_TYPE* if *pevent* is not pointing to a message queue.

OSQPend()

Description:	This function is used to wait for a message on the queue. If no message is present, *QSQPend()* will suspend the task until the queue receives a message.
C Prototype:	`void *OSQPend(OS_EVENT *pevent, UBT16U timeout, INT8U *err)`
Parameters:	*pevent* is a pointer to the event control block associated with the queue. *timeout* allows the task to resume if the queue is not pending within the specified number of clock ticks. *err* is a pointer to a variable to hold an error code.
Return:	The error codes for *OSSemPend()* are as follows: *OS_NO_ERR* if the message was received. *OS_TIMEOUT* if the message was not received by the specified timeout. *OS_ERR_PEND_ISR* if you called this function from an ISR. *OS_ERR_EVENT_TYPE* if *pevent* is not pointing to a message queue.

OSQPost()

Description:	This function sends a message to a queue. If there are tasks waiting for the message, those tasks will be made ready to run and the highest priority task receives the message. Messages are received FIFO.
C Prototype:	`INT8U OSQPost (OS_EVENT *pevent, void *msg)`
Parameters:	*pevent* is a pointer to the event control block associated with the queue. *msg* is the message sent to the queue. It is a pointer-sized message and its meaning is application specific.
Return:	*OSQPost()* returns one of three error codes. The error codes are as follows: *OS_NO_ERR* if the message was deposited in the queue successfully. *OS_Q_FULL* if the queue was already full. *OS_ERR_EVENT_TYPE* if *pevent* is not pointing to a message queue.

OSQPostFront()

Description:	This function sends a message to a queue. This function works the same as *OSQPost()* except the message is placed at the front of the queue.
C Prototype:	`INT8U OSQPostFront (OS_EVENT *pevent, void *msg)`
Parameters:	*pevent* is a pointer to the event control block associated with the queue. *msg* is the message sent to the queue. It is a pointer-sized message and its meaning is application specific.
Return:	*OSQPost()* returns one of three error codes. The error codes are as follows: *OS_NO_ERR* if the message was deposited in the queue successfully. *OS_Q_FULL* if the queue was already full. *OS_ERR_EVENT_TYPE* if *pevent* is not pointing to a message queue.

OSQQuery()

Description:	This function can be used to obtain information about a message queue. You must allocate an *OS_Q_DATA* data structure to receive the data from the queue's event control block.
C Prototype:	`INT8U OSQQuery (OS_EVENT *pevent, OS_Q_DATA *pdata)`
Parameters:	*pevent* is a pointer to the event control block associated with the message queue. *pdata* is a pointer to a *OS_Q_DATA* structure.
Return:	*OSSemQuery()* returns one of two error codes. The error codes are as follows: *OS_NO_ERR* if the call was successful. *OS_ERR_EVENT_TYPE* if *pevent* is not pointing to a message queue.

C.2.6 Memory Allocation Functions

OSMemCreate()

Description:	This function creates and initializes a memory partition.
C Prototype:	`OS_MEM *OSMemCreate (void *addr, MEMTYPE nblks, MEMTYPE blksize, INT8U *err)`
Parameters:	*addr* is the address of the start of the memory area to be used for the blocks. *nblks* is the number of memory blocks available in the specified partition. (*MEMTYPE* is the size of the CPU memory model, which is INT16U here.) *blksize* is the size, in bytes, of each memory block.
Return:	A pointer to the created memory partition control block. If no memory control block is available, a NULL pointer is returned. *err* is a pointer to a variable that holds the error code. Error codes include the following: *OS_NO_ERR* if the memory partition was created successfully *OS_MEM_INVALID_PART* if the memory partition was not available. *OS_MEM_INVALID_BLKS* if at least two blocks were not specified. *OS_MEM_INVALID_SIZE* if the block size is smaller than a pointer variable.

OSMemGet()

Description:	This function obtains a memory block from a memory partition.
C Prototype:	`void *OSMemGet(OS_MEM *pmem, INT8U *err)`
Parameters:	*pmem* is a pointer to the memory partition control block.
Return:	A pointer to the allocated memory block. NULL if no memory blocks are available from the memory partition. *err* is set to the appropriate error code. The error codes from *OSMemGet()* are as follows: *OS_NO_ERR* if the memory block was available. *OS_MEM_NO_FREE_BLKS* if the partition did not contain any free blocks.

OSMemPut()

Description:	This function returns a memory block to a memory partition.
C Prototype:	`INT8U OSMemPut(OS_MEM *pmem, void *pblk)`
Parameters:	*pmem* is a pointer to the memory control block. *pblk* is a pointer to the memory block to be returned.
Return:	*OSMemPut* returns one of two error codes. The error codes are as follows: *OS_NO_ERR* if the memory block was returned. *OS_MEM_FULL* if the partition could not accept more memory blocks.

OSMemQuery()

Description:	This function obtains information about a memory partition. An *OS_MEM_DATA* structure must be allocated.
C Prototype:	`INT8U OSMemQuery(OS_MEM *pmem, OS_MEM_DATA *pdata)`
Parameters:	*pmem* is a pointer to the memory control block. *pdata* is a pointer to an *OS_MEM_DATA* structure.
Return:	*OSMemQueryt* returns one error code, *OS_NO_ERR*.

C.2.7 System Functions

OSInit()

Description:	This function is used to initialize the internals of µC/OS-II and must be called prior to creating any µC/OS-II object, and prior to calling *OSStart()*.
C Prototype:	`void OSInit(void);`
Parameters:	None
Return:	None

OSIntEnter()

Description:	This function is used to notify µC/OS-II that you are about to service an interrupt service routine (ISR). This allows µC/OS-II to keep track of interrupt nesting.
C Prototype:	`void OSIntEnter(void);`
Parameters:	None
Return:	None
Notes:	1. Your ISR can directly increment *OSIntNesting* without calling this function because *OSIntNesting* has been declared "global." You *must* be sure that the increment is performed "indivisibly" by your processor to ensure proper access to this critical resource. 2. You *must* still call *OSIntExit()* even if you increment *OSIntNesting* directly. 3. You *must* invoke *OSIntEnter()* and *OSIntExit()* in pairs. In other words, for every call to *OSIntEnter()* at the beginning of the ISR, you *must* have a call to *OSIntExit()* at the end of the ISR.

OSIntExit()

Description:	This function is used to notify µC/OS-II that you have completed servicing an ISR. When the last nested ISR has completed, µC/OS-II will call the scheduler to determine whether a new high-priority task is ready to run.
C Prototype:	`void OSIntExit (void)`
Parameters:	None
Return:	None
Notes:	You *must* invoke *OSIntEnter()* and *OSIntExit()* in pairs. In other words, for every call to *OSIntEnter()* at the beginning of the ISR, you *must* have a call to *OSIntExit()* at the end of the ISR.

OSStart()

Description:	This function starts the µC/OS multitasker.
C Prototype:	`void OSStart(void)`
Parameters:	None
Return:	None
Notes:	*OSInit()* must be called before *OSStart()*.

OS_ENTER_CRITICAL()

Description:	A macro that disables interrupts.
C Prototype:	Macro
Parameters:	None
Return:	None

OS_EXIT_CRITICAL()

Description:	A macro that enables interrupts.
C Prototype:	Macro
Parameters:	None
Return:	None

References

► MOTOROLA DOCUMENTATION

CPU12 Reference Manual, CPU12RM/AD, Motorola.

M68EVB912B32 Evaluation Board User's Manual, M68EVB912B32UM/D, Motorola.

M68HC11 Reference Manual, M68HC11RM/AD, Motorola.

M68HC11F1 Technical Data, MC68HC11F1/D, Motorola.

MC68HC812A4 Technical Data, MC68HC812A4/D, Motorola.

MC68HC912B32 Technical Data, MC68HC912B32/D, Motorola.

► REFERENCE TEXTS

Ball, Stuart R., *Debugging Embedded Microprocessor Systems.* Woburn, MA: Butterworth-Heinemann, 1998.

Cady, Fredrick M., *Microcontrollers and Microcomputers, Principles of Software and Hardware Engineering.* New York: Oxford University Press, 1997.

Cady, Fredrick M., *Software and Hardware Engineering, Motorola M68HC11.* New York: Oxford University Press, 1997.

Cady, Fredrick M., *Software and Hardware Engineering, Motorola M68HC12.* New York: Oxford University Press, 2000.

685

Ganssle, Jack G., *The Art of Programming Embedded Systems*. San Diego, CA: Academic Press, 1992.

Kernighan, Brian W., and Dennis M. Ritchie, *The C Programming Language, ANSI C*. Upper Saddle River, NJ: Prentice-Hall, 1988.

Labrosse, Jean J., *Embedded Systems Building Blocks, Complete and Ready-to-Use Modules in C*. Lawrence, KS: R&D Books, 2000.

Labrosse, Jean J., *MicroC/OS-II, The Real-Time Kernel*. Lawrence, KS: R&D Books, 1999.

Laplante, Phillip A., *Real-Time Systems Design and Analysis, An Engineer's Handbook*. Piscataway, NJ: IEEE Press, 1997.

McConnell, Steve, *Code Complete*. Redmond WA: Microsoft Press, 1993.

Oualline, Steve, *Practical C Programming*. Sebastopol, CA: O'Reilly & Associates, Inc., 1997.

Perry, John W., *Advanced C Programming by Example*. Boston: PWS Publishing Company, 1998.

Slater, Michael, *Microprocessor-Based Design, A Comprehensive Guide to Effective Hardware Design*. Upper Saddle River, NJ: Prentice-Hall, 1989.

Spasov, Peter, *Microcontroller Technology, The 68HC11*. Upper Saddle River, NJ: Prentice-Hall, 1999.

Wakerly, John F., *Digital Design, Principles and Practices*. Upper Saddle River, NJ: Prentice-Hall, 2000.

► JOURNALS

Circuit Cellar, The Magazine for Computer Applications, Circuit Cellar Inc.

Embedded Systems Programming, CMP Media Inc.

Microprocessor Report, Cahners Electronics Group.

► WORLD WIDE WEB SITES

eet.etec.wwu.edu	Western Washington University, Electronics Engineering Technology Program
www.introl.com	Introl Corporation
www.mot-sps.com	Motorola Semiconductor Products Sector
www.noral.com	Noral Micrologics, Inc.
www.seattlerobotics.org	Encoder, Seattle Robotics Society
www.uCOS-II.com	Micrium, Inc. (MicroC/OS)

Index

! (C logical NOT), 448, 519
!= (C not equal-to), 519
(immediate addressing), 65–66
(C preprocessor), 444
$ (hex number), 26, 41
$ (current address), 42
% (binary number), 41
& (C bitwise AND), 448
& (decimal number), 41
& (address of C object), 500
&& (C logical AND), 448, 519
* (current address), 42
* (contents of pointer), 500
: (label suffix), 41
; (C end of statement), 444
// (C++ comment), 445
< (force direct), 66, 67
< (C less-than), 519
<= (C less-than or equal-to), 519
= (C assignment), 449
== (C relation), 449, 519
> (force extended), 66, 67
> (C greater-than), 519
>= (C greater-than or equal-to), 519
@ (octal number), 41
\ (C escape), 447
| (C bitwise OR), 448
|| (C logical OR), 448, 519
~ (C bitwise NOT), 448
6502, 15
74HC138, 426
74HC165, 364
74HC595, 361
8048, 15
8085, 15

\a, 447
ABA, 92
absolute addressing, 66
absolute locations, 131
    data at, 496–498
    pointers to, 501–503

abstraction, 513–514
ABX, 92, 97–98
ABY, 92, 97–98
ACCA, 63
ACCB, 63
ACCD, 63
access by name, 156
access by reference, 156
access time, 9, 420
accessibility, 9
accumulator, 63
ADCA, 92
ADCB, 92
ADDA, 92
ADDB, 92
ADDD, 92
addition, 92–98
    BCD, 96–97
    signed, 94, 95
    unsigned, 93, 94
address bus, 5–7
addressing modes, 64–76
ADR0-ADR7, 373
aliasing error, 368
analog signals, 13
analog-to-digital convertors (ADC), 13,
    366–376
    configuration, 370–372
    conversion modes, 372
    dynamic range, 368–369
    signal-to-noise ratio, 368
ANDA, 107
ANDB, 107
ANDCC, 107
application-specific, 2, 15
architectural design, 18
arrays, 503–506
AS, 6
as12, 47–48
ASCII, 46, 180
    BCD conversions, 186
    conversions, 180–184

hex conversions, 186–188
    table, 651
ASL, 105
ASLA, 105
ASLB, 105
ASR, 105
ASRA, 105, 221
ASRB, 105
assembler, 27–28, 32, 47–52, 461
    as12, 47–48
    directives, 42–47
        dc, 46
        ds, 47
        equ, 43
        fcb, 44
        fcc, 45
        fdb, 44–45
        org, 42
        rmb, 47
        section, 462
    listing, 27–28, 48–51
    syntax, 40–42
assembly language, 27–28, 440–441
    comments, 37, 39
    labels, 41
    line syntax, 41
    mnemonics, 27
    source organization, 37–40
    strings, 45–46
    symbols, 42
    variables, 46–47
asynchronous bus, 7
asynchronous serial ports, 13, 349–350
ATDCTL2, 370
ATDCTL4, 370
ATDCTL5, 370
ATDSTAT, 370
auto-increment/decrement addressing,
    72–73
automatic variables, 494

\b, 447
background, 249, 574

background debug system, 410–416
    basic operation, 414–416
    hardware configuration, 413
    memory maps, 412
.base, 463
Basic I/O module, 651–671
battery-backed RAM (BBRAM), 10
battery source, 384
BCC, 116
BCD, 180
    ASCII conversion, 186
    binary conversions, 188–194
    hex conversion, 184–185
    packed, 184
    unpacked, 184
BCLR, 111–112
BCS, 116
BDM, *See* background debug system
BEQ, 116
beta testing, 22
BGE, 118
BGT, 118
BHI, 116
BHS, 116
bidirectional, 6,7
big-endian, 45
binary conversions, 188–194
binary machine code, *See* machine
    language
BIOS, 11
bit conditional branches, 119–120
bit manipulation
    assembly, 107–109, 111–112
    C, 510-513
bit testing
    assembly, 109–111
    C, 510
BITA, 111
BITB, 111
BKGD, 395, 410–411
BLE, 118
BLO, 116
BLS, 116
BLT, 118
BMI, 116
BNE, 116
Boolean logic, 106–109
    operators, 106
bootloader, 11
BPL, 116
BR (breakpoint), 55
BRA, 116
branches, 113, 114–120
BRCLR, 120
break, 533–535

breakpoints, 55–56, 416
BRN, 116
BRSET, 120
BSET, 111–112
BSR, 121
.bss, 463
btod(), 192–194, 652, 661
buffer, 624,628
build errors, 32
build process, 31
build time, 33
bus acknowledge, 6
bus collision/contention, 421
bus request, 6
bus system, 3–9
    address bus, 5–7
    asynchronous, 7
    control bus, 5–7
    data bus, 5–7
    multiplexed, 6,8–9
    sharing, 6
    synchronization, 6,7
    timing, 434–437
BVC, 116
BVS, 116
bwtod(), 652, 662

C, carry flag, 64
C, programming language, 28–30,
    440–454
    comments, 445
    constants, 446–447
    identifiers and keywords, 446
    operators, 447–450
    preprocessor commands, 450–451
    tokens, 444
    white space, 444
CALL, 76, 121
call-by-name, 163–164
call-by-reference, 164
call-by-value, 163–164
calling conventions, 544–545
carry flag (C), 64
case construct, 148, 523–525
cast, 493–494, 498
CBA, 103
CCR, 63–64, 110–111
central processing unit (CPU), *See*
    microprocessor
ceramic resonators, 385
CFORC, 315–316
char, 486
chip select logic, 5, 7, 8, 421–428
CLC, 111
clearing bits, 108

CLI, 111
clock monitor exception, 389–390
clocks, 384–386
CLR, 85
CLRA, 85
CLRB, 85
CLV, 111
CMPA, 103
CMPB, 103
code reviews, 21
code segments, 30–31
cohesion, 162
COM, 107
COMA, 107
COMB, 107
compiler, 29–30, 32, 460–461
computer BIOS, 11
computer operating properly timer
    (COP), 389, 390–391
computer systems, 14
concept, 16
condition code register, 63–64
conditional branches, 115–120
conditional expressions, 519–521
    assignments in, 520–521
    boolean reduction of, 528–529
    nesting vs. complex, 526–527
    simplifying, 527–528
const modifier, 488, 496
.const section, 463
constant pointers, 502–503
constants
    defined, 450
    equated, 43
    stored, 43–46, 496
continue, 533, 535–536
control bus, 5–7
control constructs, 143–155, 518–536
cooperative multitasking, 261, 263–267
cooperative tasks, 263
COPCTL, 390
COPRST, 390
coroutines, 581–582
coupling, 162–163
CPD, 103
CPS, 103
CPU load, 233, 243, 248, 252, 267,
    321, 333, 568, 574
CPU11, 14
    addressing modes, 64, 68–69
    register set, 62–63
    stack, 89–90, 170
CPU12, 14
    addressing modes, 64–76
    analog-to-digital converter, 369–372

instruction set, 77, 80–132
instruction timing, 77–78, 175
interrupts, 274–288
pulse accumulator, 335–338
pulse-width modulator, 339–347
register set, 62–63
serial communications interface
    (SCI), 13, 350–355
serial peripheral interface (SPI),
    13, 356–366
stack, 86–89
standard timer module, 312–317
CPX, 103
CPY, 103
critical regions, 284–285, 615
cross assembler, *See* assmebler
cross compiler, *See* compiler
crystals, 385
CSCTL0-CSCTL1, 429
CSSTR0-CSSTR1, 429
custom integrated circuit, 15

DAA, 92, 97
data bus, 5–7
data direction register, 12
data object, 155–157, 485–486
data register, 12
.data section, 463
data transfer instructions, 80–85
data types, 484–494
    ANSI-C, 485–494
    in assembly code, 484
    checking, 484–485
    conversion, 493–494
    defining, 490–492
    fundamental, 486–487
    modifiers, 487–490
        scope, 489–490
        storage class, 487–489
D-Bug12 monitor program, 52–59,
    278–280, 289
    BR, 55, 414
    DEVICE, 414
    FBULK, 415
    FLOAD, 415
    G, 54
    interrupts with, 278–280
    LOAD, 54
    MD, 26, 55, 414
    MM, 55
    RD, 55
    RM, 55
    TRACE, 57
    USEHBR, 416
DDRP, 296

deallocation, 161
debugger, 31–32
debugging, 32–33, 52–59, 289–293
    helper hardware and software,
        290–293
    source level debugging, 475–481
DEC, 104
DECA, 104
DECB, 104
decimal output routines, 200–202
decode logic, *See* chip select logic
decrement, 104
decstob(), 205–207, 652, 667–668
#define, 450, 542
define constant (dc), 46
define storage (ds), 47
delay routines
    software, 174–180
demo1.s12, 37
demotion, 494
demultiplex, 9
DES, 104
DEX, 104
DEY, 104
digital-to-analog convertors (DAC), 13
direct addressing, 66–68
division, 220–229
    by repeated subtraction, 221
    result size, 220
    by shifting, 221–222
do-while construct, 148–149, 529–530
DORMANT, 604
double, 486
/DTACK, 6
dual-boot system, 408–410
duty cycle, 32, 343

E-clock, 6
edge detection, 240–241
EDIV, 224
EDIVS, 224
effective address (EA), 64
electricaly erasable PROM (EEPROM),
    10–11
#elif, 451
#else, 451
embedded systems, 1–3, 14–15
    development 15–22
EMUL, 216
EMULS, 216
endian, 45
#endif, 451
enumerated types, 509
EORA, 107
EORB, 107

EPROM, 10–11, 406–408
equate (equ), 43
evaluation boards, 8, 52–53, 378–379
event, 237
event-driven I/O, 237–244
    hardware based, 244–249
    interrupt driven, 249–253
event flags, 239, 245
event loops
    free-running, 238
    in MicroC/OS-II, 607
    timed, 253–256
event response time, *See* response time
event time, 242, 255
exception vectors, 403, 404
exceptions, 275
executable object file, 461
execution time, 77–78, 175
EXG, 83–84
expanded mode, 4, 383, 396–397
EXTAL, 385
extended addressing, 66–68
extern modifier, 488, 497, 539, 554

\f, 447
FALSE, 519
fanout, 299
fault tolerance, 389
FDIV, 224
final product development, 406–416,
    563–564
firmware, 10
first-in-first-out (FIFO), 9, 635
fixed-point numbers, 208–209
    errors in, 213–216
    signed mixed numbers, 211–213
    unsigned binary fractions, 209–210
    unsigned mixed numbers, 210–211
flash EEPROM, 10
float, 486
floating-point numbers, 208
flow diagrams, 138–141
for construct, 151–153, 530–531
foreground, 249, 574
form constant byte (fcb), 44
form constant character (fcc), 45
form double byte (fdb), 44–45
fundamental data types, 486–487
free-running counter, 312
free-running cyclic scheduler, 575–578
free-running event loops, 238
full address decoding, 423–425
functions, 536–548
    calling conventions, 544–545
    declaring and defining, 538–540

interrupt service routines, 546–548
    passing parameters, 540–542
    written in assembly, 543–546

G, 54
gadfly loops, *See* event loops
general purpose I/O (GPIO), 12–13,
    296–306
    interfacing, 298–303
    power dissipation, 303–304
    timing, 304–306
generic pointers, 500–501
getchar(), 195, 653, 656–657
getstrg(), 197–199, 653, 659–661
Gbyte, 6
global interrupt mask (I), 64
global variables, 158–160, 489–490,
    614–616
    for intertask communication, 171,
      614–616
goto, 533,536

half-carry flag (H), 63
hardware breakpoints, 416
hardware development, 15–22
hardware reviews, 21
hc912b32.h, 490
header files, 451,454
hex machine code, 26
hexidecimal
    ASCII conversions, 186–188
    output routines, 199–200
hexstobw(), 203–205, 653, 666–667
high impedence state, 7
high-level languages, 28–30
hold time, 435, 436, 437
HPRIO, 281–282
htoa(), 187, 653, 658–659
htob(), 187–188, 654, 665–666
HyperTerminal, 53

I, global interrupt mask, 64
identifiers, 446
IDIV, 224
IDIVS, 224
#if, 451
if construct, 144, 518–519
if-else construct, 144–146, 521–522
if-else if-else construct, 146–148,
    522–523
#ifdef, 451
#ifndef, 451
ihex, 48
immediate addressing, 65
implied addressing, *See* inherent
    addressing

import, 473
in-circuit emulator, 59
INC, 104
INCA, 104
INCB, 104
#include, 451–454
include files, *See* header files
includes.h, 452–454, 550, 551–553,
    601–603
increment, 104
index registers, 63
indexed addressing, 68–75
    auto-increment/decrement, 72–73
    constant offset, 70–71
    CPU11, 68–69
    CPU12, 69–70
    indirect, 74–75
    register offset, 73–74
indirect index addressing, 74–75
information hiding, 622–624
inherent addressing, 64–65
.init, 463
INITEE, 398
INITRG, 398
INITRM, 398
INPUT(), 654, 671
input capture, 246–247, 250–251,
    329–335
    event settings, 330
    to measure frequency, 333–335
    to measure pulse width, 331–332
INS, 104
int, 486
interfacing, 298–303
    loads, 299–303
    no load, 299
    reduced drive, 303
__interrupt modifier, 539,547
interrupts, 249–250, 274–288
    external, 285–287
    latency, 280–281, 283
    masks, 277–278
    priorities, 281–284
    processing, 275–278
    software, 287
    sources, 274–275
    stack frame, 277
interrupt masks, 64
interrupt service routine (ISR), 249
    in C, 546–548
intertask signals, 232
Introl-CODE
    assembler, 27, 47–52, 461
    calling conventions, 544–545

.code file, 461
    compiler, 460–461
    ihex, 48
    interrupt vectors, 564
    linker, 461
    linker command file, 471–473
    register configuration, 563
    section mapping, 461–463
inverting bits, 109
INX, 104
INY, 104
I/O devices, 3, 12–13
IRQ, 285–287
isdigit(), 180–181, 654, 669
ishex(), 181–183, 654, 664–665
IX, 63
IY, 63

jitter, 579
JMP, 113
JSR, 121

kernels, 262, 570
    cooperative, 263, 572–573, 574–591
      design, 574–591
      free-running cyclic scheduler,
        575–578
      time-slice cyclic scheduler,
        578–591
    preemptive, 262–263, 571–572,
      595–597
keywords, 446

languages, programming
    choice of, 19
    types of, 24–30
last-in-first-out (LIFO), 86, 124, 635
latency, 280–281, 283
LBCC, 118
LBCS, 118
LBEQ, 118
LBGE, 118
LBGT, 118
LBHI, 118
LBHS, 118
LBLE, 118
LBLO, 118
LBLS, 118
LBLT, 118
LBMI, 118
LBNE, 118
LBPL, 118
LBRA, 118
LBRN, 118
LBVC, 118
LBVS, 118

LDAA, 81
LDAB, 81
LDD, 81
LDS, 81
LDX, 81
LDY, 81
LEAS, 81,82
LEAX, 81,82
LEAY, 81,82
level detection, 238–239
libraries, 19, 455, 463–464
libucosii.a12, 597
light emitting diodes (LEDs), 301–302
linear address space, 5
linker, 32, 461
linker command file, 471–473
listing, 27–28, 465–468
    absolute, 467–468
    relocatable, 465–467
little-endian, 45
LOAD, 54
load time, 33
local variables, 160–162, 494
logic analyzers, 58–59, 290
logical operators, 519
long, 486
long branches, 118–119
lookup tables, 525–526
low-voltage detection, 392–394
LSL, 105
LSLA, 105
LSLB, 105
LSLD, 105
LSR, 105
LSRA, 105, 221
LSRB, 105
LSRD, 105

M68EVB912B32 evaluation board, 8,
    52–53
    as BDM debug POD, 413
    memory map, 54
machine language, 24–26, 28
macros, 30, 450, 542–543
main(), 443, 537, 557, 597–600
make, 473–475
Makefile, 473–475
manchester encoding, 348
manual reset switches, 393–394
map file, 468–471
mask, 109, 111, 119–120
mask programming, 10
masks, 277–278
MAX6314, 393–394
Mbyte, 6

MC34064, 384, 393
MC6801, 15
MC6805, 14
MC6809, 15
MC68HC11, 14–15
    address space, 5
    data bus, 6
    E-series, 8
MC68HC11F1, 7,14
MC68HC12
    data bus, 6
MC68HC812A4, 7, 14, 418
    bus cycles, 419–421
    bus timing, 434–437
    memory map priorities, 430
    programmable chip select,
        428–434
MC68HC912B32, 12, 14
    clocks, 384–386
    operating modes, 383, 394–399
    power supply, 379–382
    reset exceptions, 386–394
MD, 26
memory devices, 3,9–11
    accessibility, 9
    primary, 9
    programmability, 9,10
    random access, 9
    read-write memory, 9
    types, 9–11
    uses, 11
    volatility, 9,10
    writability, 9–10
memory map, 7–8, 396, 397–399,
    421–423
microcomputer, 3–4
microcontroller, 3,12
    choice of, 18–19
MicroC/OS-II, 594–646
    configuration, 672–675
    interrupt service routines in,
        609–610
    interrupt vectors, 674–675
    intertask communications, 614–638
        using global variables,
            614–616, 624–631
        message mailboxes, 631–635
        Message queues, 635–638
        semaphores, 616–624,
            624–631
    program organization, 597–603
    tasks
        priorities, 608–609
        stacks, 607–608
        states, 604

switching, 603–604
variables, 608
timers, 610–613
microprocessor, 3–5, 14–15
    choice of, 18–19
    for desktop computers, 4, 6, 14
    programming model, 36
MISC, 398,429
MISO, 356
mixed-signal oscilloscope, 59,
    290–293
mnemonics, 27
.mod1/.mod2, 463
MODA, 394–395
MODB, 394–395
MODE, 395
modular programming, 19, 454–475,
    548–557
    build process, 460–461
        command-line interface,
            471–475
        generated files, 465–471
    demo1 project, 456
    dependencies, 460
    file organization, 549–550
    source files, 455–460
    startup module, 458–459
modules, 30, 442, 548–557
monitor dependent programs, 31
MOSI, 356
MOVB, 85–86
MOVW, 85–86
MUL, 216, 218
multiplexed bus system, 6, 8–9
multiplication, 216–220
    by repeated addition, 217
    by shift-and-add, 217
    by shifting, 217–218
mutual exclusion, 580

\n 447
NEG, 98
NEGA, 98
NEGB, 98
negative flag (N), 63
noise immunity, 258, 261
noise margin, 298–299
non-reccurring costs (NRE), 18
NOP, 177
Noral Flex debugger, 478–481
normal single-chip mode, 395–396
NULL, 501
null pointers, 501
null-terminated strings, 506
number syntax, 41

object code, 25,32
object relocatable code, *See* relocatable code
OC7D, 326
OC7M, 326
OCDelay module, 551, 555–557
one-time programmable (OTP), 10
opcodes, 25–26, 27
open source software, 20
operands, 25–26,27
ORAA, 107
ORAB, 107
ORACC, 107
orgin(org), 42
os_cfg.h, 672–674
oscillators, *See* clocks
OS_ENTER_CRITICAL(), 684
OS_EXIT_CRITICAL(), 684
OSInit(), 600, 683
OSIntEnter(), 609, 683
OSIntExit(), 604, 609–610, 683
OSIntNesting(), 609
OS_LOWEST_PRIO, 608–609
OS_MAX_TASKS, 608
OSMboxAccept(), 632–633,678
OSMboxCreate(), 632, 679
OSMboxPend(), 605, 632–633, 679
OSMboxPost(), 632–633, 679
OSMboxQuery(), 632, 680
OSMemCreate(), 682
OSMemGet(), 682
OSMemPut(), 682
OSMemQuery(), 682
OSQCreate(), 636, 680
OSQFlush(), 636, 680
OSQPend(), 605, 638, 680
OSQPost(), 636, 681
OSQPostFront(), 636, 681
OSQQuery(), 636, 681
OSSemAccept(), 616, 677
OSSemCreate(), 616–617, 677
OSSemPend(), 601, 605, 607, 616–617, 678
OSSemPost(), 616–617, 678
OSSemQuery(), 616, 678
OSStart(), 600, 683
OSTaskCreate(), 598, 604, 675
OSTaskDel(), 604
OSTaskResume(), 675
OSTaskSuspend(), 675
OSTickInit(), 601
OSTimeDly(), 605, 606–607, 611–613, 676
OSTimeDlyHMSM(), 605, 611, 676
OSTimeDlyResume(), 611, 676

OSTimeGet(), 611, 677
OSTimeSet(), 611, 677
outcrlf(), 654, 659
outdecb(), 200–201, 654, 663
outdecw(), 201–202, 655, 663–664
outhexb(), 199–200, 655, 657–658
outhexw(), 200, 655, 658
output compare, 317–329
    actions, 318
    for generating square-wave, 320–321
        output compare 7 (OC7), 326–329
    for pulse-width modulation, 321–329
    for periodic timer events, 318–320
oscilloscope, 58
overclocking, 386
overflow flag(V), 64

PACNT, 335–336
PACTL, 312, 314, 335–336
PAFLG, 335–336
page, 5
paged memory, 5
parallel I/O, *See* general purpose I/O
partial address decoding, 425–428
    images, 428
passing parameters, 163–171
    in C functions, 540–542
    using registers, 164–165
    using the stack, 166–171
PC, 63
PCI bus, 8
pcr (program counter relative), 129–130
PEAR, 429
periodic events, 237
periodic interrupts, 13, 250
persistence, 157, 487
phasing errors, 68
pointers, 498–503
    to absolute locations, 501–503
    and arrays, 504–505
    constant, 502–503
    generic, 500–501
    initialization, 499
    null, 501
    operations, 500
    size in C, 499
polling loops, *See* event loops
PORTA, 298
portability, 513–514, 548
PORTB, 298
PORTF, 429
PORTP, 296
position dependent code, 128

position independence, 128–132
power dissipation, 303–304, 382–384
power-on reset, 388–389
power supply, 379–382
PPAGE, 76
preemption, 604–605
preemptive kernel, 262–263, 571–572, 595–597
primary memory, 9
printf(), 194
priority-based scheduler, 576–578
private, 456
program, 4
program counter, 63
program time, 232
programmability, 9,10
programmable chip selects, 428–434
programmable read only memory (PROM), 10
programming conventions
    for assembly, 648–649
    for C, 649–650
programming languages, *See* languages
programming model, 36, 61–78
project directory, 455–460
pseudo-C, 141–143
pseudo-nonmaskable interrupt mask (X), 64
PSHA, 87
PSHB, 87
PSHC, 87
PSHD, 87
PSHX, 87
PSHY, 87
public, 456
PULA, 87
PULB, 87
PULC, 87
PULD, 87
pull-up resistors, 303
pulse accumulator, 335–338
    event counter mode, 336
    gated-time accumulation mode, 336
    to measure frequency, 336–338
pulse-width modulation, 321–322
    using a normal output compare, 322–326
    using OC7, 327–329
    using the pulse-width modulator, 344–347
pulse-width modulator, 339–347
    alignment, 340
    clocks, 341
    configuration, 340
PULX, 87

PULY, 87
putchar(), 195, 655, 671
putstrg(), 196, 655, 657
PWCLK, 340
PWCTL, 340
PWDTY0-PWDTY3, 342
PWEN, 342
PWPER0-PWPER3, 342
PWPOL, 342
PWSCAL0-PWSCAL1, 340

\r, 447
R/W', 6,419,420
radix, 41
random access, 9
random access memory (RAM), 9
rate monotonic scheduling (RMS), 609
RD, 55
read cycle, 419–420
read-modify-write instructions,
    100–101, 511–512
read-only memory (ROM), 10, 11
read-write memory, 9
READY, 604
real-time, 232
real-time debugging, 289–293
real-time interrupt (RTI), 308–311
real-time kernels, See kernels
real-time programming, 37, 566–592
real-time system, 231
reciprocals, 228–229
recurring costs, 18
recursive routines, 159
reduced drive, 303
reentrant code, 159–160
register addressing, See inherent
    addressing
register modifier, 488
register offset indexed addressing,
    73–74
register set, 61–62
register variable, 157
relational operators, 519
relative addressing, 75–76, 114–115
reliability, 548
relocatable code, 129–132
requirements, 16
reserve memory byte (rmb), 47
/RESET, 387, 391
reset circuitry, 391–394
reset exceptions, 386–394
response time
    for free-running event loops,
        242–243
    for hardware-based event
        detection, 249

for independent tasks, 569–570
for interrupt-based event detection,
    253, 281, 570
for a preemptive kernel, 595–597
for a priority-based scheduler,
    576–578
for a round-robin scheduler,
    575–576
for time slice scheduler, 265, 267,
    579–580, 596
for unconditional I/O, 235, 237
reviews, 21
revision control, 22
ring buffer, 628
RM, 55
ROL, 105
ROLA, 105
ROLB, 105
ROMable object code, 461
ROR, 105
RORA, 105
RORB, 105
rotate, 105–106
round-robin scheduler, 575
routines, 30, 162–163
    C routines, See functions
    cohesion, 162
    coupling, 162–163
    passing parameters, 163–171
RTC, 121
rti (return from interrupt), 275, 277, 278
RTICTL, 308–309
RTIFLG, 308–309
RTS, 121
run-time, 34
run-time errors, 32
RUNNING, 604
RS-232, 13

S, stop control bit, 64
sample rate, 366
sampling, 366
SBA, 98
SBCA, 98
SBCB, 98
SC0BDH-SC0BDL, 351–352
SC0CR1-SC0CR2, 351–352
SC0DRH-SC0DRL, 351, 353
SC0SR1-SC0SR2, 351
sci_open(), 352, 355, 670
sci_read(), 195, 354, 355, 654, 670–671
sci_write(), 195, 353, 355, 655, 671
SCK,356
scope, 157
SEC, 111

section mapping, 461–463
    for assembly source, 462
SEI, 111
semaphores, 601, 616–624
    as event flags, 617–618
    as resource keys, 618–624
    service function, 616–617
    for task synchronization, 624–631
sequences, 143
serial communications interface (SCI),
    13, 350–355
    baud-rate generation, 350–351
serial peripheral interface (SPI), 13,
    356–366
    baud-rate generation, 357
    clock format, 357–359
    for input expansion, 364–366
    for output expansion, 361–364
serial I/O, 347–350
serial ports,13
setting bits, 108
setup time, 434–435, 436
SetUserVec(), 278–280, 404
SEX, 83–84
SEV, 111
shared global variable, 158
shift, 105
short, 486
sign extension, 83–85, 95–96
signals, 232, 237
signed, 487
simulators, 59
single-chip microcomputer, 3
single-chip mode, 4, 13, 383
slice, See time slice
slicestr(), 203, 207, 656, 669–670
snippets, 30
software, 4
    construction, 31–33, 36
    design, 36–37
    development, 15–22
software interrupts, 287
source code, 25,32
source level debugging, 475–481
    manual, 476–477
    Noral Flex debugger, 478–481
source relocatable code, 128–129
SP, stack pointer, 63
SP0BR, 357
SP0CR1-SP0CR2, 357–360
SP0DR, 360
SP0SR, 360
spaghetti code, 146
special single-chip mode, 396
specifications, 16, 21

sporadic events, 237
S-Record, 51–52
/SS, 356
STAA, 82
STAB, 82
stack, 86–92
    frame, 160, 166–167
    initialization, 405
    interrupt frame, 277
    memory requirements, 11
    for passing parameters, 166–171
    pointer (SP), 63, 91, 405
    and subroutines, 124
    usage, 91–92
stand-alone programs, 31, 563
stand-alone systems, 10–11
start-stop protocol, 349
start-up code, 400–406
    for C programs, 537, 557–564
    configuration and intialization,
        405–406, 558–559
    exception vectors, 403, 404, 559
    programmable chip select, 432–433
state decomposition, 582–583
static modifier, 487–490, 538–539
static variables, 158
STD, 82
stop control bit (S), 64
stopwatch examples
    using MicroC/OS, 638–645
    using output compare interrupt,
        318–320
    using RTI interrupt, 310–311
    using time slice scheduler, 268–274,
        291–293
    using time slice scheduler in C,
        583–591
strings, 45, 196, 506–507
    conversions, 203–207
    null-terminated, 506
.strings section, 463
structures, 507–509
    bit fields, 514–516
STS, 82
STX, 82
STY, 82
SUBA, 98
SUBB, 98
SUBD, 98
subroutines, 120–128
    documentation, 125
    preserving registers during, 126
subtraction, 98–103
    BCD, 101–103
    signed, 99, 100

unsigned, 99, 100
SWI, 287
switch debouncing, 256–261
switch statement, *See* case construct
symbol table, 471
synchronous serial ports, 13, 348–349
system design, *See* architectural design

\t, 447
TAB, 83
tag, 507
TAP, 83
task decomposition, 580–583
    coroutines, 581–582
    state decomposition, 582–583,
        588–591
task switch, 604
task synchronization, 624–631
tasks, 262
    cooperative, 263
    free-running, 567
    independent, 567–568, 571–572
    in MicroC/OS, 603–609
    period, 568
    timed, 568–569
TBA, 83
TC1-TC7, 316–317
TCNT, 312–315
TCTL1, 317–318
TCTL2, 317–318
TCTL3, 329–330
TCTL4, 329–330
ten's complement subtraction, *See* sub-
    traction, BCD
terminal count bug, 154–155,531–533
terminal emulator, 53
test plan, 16
text editor, 32
.text section, 463
TFLG1, 315–316
TFLG2, 313–315
TFR, 82–83
time slice, 263, 578–579
time slice period, 579
time slice scheduler, 263–267, 578–591
    mutual exclusion in, 580
    task decomposition, 580–583
timed event loops, 253–256
timer flags, 306–307
timer prescaler, 315
timers, 13, 306–347
TIOS, 315–316
TMSK1, 315–316
TMSK2, 312, 314
tokens, 444
toupper(), 183–184, 656, 659

TPA, 83
TRACE, 57
trap, 42
tristate outputs, 7
TRUE, 519
TSCR, 312–314
TST, 103
TSTA, 103
TSTB, 103
TSX, 83
TSY, 83
TXS, 83
type definition convention, 649–650
typedef, 491–492, 508–509
TYS, 83

UART, *See* serial communications inter-
    face
UcosDemo module, 597–603
unconditional I/O detection, 234
unit testing, 21
UNIX, 455, 473
unsigned, 487
user programmed, 10

\v, 447
V, overflow flag, 64
variables
    automatic, 494
    in C, 494–496
    global, 158–160
    initialization, 34, 495–496
    local, *See* automatic
    memory requirements, 11
    register, 157
    static, 158, 494
volatile modifier, 488, 494
volatility, 9, 10
VRH, 369–370
VRL, 369–370

WAITING, 604
watchdog timer, 390
while construct, 150, 529–530
writability, 9–10
write cycle, 420–421
write-once registers, 405

X, pseudo-nonmaskable interrupt
    mask, 64
XIRQ, 285–287
XGDX, 83–84
XGDY, 83–84
XTAL, 385

zero flag(Z), 63